T0075984

High-Performance Computing in Finance

Problems, Methods, and Solutions

CHAPMAN & HALL/CRC
Financial Mathematics Series

Aims and scope:
The field of financial mathematics forms an ever-expanding slice of the financial sector. This series aims to capture new developments and summarize what is known over the whole spectrum of this field. It will include a broad range of textbooks, reference works and handbooks that are meant to appeal to both academics and practitioners. The inclusion of numerical code and concrete real-world examples is highly encouraged.

Published Titles

American-Style Derivatives; Valuation and Computation, *Jerome Detemple*
Analysis, Geometry, and Modeling in Finance: Advanced Methods in Option Pricing, *Pierre Henry-Labordère*
C++ for Financial Mathematics, *John Armstrong*
Commodities, *M. A. H. Dempster and Ke Tang*
Computational Methods in Finance, *Ali Hirsa*
Counterparty Risk and Funding: A Tale of Two Puzzles, *Stéphane Crépey and Tomasz R. Bielecki, With an Introductory Dialogue by Damiano Brigo*
Credit Risk: Models, Derivatives, and Management, *Niklas Wagner*
Engineering BGM, *Alan Brace*
Equity-Linked Life Insurance: Partial Hedging Methods, *Alexander Melnikov and Amir Nosrati*
Financial Mathematics: A Comprehensive Treatment, *Giuseppe Campolieti and Roman N. Makarov*
The Financial Mathematics of Market Liquidity: From Optimal Execution to Market Making, *Olivier Guéant*
Financial Modelling with Jump Processes, *Rama Cont and Peter Tankov*
High-Performance Computing in Finance: Problems, Methods, and Solutions, *M.A.H. Dempster, Juho Kanniainen, John Keane, and Erik Vynckier*
Interest Rate Modeling: Theory and Practice, *Lixin Wu*
Introduction to Credit Risk Modeling, Second Edition, *Christian Bluhm, Ludger Overbeck, and Christoph Wagner*
An Introduction to Exotic Option Pricing, *Peter Buchen*
Introduction to Risk Parity and Budgeting, *Thierry Roncalli*

Introduction to Stochastic Calculus Applied to Finance, Second Edition,
Damien Lamberton and Bernard Lapeyre

Model-Free Hedging: A Martingale Optimal Transport Viewpoint,
Pierre Henry-Labordère

Monte Carlo Methods and Models in Finance and Insurance, *Ralf Korn, Elke Korn, and Gerald Kroisandt*

Monte Carlo Simulation with Applications to Finance, *Hui Wang*

Nonlinear Option Pricing, *Julien Guyon and Pierre Henry-Labordère*

Numerical Methods for Finance, *John A. D. Appleby, David C. Edelman, and John J. H. Miller*

Option Valuation: A First Course in Financial Mathematics, *Hugo D. Junghenn*

Portfolio Optimization and Performance Analysis, *Jean-Luc Prigent*

Quantitative Finance: An Object-Oriented Approach in C++, *Erik Schlögl*

Quantitative Fund Management, *M. A. H. Dempster, Georg Pflug, and Gautam Mitra*

Risk Analysis in Finance and Insurance, Second Edition, *Alexander Melnikov*

Robust Libor Modelling and Pricing of Derivative Products, *John Schoenmakers*

Stochastic Finance: An Introduction with Market Examples, *Nicolas Privault*

Stochastic Finance: A Numeraire Approach, *Jan Vecer*

Stochastic Financial Models, *Douglas Kennedy*

Stochastic Processes with Applications to Finance, Second Edition,
Masaaki Kijima

Stochastic Volatility Modeling, *Lorenzo Bergomi*

Structured Credit Portfolio Analysis, Baskets & CDOs, *Christian Bluhm and Ludger Overbeck*

Understanding Risk: The Theory and Practice of Financial Risk Management,
David Murphy

Unravelling the Credit Crunch, *David Murphy*

Proposals for the series should be submitted to one of the series editors above or directly to:
CRC Press, Taylor & Francis Group
3 Park Square, Milton Park
Abingdon, Oxfordshire OX14 4RN
UK

High-Performance Computing in Finance

Problems, Methods, and Solutions

Edited by
M. A. H. Dempster
Juho Kanniainen
John Keane
Erik Vynckier

CRC Press
Taylor & Francis Group
Boca Raton London New York

CRC Press is an imprint of the
Taylor & Francis Group, an **informa** business
A CHAPMAN & HALL BOOK

MATLAB® is a trademark of The MathWorks, Inc. and is used with permission. The MathWorks does not warrant the accuracy of the text or exercises in this book. This book's use or discussion of MATLAB® software or related products does not constitute endorsement or sponsorship by The MathWorks of a particular pedagogical approach or particular use of the MATLAB® software.

CRC Press
Taylor & Francis Group
6000 Broken Sound Parkway NW, Suite 300
Boca Raton, FL 33487-2742

Printed on acid-free paper

International Standard Book Number-13: 978-1-4822-9966-3 (Hardback)

Library of Congress Cataloging-in-Publication Data

Names: Dempster, M. A. H. (Michael Alan Howarth), 1938- editor. | Kanniainen, Juho editor. | Keane, John. editor. | Vynckier, Erik. editor.
Title: High-performance computing in finance : problems, methods, and solutions / [edited by] M.A.H. Dempster [and three others].
Description: Boca Raton, FL : CRC Press, 2018.
Identifiers: LCCN 2017052035| ISBN 9781482299663 (hardback) | ISBN 9781315372006 (ebook)
Subjects: LCSH: Finance--Mathematical models. | Finance--Data processing.
Classification: LCC HG106 .H544 2018 | DDC 332.01/5118--dc23
LC record available at https://lccn.loc.gov/2017052035

Visit the Taylor & Francis Web site at
http://www.taylorandfrancis.com

and the CRC Press Web site at
http://www.crcpress.com

Contents

Editors

Michael Dempster is Professor Emeritus, Centre for Financial Research, University of Cambridge. He has held research and teaching appointments at leading universities globally and is founding editor-in-chief of *Quantitative Finance.* His numerous papers and books have won several awards, and he is Honorary Fellow of the IFoA, Member of the Academia dei Lincei, and managing director of Cambridge Systems Associates.

Juho Kanniainen is Professor of Financial Engineering at Tampere University of Technology, Finland. He has served as coordinator of two international EU-programs: HPC in Finance (www.hpcfinance.eu) and Big Data in Finance (www.bigdatafinance.eu). His research is broadly in quantitative finance, focusing on computationally expensive problems and data-driven approaches.

John Keane is Professor of Data Engineering in the School of Computer Science at the University of Manchester, UK. As part of the UK government's Foresight Project, The Future of Computer Trading in Financial Markets, he co-authored a commissioned economic impact assessment review. He has been involved in both the EU HPC in Finance and Big Data in Finance programs. His wider research interests are data and decision analytics and related performance aspects.

Erik Vynckier is board member of Foresters Friendly Society, partner of InsurTech Venture Partners, and chief investment officer of Eli Global, following a career in banking, insurance, asset management, and petrochemical industry. He co-founded EU initiatives on high performance computing and big data in finance. Erik graduated as MBA at London Business School and as chemical engineer at Universiteit Gent.

Contributors

John Ashley
NVIDIA
Santa Clara, California

Brian Boucher

Luca Capriotti
Quantitative Strategies
Investment Banking Division
and
Department of Mathematics
University College London
London, United Kingdom

Álvaro Cartea
Mathematical Institute
Oxford-Man Institute of Quantitative Finance
University of Oxford
Oxford, United Kingdom

Omar Andres Carmona Cortes
Computation Department
Instituto Federal do Maranhão
São Luis, Brazil

M. A. H. Dempster
Centre for Financial Research
University of Cambridge
and
Cambridge Systems Associates
Cambridge, United Kingdom

Jens Deussen
Department of Computer Science
RWTH Aachen University
Germany

Ryan Donnelly
Swiss Finance Institute
École Polytechnique Fédérale de Lausanne
Switzerland

Jacques du Toit
The Numerical Algorithms Group Ltd.
United Kingdom

Christina Erlwein-Sayer
OptiRisk Systems

Georgi Gaydadjiev

Mark Gibbs
Quantitative Research
FINCAD

Michael B. Giles
Mathematical Institute
University of Oxford
Oxford, United Kingdom

James B. Glattfelder
Department of Banking and Finance
University of Zurich
Zurich, Switzerland

Peter Goddard
1QBit
Vancouver, Canada

Anton Golub
Lykke Corporation
Zug, Switzerland

Russell Goyder
Quantitative Research
FINCAD

Jon Gregory

Jonathan Hüser
Department of Computer Science
RWTH Aachen University
Germany

Sergey Ivliev
Lykke Corporation
Switzerland

and

Laboratory of Cryptoeconomics and Blockchain Systems
Perm State University
Russia

Sebastian Jaimungal
Department of Statistical Sciences
University of Toronto
Canada

Mark Joshi
Department of Economics
University of Melbourne
Melbourne, Australia

Christian Kahl
Quantitative Research
FINCAD

Juho Kanniainen
Tampere University of Technology
Tampere, Finland

Jacky Lee
Quantitative Strategies
Investment Banking Division
New York

Binghuan Lin
Tampere University of Technology
Tampere, Finland

Alexander Lipton
Stronghold Labs
Chicago, Illinois

and

MIT Connection Science and Engineering
Cambridge, Massachusetts

Stefanus C. Maree
Centrum Wiskunde & Informatica
Amsterdam, The Netherlands

Juan Ivan Martin
International Air Transport Association

Douglas McLean
Moody's Analytics
Edinburgh, Scotland, United Kingdom

Elena A. Medova
Centre for Financial Research
University of Cambridge
and
Cambridge Systems Associates
Cambridge, United Kingdom

Oskar Mencer

Andrew Milne
1QBit
Vancouver, Canada

Gautam Mitra
OptiRisk Systems and Department of Computer Science
UCL, London, United Kingdom

Yulia Mizgireva
Lykke Corporation
Switzerland

and

Laboratory of Cryptoeconomics and Blockchain Systems
Perm State University
Russia

and

Department of Mathematics
Ariel University
Israel

Uwe Naumann
The Numerical Algorithms Group Ltd.
United Kingdom

Richard B. Olsen
Lykke Corporation
Zug, Switzerland

Cornelis W. Oosterlee
Centrum Wiskunde & Informatica
Amsterdam, The Netherlands

and

Delft Institute of Applied Mathematics
Delft University of Technology
Delft, The Netherlands

Luis Ortiz-Gracia
Department of Econometrics
University of Barcelona
Barcelona, Spain

Igor Osmolovskiy
Cambridge Systems Associates
Cambridge, United Kingdom

Andrew Rau-Chaplin
Faculty of Computer Science
Dalhousie University
Halifax, NS, Canada

Christoph Reisinger
Mathematical Institute

and

Oxford-Man Institute of Quantitative Finance
University of Oxford
United Kingdom

Gary Robinson

Jonathan Rosen
Quantitative Research
FINCAD

Mark Rounds
1QBit
Vancouver, Canada

Peter Schober
Goethe University Frankfurt
Chair of Investment
Portfolio Management and Pension Finance
Frankfurt, Hesse, Germany

Lukasz Szpruch
School of Mathematics
University of Edinburgh
Edinburgh, United Kingdom

Philipp Ustinov
Cambridge Systems Associates
Cambridge, United Kingdom

Cristiano Arbex Valle
OptiRisk Systems
United Kingdom

Rainer Wehkamp
Techila Technologies Limited
Tampere, Finland

Rasmus Wissmann
Mathematical Institute
University of Oxford
Oxford, United Kingdom

Xiang Yu
OptiRisk Systems
United Kingdom

Introduction

As lessons are being learned from the recent financial crisis and unsuccessful stress tests, demand for superior computing power has been manifest in the financial and insurance industries for reliability of quantitative models and methods and for successful risk management and pricing. From a practitioner's viewpoint, the availability of high-performance computing (HPC) resources allows the implementation of computationally challenging advanced financial and insurance models for trading and risk management. Researchers, on the other hand, can develop new models and methods to relax unrealistic assumptions without being limited to achieving analytical tractability to reduce computational burden. Although several topics treated in these pages have been recently covered in specialist monographs (see, e.g., the references), we believe this volume to be the first to provide a comprehensive up-to-date account of the current and near-future state of HPC in finance.

The chapters of this book cover three interrelated parts: (i) Computationally expensive financial problems, (ii) Numerical methods in financial HPC, and (iii) HPC systems, software, and data with financial applications. They consider applications which can be more efficiently solved with HPC, together with topic reviews introducing approaches to reducing computational costs and elaborating how different HPC platforms can be used for different financial problems.

Part I offers perspectives on computationally expensive problems in the financial industry.

In Chapter 1, Jonathan Rosen, Christian Kahl, Russell Goyder, and Mark Gibbs provide a concise overview of computational challenges in derivative pricing, paying special attention to counterparty credit risk management. The incorporation of counterparty risk in pricing generates a huge demand for computing resources, even with vanilla derivative portfolios. They elaborate possibilities with different computing hardware platforms, including *graphic processing units* (GPU) and *field-programmable gate arrays* (FPGA). To reduce hardware requirements, they also discuss an algorithmic approach, called algorithmic differentiation (AD), for calculating sensitivities.

In Chapter 2, Gautam Mitra, Christina Erlwein-Sayer, Cristiano Arbex Valle, and Xiang Yu describe a method for generating daily trading signals to construct second-order stochastic dominance (SSD) portfolios of exchange-traded securities. They provide a solution for a computationally (NP) hard optimization problem and illustrate it with real-world historical data for the FTSE100 index over a 7-year back-testing period.

In Chapter 3, Anton Golub, James B. Glattfelder, and Richard B. Olsen introduce an event-based approach for automated trading strategies. In their methodology, in contrast to the usual continuity of physical time, only events (interactions) make the system's clock tick. This approach to designing automated trading models yields an algorithm that possesses many desired features and can be realized with reasonable computational resources.

In Chapter 4, Álvaro Cartea, Ryan Donnelly, and Sebastian Jaimungal consider the optimal liquidation of a position using limit orders. They focus on the question of how the misspecification of a model affects the trading strategy. In some cases, a closed-form expression is available, but additional relevant features in the framework make the model more realistic at the cost of not having closed-form solutions.

In Chapter 5, Douglas McLean discusses challenges associated with economic scenario generation (ESG) within an insurance context. Under Pilar 1 of the Solvency 2 directive, insurers who use their own models need to produce multi-year scenario sets in their asset and liability modeling systems, which is a computationally hard problem. McLean provides illustrative examples and discusses the aspects of high-performance computing in ESG as well.

Part II focuses on numerical methods in financial high-performance computing (HPC).

First, in Chapter 6, Christoph Reisinger and Rasmus Wissmann consider finite difference methods in derivative pricing. Numerical methods with partial differential equations (PDE) perform well with special cases, but computational problems arise with high-dimensional problems. Reisinger and Wissmann consider different decomposition methods, review error analysis, and provide numerical examples.

In Chapter 7, Mike Giles and Lukasz Szpruch provide a survey on the progress of the multilevel Monte Carlo method introduced to finance by the first author. Multilevel Monte Carlo has now become a widely applied variance reduction method. Giles and Szpruch introduce the idea of multilevel Monte Carlo simulation and discuss the numerical methods that can be used to improve computational costs. They consider several financial applications, including Monte Carlo Greeks, jump-diffusion processes, and the multidimensional Milstein scheme.

In Chapter 8, Stef Maree, Luis Ortiz-Gracia, and Cornelis Oosterlee discuss Fourier and wavelet methods in option pricing. First, they review different methods and then numerically show that the COS and SWIFT (Shannon wavelet inverse Fourier technique) methods exhibit exponential convergence. HPC can be used in the calibration procedure by parallelizing option pricing with different strikes.

In Chapter 9, Michael Dempster, Elena Medova, Igor Osmolovskiy, and Philipp Ustinov consider a three-factor Gaussian yield curve model that is used for scenario simulation in derivative valuation, investment modeling, and asset-liability management. The authors propose a new approximation of the

Black (1995) correction to the model to accommodate nonnegative (or negative lower bounded) interest rates and illustrate it by calibrating yield curves for the four major currencies, EUR, GBP, USD, and JPY, using the unscented Kalman filter and estimating 10-year bond prices both in and out of sample. Calibration times are comparable to those for the original model and with cloud computing they can be reduced from a few hours to a few minutes.

In Chapter 10, Uwe Naumann, Jonathan Hüser, Jens Deussen, and Jacques du Toit review the concept of *algorithmic differentiation* (AD) and *adjoint algorithmic differentiation* (AAD), which has been gaining popularity in computational finance over recent years. With AD, one can efficiently compute the derivatives of the primal function with respect to a specified set of parameters, and therefore it is highly relevant for the calculation of option sensitivities with Monte Carlo. The authors discuss aspects of implementation and provide three case studies.

In Chapter 11, Luca Capriotti and Jacky Lee consider adjoint algorithmic differentiation (AAD) by providing three case studies for real-time risk management: interest rate products, counterparty credit risk management, and volume credit products. AAD is found to be extremely beneficial for the case applications as it is speeding up by several orders of magnitude the computation of price sensitivities both in the context of Monte Carlo applications and for applications involving faster numerical methods.

In Chapter 12, Omar Andres Carmona Cortes and Andrew Rau-Chaplin use evolutionary algorithms for the optimization of reinsurance contracts. They use population-based incremental learning and differential evolution algorithms for the optimization. With a case study they demonstrate the parallel computation of an actual contract problem.

In Chapter 13, which ends Part II, Sergey Ivliev, Yulia Mizgireva, and Juan Ivan Martin consider implementation of blockchain technologies for the clearing and settlement procedure of the IATA Clearing House. They develop a simulation model to evaluate the industry-level benefits of the adoption of blockchain-based industry money for clearing and settlement.

Part III considers different computational platforms, software, and data with financial applications.

In Chapter 14, Peter Schober provides a summary of *supercomputing*, including aspects of hardware platforms, programming languages, and parallelization interfaces. He discusses supercomputers for financial applications and provides case studies on the pricing of basket options and optimizing life cycle investment decisions.

In Chapter 15, Oskar Mencer, Brian Boucher, Gary Robinson, Jon Gregory, and Georgi Gaydadjiev describe the concept of *multiscale data-flow computing*, which can be used for special-purpose computing on a customized architecture, leading to increased performance. The authors review the data-flow paradigm, describe Maxeler data-flow systems, outline the data-flow-oriented programming model that is used in Maxeler systems, and discuss how to develop data-flow applications in practice, and how to improve their performance.

Additionally, they provide a case study to estimate correlations between a large number of security returns. Other financial applications including interest rate swaps, value-at-risk, option pricing, and credit value adjustment capital are also considered.

In Chapter 16, John Ashley and Mark Joshi provide grounding in the underlying computer science and system architecture considerations needed to take advantage of future computing hardware in computational finance. They argue for a parallelism imperative, driven by the state of computer hardware and current manufacturing trends. The authors provide a concise summary of system architecture, consider parallel computing design and then provide case studies on the LIBOR market model and the Monte Carlo pricing of early exercisable Bermudan derivatives.

In Chapter 17, Binghuan Lin, Rainer Wehkamp, and Juho Kanniainen review cloud computing for financial applications. The section is written for practitioners and researchers who are interested in using cloud computing for various financial applications. The authors elaborate the concept of cloud computing, discuss suitable applications, and consider possibilities and challenges of cloud computing. Special attention is given to an implementation example with Techila middleware and to case studies on portfolio optimization.

In Chapter 18, Alexander Lipton introduces *blockchains* (BC) and *distributed ledgers* (DL) and describes their potential applications to money and banking. He presents historical instances of BL and DL. The chapter introduces the modern version of the *monetary circuit* and how it can benefit from the BL and DL framework. Lipton shows how central bank-issued digital currencies can be used to move away from the current fractional reserve banking toward narrow banking.

In Chapter 19, the last chapter in the book which points to the future, Andrew Milne, Mark Rounds and Peter Goddard consider feature selection for credit scoring and classification using a *quantum annealer*. Quantum computing has received much attention, but is still in its infancy. In this chapter, with the aid of the 1QBit Quantum-Ready Software Development Kit, the authors apply quantum computing using the DWave quantum *simulated annealing* machine to a well-known financial problem involving massive amounts of data in practice, *credit scoring and classification*. They report experimental results with German credit data.

For further in-depth background the reader should consult the bibliography.

Acknowledgments

This book arose in part from the four-year EU Marie Curie project *High-Performance Computing in Finance* (Grant Agreement Number 289032, www.hpcfinance.eu), which was recently completed. We would like to thank all the partners and participants in the project and its several public events, in particular its supported researchers, many of whom are represented in these pages. We owe all our authors a debt of gratitude for their fine contributions and for enduring a more drawn out path to publication than we had originally envisioned. We would also like to express our gratitude to the referees and to World Scientific and Emerald for permission to reprint Chapters 7[1] and 18[2], respectively. Finally, without the expertise and support of the editors and staff at Chapman & Hall/CRC and Taylor & Francis this volume would have been impossible. We extend to them our warmest thanks.

Michael Dempster
Juho Kanniainen
John Keane
Erik Vynckier
Cambridge, Tampere, Manchester
August 2017

Bibliography

1. Foresight: The Future of Computer Trading in Financial Markets. 2012. Final Project Report, Government Office for Science, London. https://www.gov.uk/government/publications/future-of-computer-trading-in-financial-markets-an-international-perspective

2. De Schryver, C., ed. 2015. *FPGA Based Accelerators for Financial Applications.* Springer.

3. Delong, Ł. 2013. *Backward Stochastic Differential Equations with Jumps and Their Actuarial and Financial Applications.* London: Springer.

4. Zopounidis, C. and Galariotis, E. 2015. *Quantitative Financial Risk Management: Theory and Practice.* New York, NY: John Wiley & Sons.

5. Uryasev, S. and Pardalos, P. M., eds. 2013. *Stochastic Optimization: Algorithms and Applications.* Vol. 54. Berlin: Springer Science & Business Media.

6. John, K. 2016. *WHPCF'14, Proceedings of the 7th Workshop on High Performance Computational Finance.* Special issue, Concurrency and Computation Practice and Experience 28(3).

7. John, K. 2015. *WHPCF'15, Proceedings of the 8th Workshop on High Performance Computational Finance, SC'15 The International Conference for High Performance Computing, Networking, Storage and Analysis.* New York: ACM.

[1]Which appeared originally in *Recent Developments in Computational Finance*, World Scientific (2013).

[2]Which will appear in Lipton, A., 2018. Blockchains and distributed ledgers in retrospective and perspective. *Journal of Risk Finance*, 19(1).

Part I

Computationally Expensive Problems in the Financial Industry

Chapter 1

Computationally Expensive Problems in Investment Banking

Jonathan Rosen, Christian Kahl, Russell Goyder, and Mark Gibbs

CONTENTS

1.1 Background

Financial instruments traded on markets are essentially contractual agreements between two parties that involve the calculation and delivery of quantities of monetary currency or its economic equivalent. This wider definition of financial investments is commonly known as financial derivatives or options, and effectively includes everything from the familiar stocks and bonds to the most complex payment agreements, which also include complicated mathematical logic for determining payment amounts, the so-called *payoff* of the derivative.

In the early days of derivatives, they were thought of more like traditional investments and treated as such on the balance sheet of a business. Complexity mainly arose in the definition and calculation of the option payoff, and applying theoretical considerations to price them. It was quickly discovered that probabilistic models, for economic factors on which option payoffs were calculated, had to be quite restrictive in order to produce computationally straightforward problems in pricing the balance sheet fair mark-to-market value. The development of log-normal models from Black and Scholes was quite successful in demonstrating not only a prescient framework for derivative pricing, but also the importance of tractable models in the practical application of risk-neutral pricing theory, at a time when computational facilities were primitive by modern standards. Developments since then in quantitative finance have been accompanied by simultaneous advancement in computing power, and this has opened the door to alternative computational methods such as Monte Carlo, PDE discretization, and Fourier methods, which have greatly increased the ability to price derivatives with complex payoffs and optionality.

Nevertheless, recent crises in 2008 have revealed the above complexities are only part of the problem. The state of credit worthiness was eventually to be revealed as a major influence on the business balance sheet in the event that a contractual counterparty in a derivatives contract fails to meet the terms for payment. It was recognized that market events could create such a scenario due to clustering and tail events. In response to the explosion of credit derivatives and subsequent global financial crisis, bilateral credit value adjustments (CVAs) and funding cost adjustments were used to represent the impact of credit events according to their likelihood in the accounting balance sheet. Credit value adjustments and funding adjustments introduced additional complexity into the business accounting for market participants. While previously simple trades only required simple models and textbook formulas to value, the CVA is a portfolio derivative problem requiring joint modeling of many state variables and is often beyond the realm of simple closed-form computation.

Meanwhile, the controversial decision to use tax payer money to bail out the financial institutions in the 2008 crisis ignited a strong political interest to introduce regulation that requires the largest investors to maintain capital holdings that meet appropriate thresholds commensurate with the financial risk present in their balance sheet. In recent years, there have been an adoption of capital requirements globally, with regional laws determining methods and criteria, for the calculation of regulatory capital holdings. The demand placed on large market participants to apply additional value adjustments for tax and capital funding costs requires modeling these effects over the lifetime of the investment, which has introduced a large systemic level of complexity in accounting for the financial investment portfolio, even when it is composed of very simple investments.

1.1.1 Valuation requirements

1.1.1.1 Derivatives pricing and risk

In the early days of option trading, accounting principles were completely decoupled from financial risk measures commonly known today as *market risk*. Investors holding derivatives on their balance sheets were mainly concerned with the ability to determine the fair market value. In the early derivative markets, the role of central exchanges was quite limited, and most derivative contracts directly involved two counterparties; these trades were known as over-the-counter (OTC). As derivative payoffs became more closely tailored to individual investors, the theoretical pricing of exotic trades was increasingly complex, and there was no possibility for liquid quotes for investor holdings, meaning theoretical risk-neutral considerations became paramount to the accounting problem of mark-to-market balance sheets.

The innovation in this area began with the seminal work of Black and Scholes, leading to tractable and computationally efficient theoretical derivative pricing formulas, which were widely adopted and incorporated into trading and accounting technology for derivative investors. An example is formula for the (at expiry) value V_t of a call option struck at K on a stock whose expected value at expiry t is F_t,

$$V_t = F_t \Phi(d_1) - K\Phi(d_2) \tag{1.1}$$

where Φ denotes the standard normal distribution and

$$\begin{aligned} d_1 &= \frac{1}{\sigma\sqrt{t}}\left[\ln\left(\frac{F_t}{K}\right) + \frac{\sigma^2}{2}t\right] \\ d_2 &= d_1 - \sigma\sqrt{t} \end{aligned} \tag{1.2}$$

where σ is the volatility parameter in the log-normal distribution of F_t assumed by the model. The number of analytic formulas that could be deduced was restricted by the complexity of the payoff and the sophistication of the dynamic model being used. Payoffs that require many looks at underlying factors and allow periodic or continuous exercise can greatly complicate or even prohibit the ability to derive a useful analytic formula or approximation for pricing and risk.

Meanwhile, the state-of-the-art in computation was becoming more advanced, leading to modern numerical techniques being applied to derivative pricing problems. Considering the fundamental theorem of asset pricing in conjunction with the theorem of Feynman-Kac, one arrives at an equivalence of the expectation of a functional of a stochastic process to partial-integro differential equation, which can be further transformed analytically using Fourier method-based approaches leading to a system of ordinary differential equations. All three might allow for further simplifications leading to closed-form solutions or need to be solved numerically.

Partial (integro-) differential equations: Partial differential equations form the backbone of continuous time-pricing theory, and a numerical approximation to the solution can be achieved with finite difference methods as well as many other techniques. However, this approach suffers from the so-called *curse of dimensionality*, very similar to the case of multivariate quadrature in that the stability of this approach breaks down for models with many underlying factors.

Fourier methods: The mathematical formulation of a conjugate variable to time can be used to greatly simplify convolution integrals that appear in derivative pricing problems by virtue of the Plancharel equality. The use of a known characteristic function of a dynamic process allows fast numerical derivative pricing for options with many looks at the underlying factor.

Monte Carlo simulation: Monte Carlo methods offer a very generic tool to approximate the functional of a stochastic process also allowing to deal effectively with path-dependent payoff structures. The most general approach to derivative pricing is based on pathwise simulation of time-discretized stochastic dynamical equations for each underlying factor. The advantage is in being able to handle any option payoff and exercise style, as well as enabling models with many correlated factors. The disadvantage is the overall time-consuming nature and very high level of complexity in performing a full simulation.

Besides dealing with the complexity of the option payoff, the Black–Scholes formula made use of a single risk-free interest rate and demonstrated that in the theoretical economy, this rate had central importance for the time value of money and the expected future growth of risk-neutral investment strategies analogous to single currency derivatives. This means a single discount curve per currency was all that was needed, which by modern standards led to a fairly simple approach in the pricing of derivatives. For example, a spot-starting interest rate swap, which has future cash flows that are calculated using a floating rate that must be discounted to present value, would use a single curve for the term structure of interest rates to both calculate the risk-neutral implied floating rates and the discount rates for future cash flows.

However, the turmoil of 2008 revealed that collateral agreements were of central importance in determining the relevant time value of money to use for discounting future cash flows, and it quickly became important to separate discounting from forward rate projection for collateralized derivatives. The subsequent computational landscape required building multiple curves in a single currency to account for institutional credit risk in lending at different tenors.

On top of this problem is the risk calculation, which requires the sensitivity of the derivative price to various inputs and calculated quantities inside the numerical calculation. A universal approach to this is to add a small amount to each quantity of interest and approximate the risk with a finite difference calculation, known as *bumping*. While this can be applied for all numerical and closed-form techniques, it does require additional calculations which themselves can be time-consuming and computationally expensive. The modern view is that by incorporating analytic risk at the software library level, known as *algorithmic differentiation* (AD), some libraries such as our own produce analytic risk for all numerical pricing calculations, to be described in Section 1.3.2.

1.1.1.2 Credit value adjustment/debit value adjustment

Since the credit crisis in 2008, counterparty credit risk of derivative positions has become an increasingly important subject. The IFRS 9 standard required a fair value option for derivative investment accounting (Ramirez, 2015) and originally proposed to include CVA for application to hedge accounting, to represent the effect of an entity's risk management activities that use financial instruments to manage market risk exposures that could affect overall profit or loss. IFRS 13 set out requirements intended to account for the risk that the counterparty of the financial derivative or the entity will default before the maturity/expiration of the transaction and will be unable to meet all contractual payments, thereby resulting in a loss for the entity or the counterparty, which required accounting for CVA and a *debit value adjustment* (DVA) in the balance sheet as nonperformance risk. However, IFRS 9 does not provide direct guidance on how CVA or DVA is to be calculated, beyond requiring that the resulting fair value must reflect the credit quality of the instrument.

There are a lot of choices to be made when computing CVA, including the choice of model used and the type of CVA to be computed. In addition, there are further decisions on whether it is unilateral or bilateral (Gregory, 2009), what type of closeout assumptions to make, how to account for collateral, and considerations for including first-to-default, which will be discussed later. The variety of possible definitions and modeling approaches used in CVA calculation has led to discrepancies in valuation across different financial institutions (Watt, 2011). There are a variety of ways to determine CVA, however, it is often a computationally expensive exercise due to the substantial number of modeling factors, assumptions involved, and the interaction among these assumptions. More specifically, CVA is a complex derivative pricing problem on the entire portfolio, which has led to substantially increasing the complexity of derivative pricing.

As a measure of the exposure of a portfolio of products to counterparty default, and if such an event occurs within some time interval, the expected loss is the positive part of the value of the remainder of the portfolio after the event. This is still in the realm of the familiar risk-neutral assumption of derivative pricing in the form of an adjustment to the fair value, whereby conditioning on the credit event explicitly and accumulating the contribution from each time interval over the life of the portfolio, an amount is obtained by which the value of the portfolio can be modified in order to account for the exposure to counterparty default. This can be expressed formally as

$$\text{CVA} = E\left[\int_0^T \max[\hat{V}(t), 0] D(t)\,(1 - R(t))\, I_{\tau \in (t, t+dt)}\right] \tag{1.3}$$

where $R(t)$ is the recovery rate at time t, $\hat{V}(t)$ the value at time t of the remainder of the underlying portfolio whose value is $V(t)$, $D(t)$ the discount factor at time t, τ the time of default, and $I_{\tau \in (t,t+dt)}$ is a default indicator, evaluating to 1 if τ lies between time t and $t + dt$ and 0 otherwise. Note that Equation 1.3 does not account for default of the issuer, an important distinction which is in particular relevant for regulatory capital calculation purposes (Albanese and Andersen, 2014). Including issuer default is commonly referred to as first-to-default CVA (FTDCVA).

The first step in proceeding with computation of CVA is to define a time grid over periods in which the portfolio constituents expose either party to credit risk. Next, the present value of the exposure upon counterparty default is calculated at each point in time. Often this involves simplifying assumptions, such that the risk-neutral drift and volatility are sufficient to model the evolution of the underlying market factors of the portfolio, and similarly that counterparty credit, for example, is modeled as jump-to-default process calibrated to observable market CDS quotes to obtain corresponding survival probabilities.

However, in practically all realistic situations, the investment portfolio of OTC derivatives will contain multiple trades with any given counterparty. In these situations, the entities typically execute netting agreements, such as the ISDA master agreement, which aims to consider the overall collection of trades as a whole, such that gains and losses on individual positions are offset against one another. In the case that either party defaults, the settlement agreement considers the single net amount rather than the potentially many individual losses and gains. One major consequence of this is the need to consider CVA on a large portfolio composed of arbitrary groups of trades entered with a given counterparty. As Equation 1.3 suggests, the CVA and DVA on such a portfolio cannot be simply decomposed into CVA on individual investments, thus it is a highly *nonlinear* problem which requires significant computational facilities in order to proceed.

Quantifying the extent of the risk mitigation benefit of collateral is important, and this usually requires modeling assumptions to be made. Collateral modeling often incorporates several mechanisms which in reality do not lead to perfect removal of credit default risk, which can be summarized as follows (Gregory, 2015).

Granularity of collateral posting: The existence of thresholds and minimum transfer amounts, for example, can lead to over- and under-collateralization, which can provide deficits and surpluses for funding under certain collateral agreements or Credit Support Annexes (CSAs).

Delay in collateral posting: The operational complexities of modern CSA management involve some delay between requesting and receiving collateral, including the possibility of collateral dispute resolution.

Cash equivalence of collateral holdings: In situations where collateral includes assets or securities that are not simply cash in the currency in which exposure is assessed, the potential variation of the value of the collateral holdings must be modeled over the life of the investment.

Given the arbitrary nature of the investment portfolio's composition and the nonlinear nature of CVAs, the computation of CVA typically requires numerical methods beyond the scope of closed-form computation. Monte Carlo simulation approaches are common. There is a framework for Monte Carlo copula simulation described by Gibbs and Goyder (2013a). While simulation approaches are highly general and well suited to the CVA problem, they are known to be significantly more complex to define than closed-form methods, often increasing calculation times by orders of magnitude, not to mention the difficulty in setting up simulation technology in the most efficient and robust ways possible for business accounting operations.

DVA, as a balance sheet value adjustment required under IFRS 9, is defined analogously to CVA except it represents the complimentary adjustment for risk exposure in the financial portfolio to the credit of the entity. DVA reduces the value of a liability, recognizing reduced losses when an entity's own creditworthiness deteriorates. Some criticism of this balance sheet adjustment points to the difficulty in realizing profits resulting from an entity's default when closing out financial positions, and the possibility of significant volatility in profit or loss in periods of credit market turmoil.

CVA and DVA introduce significant model dependence and risk into the derivative pricing problem, as it is impacted heavily by dynamic features such as volatilities and correlations of the underlying portfolio variables and counterparty credit spreads. Some open-ended problems suggest that the modeling complexity can increase significantly in times of market stress. Clustering and tail risk suggest non-Gaussian simulation processes and copulae of many state variables are necessary to accurately reflect credit risk in times of market turmoil. Wrong way risk can also be an important effect, which is additional risk introduced by correlation of the investment portfolio and default of the counterparty. Additionally, the incorporation of first-to-default considerations can have significant effects (Brigo et al., 2012). The sum of these complexities introduced by IFRS 9 led to a continuous need for computationally expensive credit risk modeling in derivative pricing.

1.1.1.3 Funding value adjustment

A more recently debated addition to the family of value adjustments is the *funding value adjustment* (FVA). Until recently, the conceptual foundations of FVA were less well understood relative to counterparty risk (Burgard and Kjaer, 2011), though despite the initial debate, there is consensus that funding is relevant to theoretical replication and reflects a similar dimension to the CVA specific to funding a given investment strategy. This is simply a way of saying that funding a trade is analogous to funding its hedging strategy, and this consideration leads to the practical need for FVA in certain situations. Some important realities of modern financial markets that involve funding costs related to derivative transactions are as follows (Green, 2016).

1. Uncollateralized derivative transactions can lead to balance sheet investments with effective loan or deposit profiles. Loan transfer pricing within banks incorporates these exposure profiles in determining funding costs and benefits for trading units.

2. The details of CSA between counterparties affect the availability of posted funds for use in collateral obligations outside the netting set. The presence or absence of the ability to *rehypothecate* collateral is critical to properly account for funding costs of market risk hedges.

3. The majority of the time derivatives cannot be used as collateral for a secured loan like in the case of a bond. This reflects the lack of *repo market* for derivatives, which means that funding is decided at the level of individual investment business rather than in the market, requiring accounting for detailed FVA.

Central to FVA is the idea that funding strategies are implicit in the derivatives portfolio (essentially by replication arguments), however, the funding policy is typically specific to each business, and so incorporating this in a value adjustment removes the symmetry of one price for each transaction. This leads to a general picture of a balance sheet value which differs from the mutually agreed and traded price. This lack of symmetry in the business accounting problem is one of the reasons that FVA has yet to formally be required on the balance sheet through regulation, though IFRS standards could change in the near future to account for the increasing importance of FVA (Albanese and Andersen, 2014; Albanese et al., 2014).

The details of funding costs are also linked to collateral modeling, which also has strong impact on CVA/DVA, thus in many cases it becomes impossible to clearly separate the FVA and CVA/DVA calculations in an additive manner. Modern FVA models are essentially exposure models which derive closely from CVA, and fall into two broad classes, namely expectation approaches using Monte Carlo simulation and replication approaches using PDEs (Burgard and Kjaer, 2011; Brigo et al., 2013).

1.1.2 Regulatory capital requirements

Legal frameworks for ensuring that financial institutions hold sufficient surplus capital, commensurate with their financial investment risk, have become increasingly important on a global scale. Historically speaking, regulatory capital requirements (RCR) were not a major consideration for the front-office trading desk, because the capital holdings were not included in derivative pricing. Indeed, the original capitalization rules of the 1988 Basel Capital Accord lacked sensitivity to counterparty credit risk, which was acknowledged as enabling the reduction of RCR without actually reducing financial risk-taking.

A major revision to the Basel framework came in 2006 (*Basel II*) to address the lack of risk sensitivity and also allowed for supervisory approval for banks to use internal models to calculate counterparty default risk, based on calibrated counterparty default probability and loss-given-default. The new capital rules led to a sizable increase in RCR directly associated with the front-office derivatives trading business, led to a subsequent need to focus on the risks associated with derivatives trading, and for the careful assessment and modeling of RCR on the trading book. The inflated importance of regulatory capital for derivatives trading has led to the creation of many new front-office capital management functions, whose task is to manage the capital of the bank deployed in support of trading activities.

One serious shortcoming of Basel II was that it did not take into account the possibility of severe mark-to-market losses due to CVA alone. This was acknowledged in the aftermath of the 2008 financial crisis, where drastic widening of counterparty credit spreads resulted in substantial balance sheet losses due to CVA, even in the absence of default. The revised Basel III framework proposed in 2010 significantly increases counterparty risk RCR, primarily by introducing an additional component for CVA, which aims to capitalize the substantial risk apparent in CVA accounting volatility. The need to accurately model capital holdings for the derivatives trading book, which in turn requires accurate modeling of risk-neutral valuation including counterparty credit risk value adjustments, potentially introduces a much greater level of complexity into the front-office calculation infrastructure. This has led to a dichotomy of RCR calculation types with varying degrees of complexity, applicable for different institutions with respective degrees of computational sophistication.

To address the widening gap between accurate and expensive computation and less sophisticated investors, a family of simpler calculation methods has been made available. These approaches only take as input properties of the trades themselves and thus lack the risk sensitivity that is included in the advanced calculation methods of RCR for counterparty default and CVA risks. There is a hurdle for model calculations, involving backtesting on PnL attribution, which is put in place so that special approval is required to make use of internal model calculations. The benefit of advanced calculation methods is improved accuracy compared with standard approaches, and greater risk sensitivity in RCR, allowing more favorable aggregation and better recognition of hedges. These factors typically allow advanced methods to provide capital relief, resulting in a competitive advantage in some cases.

1.1.2.1 Calculation of market risk capital

Under Basel II, a standardized method for market risk capital was introduced, which is based on a series of formulas to generate the trading book capital requirement, with different approaches to be used for various categories of assets. The formulas do not depend on market conditions, only properties of the trades themselves, with overall calibration put in place by the Basel Committee to achieve adequate capital levels.

For banks approved to use internal risk models under the *internal model method* (IMM) for determining the trading book capital requirement, this is typically based on Value-at-Risk (VaR) using one-sided 99% confidence level based on 10-day interval price shock scenarios with at least 1 year of time-series data. In Basel III, it is required to include a period of significant market stress relevant for the bank to avoid historical bias in the VaR calculation. While VaR calculation can be presented as a straightforward percentile problem of an observed distribution, it is also nonlinear and generally requires recalculating for each portfolio or change to an existing portfolio. In addition, the extensive history of returns that must be calculated and stored, as well as the definition of suitable macroeconomic scenarios at the level of pricing model inputs both represent significant challenges around implementing and backtesting of VaR relating to computational performance, storage overhead, and data management.

There has been criticism of VaR as an appropriate measure of exposure since it is not a *coherent* risk measure due to lack of subadditivity. This diminished recognition of diversification has been addressed in a recent proposal titled *The Fundamental Review of the Trading Book* (FRTB), part of which is the replacement of VaR with expected shortfall (ES), which is less intuitive than VaR but does not ignore the severity of large losses. However, theoretical challenges exist since ES is not elicitable, meaning that backtesting for regulatory approval is a computationally intensive endeavor which may depend on the risk model being used by the bank (Acerbi and Szkely, 2014).

1.1.2.2 Credit risk capital

The Basel II framework developed a choice of methods for exposure-at-default (EAD) for each counterparty netting set, both a standardized method and an advanced method. A serious challenge was posed to define EAD, due to the inherently uncertain nature of credit exposure driven by market risk factors, correlations, and legal terms related to netting and collateral. Consistently with previous methods, SA-CCR treats the exposure at default as a combination of the mark-to-market and the potential future exposure. If a bank receives supervisory authorization, they can use advanced models to determine capital requirements under the IMM, which is the most risk-sensitive approach for EAD calculation under the Basel framework. EAD is calculated at the netting set level and allows full netting and collateral modeling. However to gain IMM approval takes significant time and requires a history of backtesting which can be very costly to implement.

During the financial crisis, roughly two-thirds of losses attributed to counterparty credit were due to CVA losses and only about one-third were due to actual defaults. Under Basel II, the risk of counterparty default and credit migration risk was addressed, for example, in credit VaR, but mark-to-market losses due to CVA were not. This has led to a Basel III proposal that considers fairly severe capital charges against CVA, with relatively large multipliers on risk factors used in the CVA RCR calculation, compared with similar formulas used in the trading book RCR for market risk.

The added severity of CVA capital requirements demonstrates a lack of alignment between the trading book and CVA in terms of RCR, which is intentional to compensate for structural differences in the calculations. As CVA is a nonlinear portfolio problem, the computational expense is much greater, often requiring complex simulations, compared with the simpler valuations for instruments in the trading book. The standard calculations of risk involve revaluation for bumping each risk factor. The relative expense of CVA calculations means that in practice fewer risk factors are available, and thus capital requirements must have larger multipliers to make up for the lack of risk sensitivity.

However, one technology in particular has emerged to level the playing field and facilitate better alignment between market risk and CVA capital requirements. AD, described in Section 1.3.2, reduces the cost of risk factor calculations by eliminating the computational expense of adding additional risk factors. With AD applied to CVA pricing, the full risk sensitivity of CVA capital requirements can be achieved without significantly increasing the computational effort, although the practical use of AD can involve significant challenges in the software implementation of CVA pricing.

1.1.2.3 Capital value adjustment

The ability to account for future capital requirements is a necessity for the derivatives trading business which aims to stay capitalized over a certain time horizon. The complexity of the funding and capitalization of the derivatives business presents a further need to incorporate funding benefits associated with regulatory capital as well as capital requirements generated from hedging trades into the derivative pricing problem as a capital value adjustment (KVA). As in the case of CVA and FVA, the interconnected nature of value adjustments means that it is not always possible to fully separate KVA from FVA and CVA in an additive manner. Recent work (Green, 2016) has developed several extensions to FVA replication models to incorporate the effects associated with KVA in single asset models.

The current and future landscape of RCR appears to be driven by calculations that are risk sensitive, and thus highly connected to the valuation models and market risk measures. This is a complex dynamic to include in the valuation adjustment since it links mark-to-market derivatives accounting with exogenous costs associated with maintaining sufficient capital reserves that, in turn, depend on market conditions, analogously to counterparty default and operational funding costs. Challenges arise due to the use of real-world risk measures to calculate exposures, which are at odds with arbitrage-free derivative pricing theory. This requires sophisticated approximation of the relationship between real-world risk measures used in regulatory EAD calculation with the market-calibrated exposure profile used in the risk-neutral CVA pricing (Elouerkhaoui, 2016).

Another significant computational challenge for KVA is related to structural misalignment of the capital requirement formulas with the hedge amounts determined by internal models. Under the current regulatory framework, there is still a possibility of reducing capital requirements by increasing risk as determined by internally approved models. Optimal hedge amounts must be carefully determined so as to maximize the profitability of the derivatives business within a certain level of market and counterparty risk, while simultaneously reducing the capital requirements associated with hedging activity. The overall picture is thus pointing towards a growing need for accuracy in KVA as a part of derivatives hedge accounting, despite the current lack of any formal requirement for KVA in the IFRS standards.

1.2 Trading and Hedging

Historically, counterparty credit risk and capital adjustments only played a minor role within financial institutions often delegated away from front-office activity and thus covered by departments such as risk and finance. Given the relatively low capital requirements prior to the financial crisis, the profit centers were simply charged on a determined frequency ranging from monthly to yearly against their relevant capital usage, but the impact has been negligible for almost all trading activity decisions. This also meant that front-office model development followed a rapid path of innovation whilst counterparty credit risk and capital modeling was only considered a needed by-product of operation. Realistically, only CVA did matter for trades against counterparties where no collateral agreement has been in place, such as corporates, for which the bank could end up in the situation of lending a significant amount of money which is already the case for a vanilla interest rate swap. Given credit spreads of almost zero for all major financial institutions, neither DVA nor FVA was worth considering.

Given the strong emphasis of both accounting and regulation on xVA related charges ranging from CVA/DVA to FVA and even KVA, it is of fundamental importance to have accurate pre- and post-trade information available. Both are a matter of computational resources albeit with a slightly different emphasis. Whilst calculation speed is critical for decision making in pretrade situations, post-trade hedging and reporting is predominantly a matter of overall computational demand only limited by the frequency of reporting requirements, which in case of market risk require a daily availability of fair values, xVA and their respective risk sensitivities. A similar situation holds for hedging purposes where portfolio figures don't always need to be available on demand but calculations on a much coarser frequency are sufficient. It is important to point out in particular that in periods of stress, the ability to generate accurate post-trade analysis becomes more important as volatility markets will also make risk and hedging figures fluctuate more and thus potentially require to be run on a higher frequency in order to make decision ranging from individual hedges all the way down to unwinding specific business activity, closing out of desks to raising more regulatory capital. Since it is not economical to just warehouse idle computation resource for the event of a crisis, one would either have to readjust their current usage where possible or rely on other scaling capabilities. A point we are investigating in more detail is in Section 1.3.1.

1.3 Technology

There are a multitude of technology elements supporting derivative pricing ranging from hardware (see Section 1.3.1) supporting the calculation to software technology such as algorithmic adjoint differentiation (see Section 1.3.2).

In the discussion below, we focus predominantly on the computational demand related to derivative pricing and general accounting and regulatory capital calculations, and less on the needs related to high-frequency applications including the collocation of hardware on exchanges.

1.3.1 Hardware

Choice of hardware is a critical element in supporting the pricing, risk management, accounting and regulatory reporting requirements of financial institutions. In the years prior to the financial crisis, the pricing and risk management have typically been facilitated by in-house developed pricing libraries using an object-oriented programming language (C++ being a popular choice) on regular CPU hardware and respective distributed calculation grids.

1.3.1.1 Central processing unit/floating point unit

Utilizing an object-oriented programming language did allow for a quick integration of product, model, and pricing infrastructure innovation into production which is a critical element for a structured product business to stay ahead of competition. Even with product and model innovation being of less importance, postcrisis being able to reutilize library components aids to the harmonization of pricing across different asset classes and products which is in particular important in light of regulatory scrutiny. C++ has been the programming language of choice as one could make full benefit from hardware development on the floating-point-unit writing performance critical code very close to the execution unit. This is particularly important with CPU manufacturers utilizing the increase in transistor density to allow for greater vectorization of parallel instructions (SSEx, AVX, AVX2, ...) going hand in hand with the respective compiler support.

1.3.1.2 Graphic processing unit

Graphic processing units (GPUs) lend themselves very naturally to the computationally expensive task in the area of computational finance. The design of GPUs offers to take the well-established paradigm of multithreading to the next level, offering advantages in scale and also importantly energy usage. In particular, Monte Carlo simulations lend themselves very naturally to multithreading tasks due to their inherent parallel nature, where sets of paths can be processed efficiently in parallel. Thus, GPUs offer a very powerful solution for well-encapsulated high-performance tasks in computational finance. One of the key difficulties utilizing the full capabilities of GPUs is the integration into the analytics framework as most of the object management, serialization and interface functionality need to be handled by a higher level language such as C++, C#, Java, or even Python. All of those offer seamless integration with GPUs where offloading of task only requires some simple data transfer and dedicated code running on the GPU. However, this integration task does make the usage of GPUs a deliberate choice as one typically has to support more than one programming language and hardware technology. Additionally, upgrades to GPUs would make a reconfiguration or even code rewrite not unlikely to fully utilize the increased calculation power, which is no different to the greater level of vectorizations on CPUs.

1.3.1.3 Field programmable gate array

Field programmable gate arrays (FPGAs) offer high-performance solutions close to the actual hardware. At the core of FPGAs are programmable logic blocks which can be "wired" together using hardware description languages. At the very core, the developer is dealing with simple logic gates such as AND and XOR. Whilst both CPUs and GPUs have dedicated floating point units (FPUs) specifically designed to sequentially execute individual tasks at high speed, FPGAs allow systems to take advantage of not only a high level of parallelism similar to multithreading, but in particular creating a pipe of calculation tasks. At any clock cycle (configurable by the FPGA designer), the FPGA will execute not only a single task but every instruction along the calculation chain is executed. Monte Carlo simulation on FPGA can simultaneously execute the tasks of uniform random number generation, inverse cumulative function evaluation, path construction, and payoff evaluation—and every sub-step in this process will also have full parallelization and pipelining capacity. Therefore even at a far lower clock speed, one can execute many tasks along a "pipe" with further horizontal acceleration possible. This allows a far higher utilization of transistors compared to CPUs/GPUs, where most parts of the FPU are idle as only a single instruction is dealt with. One of the core difficulties of FPGAs is that one not only needs to communicate data between the CPU and the FPGA but the developer also needs to switch programming paradigm. Both CPUs and GPUs can handle a new set of tasks with relative ease by a mere change of the underlying assembly/machine code, whilst the FPGA requires much more dedicated reconfiguration of the executing instructions. Whilst FPGAs would offer even greater calculation speedups and in particular a significant reduction in energy usage it is yet to be seen whether they provide a true competitor to traditional hardware and software solutions. That being said, FPGAs do have a widespread usage in the context of high-frequency trading; in particular, colocated on exchange premises as they offer the ultimate hardware acceleration for a very dedicated task.

1.3.1.4 In-memory data aggregation

Significant increase in in-memory data storage capacity opened new capabilities for the calculation and aggregation of risk, counterparty and regulatory data. Calculation intensive computations such as Monte Carlo simulation for counterparty credit risk used to require data aggregation as an in-built feature for the efficiency of the algorithm where one had to discard a significant amount of intermediate calculation results such as the realization of an individual index at any time-discretization point. Thus, any change in the structure of the product (or counterparty, or collateral agreement) required a full rerun of the Monte Carlo simulation recreating the same intermediate values. Whilst storage capacity made this the only viable solution a few years ago, memory storage on 64-bit system does allow for consideration of more efficient alternatives. In-memory data aggregation in the context of computational finance does provide the distinctive advantage that more business decisions can be made in real time or further high-level optimization to be deployed such as best counterparty to trade with considering collateral cost, counterparty credit risk and regulatory capital requirements even keeping the product features the same.

1.3.2 Algorithmic differentiation

While the impact of hardware choices on computational performance is critical, that of software choices can be overlooked. Within software, optimization efforts tend to fall into at least two categories: first, the application of profiling tools to optimize at a low level, often very close to the hardware, leveraging features such as larger processor caches and vectorization within FPUs, and second, the choice of algorithms to perform the required calculations. We might consider a third category to be those concerns relevant to distributed computation: network latency and bandwidth, and marshaling between wire protocols and different execution environments in a heterogeneous system. For optimal performance, a holistic approach is required where both local and global considerations are taken into account.

The impact of algorithm choice on performance is perhaps nowhere more dramatic than in the context of calculating sensitivities, or greeks, for a portfolio. Current proposed regulatory guidance in the form of ISDA's Standard Initial Margin Model (SIMM) and the market risk component of the FRTB both require portfolio sensitivities as input parameters, and sensitivity calculation is a standard component of typical nightly risk runs for any trading desk interested in hedging market risk. The widespread approach to calculating such sensitivities is that of finite difference, or "bumping," in which each risk factor (typically a quote for a liquid market instrument) is perturbed, or bumped, by a small amount, often a basis point, and the portfolio revalued in order to assess the effect of the change. This process is repeated for each risk factor of interest, although in order to minimize computational cost, it is common to bump collections of quotes together to measure the effect of a parallel shift of an entire interest rate curve or volatility surface, for example.

Owing to the linear scaling of the computational burden of applying finite difference to a diverse portfolio with exposure to many market risk factors, sensitivity calculation has traditionally been expensive. This expense is mitigated by the embarrassingly parallel nature of the problem, which means long-running calculations can be accelerated through investment in computational grid technology and the associated software, energy and infrastructure required to support it.

With an alternative algorithmic approach, however, the need for portfolio revaluation can be eliminated, along with the linear scaling of cost with number of risk factors. This approach is called algorithmic differentiation (AD). While popularized in finance relatively recently (Sherif, 2015), AD is not new. It has been applied in a wide range of scientific and engineering fields from oceanography to geophysics since the 1970s (Ng and Char, 1979; Galanti and Tziperman, 2003; Charpentier and Espindola, 2005). It is based on encoding the operations of differential calculus for each fundamental mathematical operation used by a given computation, and then combining them using the chain rule when the program runs (Griewank, 2003). Consider the function $h : \mathbb{R}^m \rightarrow \mathbb{R}^n$ formed by composing the functions $f : \mathbb{R}^p \rightarrow \mathbb{R}^n$ and $g : \mathbb{R}^m \rightarrow \mathbb{R}^p$ such that

$$\begin{aligned} \vec{y} = h(\vec{x}) &= f(\vec{u}) \\ \vec{u} &= g(\vec{x}) \end{aligned} \tag{1.4}$$

for $\vec{x} \in \mathbb{R}^m$, $\vec{u} \in \mathbb{R}^p$, and $\vec{y} \in \mathbb{R}^n$. The chain rule allows us to decompose the $n \times m$ Jacobian J of h as follows:

$$J_{ij} = \left[\frac{\partial h(\vec{x})}{\partial \vec{x}}\right]_{ij} = \frac{\partial y_i}{\partial x_j} = \sum_{k=1}^{p} \frac{\partial y_i}{\partial u_k}\frac{\partial u_k}{\partial x_j} = \sum_{k=1}^{p} \left[\frac{\partial f(\vec{u})}{\partial \vec{u}}\right]_{ik} \left[\frac{\partial g(\vec{x})}{\partial \vec{x}}\right]_{kj}$$
$$= \sum_{k=1}^{p} A_{ik} B_{kj} \tag{1.5}$$

where A and B are the Jacobians of f and g, respectively. There is a straightforward generalization to an arbitrary number of function compositions, so we choose a single composition here for convenience and without loss of generality. Suppose that a computer program contains implementations of f and g explicitly, with h formed implicitly by supplying the output of g to a call to f.

1.3.2.1 Implementation approaches

AD defines two approaches to computing J; forward (or tangential) and reverse (or adjoint) accumulation (Rall and Corliss, 1996). In forward accumulation, B is computed first, followed by A. In other words, the calculation tree for the operation performed by h is traversed from its leaves to the root. The computational cost of such an approach scales linearly with the number of leaves because the calculation needs to be "seeded," that is repeatedly evaluated with $\vec{x} = \text{diag}(\delta_{jq})$ on the qth accumulation. This is well suited to problems where $n \gg m$, because it allows all n rows in J to be computed simultaneously for the qth accumulation. Such problems are so hard to find in the context of financial derivative valuation that it is very rare to read of any application of forward accumulation in the literature. The main exception to this rule may be in calibration, where the sensitivity of several instrument values to an often much smaller number of model parameters is useful for gradient descent optimizers.

In contrast, reverse accumulation is most efficient when $n \ll m$. It consists of two stages—a "forward sweep," where the relevant partial derivatives (termed "work variables") are formed, and then a "backward sweep" where the relevant products of partial derivatives are added into each element of A and B, which can then be multiplied to obtain the full Jacobian J. The result is a single calculation that calculates exact (to machine precision rather than approximations at an accuracy determined by the size of perturbation) derivatives to every risk factor to which the portfolio is exposed. This scales linearly with complexity of the portfolio (the number of links in the chain), but is effectively constant with respect to the number of risk factors. In practice, the cost of obtaining sensitivities through AD is typically a factor of between 2 and 4 times the cost of valuing the portfolio.

Often, the assembly of differentiation operations into the full chain for a given calculation is performed automatically by using overloaded operators on basic data types that record the required operations for later replay, and so AD is also termed Automatic Differentiation. In the AD literature, one finds two main approaches to implementation, whether forward or reverse accumulation, for the function h. Both approaches emphasize the problem of adding a differentiation capability to an existing codebase that computes the value alone, as opposed to designing AD into a codebase from the beginning. The first method, source code transformation, is based on a static analysis of the source code for the function h. New source code is generated for a companion function that computes J, then both are compiled (or interpreted).

The second method, operator overloading, requires a language where basic types can be redefined and operators can be overloaded (e.g., C++, C# or Java), so that existing code that performs these operations will also trigger the corresponding derivative calculations. Forward accumulation is easier to implement in an operator overloading approach than reverse. A common technique for reverse accumulation is to generate a data structure commonly termed the "tape" that records the relevant set of operations, then interpret that tape in order to obtain the desired derivatives.

1.3.2.2 Performance

Source code transformation is a disruptive technique that complicates build processes and infrastructure. In practice, operator overloading is more popular, but it suffers from high storage costs and long run times, with the result that numerous implementation techniques have been devised to mitigate the performance challenges inherent with AD and it remains an active area of research (e.g., Faure and Naumann, 2002). The fundamental barrier to performance in AD is the granularity at which the chain rule is implemented. When using tools to apply operator overloading to an existing codebase, the granularity is defined by the operators that are overloaded, which is typically very fine. Any per-operator overhead is multiplied by the complexity of the calculation, and for even modestly sized vanilla portfolios, this complexity is considerable. If instead the Jacobians can be formed at a higher level, so that each one combines multiple functional operations (multiple links in the chain, such as f and g together), then per-operator overhead is eliminated for each group and performance scales differently (Gibbs and Goyder, 2013b).

In addition, it is not necessary to construct an entire Jacobian at any level in a calculation and hold it in memory as a single object. To illustrate, suppose Equation 1.4 takes the concrete form

$$h(x, y, z) = \alpha x + \beta yz = f(x, g(y, z)) \tag{1.6}$$

where $f(a, b) = \alpha a + \beta b$ and $g(a, b) = ab$ where all variables are $\in \mathbb{R}$ and α and β are considered constants. By differentiating,

$$\begin{aligned} \mathrm{d}h &= \frac{\partial h}{\partial x}\,\mathrm{d}x + \frac{\partial h}{\partial z}\,\mathrm{d}y + \frac{\partial h}{\partial z}\,\mathrm{d}z \\ &= \frac{\partial f}{\partial x}\,\mathrm{d}x + \frac{\partial f}{\partial g}\frac{\partial g}{\partial y}\,\mathrm{d}y + \frac{\partial f}{\partial g}\frac{\partial g}{\partial z}\,\mathrm{d}z \\ &= \alpha\,\mathrm{d}x + \beta z\,\mathrm{d}y + \beta y\,\mathrm{d}z. \end{aligned} \tag{1.7}$$

The Jacobians are $A = (\alpha, \beta)$ and $B = (b, a)$, however, in the chain rule above, only factors such as $(\partial f/\partial g)(\partial g/\partial y)$ are required to determine the sensitivity to a given variable, such as y. These factors comprise subsets of each Jacobian and turn out to be exactly those subsets that reside on the stack when typically implementations of hand-coded AD are calculating the sensitivity to a given risk factor (Gibbs and Goyder, 2013b).

Implementing AD "by hand" in this way also enables a number of optimizations which further reduce both storage costs and increase speed of computation. Deeply recursive structures arise naturally in a financial context, in path-dependent payoffs such as those found in contracts with accumulator features, and in curve bootstrapping algorithms where each maturity of, say, expected Libor depends on earlier maturities. Naive implementations of AD suffer from quadratic scaling in such cases, whereas linear performance is possible by flattening each recursion.

In Monte Carlo simulations, only the payoff and state variable modeling components of a portfolio valuation change on each path. Curves, volatility surfaces, and other model parameters do not, and so it is beneficial to calculate sensitivities per path to state variables and only continue the chain rule through the relationship between model parameters and market data (based on calibration) once. In contrast, for certain contracts such as barrier options, it is common to evaluate a volatility surface many times, which would result in a large Jacobian. Instead, performance will be improved if the chain rule is extended on each volatility surface evaluation through the parameters on which the surface depends. This set of parameters is typically much smaller than the number of volatility surface evaluations.

While the value of implementing AD by hand is high, so is the cost. A good rule of thumb is that implementing AD costs, on average, is approximately the same as implementing the calculation whose sensitivities are desired. The prospect of doubling investment in quantitative development resources is unwelcome for almost any financial business, and consequently implementations of AD tend to be patchy in their coverage, restricted to specific types of trade or other bespoke aspects of a given context.

1.3.2.3 Coverage

When evaluating value, risk, and similar metrics, it is important to have a complete picture. Any hedging, portfolio rebalancing, or other related decision making that is performed on incomplete information will be exposed to the slippage due to the difference between the actual position and the reported numbers used. Whilst some imperfections, such as those imposed by model choice, are inevitable, there is no need or even excuse to introduce others through unnecessary choices such as an incomplete risk implementation. This can be easily avoided by adopting a generic approach that enables the necessary computations to be easily implemented at every step of the risk calculation.

For complex models, the chain rule of differentiation can and does lead to the mixing of exposures to different types of market risk. For example, volatility surfaces specified relative to the forward are quite common and lead to contributions of the form $(\partial C/\partial\sigma)((\partial\sigma(F, K))/\partial F)$ to the delta of an option trade through the "sticky smile." This mixing obviously increases with model complexity and consequently any generic approach has to confront it directly. Imposing the simple requirement that all first-order derivatives, rather than just some targeted towards a specific subset of exposures, are always calculated as part of the AD process allows an implementation to avoid missing these risk contributions.

It is possible to separate valuation and related calculations into a chain of logical steps with a high level of commonality and reusability for each link in the chain. In general, the steps are the calibration of model parameters to input observables such as market data quotes, the setup of simulation or similar calculations, specific pricing routines for classes of financial contracts, and portfolio-level aggregation of the values calculated per-trade. These links in the chain can be separately built up and independently maintained to form a comprehensive analytic library that can then be used to cover essentially all valuation, risk, and related calculations.

Within each of these steps, the dependency tree of calculations has to be checked and propagated through the calculation. By carefully defining how these steps interact with each other, a high level of reuse can be achieved. As well as enabling the use of pricing algorithms within model calibration, a key consideration when setting up complex models, the nested use of such calculations is important for analyzing the important class of contracts with embedded optionality. In particular, the presence of multiple exercise opportunities in American and Bermudan options necessitates some form of approximation of future values when evaluating the choices that can be made at each of these exercise points. Typically, this approximation is formed by regressing these future values against explanatory variables known at the choice time. Standard methods such as maximum likelihood determine the best fit through minimizing some form of error function; the act of minimization means that first-order derivatives are by construction zero. Note that this observation applies on average to the particular values (Monte Carlo paths, or similar) used in the regression. Subsequent calculations are exposed to numerical noise when a different set of values are used in an actual calculation, and this noise extends to the associated risk computations.

1.4 Conclusion

This chapter provides a comprehensive overview of computational challenges in the history of derivative pricing. Before the 2008 credit crisis, the main driver of computational expense was the complexity of sophisticated contracts and associated market risk management. In the wake of the crisis, counterparty credit risk came to the fore, while contracts simplified. Paradoxically, incorporating counterparty risk into pricing for even vanilla derivative portfolios presents computational problems more challenging than for the most complex precrisis contracts.

In response to this challenge, investment in specialized computing hardware such as GPU and FPGA, along with the rising tide of commodity compute power available in clouds, rose sharply in the last decade. However, specialized hardware requires specialized software development to realize its power, which in turn limits the flexibility of systems and the speed at which they are able to cope with changing market and regulatory requirements. Commodity hardware, particularly if managed in-house, can incur nontrivial infrastructure and maintenance costs.

For calculating sensitivities, a key input to both market risk management and regulatory calculations for initial margin and capital requirements, a software technique called AD offers the potential to reduce hardware requirements dramatically. While popularized only recently, AD is an old idea. Tools for applying AD to existing codebases exist but suffer from performance drawbacks such that the full potential of the technique, for both performance and coverage of contract and calculation types, is only realized when AD is incorporated into a system's architecture from the beginning.

References

Acerbi, C. and Szkely, B.: Back-testing expected shortfall. *Risk Magazine* (December 2014).

Albanese, C. and Andersen, L.: *Accounting for OTC Derivatives: Funding Adjustments and the Re-Hypothecation Option.* Available at SSRN: https://ssrn.com/abstract=2482955 or http://dx.doi.org/10.2139/ssrn.2482955, accessed March 10, 2017. 2014.

Albanese, C., Andersen, L., and Stefano, I.: *The FVA Puzzle: Accounting, Risk Management and Collateral Trading.* Available at SSRN: https://ssrn.com/abstract=2517301 or http://dx.doi.org/10.2139/ssrn.2517301, accessed March 10, 2017. 2014.

Brigo, D., Buescu, C., and Morini, M.: Counterparty risk pricing: Impact of closeout and first-to-default times. *International Journal of Theoretical and Applied Finance*, 15(6): 2012.

Brigo, D., Morini, M., and Pallavicini, A.: *Counterparty Credit Risk, Collateral and Funding.* John Wiley & Sons, UK, 2013.

Burgard, C. and Kjaer, M.: Partial differential equation representations of derivatives with bilateral counterparty risk and funding costs. *The Journal of Credit Risk*, 7:75–93, 2011.

Charpentier, I. and Espindola, J.M.: A study of the entrainment function in models of Plinian columns: Characteristics and calibration. *Geophysical Journal International*, 160(3):1123–1130, 2005.

Elouerkhaoui, Y.: From FVA to KVA: Including cost of capital in derivatives pricing. *Risk Magazine* (March 2016).

Faure, C. and Naumann, U.: Minimizing the tape size. In Corliss G., Faure C., Griewank A., Hascoët L., and Naumann U., editors, *Automatic Differentiation of Algorithms: From Simulation to Optimization*, Computer and Information Science, Chapter 34, pages 293–298. Springer, New York, NY, 2002.

Galanti, E. and Tziperman, E.: A midlatitude-enso teleconnection mechanism via baroclinically unstable long rossby waves. *Journal of Physical Oceanography*, 33(9):1877–1888, 2003.

Gibbs, M. and Goyder, R.: Automatic Numeraire Corrections for Generic Hybrid Simulation. Available at SSRN: https://ssrn.com/abstract=2311740 or http://dx.doi.org/10.2139/ssrn.2311740, accessed August 16. 2013a.

Gibbs, M. and Goyder, R.: *Universal Algorithmic Differentiation*^{TM} *in the F3 Platform*. Available at: http://www.fincad.com/resources/resource-library/whitepaper/universal-algorithmic-differentiation-f3-platform, accessed March 10, 2017. 2013b.

Green, A.: *XVA: Credit, Funding and Capital Valuation Adjustments*. John Wiley & Sons, UK, 2016.

Gregory, J.: Being two faced over counterparty credit risk. *Risk*, 22(2):86–90, 2009.

Gregory, J.: *The xVA Challenge*. John Wiley & Sons, UK, 2015.

Griewank, A.: A mathematical view of automatic differentiation. In Arieh I., editor, *Acta Numerica*, volume 12, pages 321–398. Cambridge University Press, 2003.

Ng, E. and Char, B. W.: Gradient and Jacobian computation for numerical applications. In Ellen Golden V., editor, *Proceedings of the 1979 Macsyma User's Conference*, pages 604–621. NASA, Washington, D.C., June 1979.

Rall, L. B. and Corliss, G. F.: An introduction to automatic differentiation. In Berz M., Bischof C. H., Corliss G. F., and Griewank A., editors, *Computational Differentiation: Techniques, Applications, and Tools*, pages 1–17. SIAM, Philadelphia, PA, 1996.

Ramirez, J.: *Accounting for Derivatives: Advanced Hedging under IFRS 9*. John Wiley & Sons, UK, 2015.

Sherif, N.: Chips off the menu. *Risk Magazine*, 27(1): 12–17, 2015.

Watt, M.: Corporates rear CVA charge will make hedging too expensive. *Risk*, October Issue 2011.

Chapter 2

Using Market Sentiment to Enhance Second-Order Stochastic Dominance Trading Models

Gautam Mitra, Christina Erlwein-Sayer, Cristiano Arbex Valle, and Xiang Yu

CONTENTS

2.1 Introduction and Background

We propose a method of computing (daily) trade signals; the method applies a second-order stochastic dominance (SSD) model to find portfolio weights within a given asset universe. In our report of the "use cases" the assets are the constituents of the exchange traded securities of different major indices such as Nikkei, Hang Seng, Eurostoxx 50, and FTSE 100 (see Sections 2.6 and 2.7). The novelty and contribution of our research are fourfold: (1) We introduce a model of "enhanced indexation," which applies the SSD criterion. (2) We then improve the reference, that is, the benchmark distribution using a tilting method; see Valle, Roman, and Mitra (2017; also Section 2.4.3). This is achieved using impact measurement of market sentiment; see Section 2.4.1. (3) We embed the static SSD approach within a dynamic framework of "volatility pumping" with which we control drawdown and the related temporal risk measures. (4) Finally, we apply a "branch and cut" technique to speed up the repeated solution of a single-period stochastic integer programming (SIP) model instances. Our models are driven by two sets of time-series data, which are first the market price (returns) data and second the news (meta) data; see Section 2.2 for a full discussion of these.

2.1.1 Enhanced indexation applying SSD criterion

SSD has a well-recognized importance in portfolio selection due to its connection to the theory of risk-averse investor behavior and tail-risk minimization. Until recently, stochastic dominance models were considered intractable, or at least very demanding from a computational point of view. Computationally tractable and scalable portfolio optimization models that apply the concept of SSD were proposed recently (Dentcheva and Ruszczyński 2006; Roman, Darby-Dowman, and Mitra 2006; Fábián, Mitra, and Roman, 2011a). These portfolio optimization models assume that a benchmark, that is, a desirable "reference" distribution is available and a portfolio is constructed, whose return distribution dominates the reference distribution with respect to SSD. Index tracking models also assume that a reference distribution (that of a financial index) is available. A portfolio is then constructed, with the aim of replicating, or tracking, the financial index. Traditionally, this is done by minimizing the tracking error, that is, the standard deviation of the differences between the portfolio and index returns. Other methods have been proposed (for a review of these methods, see Beasley et al. 2003; Canakgoz and Beasley 2008).

The passive portfolio strategy of index tracking is based on the well-established "Efficient Market Hypothesis" (Fama, 1970), which implies that financial indices achieve the best returns over time. Enhanced indexation models are related to index tracking in the sense that they also consider the return distribution of an index as a reference or benchmark. However, they aim to outperform the index by generating "excess" return (DiBartolomeo, 2000; Scowcroft and Sefton, 2003). Enhanced indexation is a very new area of research and there is no generally accepted portfolio construction method in this field (Canakgoz and Beasley, 2008). Although the idea of enhanced indexation was formulated as early as 2000, only a few enhanced indexation methods were proposed later in the research community; for a review of this topic see Canakgoz and Beasley (2008). These methods are predominantly concerned with overcoming the computational difficulty that arises due to restriction on the cardinality of the constituent assets in the portfolios. Not much consideration is given to answering the question if they do attain their stated purpose, that is, achieve return in excess of the index.

In an earlier paper (Roman, Mitra, and Zverovich, 2013), we have presented extensive computational results illustrating the effective use of the SSD criterion to construct "models of enhanced indexation." SSD dominance criterion has been long recognized as a rational criterion of choice between wealth distributions (Hadar and Russell 1969; Bawa 1975; Levy 1992). Empirical tests for SSD portfolio efficiency have been proposed in Post (2003) and Kuosmanen (2004). In recent times, SSD choice criterion has been proposed (Dentcheva and Ruszczynski 2003, 2006, Roman et al. 2006) for portfolio construction by researchers working in this domain. The approach described in Dentcheva and Ruszczynski (2003, 2006) first considers a reference (or benchmark) distribution and then computes a portfolio, which dominates the benchmark distribution by the SSD criterion. In Roman et al. (2006), a multi-objective optimization model is introduced to achieve SSD dominance. This model is both novel and usable since, when the benchmark solution itself is SSD efficient or its dominance is unattainable, it finds an SSD-efficient portfolio whose return distribution comes close to the benchmark in a satisfying sense. The topic continues to be researched by academics who have strong interest in this approach: Dentcheva and Ruszczynski (2010), Post and Kopa (2013), Kopa and Post (2015), Post et al. (2015), Hodder, Jackwerth, and Kolokolova (2015), Javanmardi and Lawryshy (2016). Over the last decade, we have proposed computational solutions and applications to large-scale applied problems in finance (Fábián et al. 2011a, 2011b; Roman et al. 2013; Valle et al. 2017).

From a theoretical perspective, enhanced indexation calls for further justification. The efficient market hypothesis (EMH) is based on the key assumption that security prices fully reflect all available information. This hypothesis, however, has been continuously challenged; the simple fact that academics and practitioners commonly use "active," that is, non-index tracking strategies, vindicates this claim. An attempt to reconcile the advocates and opponents of the EMH is the "adaptive market hypothesis" (AMH) put forward by Lo (2004). AMH postulates that the market "adapts" to the information received and is generally efficient, but there are periods of time when it is not; these periods can be used by investors to make profit in excess of the market index. From a theoretical point of view, this justifies the quest for techniques that seek excess return over financial indices. In this sense, enhanced indexation aims to discover and exploit market inefficiencies. As set out earlier, a common problem with the index tracking and enhanced indexation models is the computational difficulty which is due to cardinality constraints that limit the number of stocks in the chosen portfolio. It is well known that most index tracking models naturally select a very large number of stocks in the composition of the tracking portfolio. Cardinality constraints overcome this problem, but they require introduction of binary variables and thus the resulting model becomes much more difficult to solve. Most of the literature in the field is concerned with overcoming this computational difficulty. The good in-sample properties of the return distribution of the chosen portfolios have been underlined in previous papers: Roman, Darby-Dowman, and Mitra (2006), using historical data; Fábián et al. (2011c), using scenarios generated via geometric Brownian motion.

However, it is the actual historical performance of the chosen portfolios (measured over time and compared with the historical performance of the index) that provides empirical validation of whether the models achieved their stated purpose of generating excess return.

We also investigate aspects related to the practical application of portfolio models in which the asset universe is very large; this is usually the case in index tracking and enhanced indexation models. It has been recently shown that very large SSD-based models can be solved in seconds, using solution methods which apply the cutting-plane approach, as proposed by Fábián et al. (2011a). Imposing additional constraints that add realism (e.g., cardinality constraints, normally required in index tracking) increase the computational time dramatically.

2.1.2 Revising the reference distribution

The reference distribution, that is, the distribution of the financial index under consideration is "achievable" since there exists a feasible portfolio that replicates the index. Empirical evidence (Roman et al. 2006, 2013; Post and Kopa 2013) further suggests that in most cases this distribution can be SSD dominated; this in turn implies that the distribution is not SSD efficient. When there exists a scope of improvement we set out to compute an improved distribution as our revised "benchmark." We achieve this by using a tilting method and the improvement step is triggered by the impact measure of the market sentiment. For a full description of this method, see Valle, Roman, and Mitra (2017; also Section 2.4.3) and Mitra, Erlwein-Sayer, and Roman (2017).

2.1.3 Money management via "volatility pumping"

Similar to the Markowitz model, SSD works as a "single period SP," that is, a static or myopic framework. Whereas volatility or tail risks such as value at risk (VAR) and conditional value at risk (CVAR) can be controlled, these are static measures of risk. In contrast "maximum drawdown" and "days to recovery" are dynamic risk measures, which are considered to be more important performance measures by the active trading and fund management community. In order to control drawdown and the related temporal risk measures we have resorted to money management via "volatility pumping." This is based on the eponymous criterion called "Kelly Criterion" of John Kelly, extended and much refined by Ed Thorp. Ed Thorp first applied this in the setting of "Black Jack" (see Thorp 1966) and subsequently in trading and fund management (see Thorp 1967). The underlying principles are lucidly explained by David Luenberger (1997; see Chapter 15) who has coined the term "volatility pumping."

2.1.4 Solution methods for SIP models

Computing "long only" SSD portfolios (strategies) requires solution of an LP; this can be a computationally tractable problem, yet it may require including a large number of constraints or cuts which are generated dynamically in the solution process (see Fábián et al. 2011a, 2011b). To apply "long–short" strategy leads to further computational challenge of solving stochastic integer programs (SIP) made up of binary variables. Given n assets, the model requires $2n$ binary decision variables and $2n$ continuous variables (200 binary and 200 continuous for the components of FTSE100, for instance). As such the model is relatively small, however, if we consider all the cuts which are dynamically added (see Equation 2.7), the model size grows exponentially as a huge number of constraints are generated: one for every subset of S with cardinality $s = 1, \ldots, S$. We refer the readers to the formulation set out in Section 2.4.2 (see Equations 2.1 through 2.8).

2.1.5 Guided tour

In Section 2.2, we discuss the two time-series data sets with which our models are instantiated. In Section 2.3, we describe the general system architecture and in Section 2.4, we set out in detail the relevant models and methods to find SSD optimal portfolios. We present our strategies and the concepts that underpin these in Section 2.5. The results of out-of-sample back tests are presented and discussed in Section 2.6.

TABLE 2.1: Description of all the data fields for a company in the market data

Data field	Field name	Description
1	##RIC	Reuters instrument code individually assigned to each company
2	Date	In the format DD-MM-YYYY
3	Time	In the format hh:mm:ss, given to the nearest minute
4	GMT Offset	Difference from Greenwich mean time
5	Type	Type of market data
6	Last	Last prices for the corresponding minute

2.2 Data

Our modelling architecture uses two streams of time-series data: (i) market data which is given on a daily frequency, and (ii) news metadata as supplied by Thomson Reuters. A detailed description of these datasets is given below.

2.2.1 Market data

In these set of experiments, daily prices for each asset have been used to test the trading strategy. The data fields of the market data are set out in Table 2.1. The trading strategy (in Section 2.5) is tested on the entire asset universe of FTSE100.

2.2.2 News meta data

News analytics data are presented in the well-established metadata format whereby a news event is given tags of relevance, novelty and sentiment (scores) for a given individual asset. The analytical process of computing such scores is fully automated from collecting, extracting, aggregating, to categorizing and scoring. The result is a numerical score assigned to each news article for each of its different characteristics. Although we use intraday news data, for a given asset the number of news stories, hence data points, are variable and do not match the time frequencies of market data. The attributes of news stories used in our study are relevance and sentiment.

The news metadata for the chosen assets were selected under the filter of relevance score, that is, any news item that had a relevance score under the value of 100 was ignored and not included in the data set. This ensured with a high degree of certainty that the sentiment scores to be used are indeed focused on the chosen asset and is not just a mention in the news for comparison purposes, for example.

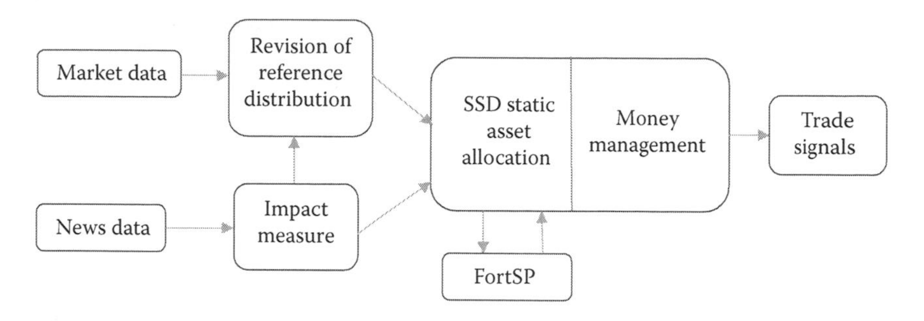

FIGURE 2.1: System architecture.

2.3 Information Flow and Computational Architecture

News analytics in finance focus on improving IT-based legacy system applications. These improvements come through research and development directed to automated/semi-automated programmed trading, fund rebalancing and risk control applications.

The established good practice of applying these analytics in the traditional manual approach is as follows. News stories and announcements arrive synchronously and asynchronously. In the market assets' (stocks, commodities, FX rates, etc.) prices move (market reactions). The professionals digest these items of information and accordingly make trading decisions and re-compute their risk exposures.

The information flow and the (semi-) automation of the corresponding IS architecture is set out in Figure 2.1. There are two streams of information which flow simultaneously, news data and market data. Pre-analysis is applied to news data; it is further filtered and processed by classifiers to relevant metrics. This is consolidated with the market data of prices, and together they constitute the classical datamart which feed into the SSD asset allocation model. A key aspect of this application is that it sets out to provide technology-enabled support to professional decision makers.

The SSD asset allocation optimization model is solved with a specialized solver and its output feeds the recommended traded signals.

The computational system is run in an Intel(R) Core(TM) i5-3337U CPU@1.80 GHz with 6 GB of RAM and Linux as operating system. The back testing framework was written in R and the required data are stored in a SQLite database. The SSD solver is invoked by the R framework and it was written in C++ with FortSP and CPLEX 12.6 as mixed-integer programming solvers.

2.4 Models

2.4.1 Impact measure for news

In our analytical model for news we introduce two concepts, namely, (i) sentiment score and (ii) impact score. The sentiment score is a quantification of the mood (of a typical investor) with respect to a news event. The impact score takes into consideration the decay of the sentiment of one or more news events and how after aggregation these impact the asset behavior.

2.4.1.1 Sentiment score

Thomson Reuters' news sentiment engine analyzes and processes each news story that arrives as a machine readable text. Through text analysis and other classification schemes, the engine then computes for each news event: (i) relevance, (ii) entity recognition, and (iii) sentiment probabilities, as well as a few other attributes (see Mitra and Mitra, 2011). A news event sentiment can be positive, neutral and negative and the classifier assigns probabilities such that all three probabilities sum to one.

We turn these three probabilities into a single sentiment score in the range +50 to −50 using the following equation:

$$Sent = 100 * \left(Prob(\text{positive}) + \frac{1}{2} Prob(\text{neutral}) \right) - 50 \tag{2.1}$$

where $Sent$ denotes a single transformed sentiment score. We find that such a derived single score provides a relatively better interpretation of the mood of the news item. Thus, the news sentiment score is a relative number that describes the degree of positivity and negativity in a piece of news. During the trading day, as news arrives it is given a sentiment value. Given that $-50 \leq Sent \leq 50$, for a given news item k at the time bucket t_k, we define $PNews(k, t_k)$ and $NNews(k, t_k)$ as the sentiments of the kth news (see the following section).

2.4.1.2 Impact score

It is well known from research studies that news flow affects asset behavior (Patton and Verardo, 2012; Mitra, Mitra, and diBartolomeo, 2009). Therefore, the accumulation of news items as they arrive is important. Patton and Verardo (2012) noticed decay in the impact of news on asset prices and their betas on a daily timescale and further determine the complete disappearance of news effects within 2–5 days. Mitra, Mitra, and diBartolomeo (2009) created a composite sentiment score in their volatility models after initial experiments revealed no effect on volatility predictions with sentiment alone; the decay period in this study was over 7 days.

To compute the impact of news events over time, we first find an expression which describes the attenuation of the news sentiment score. The impact of a news item does not solely have an effect on the markets at the time of release; the impact also persists over finite periods of time that follow. To account for this prolonged impact, we have applied an attenuation technique to reflect the instantaneous impact of news releases and the decay of this impact over a subsequent period of time. The technique combines exponential decay and accumulation of the sentiment score over a given time bucket under observation. We take into consideration the attenuation of positive sentiment to the neutral value and the rise of negative sentiment also to the neutral value and accumulate (sum) these sentiment scores separately. The separation of the positive and negative sentiment scores is only logical as this avoids cancellation effects. For instance, cancellation reduces the news flow and an exact cancellation leads to the misinterpretation of no news.

News arrives asynchronously; depending on the nature of the sentiment it creates, we classify these into three categories, namely: positive, neutral, and negative. For the purpose of deriving impact measures, we only consider the positive and negative news items.

Let,

POS denote the set of news with positive sentiment value $Sent > 0$

NEG denote the set of news with negative sentiment value $Sent < 0$

$PNews(k, t_k)$ denote the sentiment value of the kth positive news arriving at time bucket t_k, $1 \leq t_k \leq 630$ and $k \in POS$; $PNews(k, t_k) > 0$

$NNews(k, t_k)$ denote the sentiment value of the kth negative news arriving at time bucket t_k, $1 \leq t_k \leq 630$ and $k \in NEG$; $NNews(k, t_k) < 0$

Let λ denote the exponent which determines the decay rate. We have chosen λ such that the sentiment value decays to half the initial value in a 90 minute time span. The cumulated positive and negative sentiment scores for one day are calculated as

$$PImpact(t) = \sum_{\substack{k \in POS \\ t_k \leq t}} PNews(k, t_k)e^{-\lambda(t-1)} \tag{2.2}$$

$$NImpact(t) = \sum_{\substack{k \in NEG \\ t_k \leq t}} NNews(k, t_k)e^{-\lambda(t-1)} \tag{2.3}$$

In Equations 2.2 and 2.3 for intraday $PImpact$ and $NImpact$, t is in the range, $t = 1, \ldots, 630$. On the other hand for a given asset all the relevant news items which arrived in the past, in principle, have an impact for the asset. Hence, the range of t can be widened to consider past news, that is, news which are 2 or more days "old."

A bias exists for those companies with a bigger market capitalization because they are covered more frequently in the news. For the smaller companies within a stock market index, their press coverage will be less and therefore, there is fewer data points to work with.

2.4.2 Long–short discrete optimization model based on SSD

Let $X \subset \mathbb{R}^n$ denote the set of the feasible portfolios and assume that X is a bounded convex polytope. A portfolio x^* is said to be SSD-efficient if there is no feasible portfolio: $x \in X$ such that $R_x \geq_{ssd} R_{x*}$.

Recently proposed portfolio optimization models based on the concept of SSD assume that a reference (benchmark) distribution R^{ref} is available. Let $\hat{\tau}$ be the tails of the benchmark distribution at confidence levels $(1/S, \ldots, S/S)$; that is, $\hat{\tau} = (\hat{\tau}_1, \ldots, \hat{\tau}_S) = (\mathrm{Tail}_{\frac{1}{S}} R^{ref}, \ldots, \mathrm{Tail}_{\frac{S}{S}} R^{ref})$.

Assuming equiprobable scenarios as in Roman et al. (2006, 2013) and Fábián et al. (2011a, 2011b), the model in Fábián et al. (2011b) optimizes the worst difference between the "scaled" tails of the benchmark and of the return distribution of the solution portfolio; the "scaled" tail is defined as $(1/\beta)\mathrm{Tail}_\beta(R)$. $V = \min\limits_{1 \leq s \leq S} (1/S)(\mathrm{Tail}_{s/S}(R_x) - \hat{\tau}_s$ represents the worst partial achievement of the differences between the scaled tails of the portfolio return and the scaled tails of the benchmark. The scaled tails of the benchmark are $\left(\frac{S}{1}\hat{\tau}_1, \frac{S}{2}\hat{\tau}_2, \ldots, \frac{S}{S}\hat{\tau}_S\right)$.

When short-selling is allowed, the amount available for purchases of stocks in long positions is increased. Suppose we borrow from an intermediary a specified number of units of asset $i (i = 1, \ldots, n)$ corresponding to a proportion x_i^- of capital. We sell them immediately in the market and hence have a cash sum of $(1 + \Sigma_{j=1}^n x_i^-)C$ to invest in long positions; where C is the initial capital available.

In long–short practice, it is common to fix the total amount of short-selling to a pre-specified proportion α of the initial capital. In this case, the amount available to invest in long positions is $(1+\alpha)C$. A fund that limits their exposure with a proportion $\alpha = 0.2$ is usually referred to as a 120/20 fund. For modelling this situation, to each asset $i \in \{1, \ldots, n\}$ we assign two continuous nonnegative decision variables x_i^+, x_i^-, representing the proportions invested in long and short positions in asset i, and two binary variables z_i^+, z_i^- that indicate whether there is investment in long or short positions in asset i. For example, if 10% of the capital is shortened in asset i, we write this as $x_i^+ = 0, x_i^- = 0.1, z_i^+ = 0, z_i^- = 1$. We also assigned a decision variable V defined as above (worst partial achievement).

Using a cutting-plane representation Fábián et al. (2011a), the scaled long/short formulation of the achievement-maximization problem is written as

$$\max V \tag{2.4}$$

subject to

$$\sum_{i=1}^{n} x_i^+ = 1 + \alpha \tag{2.5}$$

$$\sum_{i=1}^{n} x_i^- = \alpha \tag{2.6}$$

$$x_i^+ \leq (1+\alpha)z_i^+ \qquad \forall i \in N \tag{2.7}$$

$$x_i^- \leq \alpha z_i^- \qquad \forall i \in N \tag{2.8}$$

$$z_i^+ + z_i^- \leq 1 \qquad \forall i \in N \tag{2.9}$$

$$\frac{s}{S}V + \hat{\tau}_s \leq \frac{1}{S}\sum_{j \in J_s}\sum_{i=1}^{n} r_{ij}(x_i^+ - x_i^-) \quad \forall J_s \subset \{1,\ldots,S\}, |J_s| = s, s = \{1,\ldots,S\} \tag{2.10}$$

$$V \in \mathbb{R}, x_i^+, x_i^- \in \mathbb{R}^+, z_i^+, z_i^- \in \{0,1\} \quad \forall i \in N \tag{2.11}$$

2.4.3 Revision of reference distribution

In recent research, the most common approach is to set the benchmark as the return distribution of a financial index. This is natural since discrete approximations for this choice can be directly obtained from publicly available historical data, and also due to the meaningfulness of interpretation; the common practice is to compare the performance of a portfolio with the performance of an index. Applying the SSD criterion, we may construct a portfolio that dominates a chosen index yet there is no guarantee that this portfolio will have the desirable properties that an informed investor is looking for. Set against this background, we propose a method of reshaping a given reference distribution and compute a synthetic (improved) reference distribution. It may not be possible to SSD dominate such a reference distribution; in these cases, the closest SSD efficient portfolio is constructed by our model (see Roman et al., 2006). To clarify what we mean by improved reference distribution, let us consider the upper (light grey) density curve of the original reference distribution in Figure 2.2. In this example, the reference distribution is nearly symmetrical and has a considerably long left-tail. The lower (black) curve in Figure 2.2 represents the density curve of what we consider to be an improved reference distribution. Desirable properties include a shorter left tail (reduced probability of large losses), and a higher expected return which translates into higher skewness. A smaller standard deviation is not necessarily desirable, as it might limit the upside potential of high returns. Instead, we require the standard deviation of the new distribution to be within a specified range from the standard deviation of the original distribution.

We would like to transform the original reference distribution into a synthetic reference distribution given target values for the first three statistical moments (mean, standard deviation, and skewness). In a recent paper, we have developed an approximate method of finding the three moments of the improved distribution. This method solves a system of nonlinear equations using the Newton–Raphson iterations. For a full description of this approach, see Valle et al. (2017).

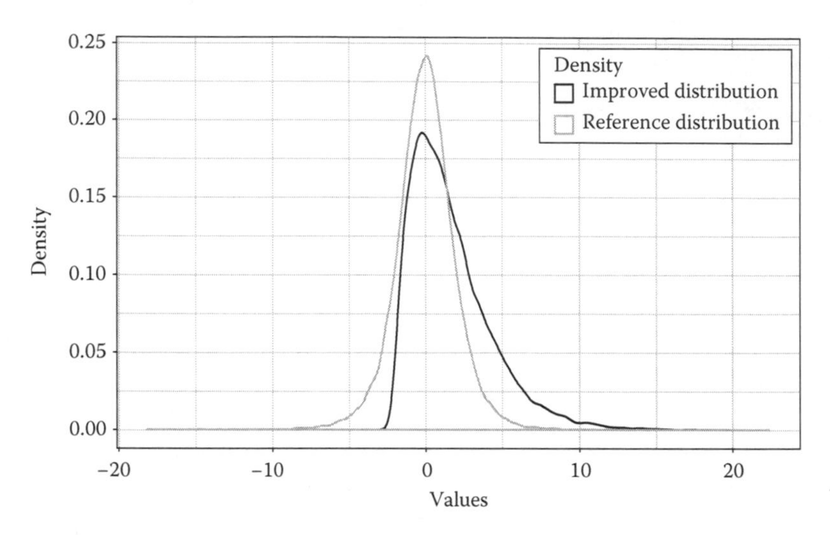

FIGURE 2.2: Density curves for the original and improved reference distributions.

In a recent research project, we have focused on this problem further. Essentially, our aim is to find a reference distribution with target values μ_T, σ_T and γ_T. We standardize our observed index distribution and perform a shift toward the target values. Since we wish to increase the upside potential this can be stated as maximizing the skewness, that is, the third moment, subject to constraints. This is formulated as a non-linear optimization problem. The proposed approach (Mitra et al. 2017) sets out to solve an NLP in which the constraints are set between the three moments of an index distribution (these are known, hence they are the parameters) and those of the target distribution (these are therefore the decision variables) of the NLP model. The solution of the NLP leads to an improved reference distribution. This in turn is utilized within the SSD optimization to achieve an enhanced portfolio performance.

2.5 Trading Strategies

In this study, we present three different trading strategies. The first strategy will be the base case which takes into account all available assets and forms the portfolio composition based on the SSD methodology stated above. The second and third strategies also employ the SSD methodology, but restrict the available asset universe. In the following, we briefly describe the used strategies.

2.5.1 Base strategy

2.5.1.1 Strategy A: Full asset universe

This strategy basically utilizes the concept of SSD. Whenever new portfolio weights are needed, we call the SSD optimization to determine these optimal portfolio weights. The computation of these weights depends on recent market data as well as on the market regime.

Generally, all assets in the index are considered potential portfolio constituents and the SSD optimization decides which assets will be in the portfolio (and are chosen either long and short) and which assets are not considered.

2.5.2 Using relative strength index as a filter

2.5.2.1 Strategy B: Asset filter relative strength index

Our second trading strategy restricts the asset universe for our SSD tool. We employ the relative strength index (RSI), which is a technical indicator and is rather standard in technical analysis of financial assets. RSI was introduced by J. Welles Wilder (1978) and has since gained in popularity with technical traders. It is a momentum key figure and analyzes the performance of an asset over an average period of 14 days. RSI is the ratio of medium gains to losses. It highlights overbought or oversold assets and indicates an imminent market reversal.

The RSI takes on values between 0 and 100 and is calculated through

$$RSI = 100 - \frac{100}{1 + RS}$$

where RS denotes the relative strength and is computed by dividing the average gains by the average losses, that is,

$$RS = \frac{EMA(Gain_t)}{EMA(Loss_t)}$$

Here, we calculate the averages as exponential moving averages. In fact, the average loss is computed as

$$EMA(Loss_t) = \frac{1}{n}\sum_{i=0}^{n} \alpha(i) L_{t-i}$$

where $\alpha(i) = \lambda e^{-\lambda * i}$ is the exponential weight, X_t is the price of the asset and $L_t = (X_t - X_{t-days}) I_{\{X_t - X_{t-days} < 0\}}$ is the loss at time t with *days* an offset of chosen number of days. Analogous, we calculate the average gain as

$$EMA(Gain_t) = \frac{1}{n}\sum_{i=0}^{n} \alpha(i) G_{t-i},$$

where $G_t = (X_t - X_{t-days}) I_{\{X_t - X_{t-days} \geq 0\}}$ is the gain at time t. Typical RSI values are calculated with average gains and losses over a period of 14 days, often called the lookback period. In the literature, the RSI is considered to highlight overbought assets, that is when the RSI is above the threshold of 70, and oversold, that is when it is below the threshold of 30.

In our application, we compute the RSI value for each available asset and flag assets as potential long candidates, if the RSI is below 30 and as potential short candidates, if the RSI is above 70. If, on the other hand, the RSI value is between 30 and 70, the asset is not considered to be a portfolio constituent. By doing so, we restrict the available asset universe compared to the above base-case strategy A.

2.5.3 Using relative strength index and impact as filters

2.5.3.1 Strategy C: Asset filter relative strength index and impact

For this strategy, we further restrict the available asset universe. As a first step, we apply the asset filter as in strategy B, which gives us two potential set of stocks out of which we can choose long and short exposures. In the next step, we remove a potential long asset, if the corresponding news impact is negative. Analogously, we remove potential short assets, if the corresponding impact is positive. This strategy therefore combines the momentum strategy with news sentiment. The news impact gives an additional lead whether the RSI indicator picks up the current market information. If the current market sentiment seems to contradict the indicator, the asset is not included in the potential asset universe. The impact reflects the current state of the market and therefore improves further the stock filtering step.

Since we only remove assets as potential portfolio constituents, the asset universe for strategy C is smaller or equal to the asset universe for strategy B. The reduction of the asset universe is particularly vital for a large asset universe, for example, components of NIKKEI 250. The reduction of the number of assets further accelerates the determination of the portfolio composition. Further aspects of momentum and value strategies including news sentiment are discussed in Valle and Mitra (2013).

2.5.4 A dynamic strategy using money management

A major performance criterion in the fund management industry is to measure "maximum draw down" and "days to recovery." Thorp and his associates had formally analyzed the problem of gambling (Thorp, 1966) and stock market trading (Thorp and Kassouf, 1967) and had derived strategies which are "winning propositions." The central plank to these is the celebrated "Kelly Strategy (KS)." Luenberger (1997) presents a lucid explanation of this and has coined the term "volatility pumping." Whereas SSD is an otherwise static single-period asset allocation model, we have adopted a simplified version of the KS and apply it to construct our daily trading signals. As a consequence, we are able to control the dynamic risk measures such as draw down. This we have termed "money management." Thus, portfolios are rebalanced using

SSD criterion and also applying the principle of money management. Since our SSD model has long–short positions in risky assets we also determine the long and short exposures dynamically but staying within regulatory regimes such as "Regulation T" stipulated by the US regulators. Thus if the strategy is to have (100 + alpha)/alpha; that is, (100 + alpha) long and (alpha) short exposures, then we simply control this adaptively by a limit (alpha-max) such that (alpha) <= (alpha-max). The actual settings of the money management parameter and the parameter "alpha-max" in our experiments are discussed in Section 2.7 where we report results of our experiments.

2.6 Solution Method and Processing Requirement

2.6.1 Solution of LP and SIP models

The asset allocation decision involves solving a single-stage integer stochastic program based on a set of S discrete scenarios (the formulation introduced in Section 2.4.2). Given n assets, the model requires $2n$ binary decision variables and $2n$ continuous variables (200 binary and 200 continuous for the components of FTSE100, for instance). As such the model is relatively small, however, if we consider all the cuts which are dynamically added (see Equation 2.10), the model size grows exponentially as a huge number of constraints are generated: one for every subset of S with cardinality $s = 1, \ldots, S$.

We have therefore designed the algorithm that solves this formulation set out in Equations 2.4 through 2.11 to process the model rapidly and efficiently. This is particularly important because our trade signals are generated for real-time streaming data. A delay in generating the trade signals may not be able to use the correct prices at the time of executing the trades. Moreover, in a simulation environment (back testing), we need to rerun the strategy repeatedly, whenever a rebalance is required, and so a simple simulation can be very time-consuming.

To overcome the difficulties caused by the exponential number of constraints (2.10), we generate these constraints dynamically *in situ*. This strategy ensures that the actual number of cuts generated to solve the problem is much less than its theoretical upper limit. We have further implemented a branch-and-cut algorithm (Padberg and Rinaldi 1991), which extends the basic branch-and-bound strategy by generating new cuts prior to branching on a partial solution.

We begin processing the model with a single constraint (2.10), otherwise the formulation is unbounded as V can take an infinite value. At every node of the branch-and-bound tree, we analyze the linear relaxation solution found at that node to check whether any constraint (2.10) has been violated. For that, we employ the (polynomial) separation algorithm proposed by Fábián et al. (2011a), which works as follows.

Given a candidate solution, we sort the scenarios in ascending order of portfolio return (from worst to best). Then, for $s = 1, \ldots, S$, we compute the value of the logical variables corresponding to the constraint (2.10) for a set J_s composed of the s worst scenarios. We then add the most violated constraint (if any exists) to the model and resolve that node. A solution is only accepted as a candidate for branching whenever it satisfies all the constraints of type (2.10).

The branch-and-cut algorithm is written in C++ and the back testing framework is written in R (2016). We have used CPLEX 12.6 (see IBM ILOG CPLEX 2016) as the mixed-integer programming solver. For the tests reported in this paper, we impose a time limit of 30 seconds each time we process a problem instance; that is, if within 30 seconds the algorithm is unable to prove optimality, we retrieve the best integer solution found so far.

In order to give an indication of the computational effort required, we set out the below processing times for back testing (simulation) runs for the following data sets. We report the test results for FTSE 100 and NIKKEI 225 (each containing 100 and 225 assets, respectively). In each case, the asset universe is composed of the constituents of the index and the index future (also included as a tradable asset). The experiments were run on an Intel(R) Core(TM) i7-3770@3.40 GHz with 8 GB RAM.

The simulation adopts successive rebalancing over time at a specified frequency (for instance every 5 days). We set $S = 1200$ and we run the simulation throughout a seven-year period and the computational times are shown in Table 2.2.

The number of trading days differs between FTSE 100 and NIKKEI 225 due to different holidays in each market. For FTSE 100, 307 rebalances are computed whenever a 5-day rebalancing frequency is selected, and 1531 if the rebalance frequency is every day, which is one less than the total number of trading days as we close all portfolio positions in the last day.

For the FTSE 100, the time required to perform the simulation was about 32 minutes for a 5 days rebalancing frequency and about 155 minutes for daily rebalance. Naturally, NIKKEI 225 requires more computational time due to larger models, at roughly 5 hours for daily rebalancing. The vast majority of rebalances were solved within the time limit set: all for FTSE 100 and about 94% for NIKKEI 225. On average, each FTSE 100 rebalance was solved in 6 seconds and each NIKKEI 225 rebalance in about 11.5 seconds.

Given these results, we believe that the algorithm described here is appropriate for back testing virtual trading and finally live trading.

TABLE 2.2: Backtesting

Asset universe	Back testing period	Number of trading days	Rebalancing frequency (days)	Number of rebalances	CPU time (s)	Rebalances solved within 30s
FTSE100	January 1, 2010	1532	5	307	1903.4	307
			1	1531	9339.1	1531
NIKKEI225	–	1536	5	308	3544.5	287
	October 26, 2016		1	1535	17624.5	1445

2.6.2 Scale-Up to process larger models

In the example above, we have considered a universe of up to 225 assets and a set of 1200 discrete scenarios. Naturally, the model can grow substantially if:

- A larger asset universe is considered, such as components of the S&P 500 US or S&P1200 Global indices.
- A higher number of scenarios are taken into account, especially if we consider higher-frequency trading where Δt is in minutes or seconds instead of daily.

In a higher-frequency setting, we may also be unable to afford a 30 second rebalance time as decisions must be made quickly. As higher performance becomes a priority, a natural alternative is to consider parallelizing the branch-and-cut algorithm described earlier. A branch-and-cut is a combination of branch-and-bound and cutting-plane techniques; we describe the challenges in parallelizing them below.

One of the most critical points in a branch-and-bound algorithm is the fathoming of tree branches—which aims to reduce the (exponential) search space of enumerated solutions. Fathoming depends on finding strong lower and upper bounds. In a maximization problem, a lower bound represents the best-known solution for the problem so far and an upper bound represents a limit on the value of the global optimal solution. The upper bound is obtained by solving linear relaxations of the problem—that is, the original formulation without integrality constraints.

Nodes in the branch-and-bound tree are placed in a priority queue and solved sequentially and independently. This is a natural setting for parallelization. The challenge however is sharing, among concurrent nodes, the information on upper and lower bounds in order to reduce the amount of redundant work. For instance, an upper bound found after solving a node could allow another node to be fathomed. If this node is already being solved by a concurrent processor, we have a waste of processing power. Some redundant work is unavoidable, but a parallel branch-and-bound must reduce these inefficiencies.

The generation of cuts in the cutting-plane part of the algorithm is another critical factor. We remind the reader that the full description of the SSD formulation presented in Section 2.4.2 requires an exponential number of constraints (2.10), and it is effectively impossible to have them all represented in memory. As such we start with a relaxed version of the problem and, at each node, we run the separation algorithm described earlier to search for any violated constraint(s) that must be included in the model. Any cut identified in one node is valid for all other nodes in the priority queue.

When we have concurrency, parallel nodes may waste computational power in finding redundant cuts; a lower bound cannot be accepted if it violates a constraint (2.10), as that lower bound is not valid for the original model. So a parallel node may have to run many executions of the separation algorithm until it finds no violated cut for that linear relaxation solution. Perhaps, if that node was executed after a previous one, the cut found before could have prevented several of these executions.

Thus in an efficient implementation of branch-and-cut algorithm parallel processors share a collection of information which comprise current bounds, pending node priorities, and violated cuts. This is a suitable setting for the classical *master-slaves*, or *centralized control*, strategy. In this case there would be a dedicated master process handling the queue, fathoming nodes, updating priorities and controlling cuts, all based on the arrival of asynchronous information. Slave processes would be responsible for solving the highest priority pending tree nodes. Thus, the master process maintains global knowledge and controls the entire search, while slave processes receive pending nodes from the master processor, solve their linear relaxations and attempt to find new violated cuts. Finally, the slave returns information to the master processor.

The higher the number of cuts in the model, the slower is the resolution of a linear relaxation in a particular node. In a branch-and-cut setting, the master processor should also maintain a *cut pool* with a list of previously found cuts. The master processor must handle a few tasks, listed below:

- Check the cut pool in order to identify "tight" and "slack" cuts, that is, the master processor must identify cuts that are likely (unlikely) to change the value of a linear relaxation solution.

- Choose which cuts should be provided to a slave processor. A properly implemented cut pool may prevent very large and time-consuming linear relaxations.

- Receive newly identified cuts from slave processors and add them to the pool.

Overall, some redundant work is unavoidable: a node may be solved before another processor realizes it could be fathomed, or a node may search for violated constraints which would not be necessary if it had information about other previously identified constraints. However, if a master processor properly implements the policies described above, the branch-and-cut implementation could benefit from parallelization and be able to solve larger models. Other strategies for parallelization may be considered here, a more thorough discussion of this technique can be seen in Chapters 1 and 3 of Talbi (2006).

2.7 Results

We use real-world historical daily data (adjusted closing prices) taken from the universe of assets defined by the Financial Times Stock Exchange

100 (FTSE100) index over the period October 9, 2008 to November 1, 2016 (1765 trading days). The data were collected from Thomson Reuters Data Stream platform and adjusted to account for changes in index composition. This means that our models use no more data than was available at the time, removing susceptibility to the influence of survivor bias. For each asset, we compute the corresponding daily rates of return. The original benchmark distribution is obtained by considering the historical daily rates of return of FTSE100 during the same time period.

The methodology we adopt is successive rebalancing over time with recent historical data as scenarios. We start from the beginning of our data set. Given in-sample duration of S days, we decide a portfolio using data taken from an in-sample period corresponding to the first $S+1$ days (yielding S daily returns for each asset). The portfolio is then held unchanged for an out-of-sample period of 5 days. We then rebalance (change) our portfolio, but now using the most recent S returns as in-sample data. The decided portfolio is then again held unchanged for an out-of-sample period of 5 days, and the process repeats until we have exhausted all of the data. We set $S = 1200$; the total out-of-sample period spans slightly more than 6 years (October 1, 2010 to October 26, 2010).

Once the data have been exhausted we have a time series of 1532 portfolio return values for out-of-sample performance, here from period 1201 (the first out-of-sample return value, corresponding to January 1, 2010) until the end of the data.

Portfolios are rebalanced every 5 days, for each experiment we solve 307 instances of the long–short SSD formulation described in Section 2.4.2, each corresponding to a single rebalance.

For every experiment, we set $\alpha = 0.2$, that is, portfolios can have a long–short exposure of up to 120/20. The strategies below also apply a money management technique. That is, at every day, the percentage of the portfolio mark-to-market value invested in risky assets is fixed at 75%, the remaining 25% being invested in a risk-free investment of 2% a year. Hence, the SSD strategy itself is rebalanced every 5 days in order to bring the portfolio to desired proportions.

Figure 2.3 shows portfolio paths for the three different strategies A, B, and C as well as the Financial Times Stock Exchange 100 Index (FTSE100) over the period from October 1, 2010 to October 26, 2016. The strategies all invest in a subset of the companies listed on the FTSE100, where the actual asset universe is defined by the asset universe filter stated above. The FTSE100 index is shown in solid black; the strategies A, B, and C are shown in dashed dark-grey, dashed light-grey and solid light-grey, respectively.

All strategies outperform the FTSE100 index in the period considered. We can also see that strategies B and C, where we filter the asset universe, outperformed strategy A. Table 2.3 shows selected performance statistics, namely:

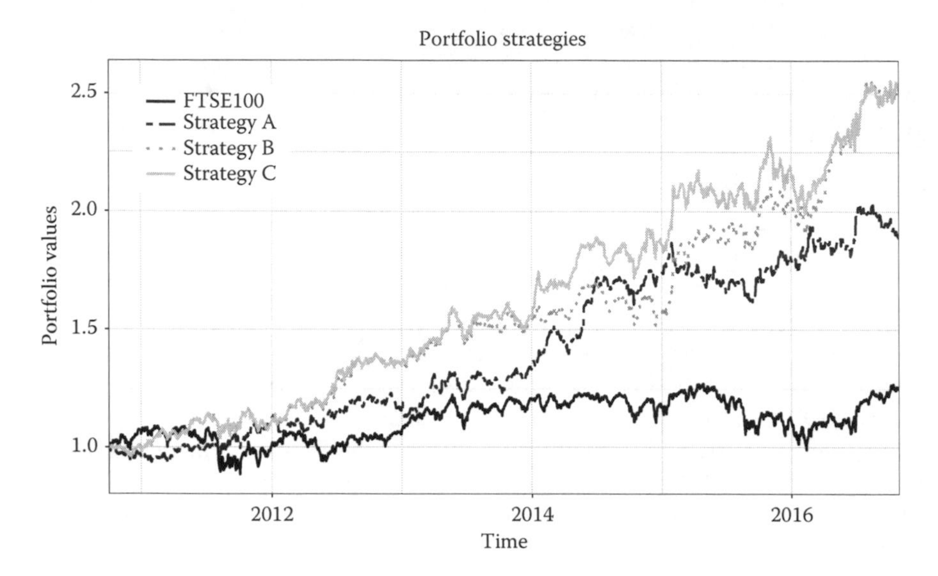

FIGURE 2.3: Portfolio strategies

TABLE 2.3: Performance measures

Portfolio	Final value	Excess RFR (%)	Sharpe ratio	Sortino ratio	Max draw-down (%)	Max. rec. days	Beta	Av. turnover
FTSE100	1.24	1.68	0.10	0.14	22.06	481		
Strategy A	1.88	9.01	0.70	1.01	13.40	264	0.44	19.02
Strategy B	2.47	14.03	0.99	1.48	11.40	165	0.57	34.42
Strategy C	2.51	14.37	1.00	1.48	14.24	125	0.58	34.26

- *Final value:* Normalized final value of the portfolio at the end of the out-of-sample period.
- *Excess over RFR (%)*: Annualized excess return over the risk-free rate. For FTSE100, we used a yearly risk-free rate of 2%.
- *Sharpe ratio*: Annualized Sharpe ratio of returns.
- *Sortino ratio*: Annualized Sortino ratio of returns.
- *Max drawdown (%)*: Maximum peak-to-trough decline (as percentage of the peak value) during the entire out-of-sample period.
- *Max recovery days*: Maximum number of days for the portfolio to recover to the value of a former peak.
- *Beta*: Portfolio beta when compared to the FTSE100 index.
- *Average turnover*: Average turnover per day as a percentage of portfolio mark-to-market.

Both strategies B and C had higher returns and quicker recovery rates when compared to strategy A. Strategies B and C have very similar statistics. Overall, reducing the asset universe via strategies B and C allows us to improve returns and reduce our risk exposure. However, that comes at a cost of a higher correlation to the market itself (a higher beta) and a higher average turnover. The latter is due to the asset universe in two consecutive rebalances being potentially different; in such cases we may need to liquidate current positions in assets that are not included in the current asset universe.

Our back testing results for NIKKEI 250 are in line with the FTSE results stated above. Strategies B and C lead to improved performance measures, the consideration of technical analysis combined with news sentiment results to desired portfolio properties.

2.8 Conclusion and Discussions

For the FTSE 100 and for the 7-year back testing period under consideration, the three strategies described earlier lead to progressive improvement on the performance of a passive index fund. In our model, we apply the SSD criterion to find an optimal static portfolio. For this, the actual data sets of constituents of the FTSE100 index are used as an input to our optimization system. The SSD optimization finds an optimal portfolio with regards to a reference distribution derived from the underlying index. This revision of the return distribution leads to an improved portfolio with a shorter left tail. Money management-based rebalancing is then applied to control the maximum drawdown of the derived portfolio. We investigated three trading strategies: the first gives all available index constituents to the SSD optimization; in the second strategy, the available asset universe for SSD optimization in terms of long–short assets was limited to assets with an RSI value above/below certain thresholds. Furthermore, sentiments covering news items were introduced to the third strategy to filter out contradicting sentiments to RSI choices. The third strategy performed best for most of our key figures. An enhancement of SSD portfolio choices through sentiment data feeds is achieved in this combined static and temporal setting. We have planned further refinement of our strategies which include the detection of market regimes. We propose to use the identification of bull or bear regime to derive a market exposure strategy which assigns limits to the long–short partition.

Acknowledgments

We gratefully acknowledge the contribution of Tilman Sayer to this work. While working at OptiRisk, Tilman participated in this project and contributed to the development of some of the trading strategies reported in this paper.

References

Bawa, V. S. Optimal rules for ordering uncertain prospects. *Journal of Financial Economics*, 2(1):95–121, 1975.

Beasley, J. E., Meade, N., and Chang, T. J. An evolutionary heuristic for the index tracking problem. *European Journal of Operational Research*, 148(3):621–643, 2003.

Canakgoz, N. A. and Beasley, J. E. Mixed-integer programming approaches for index tracking and enhanced indexation. *European Journal of Operational Research*, 196(1):384–399, 2008.

Dentcheva, D. and Ruszczynski, A. Optimization with stochastic dominance constraints. *SIAM Journal on Optimization*, 14(2):548–566, 2003.

Dentcheva, D. and Ruszczyñski, A. Portfolio optimization with stochastic dominance constraints. *Journal of Banking & Finance*, 30(2):433–451, 2006.

Dentcheva, D. and Ruszczynski, A. Inverse cutting plane methods for optimization problems with second order stochastic dominance constraints. *Optimization: A Journal of Mathematical Programming and Operations Research*, 59(3):323–338, 2010.

DiBartolomeo, D. *The Enhanced Index Fund as an Alternative to Indexed Equity Management*. Northfield Information Services, Boston, 2000.

Fábián, C. I., Mitra, G., and Roman, D. Processing second-order stochastic dominance models using cutting-plane representations. *Mathematical Programming*, 130(1):33–57, 2011a.

Fábián, C. I., Mitra, G., and Roman, D. An enhanced model for portfolio choice with SSD criteria: A constructive approach. *Quantitative Finance*, 11(10):1525–1534, 2011b.

Fábián, C. I., Mitra, G., and Roman, D. Portfolio choice models based on second-order stochastic dominance measures: An overview and a computational study. In: Bertocchi M., Consigli G., Dempster M. (eds). *Stochastic Optimization Methods in Finance and Energy. International Series in Operations Research & Management Science*, Springer, New York, NY. vol. 163 pp. 441–469, 2011c.

Fama, E. F. Efficient capital markets: A review of theory and empirical work. *The Journal of Finance*, 25(2):383–417, 1970.

Hadar, J. and Russell, W. R. Rules for ordering uncertain prospects. *The American Economic Review*, 59(1):25–34, 1969.

Hodder, J. E., Jackwerth, J. C., and Kolokolova, O. Improved portfolio choice using second-order stochastic dominance. *Review of Finance*, 19(1):1623–1647, 2015.

IBM ILOG CPLEX Optimizer. Available from https://www-01.ibm.com/software/commerce/optimization/cplex-optimizer/, last accessed March 5, 2017. 2016.

Javanmardi, L. and Lawryshy, Y. A new rank dependent utility approach to model risk averse preferences in portfolio optimization. *Annals of Operations Research*, 237(1):161–176, 2016.

Kopa, M. and Post, T. A general test for SSD portfolio efficiency. *OR Spectrum*, 37(1):703–734, 2015.

Kuosmanen, T. Efficient diversification according to stochastic dominance criteria. *Management Science*, 50(10):1390–1406, 2004.

Levy, H. Stochastic dominance and expected utility: Survey and analysis. *Management Science*, 38(4):555–593, 1992.

Lo, A.W. The adaptive markets hypothesis. *The Journal of Portfolio Management*, 30(5):15–29, 2004.

Luenberger, D. G. *Investment Science*. Oxford University Press, USA, 1997. ISBN: 9780195108095.

Mitra, L. and Mitra, G. Applications of news analytics in finance: A review. In *The Handbook of News Analytics in Finance, Chapter 1*, John Wiley & Sons, 2011.

Mitra, L. Mitra, G. and diBartolomeo, D. Equity portfolio risk (volatility) estimation using market information and sentiment. *Quantitative Finance*, 9(8):887–895, 2009.

Mitra, G., Erlwein-Sayer, C., and Roman, D. *Revision of benchmark distributions to enhance portfolio choice by the Second Order Stochastic Dominance criterion.* White paper under preparation OptiRisk Systems, 2017.

Padberg, M. and Rinaldi, G. A branch-and-cut algorithm for resolution of large scale of symmetric traveling salesman problem. *SIAM Review*, 33:60–100, 1991.

Patton, A. J. and Verardo, M. Does beta move with news? Firm-specific information flows and learning about profitability. *The Review of Financial Studies*, 25(9):2789–2839, 2012.

Post, T. Empirical tests for stochastic dominance efficiency. *The Journal of Finance*, 58(5):1905–1931, 2003.

Post, T., Fang, Y., and Kopa, M. Linear tests for DARA stochastic dominance. *Management Science*, 61(1):1615–1629, 2015.

Post, T. and Kopa, M. General linear formulations of stochastic dominance criteria. *European Journal of Operational Research*, 230(2):321–332, 2013.

R Core Team. *R: A Language and Environment for Statistical Computing.* R Foundation for Statistical Computing, Vienna, Austria. Available at https://www.R-project.org/, last accessed March 5, 2017.

Roman, D., Darby-Dowman, K., and Mitra, G. Portfolio construction based on stochastic dominance and target return distributions. *Mathematical Programming*, 108(2):541–569, 2006.

Roman, D., Mitra, G., and Zverovich, V. Enhanced indexation based on second-order stochastic dominance. *European Journal of Operational Research*, 228(1):273–281, 2013.

Scowcroft, A. and Sefton, J. *Enhanced indexation. In: Satchell, A. and Scowcroft, A. (eds.) Advances in Portfolio Construction and Implementation*, Butterworth-Heinemann, London, pp. 95–124, 2003.

Talbi, E. G. *Parallel Combinatorial Optimization.* John Wiley & Sons, USA, 2006.

Thorp, E. O. *Beat the Dealer: A Winning Strategy for the Game of Twenty-One.* Random House, New York, NY, 1966.

Thorp, E. O. and Kassouf, S. T. *Beat the Market: A Scientific Stock Market System.* Random House, New York, NY, 1967.

Valle, C. and Mitra, G. *News Analytics Toolkit User Manual*, OptiRisk Systems, London, UK, 2004. available online: http://www.optirisksystems.com/manuals/NAToolkit_User_Manual.pdf.

Valle, C. A., Roman, D., and Mitra, G. Novel approaches for portfolio construction using second order stochastic dominance. *Computational Management Science*, DOI 10.1007/s10287-017-0274-9, 2017.

Welles Wilder, J. *New Concepts in Technical Trading Systems.* Trend Research, UK, 1978.

Chapter 3

The Alpha Engine: Designing an Automated Trading Algorithm

Anton Golub, James B. Glattfelder, and Richard B. Olsen

CONTENTS

3.1 Introduction

The asset management industry is one of the largest industries in modern society. Its relevance is documented by the astonishing amount of assets that are managed. It is estimated that globally there are 64 trillion USD under management [1]. This is nearly as big as the world product of 77 trillion USD [2].

3.1.1 Asset management

Asset managers use a mix of analytic methods to manage their funds. They combine different approaches from fundamental to technical analysis. The time frames range from intraday, to days and weeks, and even months. Technical analysis, a phenomenological approach, is utilized widely as a toolkit to build trading strategies.

A drawback of all such methodologies is, however, the absence of a consistent and overarching framework. What appears as a systematic approach to asset management often boils down to gut feeling, as the manager chooses from a broad blend of theories with different interpretations. For instance, the choice and configuration of indicators is subject to the specific preference of the analyst or trader. In effect, practitioners mostly apply ad hoc rules which are not embedded in a broader context. Complex phenomena such as changing liquidity levels as a function of time go unattended.

This lack of consensus, or intellectual coherence, in such a dominant and relevant industry underpinning our whole society is striking. In a day and age where computational power and digital storage capacities are growing exponentially, at shrinking costs, and where there exists an abundance of machine learning algorithms and big data techniques, one would expect a more unified and comprehensive methodological and theoretical framework. To illustrate, consider the recent unexpected success of Google's AlphaGo algorithm beating the best human players [3]. This is a remarkable feat for a computer, as the game of Go is notoriously complex and players often report that they select moves based solely on intuition.

There is, however, one exception in the asset management and trading industry that relies fully on algorithmic trade generation and automated execution. Referred to under the umbrella of term "high-frequency trading," this approach has witnessed substantial growth. These strategies take advantage of short-term arbitrage opportunities and typically analyze the limit order books to jump the queue, whenever there are large orders pending [4]. While high-frequency trading results in high trade volumes the assets managed with these types of strategies are around 140 billion [5]. This is microscopic compared to the size of the global assets under management.

3.1.2 The foreign exchange market

For the development of our trading model algorithm, and the evaluation of the statistical price properties, we focus on the foreign exchange market. This market can be characterized as a complex network consisting of interacting agents: corporations, institutional and retail traders, and brokers trading through market makers, who themselves form an intricate web of interdependence. With an average daily turnover of approximately 5 trillion USD [6] and with price changes nearly every second, the foreign exchange market offers a unique opportunity to analyze the functioning of a highly liquid, over-the-counter market that is not constrained by specific exchange-based rules. These markets are an order of magnitude bigger than futures or equity markets [7].

In contrast to other financial markets, where asset prices are quoted in reference to specific currencies, exchange rates are symmetric: quotes are currencies in reference to other currencies. The symmetry of one currency against another neutralizes effects of trend, which are a significant drivers in other markets, such as stock markets. This property of symmetry makes currency markets notoriously hard to trade profitably.

We focus on the foreign exchange market for the development of our trading model algorithm. Its high liquidity and long/short symmetry make it an ideal environment for the research and development of fully automated and algorithmic trading strategies. Indeed, any profitable trading algorithm for this market should, in theory, also be applicable to other markets.

3.1.3 The rewards and challenges of automated trading

During the crisis of 2007 and 2008, the world witnessed how the financial system destabilized the real economy and destroyed vast amounts of wealth. At other times, when there are favorable economic conditions, financial markets contribute to wealth accumulation. The financial system is an integral part of the real economy with a strong cross dependency. Markets are not a closed system, where the sum of all profits and losses net out. If investment strategies contribute to market liquidity, they can help stabilize prices and reduce the uncertainty in financial markets and the economy at large. For such strategies, the investment returns can be viewed as a payoff for the value-added provided to the economy.

Liquid financial markets offer a large profit potential. The length of a foreign exchange price curve, as measured by the sum of up and down price movements of increments of 0.05%, during the course of a year, is, on average, approximately 1600%, after deducting transaction costs [8]. An investor can, in theory, earn 1600% unleveraged per year, assuming perfect foresight in exploiting this coastline length. With leverage, the profit potential is even greater. Obviously, as no investor has perfect foresight, capturing 1600% is not feasible.

However, why do most investment managers have such difficulty in earning even small returns on a systematic basis, if the profit potential is so big? Especially as traders can manage their risk with sophisticated money management rules which have the potential to turn losses into profits. Again, the question arises as to why hedge funds, who can hire the best talent in the world, find it so hard to earn consistent annual returns. For instance, the Barclay Hedge Fund Index,[1] measuring the average returns of all hedge funds (except funds of funds) in their database, reports an average yearly return of 5.035% (±4.752%) for the past 4 years. How can we develop liquidity-providing

[1] www.barclayhedge.com/research/indices/ghs/Hedge_Fund_Index.html.

investment algorithms that consistently generate positive and sizable returns? What is missing in the current paradigm?

Another key criterion of the quality of an investment strategy is the size of assets that can be deployed without a deterioration of performance. Closely related to this issue is the requirement that the strategy does not distort the market dynamics. This is, for example, the case with the trend following strategies that are often deployed in automated trading. Such strategies have the disadvantage that the investor does not know for sure how his action of following the trend amplifies the trend. In effect, the trend follower can get locked into a position that he cannot closeout without triggering a price dislocation.

Finally, any flavor of automated trading is constrained by the current computational capacities available to researchers. Although this constraint is loosening day by day due to the prowess of high-performance computing in finance, some approaches rely more on number crunching than others. Ideally, any trading model algorithm should be implementable with reasonable resources to make it useful and applicable in the real world.

3.1.4 The hallmarks of profitable trading

Investment strategies need to be fully automated. For one, the number of traded instruments should not be constrained by human limitations. Then, the trading horizons should also include intraday activity, as a condition sine qua non. Complete automation has its own challenges, because computer code can go awry and cause huge damage, as witnessed by Knight Capital, which lost 500 million USD in a matter of 30 seconds due to an operational error.[2]

Many modeling attempts fail because developers succumb to curve fitting. They start with a specific data sample and tweak their model until it makes money in simulation runs. Such trading models can disappoint from the start when going live or boast good performance only for some period of time until a regime shift occurs and the specific conditions the model was optimized for (i.e., curve fitted) disappear and losses are incurred.

Trading models need to be parsimonious and have a limited set of variables. If the models have too many variables, the parameter space becomes vast and hard to navigate. Parsimonious models are powerful, because they are easier to calibrate, assess, and understand why they perform. Moreover, investment models need to be robust to market changes. For instance, the models can be adaptive and have their behavior depend on the current market regime. Therefore, algorithmic investment strategies have to be developed on the basis of robust and consistent approaches and methods that provide a solid framework of analysis.

Financial markets are comprised of a large number of traders that take positions on different time horizons. Agent-based models can mimic the actual

[2]www.sec.gov/news/press-release/2013-222.

traders and are therefore well suited to research market behavior [9]. If agent-based models are fractal, that is, behave in a self-similar manner across time horizons and only differ with respect to the scaling of their parameters, the short-term models are a filter for the validity of the long-term models. In practice, this allows for the short-term agent-based models to be tested and validated over a huge data sample with a multitude of events. As a result, the scarcity of data available for the long-term model is not a hindrance of acceptance if it is self-similar with respect to the short-term models. In effect, the validation of the model structure for short-term models also implies a validation for the long-term models, by virtue of the scaling effects. In contrast, most standard modeling approaches are typically devised for one time horizon only and hence there are no self-similar models that complement each other.

Moreover, the modeling approach should be modular and enable developers to combine smaller blocks to build bigger components. In other words, models are built in a bottom-up spirit, where simple building blocks are assembled into more complex units. This also implies an information flow between building blocks.

To summarize, our aim is to develop trading models based on parsimonious, self-similar, modular, and agent-based behavior, designed for multiple time horizons and not purely driven by trend following action. The intellectual framework unifying these angles of attack is outlined in Section 3.3. The result of this endeavor is interacting systems that are highly dynamic, robust, and adaptive; in other words, a type of trading model that mirrors the dynamic and complex nature of financial markets. The performance of this automated trading algorithm is outlined in the following section.

In closing, it should be mentioned that transaction costs can represent real-world stumbling blocks for trading models. Investment strategies that take advantage of short-term price movements in order to achieve good performance have higher transaction volumes than longer term strategies. This obviously increases the impact of transaction costs on the profitability. As far as possible, it is advisable to use limit orders to initiate trades. They have the advantage that the trader does not have to cross the spread to get his order executed, thus reducing or eliminating transaction costs. The disadvantage of limit orders is, however, that execution is uncertain and depends on buy and sell interest.

3.2 In a Nutshell: Trading Model Anatomy and Performance

In this section, we provide an overview of the trading model algorithm and its performance. For all the details on the model, see Section 3.4, and the code can be downloaded from GitHub [10].

The Alpha Engine is a counter-trending trading model algorithm that provides liquidity by opening a position when markets overshoot and manages positions by cascading and de-cascading during the evolution of the long coastline of prices, until it closes in a profit. The building blocks of the trading model are as follows:

- An endogenous time scale called intrinsic time that dissects the price curve into directional changes and overshoots;
- Patterns called scaling laws that hold over several orders of magnitude, providing an analytical relationship between price overshoots and directional change reversals;
- Coastline trading agents operating at intrinsic events, defined by the event-based language;
- A probability indicator that determines the sizing of positions by identifying periods of market activity that deviate from normal behavior;
- Skewing of cascading and de-cascading designed to mitigate the accumulation of large inventory sizes during trending markets; and
- The splitting of directional change and, consequently, overshoot thresholds into upward and downward components, that is, the introduction of asymmetric thresholds.

The trading model is backtested on historical data comprised of 23 exchange rates:

> AUD/JPY, AUD/NZD, AUD/USD, CAD/JPY, CHF/JPY, EUR/AUD, EUR/CAD, EUR/CHF, EUR/GBP, EUR/JPY, EUR/NZD, EUR/USD, GBP/AUD, GBP/CAD, GBP/CHF, GBP/JPY, GBP/USD, NZD/-CAD, NZD/JPY, NZD/USD, USD/CAD, USD/CHF, USD/JPY.

The chosen time period is from the beginning of 2006 until the beginning of 2014, that is, 8 years. The trading model yields an unlevered return of 21.3401%, with an annual Sharp ratio of 3.06, and a maximum drawdown (computed on daily basis) of 0.7079%. This event occurs at the beginning of 2013 and lasts approximately 4 months, as the JYP weakens significantly following the Quantitative Easing program ("three arrows" of fiscal stimulus) launched by the Bank of Japan.

Figure 3.1 shows the performance of the trading model across all exchange rates. Table 3A.1 reports the monthly and yearly returns. The difference in returns among the various exchange rates is explained by volatility: the trading model reacts only to occurrences of intrinsic time events, which are functionally dependent on volatility. Exchange rates with higher volatility will have a greater number of intrinsic events and hence more opportunities for the model to extract profits from the market. This behavior can be witnessed during the

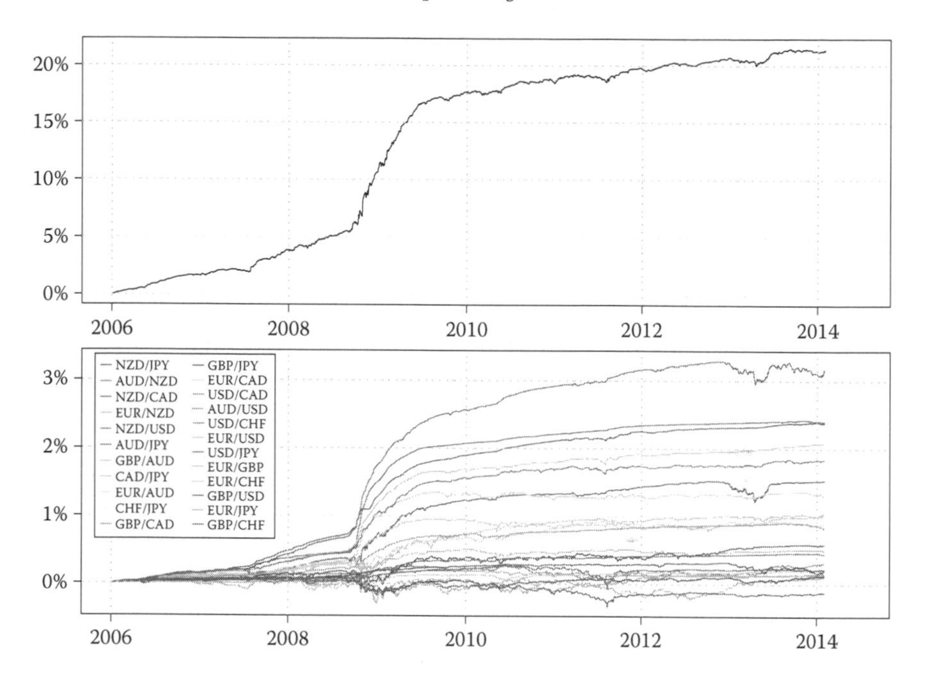

FIGURE 3.1: Daily Profit & Loss of the Alpha Engine, across 23 currency pairs, for 8 years. See details in the main text of this section and Section 3.4.

financial crisis, where its deleterious effects are somewhat counterbalanced by an overall increase in profitable trading behavior of the model, fueled by the increase in volatility.

The variability in performance of the individual currency pairs can be addressed by calibrating the "aggressiveness" of the model with respect to the volatility of the exchange rate. In other words, the model trades more frequently when the volatility is low and vice versa. For the sake of simplicity, and to avoid potential overfitting, we have excluded these adjustments to the model. In addition, we also refrained from implementing cross-correlation measures. By assessing the behavior of the model for one currency pair, information can be gained that could be utilized as an indicator which affects the model's behavior for other exchange rates. Finally, we have also not implemented any risk management tools.

In essence, what we present here is a proof of concept. We refrained from tweaking the model to yield better performance, in order to clearly establish and outline the model's building blocks and fundamental behavior. We strongly believe there is great potential for obvious and straightforward improvements, which would give rise to far better models. Nevertheless, the bare-bones model we present here already has the capability of being implemented as a robust and profitable trading model that can be run in real time. With a leverage factor of 10, the model experiences a drawdown of 7.08%

while yielding an average yearly profit of 10.05% for the last 4 years. This is still far from realizing the coastline's potential, but, in our opinion, a crucial first step in the right direction.

Finally, we conclude this section by noting that, despite conventional wisdom, it is in fact possible to "beat" a random walk. The Alpha Engine produces profitable results even on time series generated by a random walk, as seen in Figure 3B.1. This unexpected feature results from the fact that the model is dissecting Brownian motion into intrinsic time events. Now these directional changes and overshoots yield a novel context, where a cascading event is more likely to be followed by a de-cascading event than another cascading one. In detail, the probability of reaching the profitable de-cascading event after a cascade is $1 - e^{-1} \approx 0.63$, while the probability for an additional cascade is about 0.37. In effect, the procedure of translating a tick-by-tick time series into intrinsic time events skews the odds in one's favor—for empirical as well as synthetic time series. For details, see Reference 11.

In the following section, we will embark on the journey that would ultimately result in the trading model described above. For a prehistory of events, see Appendix 3A.

3.3 Guided by an Event-Based Framework

The trading model algorithm outlined in the last section is the result of a long journey that began in the early 1980s. Starting with a new conceptual framework of time, this voyage set out to chart new terrain. The whole history of this endeavor is described in Appendix 3A. In the following, the key elements of this new paradigm are highlighted.

3.3.1 The first step: Intrinsic time

We all experience time as a fundamental and unshakable part of reality. In stark contrast, the philosophy of time and the notion of time in fundamental physics challenge our mundane perception of it. In an operational definition, time is simply what instruments measure and register. In this vein, we understand the passage of time in financial time series as a set of events, that is, system interactions.

In this novel time ontology, time ceases to exist between events. In contrast to the continuity of physical time, now only interactions, or events, let a system's clock tick. Hence, this new methodology is called intrinsic time [12]. This event-based approach opens the door to a modeling framework that yields self-referential behavior which does not rely on static building blocks and has a dynamic frame of reference.

Implicit in this definition is the threshold for the measurement of events. At different resolutions, the same price series reveals different characteristics. In essence, intrinsic time increases the signal to noise ratio in a time series by filtering out the irrelevant information between events. This dissection of price curves into events is an operator, mapping a time series $x(t)$ into a discrete set of events $\Omega[x(t), \delta]$, given the directional change threshold δ.

We focus on two types of events that represent ticks of intrinsic time:

1. A directional change δ [8,13–16];

2. An overshoot ω [8,13,15].

With these events, every price curve can be dissected into components that represent a change in the price trend (directional change) and a trend component (overshoot). For a directional change to be detected, first an initial direction mode needs to be chosen. As an example, in an up mode an increasing price move will result in the extremal price being updated and continuously increased. If the price goes down, the difference between the extremal price and the current price is evaluated. If this distance (in percent) exceeds the predefined directional change threshold, a directional change is registered. Now the mode is switched to down and the algorithm continues correspondingly. If now the price continues to move in the same direction as the directional change, for the size of the threshold, an overshoot event is registered. As long as a trend persists, overshoot events will be registered. See Figure 3.2a for an illustration. Note that two intrinsic time series will synchronize after one directional change, regardless of the chosen starting direction.

As a result, a price curve is now comprised of segments, made up of a directional change event δ and one or more overshoots of size ω. This event-based

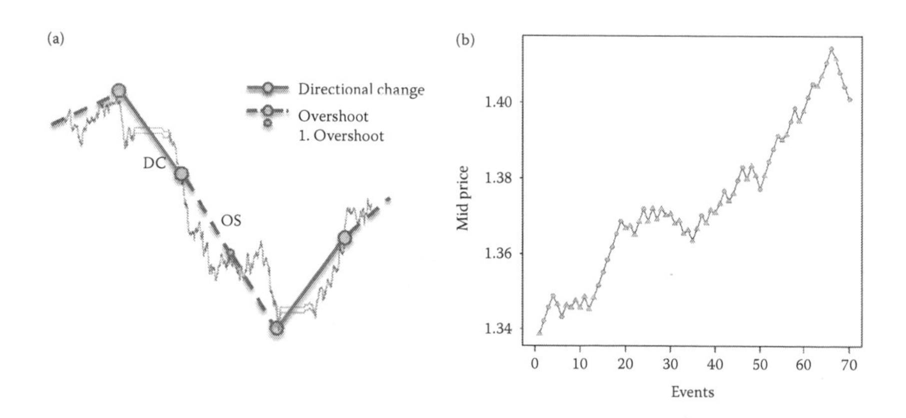

FIGURE 3.2: (a) Directional change and overshoot events. (b) A coastline representation of the EUR_USD price curve (2008-12-14 22:10:56 to 2008-12-16 21:58:20) defined by a directional change threshold $\delta = 0.25\%$. The triangles represent directional change and the bullets overshoot events.

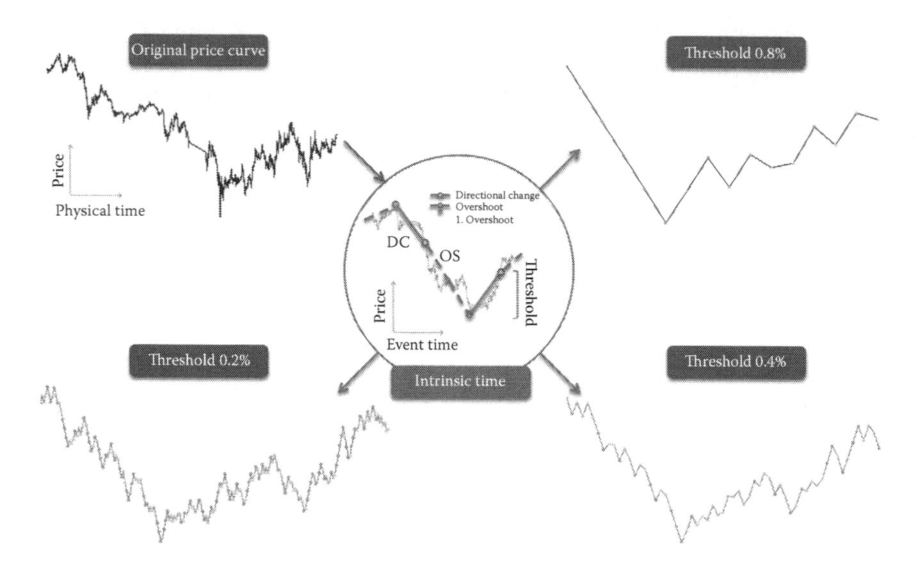

FIGURE 3.3: Coastline representation of a price curve for various directional change thresholds δ.

time series is called the coastline, defined for a specific directional change threshold. By measuring the various coastlines for an array of thresholds, multiple levels of event activity can be considered. See Figures 3.2b and 3.3. This transformed time series is now the raw material for further investigations [8]. In particular, this price curve will be used as input for the trading model, as described in Section 3.3.4. With the publication [17], the first decade came to a close.

3.3.2 The emergence of scaling laws

A validation for the introduction of intrinsic time is that this event-based framework uncovers statistical properties otherwise not detectable in the price curves, for instance, scaling laws. Scaling-law relations characterize an immense number of natural processes, prominently in the form of

1. Scaling-law distributions;
2. Scale-free networks; and
3. Cumulative relations of stochastic processes.

Scaling-law relations display scale invariance because scaling the function's argument x preserves the shape of the function $f(x)$ [18]. Measurements of scaling-law processes yield values distributed across an enormous dynamic range, and for any section analyzed, the proportion of small to large events stays constant.

Scaling-law distributions have been observed in an extraordinary wide range of natural phenomena: from physics, biology, earth and planetary sciences, economics and finance, computer science and demography to the social sciences [19–22]. Although scaling law distributions imply that small occurrences are extremely common, whereas large instances are rare, these large events occur nevertheless much more frequently compared to a normal probability distribution. Hence, scaling-law distributions are said to have "fat tails."

The discovery of scale-free networks [23,24], where the degree distributions of nodes follow a scaling-law distribution, was a seminal finding advancing the study of complex networks [25]. Scale-free networks are characterized by high robustness against random failure of nodes, but susceptible to coordinated attacks on the hubs.

Scaling-law relations also appear in collections of random variables. Prominent empirical examples are financial time series, where one finds scaling laws governing the relationship between various observed quantities [16,17,26]. The introduction of the event-based framework leads to the discovery of a series of new scaling relations in the cumulative relations of properties in foreign exchange time series [8]. In detail, of the 18 novel scaling-law relations (of which 12 are independent), 11 relate to directional changes and overshoots.

One notable observation was that, on average, a directional change δ is followed by an overshoot ω of the same magnitude

$$\langle\omega\rangle \approx \delta. \tag{3.1}$$

This justifies the procedure of dissecting the price curve into directional change and overshoot segments of the same size, as seen in Figures 3.2 and 3.3. In other words, the notion of the coastline is statistically validated.

Scaling laws are a hallmark of complexity and complex systems. They can be viewed as a universal "law of nature" underlying complex behavior in all its domains.

3.3.3 Trading models and complexity

A complex system is understood as being comprised of many interacting or interconnected parts. A characteristic feature of such systems is that the whole often exhibits properties not obvious from the properties of the individual parts. This is called emergence. In other words, a key issue is how the macro behavior emerges from the interactions of the system's elements at the micro level. Moreover, complex systems also exhibit a high level of resilience, adaptability, and self-organization. Complex systems are usually found in socio-economical, biological or physio-chemical domains.

Complex systems are usually very reluctant to be cast into closed-form analytical expressions. This means that it is generally hard to derive mathematical quantities describing the properties and dynamics of the system under study. Nonetheless, there has been a long history of attempting to understand finance from an analytical point of view [27,28].

In contrast, we let our trading model development be guided by the insights gained by studying complex systems [29]. The single most important feature is surprisingly subtle:

> Macroscopic complexity is the result of simple rules of interaction at the micro level.

In other words, what looks like complex behavior from a distance turns out to be the result of simple rules at closer inspection. The profundity of this observation should not be underestimated, as echoed in the words of Stephen Wolfram, when he was first struck by this realization [30, p. 9]:

> Indeed, even some of the very simplest programs that I looked at had behavior that was as complex as anything I had ever seen. It took me more than a decade to come to terms with this result, and to realize just how fundamental and far-reaching its consequences are.

By focusing on local rules of interactions in complex systems, the system can be naturally reduced to a set of agents and a set of functions describing the interactions between the agents. As a result, networks are the ideal formal representation of the system. Now the nodes represent the agents and the links describe their relationship or interaction. In effect, the structure of the network, i.e., its topology, determines the function of the network.

Indeed, this perspective also highlights the paradigm shift away from mathematical models towards algorithmic models, where computations and simulation are performed by computers. In other words, the analytical description of complex systems is abandoned in favor of algorithms describing the interaction of the agents. This approach has given rise to the prominent field of agent-based modeling [31–33]. The validation of agent-based models is given by their capability to replicate patterns and behavior seen in real-world complex systems by virtue of agents interacting according to simple rules.

Financial markets can be viewed as the epitome of a human-generated complex system, where the trading choices of individuals, aggregated in a market, give rise to a stochastic and highly dynamic price evolution. In this vein, a long or short position in the market can be understood as an agent. In detail, a position p_i is comprised of the set $\{\bar{x}_i, \pm g_i\}$, where $\bar{x}_i$ is the current mid (or entry price) and $\pm g_i$ represents the position size and direction.

3.3.4 Coastline trading

In a next step, we combined the event-based price curve with simple rules of interactions. This means that the agents interact with the coastline according to a set of trading rules, yielding coastline traders [34–36]. In a nutshell, the initialization of new positions and the management of existing positions in the market are clocked according to the occurrence of directional change or overshoot events. The essential elements of coastline trading are cascading

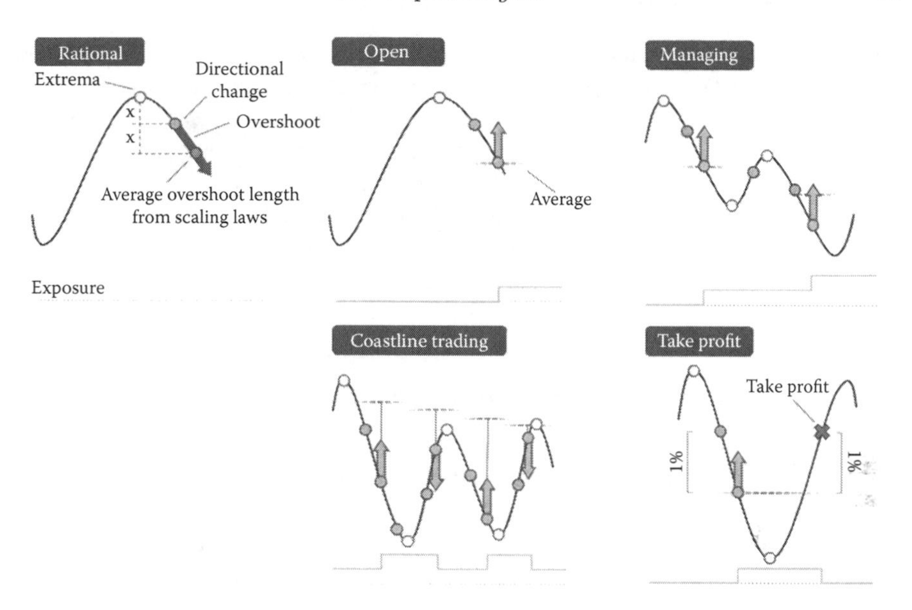

FIGURE 3.4: Simple rules: The elements of coastline trading. Cascading and de-cascading trades increase or decrease existing positions, respectively.

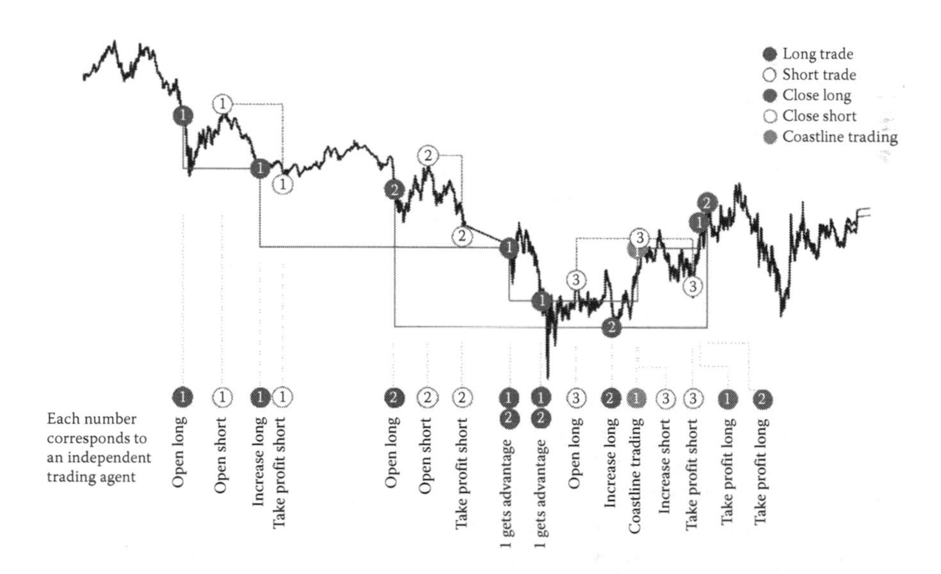

FIGURE 3.5: Real-world example of coastline trading.

and de-cascading trades. For the former, an existing position is increased by some increment in a loss, bringing the average closer to the current price. For a de-cascading event, an existing position is decreased, realizing a profit. It is important to note that because position sizes are only ever increased by the same fixed increments, coastline trading does not represent a Martingale strategy. In Figures 3.4 and 3.5, examples of such trading rules are shown.

With these developments, the second decade drew to a close. Led by the introduction of event-based time, uncovering scaling law relations, the novel framework could be embedded in the larger paradigm related to the study of complex systems. The resulting trading models were by construction, automated, agent-based, contrarian, parsimonious, adaptive, self-similar, and modular. However, there was one crucial ingredient missing, to render the models robust and hence profitable in the long term. And so the journey continued.

3.3.5 Novel insights from information theory

In a normal market regime, where no strong trend can be discerned, the coastline traders generate consistent returns. By construction, this trading model algorithm is attuned to directional changes and overshoots. As a result, so long as markets move in gentle fluctuations, this strategy performs. In contrast, during times of strong market trends, the agents tend to build up large positions which they cannot unload. Consequentially, each agent's inventory increases in size. As this usually happens over multiple threshold sizes, the overall resulting model behavior derails.

This challenge, related to trends, led to the incorporation of novel elements into the trading model design. A new feature, motivated by information theory was added. Specifically, a probability indicator was constructed. Equipped with this new tool, the challenges presented by market trends could now be tackled. In effect, the likeliness of a current price evolution with respect to a Brownian motion can be assessed in a quantitative manner.

In the following, we will introduce the probability indicator $\mathcal{L}$. This is an information theoretic value that measures the unlikeliness of the occurrence of price trajectories. As always, what is actually analyzed is the price evolution which is mapped onto the discretized price curve, which results from the event-based language in combination with the overshoot scaling law. Point-wise entropy, or surprise, is defined as the entropy for a certain realization of a random variable. Following [37], we understand the surprise of the event-based price curve being related to the transitioning probability from the current state s_i to the next intrinsic event s_j, i.e., $\mathbb{P}(s_i \rightarrow s_j)$. In detail, given a directional change threshold δ, the set of possible events is given by directional changes or overshoots. In other words, a state at "time" i is given by $s_i \in \mathcal{S} = \{\delta, \omega\}$. Given $\mathcal{S}$, we now can understand all possible transitions as happening in the stylized network of states seen in Figure 3.6. The evolution of intrinsic time can progress from a directional change to another directional change or an overshoot, which, in turn, can transit to another overshoot event ω or back to a directional change δ.

We define the surprise of the transitions from state s_i to state s_j as

$$\gamma_{ij} = -\log\mathbb{P}(s_i \rightarrow s_j), \tag{3.2}$$

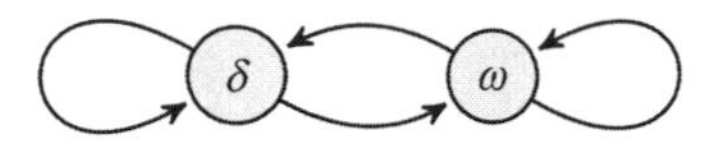

FIGURE 3.6: The transition network of states in the event-based representation of the price trajectories. Directional changes δ and overshoots ω are the building blocks of the discretized price curve, defining intrinsic time.

which, as mentioned, is the point-wise entropy that is large when the probability of transitioning from state s_i to state s_j is small and vice versa. Consequently, the surprise of a price trajectory within a time interval $[0, T]$, that has experienced K transitions, is

$$\gamma_K^{[0,T]} = \sum_{k=1}^{K} -\log\mathbb{P}(s_{i_k} \to s_{i_{k+1}}). \tag{3.3}$$

This is now a measure of the unlikeliness of price trajectories. It is a path-dependent measurement: two price trajectories exhibiting the same volatility can have very different surprise values.

Following [11], $H^{(1)}$ denotes the entropy rate associated with the state transitions and $H^{(2)}$ is the second order of informativeness. Utilizing these building blocks, the next expression can be defined as

$$\Delta = \frac{\gamma_K^{[0,T]} - K \cdot H^{(1)}}{\sqrt{K \cdot H^{(2)}}}. \tag{3.4}$$

This is the surprise of a price trajectory, centered by its expected value, that is, the entropy rate multiplied by the number of transitions, and divide it by the square root of its variance, that is, the second order of informativeness multiplied by the number of transitions. It can be shown that

$$\Delta \to \mathcal{N}(0, 1), \quad \text{for } K \to \infty, \tag{3.5}$$

by virtue of the central limit theorem [38]. In other words, for large K, Δ converges to a normal distribution. Equation 3.4 now allows for the introduction of our probability indicator $\mathcal{L}$, defined as

$$\mathcal{L} = 1 - \Theta\left(\frac{\gamma_K^{[0,T]} - K \cdot H^{(1)}}{\sqrt{K \cdot H^{(2)}}}\right), \tag{3.6}$$

where Θ is the cumulative distribution function of normal distributions. Thus an unlikely price trajectory, strongly deviating from a Brownian motion, leads to a large surprise and hence $\mathcal{L} \approx 0$. We can now quantify when markets show normal behavior, where $\mathcal{L} \approx 1$. Again, the reader is referred to [11] for more details.

We now assess how the overshoot event ω should be chosen. The standard framework for coastline trading dictates that an overshoot event occurs in the price trajectory when the price moves by δ in the overshoots' direction after a directional change. In the context of the probability indicator, we depart from this procedure and define the overshoots to occur when the price moves by $2.525729 \cdot \delta$. This value comes from maximizing the second-order informativeness $H^{(2)}$ and guarantees maximal variability of the probability indicator $\mathcal{L}$. For details, see Reference 11.

The probability indicator $\mathcal{L}$ can now be used to navigate the trading models through times of severe market stress. In detail, by slowing down the increase of the inventory of agents during price overshoots, the overall trading models exposure experiences smaller drawdowns and better risk-adjusted performance. As a simple example, when an agent cascades, that is, increases its inventory, the unit size is reduced in times where $\mathcal{L}$ starts to approach zero.

For the trading model, the probability indicator is utilized as follows. The default size for cascading is one unit (lot). If $\mathcal{L}$ is smaller than 0.5, this sizing is reduced to 0.5, and finally if $\mathcal{L}$ is smaller than 0.1, then the size is set to 0.1.

Implementing the above-mentioned measures allowed the trading model to safely navigate treacherous terrain, where it derailed in the past. However, there was still one crucial insight missing, before a successful version of the Alpha Engine could be designed. This last insight evolves around a subtle recasting of thresholds which has profound effects on the resulting trading model performance.

3.3.6 The final pieces of the puzzle

Coming back full circle, the focus was again placed on the nature of the event-based formalism. By allowing for new degrees of freedom, the trading model puzzle could be concluded. What before were rigid and static thresholds are now allowed to breathe, giving rise to asymmetric thresholds and fractional position changes.

In the context of directional changes and overshoots, an innocuous question to ask is whether the threshold defining the events should depend on the direction of the current market. In other words, does it make sense to introduce a threshold that is a function of the price move direction? Analytically

$$\delta \rightarrow \begin{cases} \delta_{\text{up}} & \text{for increasing prices;} \\ \delta_{\text{down}} & \text{for decreasing prices.} \end{cases} \tag{3.7}$$

These asymmetric thresholds now register directional changes at different values of the price curve, depending on the direction of the price movement. As a consequence $\omega = \omega(\delta_{\text{up}}, \delta_{\text{down}})$ denotes the length of the overshoot corresponding to the new upward and downward directional change thresholds. By virtue of the overshoot size scaling law

$$\langle \omega_{\text{up}} \rangle = \delta_{\text{up}}, \quad \langle \omega_{\text{down}} \rangle = \delta_{\text{down}}. \tag{3.8}$$

To illustrate, let P_t be a price curve, modeled as an arithmetic Brownian motion B_t with trend μ and volatility σ, meaning $\mathrm{d}P_t = \mu \mathrm{d}t + \sigma \mathrm{d}B_t$. Now the

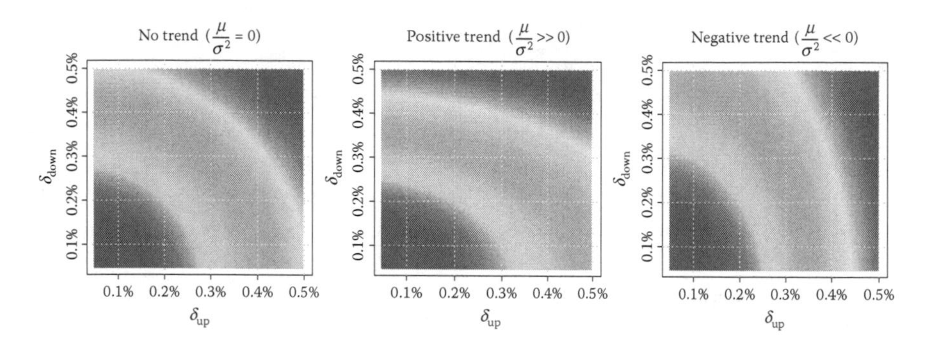

FIGURE 3.7: Monte Carlo simulation of the number of directional changes N, seen in Equation 3.9, as a function of the asymmetric directional change thresholds δ_{up} and δ_{down}, for a Brownian motion, defined by μ and σ. The left-hand panel shows a realization with no trend, while the other two panels have an underlying trend.

expected number of upward and downward directional changes during a time interval $[0, T]$ is a function

$$N = N(\delta_{\text{up}}, \delta_{\text{down}}, \mu, \sigma, [0, T]). \tag{3.9}$$

In Figure 3.7, the result of a Monte Carlo simulation is shown. For the situation with no trend (left-hand panel), we see the contour lines being perfect circles. In other words, by following any defined circle, the same number of directional changes are found for the corresponding asymmetric thresholds. Details about the analytical expressions and the Monte Carlo simulation regarding the number of directional changes can be found in Reference 39.

This opens up the space of possibilities, as up to now, only the 45-degree line in all panels of Figure 3.7 were considered, corresponding to symmetric thresholds $\delta = \delta_{\text{up}} = \delta_{\text{down}}$. For trending markets, one can observe a shift in the contour lines, away from the circles. In a nutshell, for a positive trend the expected number of directional changes is larger if $\delta_{\text{up}} > \delta_{\text{down}}$. This reflects the fact that an upward trend is naturally comprised of longer up-move segments. The contrary is true for down moves.

Now it is possible to introduce the notion of invariance as a guiding principle. By rotating the 45-degree line in the correct manner for trending markets, the number of directional changes will stay constant. In other words, if the trend is known, the thresholds can be skewed accordingly to compensate. However, it is not trivial to construct a trend indicator that is predictive and not only reactive.

A workaround is found by taking the inventory as a proxy for the trend. In detail, the expected inventory size I for all agents in normal market conditions can be used to gauge the trend: $\mathbb{E}[I(\delta_{\text{up}}, \delta_{\text{down}})]$ is now a measure of trendiness and hence triggers threshold skewing. In other words, by taking the inventory as an invariant indicator, the 45-degree line can be rotated due to the asymmetric thresholds, counteracting the trend.

A more mathematical justification can be found in the approach of what is known as "indifference prices" in market making. This method can be translated into the context of intrinsic time and agent's inventories. It then mandates that the utility (or preference) of the whole inventory should stay the same for skewed thresholds and inventory changes. In other words, how can the thresholds be changed in a way that "feels" the same as if the inventory increases or decreased by one unit? Expressed as equations

$$U(\delta_{\text{down}}, \delta_{\text{up}}, I) = U(\delta^{*}_{\text{down}}, \delta^{*}_{\text{up}}, I + 1), \tag{3.10}$$

and

$$U(\delta_{\text{down}}, \delta_{\text{up}}, I) = U(\delta^{**}_{\text{down}}, \delta^{**}_{\text{up}}, I - 1), \tag{3.11}$$

where U represents a utility function. The thresholds δ^{*}_{up}, δ^{*}_{down}, δ^{**}_{up}, and $\delta^{**}_{\text{down}}$ are "indifference" thresholds.

A pragmatic implementation of such an inventory-driven skewing of thresholds is given by the following equation, corresponding to a long position

$$\frac{\delta_{\text{down}}}{\delta_{\text{up}}} = \begin{cases} 2 & \text{if } I \geq 15; \\ 4 & \text{if } I \geq 30. \end{cases} \tag{3.12}$$

For a short position, the fractions are inverted

$$\frac{\delta_{\text{up}}}{\delta_{\text{down}}} = \begin{cases} 2 & \text{if } I \leq -15; \\ 4 & \text{if } I \leq -30. \end{cases} \tag{3.13}$$

In essence, in the presence of trends, the overshoot thresholds decrease as a result of the asymmetric directional change thresholds.

This also motivates a final relaxation of a constraint. The final ingredients of the Alpha Engine are fractional position changes. Recall that coastline trading is simply an increase or decrease in the position size at intrinsic time events. This cascading and de-cascading were done by one unit. For instance, increasing a short position size by one unit if the price increases and reaches an upward overshoot. To make this procedure compatible with asymmetric thresholds, the new cascading and de-cascading events resulting from the asymmetric threshold are now done with a fraction of the original unit. The fractions are also dictated by Equations 3.12 and 3.13. In effect, the introduction of asymmetric thresholds leads to a subdivision of the original threshold into smaller parts, where the position size is changed by subunits on these emerging threshold.

An example is shown in Figure 3.8. Assuming that a short position was opened at a lower price than the minimal price in the illustration, the directional change will trigger a cascading event. In other words, one (negative) unit of exposure (symbolized by the large arrows) is added to the existing short position. The two overshoot events in Figure 3.8a trigger identical cascading

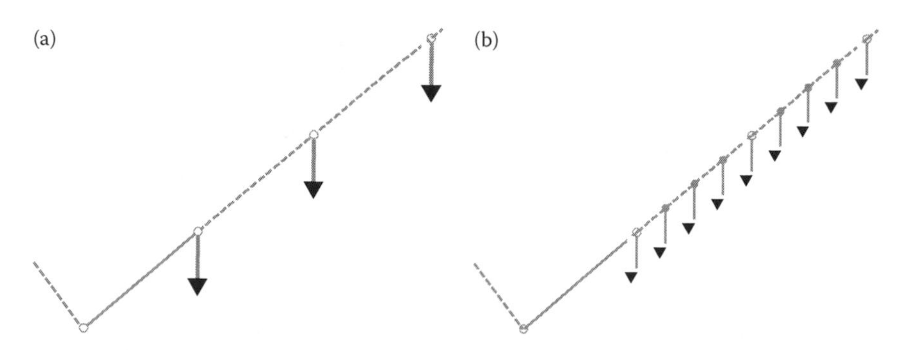

FIGURE 3.8: Cascading with asymmetric thresholds. A stylized price curve is shown in both panels. (a) The original (symmetric) setup with an upward directional change event (continuous line) and two overshoots (dashed lines). Short position size increments are shown as downward arrows. (b) The situation corresponding to asymmetric thresholds, where intrinsic time accelerates and smaller position size increments are utilized for coastline trading. See details in text.

events. In Figure 3.8b, the same events are augmented by asymmetric thresholds. Now $\omega_{\text{up}} = \omega_{\text{down}}/4$. As a result, each overshoot length is divided into four segments. The new cascading regime is as follows: increase the position by one-fourth of a (negative) unit (small arrow) at the directional change and another fourth at the first, second, and third asymmetric overshoots each. In effect, the cascading event is "smeared out" and happens in smaller unit sizes over a longer period. For the cascading events at the first and second original overshoots, this procedure is repeated.

This concludes the final chapter in the long history of the trading model development. Many insights from diverse fields were consolidated and a unified modeling framework emerged.

3.4 The Nuts and Bolts: A Summary of the Alpha Engine

All the insights gained along this long journey need to be encapsulated and translated into algorithmic concepts. In this section, we summarize in detail the trading model behavior and specify the parameters.

The intrinsic time scale dissects the price curve into directional changes and overshoots, yielding an empirical scaling law equating the length of the overshoot ω to the size of the directional change threshold δ, that is, $\langle\omega\rangle \approx \delta$. This scaling law creates an event-based language that clocks the trading model. In essence, intrinsic time events define the coastline trading behavior with its hallmark cascading and de-cascading events. In other words, the discrete price curve with occurrences of intrinsic time events triggers an increase or decrease in position sizes.

In detail, an intrinsic event is either a directional change or a move of size δ in the direction of the overshoot. For each exchange rate, we assign four coastline traders $\text{CT}_i[\delta_{\text{up/down}}(i)]$, $i = 1, 2, 3, 4$, that operate at various scales, with upward and downward directional change thresholds equaling $\delta_{\text{up/down}}(1) = 0.25\%, \delta_{\text{up/down}}(2) = 0.5\%, \delta_{\text{up/down}}(3) = 1.0\%$, and $\delta_{\text{up/down}}(4) = 1.5\%$.

The default size for cascading and de-cascading a position is one unit (lot). The probability indicator $\mathcal{L}_i$, assigned to each coastline trader, is evaluated on the fixed scale $\delta(i) = \delta_{\text{up/down}}(i)$. As a result, its states are directional changes of size $\delta(i)$ or overshoot moves of size $2.525729 \cdot \delta_i$. The default unit size for cascading is reduced to 0.5 if $\mathcal{L}_i$ is smaller than 0.5. Additionally, if $\mathcal{L}_i$ is smaller than 0.1, then the size is further reduced to 0.1.

In case a coastline trader accumulates an inventory with a long position greater than 15 units, the upward directional change threshold $\delta_{\text{up}}(i)$ is increased to 1.5 of its original size, while the downward directional change threshold $\delta_{\text{down}}(i)$ is decreased to 0.75 of its original size. In effect, the ratio for the skewed thresholds is $\delta_{\text{up}}(i)/\delta_{\text{down}}(i) = 2$. The agent with the skewed thresholds will cascade when the overshoot reaches 0.5 of the skewed threshold, that is, half of the original threshold size. In case the inventory with long position is greater than 30, then the upward directional change threshold $\delta_{\text{up}}(i)$ is increased to 2.0 of its original size and the downward directional change threshold $\delta_{\text{down}}(i)$ is decreased to 0.5. The ratio of the skewed thresholds now equals $\delta_{\text{up}}(i)/\delta_{\text{down}}(i) = 4$. The agent with these skewed thresholds will cascade when the overshoot extends by 0.25 of the original threshold, with one-fourth of the specified unit size. This was illustrated in Figure 3.8b. The changes in threshold lengths and sizing are analogous for short inventories.

This concludes the description of the trading model algorithm and the motivation of the chosen modeling framework. Recall that the interested reader can download the code from GitHub [10].

3.5 Conclusion and Outlook

The trading model algorithm described here is the result of a meandering journey that lasted for decades. Guided by an overarching event-based framework, recasting time as discrete and driven by activity, elements from complexity theory and information theory were added. In a nutshell, the proposed trading model is defined by a set of simple rules executed at specific events in the market. This approach to designing automated trading models yields an algorithm that fulfills many desired features. Its parsimonious, modular, and self-similar design results in behavior that is profitable, robust, and adaptive.

Another crucial feature of the trading model is that it is designed to be counter trend. The coastline trading ensures that positions, which are going against a trend, are maintained or increased. In this sense, the models provide liquidity to the market. When market participants want to sell, the investment strategy will buy and vice versa. This market-stabilizing feature of the model is beneficial to the markets as a whole. The more such strategies are implemented, the less we expect to see runaway markets but healthier market conditions overall. By construction, the trading model only ceases to perform in low-volatility markets.

It should be noted that the model framework presented here can be realized with reasonable computational resources. The basic agent-based algorithm shows profitable behavior for four directional change thresholds, on which the positions (agents) live. However, by adding more thresholds the model behavior is expected to become more robust, as more information coming from the market can be processed by the trading model. In other words, by increasing the model complexity the need for performant computing becomes relevant for efficient prototyping and backtesting. In this sense, we expect the advancements in high-performance computing in finance to positively impact the Alpha Engine's evolution.

Nevertheless, with all the merits of the trading algorithm presented here, we are only at the beginning. The Alpha Engine should be understood as a prototype. It is, so to speak, a proof of concept. For one, the parameter space can be explored in greater detail. Then, the model can be improved by calibrating the various exchange rates by volatility or by excluding illiquid ones. Furthermore, the model treats all the currency pairs in isolation. There should be a large window of opportunity for increasing the performance of the trading model by introducing correlation across currency pairs. This is a unique and invaluable source of information not yet exploited. Finally, a whole layer of risk management can be implemented on top of the models.

We hope to have presented a convincing set of tools motivated by a consistent philosophy. If so, we invite the reader to take what is outlined here and improve upon it.

Appendix 3A A History of Ideas

This section is a personal recount of the historical events that would ultimately lead to the development of the trading model algorithm outlined in this chapter, told by Richard B. Olsen.

The development of the trading algorithm and model framework dates back to my studies in the mid-70s and 80s. From the very start, my interests in economics were influenced by my admiration of the scientific rigor of natural sciences and their successful implementations in the real world. I argued that the resilience of the economic and political systems depends on the underlying economic and political models. Motivated to contribute to the well-being of society I wanted to work on enhancing economic theory and work on applying the models.

I first studied law at the University of Zurich and then, in 1979, moved to Oxford to study philosophy, politics, and economics. In 1980, I attended a course on growth models by James Mirrlees, who, in 1996, received a Nobel prize in economics. In his first lecture, he discussed the shortcomings of the models, such as [40]. He explained that the models are successful in explaining growth as long as there are no large exogenous shocks. But unanticipated events are inherent to our lives and the economy at large. I thus started to search for a model framework that can both explain growth and handle unexpected exogenous shocks. I spent one year studying the Encyclopedia Britannica and found my inspiration in relativity theory.

In my 1981 PhD thesis, titled "Interaction Between Law and Society," at the University of Zurich, I developed a new model framework that describes in an abstract language, how interactions in the economy occur. At the core of the new approach are the concepts of object, system, environment, and event-based intrinsic time. Every object has its system that comprises all the forces that impact and influence the object. Outside the system is its environment with all the forces that do not impact the object. Every object and system has its own frame of reference with an event-based intrinsic time scale. Events are interactions between different objects and their systems. I concluded that there is no abstract and universal time scale applicable to every object. This motivated me to think about the nature of time and how we use time in our everyday economic models.

After finishing my studies, I joined a bank working first in the legal department, then in the research group, and finally joined the foreign exchange trading desk. My goal was to combine empirical work with academic research, but was disappointed with the pace of research at the bank. In the mid-80s, there was the first buzz about start-ups in the United States. I came up with a business idea: banks have a need for quality information to increase profitability, so there should be a market for quality real-time information.

I launched a start-up with the name of Olsen & Associates. The goal was to build an information system for financial markets with real-time forecasts and trading recommendations using tick-by-tick market data. The product idea combined my research interest with an information service, which would both improve the quality of decision making in financial markets and generate revenue to fund further research. The collection of tick market data began in January 1986 from Reuters. We faced many business and technical obstacles, where data storage cost was just one of the many issues. After many setbacks, we successfully launched our information service and eventually acquired 60 big to mid-sized banks across Europe as customers.

In 1990, we published our first scientific paper [26] revealing the first scaling law. The study showed that intraday prices have the same scaling law exponent as longer term price movements. We had expected two different exponents: one for intraday price movements, where technical factors dictate price discovery, and another for longer term price movements that are influenced by fundamentals. The result took us by surprise and was evidence that there are universal laws that dictate price discovery at all scales. In 1995, we organized the first high-frequency data conference in Zurich, where we made a large sample of tick data available to the academic community. The conference was a big success and boosted market microstructure research, which was in its infancy at that time. In the following years, we conducted exhaustive research testing all possible model approaches to build a reliable forecasting service and trading models. Our research work is described in the book [17]. The book covers data collection and filtering, basic stylized facts of financial market time series, the modeling of 24-hour seasonal volatility, realized volatility dynamics, volatility processes, forecasting return and risks, correlation, and trading models. For many years, the book was a standard text for major hedge funds. The actual performance of our forecasting and trading models was, however, spurious and disappointing. Our models were best in class, but we had not achieved a breakthrough.

Back in 1995, we were selling tick-by-tick market data to top banks and created a spinoff under the name of OANDA to market a currency converter on the emergent Internet and eventually build a foreign exchange market making business. The OANDA currency converter was an instant success. At the start of 2001, we were completing the first release of our trading platform. At the same time, Olsen & Associates was a treasure store of information and risk services, but did not have cash to market the products and was struggling for funding. When the Internet bubble burst and markets froze, we could not pay our bills and the company went into default. I was able to organize a bailout with a new investor. He helped to salvage the core of Olsen & Associates with the aim of building a hedge fund under the name of Olsen Ltd and buying up the OANDA shares.

In 2001, the OANDA trading platform was a novelty in the financial industry: straight through processing, one price for everyone, and second-by-second interest payments. At the time, these were true firsts. At OANDA, a trader could buy literally 1 EUR against USD at the same low spread as a buyer of 1 million EUR against USD. The business was an instant success. Moreover, the OANDA trading platform was a research laboratory to analyze the trades of 10,000 traders, all buying and selling at the same terms and conditions, and observe their behavior patterns in different market environments. I learned hands on, how financial markets really work and discovered that basic assumptions of market efficiency that we had taken for granted at Olsen & Associates were inappropriate. I was determined to make a fresh start in model development.

At Olsen Ltd, I made a strategic decision to focus exclusively on trading model research. Trading models have a big advantage over forecasting models: the profit and losses of a trading model are an unambiguous success criterion of the quality of a model. We started with the forensics of the old model algorithms and discovered that the success and failure of a model depends critically on the definition of time and how data is sampled. Already at Olsen & Associates, we were sensitive to the issue of how to define time and had rescaled price data to account for the 24-hour seasonality of volatility, but did not succeed with a more sophisticated rescaling of time. There was one operator that we had failed to explore. We had developed a directional change indicator and had observed that the indicator follows a scaling law behavior similar to the absolute price change scaling law [16]. This scaling law was somehow forgotten and was not mentioned in our book [17]. I had incidental evidence that this operator would be successful to redefine time because traders use such an operator to analyze markets. The so-called point and figure chart replaces the x-axis of physical time with an event scale. As long as a market price moves up, the prize stays frozen in the same column. When the price moves down by a threshold bigger than the box size, the plot moves to the next column. A new column is started, when the price reverses its direction.

Then I also had another key insight of the path dependence of market prices from watching OANDA traders. There was empirical evidence that a margin call of one trader from anywhere in the world could trigger a whole cascade of margin calls in the global foreign exchange markets in periods of herding behavior. Cascades of margin calls wipe out whole cohorts of traders and tilt the market composition of buyers and sellers and skew the long-term price trajectory. Traditional time series models cannot adequately model these phenomena. We decided to move to agent-based models to better incorporate the emergent market dynamics and use the scaling laws as a framework to calibrate the algorithmic behavior of the agents. This seemed attractive, because we could configure self-similar agents at different scales.

I was adamant to build bare-bone agents and not to clutter our model algorithms with tools of spurious quality. In 2008, we were rewarded with major breakthrough: we discovered a large set of scaling laws [8]. I expected that model development would be plain sailing from thereon. I was wrong. The road of discovery was much longer than anticipated. Our hedge fund had several significant drawdowns that forced me to close the fund in 2013. At OANDA, things had also deteriorated. After raising 100 million USD for 20% of the company in 2007, I had become chairman without executive powers. OANDA turned into a conservative company and lost its competitive edge. In 2012, I left the board.

In July 2015, I raised the first seed round for Lykke, a new startup. Lykke builds a global marketplace for all asset classes and instruments on the blockchain. The marketplace is open source and a public utility. We will earn money by providing liquidity with our funds and/or customer's funds, with algorithms as described in this chapter.

TABLE 3A.1: Monthly performance of the unleveraged trading model

%	Jan	Feb	Mar	Apr	May	June	July	Aug	Sep	Oct	Nov	Dec	Year
2006	0.16	0.15	0.07	0.12	0.22	0.17	0.19	0.20	0.18	0.08	−0.00	0.04	**1.58**
2007	0.08	0.22	0.14	0.02	−0.05	−0.03	0.32	0.59	0.07	0.11	0.47	0.20	**2.03**
2008	0.24	0.07	0.05	0.50	0.26	0.09	0.26	0.16	0.66	2.22	1.27	0.98	**6.03**
2009	1.14	1.41	1.17	1.00	0.75	0.59	0.22	0.19	−0.13	0.28	0.06	0.25	**7.70**
2010	0.15	−0.34	0.24	0.14	0.30	0.17	0.27	−0.02	0.03	0.06	0.14	−0.31	**1.42**
2011	0.45	0.13	0.11	−0.16	0.04	−0.06	−0.40	0.43	0.45	−0.03	0.32	−0.03	**0.97**
2012	−0.08	0.19	0.29	0.08	−0.12	0.15	−0.20	0.23	0.10	0.13	0.12	0.11	**0.86**
2013	−0.17	−0.01	−0.10	−0.08	0.32	0.52	0.04	0.24	−0.10	0.01	−0.01	−0.16	**0.77**

Note: The P&L is given in percentages. All 23 currency pairs are aggregated.

Appendix 3B Supplementary Material

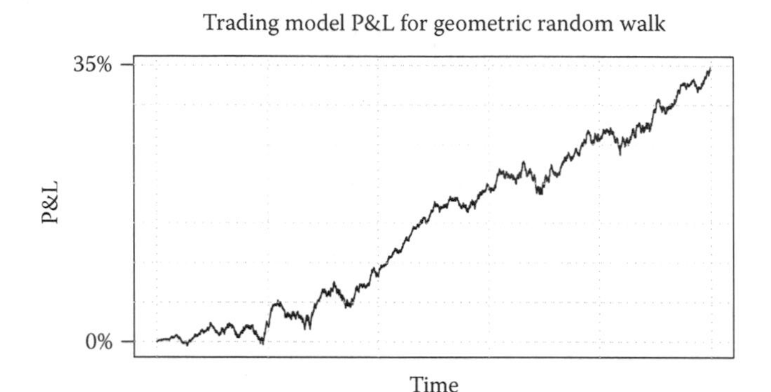

FIGURE 3B.1: Profit & Loss for a time series, generated by a geometric random walk of 10 million ticks with annualized volatility of 25%. The average of 60 Monte Carlo simulations is shown. In the limiting case, the P&L curve becomes a smooth increasing line.

References

1. Baghai, P., Erzan, O., and Kwek, J.-H. The $64 trillion question, convergence in asset management, McKinsey & Company, 2015.

2. World Bank. World development indicators database, 2015.

3. Chen, J.X. The evolution of computing: Alphago. *Computing in Science & Engineering* 18(4), 2016, pp. 4–7.

4. Bouveret, A., Guillaumie, C., Roqueiro, C.A., Winkler, C., and Nauhaus, S. High frequency trading activity in EU equity markets, European Securities and Markets Authority, 2014.

5. Roseen, T. Are quant funds worth another look? Thomson Reuters, 2016.

6. Bank of International Settlement. *Triennial Central Bank Survey of Foreign Exchange, and OTC Derivatives Markets, 2016.* Monetary and Economic Department, Basel, 2016.

7. ISDA. Central clearing in the equity derivatives market, 2014.

8. Glattfelder, J.B., Dupuis, A., and Olsen, R.B. Patterns in high-frequency fx data: Discovery of 12 empirical scaling laws. *Quantitative Finance* 11(4), 2011, pp. 599–614.

9. Doyne Farmer, J. and Foley, D. The economy needs agent-based modelling. *Nature* 460(7256), 2009, pp. 685–686.

10. The alpha engine: Designing an automated trading algorithm code. https://github.com/AntonVonGolub/Code/blob/master/code.java, 2017. Accessed: 2017-01-04. 2017.

11. Golub, A., Chliamovitch, G., Dupuis, A., and Chopard, B. Multi-scale representation of high frequency market liquidity. *Algorithmic Finance*, 5(1), 2016, pp. 3–19.

12. Müller, U.A., Dacorogna, M.M., Davé, R.D., Pictet, O.V., Olsen, R.B., and Robert Ward, J. Fractals and intrinsic time: A challenge to econometricians. *Presentation at the XXXIXth International AEA Conference on Real Time Econometrics*, 14–15 Oct 1993, Luxembourg, 1993.

13. Aloud, M., Tsang, E., Olsen, R.B., and Dupuis, A. A directional-change events approach for studying financial time series. *Economics Discussion Papers*, (2012-36), 6, 2012, pp. 1–17.

14. Ao, H. and Tsang, E. Capturing market movements with directional changes. *Working paper: Centre for Computational Finance and Economic Agents, Univ. of Essex*, 2013.

15. Bakhach, A., Tsang, E.P.K., and Ng, W. Lon. Forecasting directional changes in financial markets. *Working paper: Centre for Computational Finance and Economic Agents, Univ. of Essex*, 2015.

16. Guillaume, D.M., Dacorogna, M. M., Davé, R. R., Müller, U. A., Olsen, R. B., and V Pictet, O. From the bird's eye to the microscope: A survey of new stylized facts of the intra-daily foreign exchange markets. *Finance and Stochastics*, 1(2), 1997, pp. 95–129.

17. Gençay, R., Dacorogna, M., Muller, U.A., Pictet, O., and Olsen, R. *An Introduction to High-Frequency Finance.* Academic Press, New York, 2001.

18. Mandelbrot, B. The variation of certain speculative prices. *Journal of Business* 36(4), 1963, pp. 394–419.

19. Newman, M.E.J. Power laws, pareto distributions and Zipf's law. *Contemporary Physics* 46(5), 2005, pp. 323–351.

20. Pareto, V. Cours d'economie politique. 1897.

21. West, G.B., Brown, J.H., and Enquist, B.J. A general model for the origin of allometric scaling laws in biology. *Science* 276(5309), 1997, p. 122.

22. Zipf, G.K. *Human Behaxvior and the Principle of Least Effort.* Addison-Wesley, Reading, MA, 1949.

23. Albert, R. and Barabási, A.L. Statistical mechanics of complex networks. *Review of Modern Physics* 74(1), 2002, pp. 47–97.

24. Barabási, A.L. and Albert, R. Emergence of scaling in random networks. *Science* 1999, p. 509.

25. Newman, M.E.J. The structure and function of complex networks. *SIAM review* 45(2), 2003, pp. 167–256.

26. Müller, U.A., Dacorogna, M.M., Olsen, R.B., Pictet, O.V., Schwarz, M., and Morgenegg, C. Statistical study of foreign exchange rates, empirical evidence of a price change scaling law, and intraday analysis. *Journal of Banking & Finance* 14(6), 1990, 1189–1208.

27. Hull, J.C. *Options, Futures and Other Derivative Securities*, 9th edition. Pearson, London, 2014.

28. Voit, J. *The Statistical Mechanics of Financial Markets*, 3rd edition. Springer, Berlin, 2005.

29. Glattfelder, J.B. *Decoding Complexity.* Springer, Heidelberg, 2013.

30. Wolfram, S. *A New Kind of Science.* Wolfram Media, Champaign, 2002.

31. Andersen, J.V. and Sornette, D. A mechanism for pockets of predictability in complex adaptive systems. *EPL (Europhysics Letters)*, 70(5), 2005, p. 697.

32. Helbing, D. Agent-based modeling. In: *Social Self-Organization.* Springer, 2012, pp. 25–70.

33. Lux, T. and Marchesi, M. Volatility clustering in financial markets: A microsimulation of interacting agents. *International Journal of Theoretical and Applied Finance* 3(4), 2000, pp. 675–702.

34. Aloud, M., Tsang, E., Dupuis, A., and Olsen, R. Minimal agent-based model for the origin of trading activity in foreign exchange market. In: *2011 IEEE Symposium on Computational Intelligence for Financial Engineering and Economics (CIFEr). IEEE.* 2011, pp. 1–8.

35. Dupuis, A. and Olsen, R.B. *High Frequency Finance, Using Scaling Laws to Build Trading Models*, John Wiley & Sons, Inc., 2012, pp. 563–584.

36. Glattfelder, J.B., Bisig, T., and Olsen, R.B. R&D Strategy Document. Technical report, A Paper by the Olsen Ltd. Research Group, 2010.

37. Cover, T.M. and Thomas, J.A. *Elements of Information Theory.* John Wiley & Sons, Hoboken, 1991.

38. Pfister, H.D., Soriaga, J.B., and Siegel, P.H. On the achievable information rates of finite state ISI channels. In *Proc. IEEE Globecom.* Eds. Kurlander, D., Brown, M., and Rao, R. ACM Press, November 2001, pp. 41–50.

39. Golub, A., Glattfelder, J.B., Petrov, V., and Olsen, R.B. *Waiting Times and Number of Directional Changes in Intrinsic Time Framework*, 2017. Lykke Corp & University of Zurich Working Paper.

40. Kaldor, N. and Mirrlees, J.A. A new model of economic growth. *The Review of Economic Studies* 29(3), 1962, pp. 174–192.

Chapter 4

Portfolio Liquidation and Ambiguity Aversion

Álvaro Cartea, Ryan Donnelly, and Sebastian Jaimungal

CONTENTS

4.1 Introduction

This work considers an optimal execution problem in which an agent is tasked with liquidating a number of shares of an asset before the end of a defined trading period. The development of an algorithm which quantitatively specifies the execution strategy ensures that the liquidation is performed with the perfect balance of risk and profits, and in the modern trading environment in which human trading activity is essentially obsolete, the execution must be based on a set of predefined rules. However, the resulting set of rules depends heavily upon modeling assumptions made by the agent. If the model is misspecified, then the balance between risk and profits is imperfect and the strategy may be exposed to risks that are not captured by model features.

The design of optimal execution algorithms began in the literature with the seminal work by Almgren and Chriss (2001). This approach was generalized over time with models that include additional features. Kharroubi and Pham (2010) take a different approach to modeling the price impact of the agent's trades depending on the amount of time between them. A more direct approach in modeling features of the limit order book and an induced price impact appears in Obizhaeva and Wang (2013).

The aforementioned papers allow the agent to employ only market orders in their liquidation strategy. By allowing the agent to use limit orders instead (or in addition), the profits accumulated by the portfolio liquidation can potentially be increased. Intuitively this is possible due to acquiring better trade prices by using limit orders instead of market orders. Guéant et al. (2012) rephrase the problem as one in which all of the agent's trades are performed by using limit orders. This work is then continued in Guéant and Lehalle (2015) where more general dynamics of transactions are considered. A combination of limit order and market order strategy is developed in Cartea and Jaimungal (2015a). For an overview on these and related problems in algorithmic and high-frequency trading, see Cartea et al. (2015).

The present work is most related to Guéant et al. (2012) and Cartea and Jaimungal (2015a). The underlying dynamics we use here are identical to Guéant et al. (2012), but we use a different objective function. Our work considers a risk-neutral agent, whereas the former considers one with exponential utility. Since the purpose of this work is to investigate the effects of ambiguity aversion, we consider the risk-neutral agent in order to eliminate possible effects due to risk aversion. In addition, we also consider the inclusion of market orders in the agent's trading strategy toward the end of this work, similar to what is done in Cartea and Jaimungal (2015a).

If the agent expresses equal levels of ambiguity toward the arrival rate and fill probability of orders, then her optimal trading strategy can be found in closed form. This is of benefit to the agent if the strategy is computed and implemented in real time because there is no need to numerically solve a nonlinear PDE. Incorporating other relevant features in the framework makes the model more realistic, but this may come at the cost of not having closed-form solutions. An example is when the admissible strategies include market orders. In this case, the optimal strategies are obtained by solving a quasi-variational inequality. This requires sophisticated numerical methods and computer power to implement the strategies in real time.

The rest of the work is presented as follows. In Section 4.2, we present the dynamics of price changes and trade arrivals that our agent considers to be her reference model. The solution to the optimal liquidation is presented here in brief. In Section 4.3, we introduce the method by which our agent expresses ambiguity aversion. This amounts to specifying a class of equivalent candidate measures which she considers and a function which assigns to each candidate measure a penalization based on its deviation from the reference measure. In Section 4.4, we show how ambiguity with respect to different aspects of the reference dynamics affects the trading strategy of the agent. Section 4.5 presents some closed-form solutions for the value function and considers asymptotic behavior of the trading strategy, and Section 4.6 develops the inclusion of market orders and again analyzes how ambiguity effects the timing of market order placements.

4.2 Reference Model

4.2.1 Dynamics

The reference model is structured in a similar manner to Avellaneda and Stoikov (2008), Guéant et al. (2013), and Cartea and Jaimungal (2015b), except only the sell side of the LOB dynamics is considered. It is not necessary to consider the intensity or volume distribution of market sell orders since the agent will only be posing LOs on the sell side of the LOB.

Midprice: The midprice S_t satisfies

$$dS_t = \alpha\, dt + \sigma\, dW_t,$$

where α and $\sigma > 0$ are constants and $(W_t)_{0\leq t\leq T}$ is a standard Brownian motion.

Market Orders: Let μ be a Poisson random measure with compensator given by $\nu_{\mathbb{P}}(dy, dt) = \lambda\, F(dy)\, dt$, where $F(dy) = \kappa\, e^{-\kappa\, y}\, dy$. The number of market buy orders that have arrived up to time t is set equal to $M_t = \int_0^t \int_0^\infty \mu(dy, ds)$. The spacial dimension of the Poisson random measure μ is interpreted as the maximum price of execution of the incoming MO. The form of the compensator $\nu_{\mathbb{P}}$ implies that MOs arrive at a constant intensity λ and that the distribution of the maximum price of execution is exponential with parameter κ.

Limit Order Placement: The agent places limit orders only on the sell side of the limit order book. The distance of the limit order from the midprice is denoted by δ_t. Since the agent's LO is only lifted by an MO that has a maximum execution price greater than $S_t + \delta_t$, the number of LOs which are filled is equal to $N_t = \int_0^t \int_{\delta_s}^\infty \mu(dy, ds)$. Given that an MO arrives at time t and the agent has an LO posted at price $S_t + \delta_t$, the probability that the limit order is filled is $e^{-\kappa\, \delta_t}$. This is due to the fact that the LO is only filled if the maximal execution price of the MO is greater than the price of the LO.

The appropriate filtered probability space to consider is $(\Omega, \mathbb{F}, \mathcal{F} = \{\mathcal{F}_t\}_{0\leq t\leq T}, \mathbb{P})$ where $\mathcal{F}$ is the completion of the filtration generated by the midprice $(S_t)_{0\leq t\leq T}$ and the process $(P_t)_{0\leq t\leq T}$, where $P_t = \int_0^t \int_0^\infty y\mu(dy, ds)$.[1] The agent uses δ_t, an $\mathcal{F}_t$-predictable process, as their control.

[1] Observing the process P_t allows one to detect both the time of arrival and the maximum execution price of MOs.

Inventory: The agent's inventory is denoted by q_t, and she begins with a total of Q shares to be liquidated. Thus $q_t = Q - N_t$, which gives

$$\mathrm{d}q_t = -\mathrm{d}N_t.$$

Wealth: Immediately after an MO lifts one of the agent's LOs, her wealth increases by an amount equal to the trade value $S_t + \delta_t$. The wealth, X_t, therefore, has dynamics

$$\mathrm{d}X_t = (S_t + \delta_{t^-})\,\mathrm{d}N_t.$$

4.2.2 Optimal liquidation problem

Given that the agent's goal is to liquidate her position in a way that is financially optimal, we cease the agent's ability to trade when her inventory reaches zero. The optimization problem under the reference model is to select an LO posting strategy which maximizes expected terminal wealth

$$H(t,x,q,S) = \sup_{(\delta_s)_{t\le s\le T}\in\mathcal{A}} \mathbb{E}^{\mathbb{P}}_{t,x,q,S}\left[X_{\tau\wedge T} + q_{\tau\wedge T}\left(S_{\tau\wedge T} - \ell(q_{\tau\wedge T})\right)\right], \tag{4.1}$$

where $\tau = \inf\{t : q_t = 0\}$, T is the terminal time of the strategy, q_T final inventory, $\mathbb{E}^{\mathbb{P}}_{t,x,q,S}[\,\cdot\,]$ denotes $\mathbb{P}$ expectation conditional on $X_{t^-} = x$, $q_{t^-} = q$ and $S_t = S$, and $\mathcal{A}$ denotes the set of admissible strategies which are non-negative $\mathcal{F}_t$-predictable processes. Moreover, the function $\ell(q_T)$, with $\ell(0) = 0$ and $\ell(q)$ increasing in q, is a liquidation penalty that consists of fees and market impact costs when the agent unwinds terminal inventory. For example, $\ell(q) = \theta\, q$ represents a linear impact when liquidating q shares, where $\theta \ge 0$ is a penalty parameter. This liquidation penalty will only come into effect if the agent does not manage to liquidate her entire inventory before the end of the trading period. Note that since $q_\tau = 0$, this gives $X_\tau + q_\tau\,(S_\tau - \ell(q_\tau)) = X_\tau$, and so $H(t,x,0,S) = x$.

By standard results, the HJB equation associated with the value function H is given by

$$\partial_t H + \alpha\partial_S H + \tfrac{1}{2}\sigma^2\partial_{SS}H + \sup_{\delta\ge 0} \lambda\, e^{-\kappa\,\delta}\mathcal{D}H\mathbb{1}_{q\neq 0} = 0, \tag{4.2}$$

subject to the terminal and boundary conditions

$$H(T,x,q,S) = x + q(S - \ell(q)), \tag{4.3}$$
$$H(t,x,0,S) = x, \tag{4.4}$$

where the operator $\mathcal{D}$ acts as

$$\mathcal{D}H = H(t, x + (S+\delta), q-1, S) - H(t,x,q,S), \tag{4.5}$$

and $\mathbb{1}$ is the indicator function.

The form of the HJB equation 4.2 and the conditions 4.3 and 4.4 allow for the ansatz $H(t,x,q,S) = x + q\,S + h_q(t)$. Substituting this expression into the above HJB equation gives

$$\partial_t h_q + \alpha\, q + \sup_{\delta\geq 0} \left\{ \lambda\, e^{-\kappa\,\delta}\,(\delta + h_{q-1} - h_q) \right\} \mathbb{1}_{q\neq 0} = 0, \tag{4.6}$$

$$h_q(T) = -q\,\ell(q), \tag{4.7}$$

$$h_0(t) = 0\,. \tag{4.8}$$

4.2.3 Feedback controls

Proposition 4.1 (Optimal Feedback Controls). *The optimal feedback controls of the HJB equation 4.2 are given by*

$$\delta_q^*(t) = \left(\tfrac{1}{\kappa} - h_{q-1}(t) + h_q(t) \right)_+ , \quad q \neq 0, \tag{4.9}$$

where $(x)_+ = \max(x, 0)$.

Proof. Apply first-order conditions to the supremum term in Equation 4.6 to give 4.9. Verifying that this gives a maximizer to Equation 4.6 is a simple exercise. □

As long as the stated feedback controls remain positive, they can be substituted into Equation 4.6 to obtain a nonlinear system of equations for $h_q(t)$:

$$\partial_t h_q + \alpha\, q + \frac{\lambda}{\kappa} e^{-\kappa\left(\frac{1}{\kappa} - h_{q-1} + h_q\right)} \mathbb{1}_{q\neq 0} = 0, \tag{4.10}$$

along with terminal conditions $h_q(T) = -q\,\ell(q)$ and boundary condition $h_0(t) = 0$. This equation has a closed-form analytical solution which will be discussed in Section 4.5.

In Figure 4.1, the optimal depth to be posted by the agent is plotted for each value of inventory. There are numerous qualitative features in this figure which have intuitive explanations. First, we see that for any level of inventory, the optimal depth decreases as the time approaches maturity. When the time until the end of the trading period is longer, the agent experiences no sense of urgency in liquidating her position and would rather be patient to sell shares for high prices. But when maturity is close, the possibility of having to pay the liquidation fee for unliquidated shares is imminent. The agent is more willing to take lower prices to avoid having to pay this cost. Similarly, we also see that the depth decreases as the current level of inventory increases for the same reason; with a larger inventory holding comes a higher probability of having to pay a liquidation fee at the end of the trading period. The agent is willing to transact at lower prices to avoid this possibility. Lastly, we see

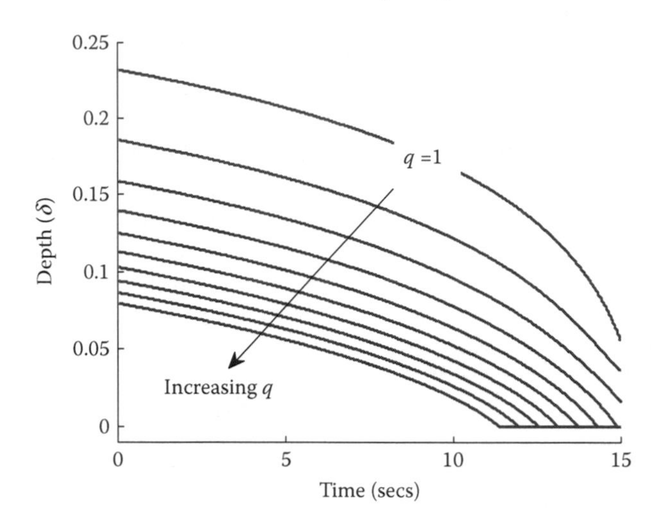

FIGURE 4.1: Optimal depth for an ambiguity neutral agent. Parameter values are $\kappa = 15$, $\lambda = 2$, $\sigma = 0.01$, $\alpha = 0$, $\ell(q) = \theta\, q$, $\theta = 0.01$, $Q = 10$, and $T = 15$ seconds.

that for times close to maturity and large levels of inventory, the agent posts a depth of zero indicating that she is willing to make essentially no profit in order to avoid the risk of terminal penalty. If it were not for the constraint $\delta_t \geq 0$, the agent would post negative depths in this region if it would mean a faster rate of order fill. This is an indication of the desire to submit a market order. A proper formulation of optimal market order submission is provided in Section 4.6.

4.3 Ambiguity Aversion

In this section, we allow the agent to acknowledge that the reference model is only an approximation of reality, but ultimately is misspecified. Adjusting one's strategy due to the desire of avoiding the risks of working with a misspecified model is called ambiguity aversion. There are many approaches to incorporating ambiguity aversion into optimal control problems, but we focus on one which allows tractable solutions in our algorithmic trading context.

The approach we consider involves the agent ranking alternative dynamics through the specification of a measure $\mathbb{Q}$ equivalent to the reference measure $\mathbb{P}$ and evaluating the performance of a trading strategy within the dynamics given by $\mathbb{Q}$. The agent also penalizes deviations of the measure $\mathbb{Q}$ from $\mathbb{P}$ which represents the cost of rejecting the reference measure in favor of the candidate. If the agent is highly confident in the specification of dynamics in the reference measure, then this cost should be large even for small deviations.

On the other hand, if the agent lacks confidence in the reference measure, then the cost should be small and even large deviations will incur a small penalty.

The optimization problem as originally posed in Equation 4.1 is altered to reflect the consideration of new dynamics:

$$\begin{aligned} H(t,x,q,S) = \sup_{(\delta_s)_{t\le s\le T}\in\mathcal{A}} \inf_{\mathbb{Q}\in\mathcal{Q}} \mathbb{E}^{\mathbb{Q}}_{t,x,q,S} \big[& X_{\tau\wedge T} + q_{\tau\wedge T}\left(S_{\tau\wedge T} - \ell(q_{\tau\wedge T})\right) \\ & + \mathcal{H}_{t,T}(\mathbb{Q}|\mathbb{P}) \big] , \end{aligned} \tag{4.11}$$

where the class of equivalent measures $\mathcal{Q}$ and the penalty $\mathcal{H}$ are specified below.

Since the agent may exhibit more confidence in the specifications of the model with respect to some aspects over others, she penalizes deviations of the model with respect to different aspects with different magnitudes. In particular, the agent wants unique penalizations toward specifications of:

- Midprice dynamics other than arithmetic Brownian motion
- MO arrival intensities other than the constant λ
- MO maximal price distributions other than exponential with parameter κ

We consider the same class of candidate measures and penalty function introduced in Cartea et al. (2014). Namely, any candidate measure is defined in terms of a Radon–Nikodym derivative

$$\frac{d\mathbb{Q}^{\alpha,\lambda,\kappa}(\eta,g)}{d\mathbb{P}} = \frac{d\mathbb{Q}^{\alpha}(\eta)}{d\mathbb{P}} \frac{d\mathbb{Q}^{\alpha,\lambda,\kappa}(\eta,g)}{d\mathbb{Q}^{\alpha}(\eta)}, \tag{4.12}$$

where the intermediate Radon–Nikodym derivatives are

$$\frac{d\mathbb{Q}^{\alpha}(\eta)}{d\mathbb{P}} = \exp\left\{ -\frac{1}{2}\int_0^T \left(\frac{\alpha-\eta_t}{\sigma}\right)^2 dt - \int_0^T \frac{\alpha-\eta_t}{\sigma}\, dW_t \right\} \tag{4.13}$$

and

$$\frac{d\mathbb{Q}^{\alpha,\lambda,\kappa}(\eta,g)}{d\mathbb{Q}^{\alpha}(\eta)} = \exp\left\{ -\int_0^T\int_0^\infty (e^{g_t(y)}-1)\,\nu_{\mathbb{P}}(dy,dt) + \int_0^T\int_0^\infty g_t(y)\,\mu(dy,dt) \right\} . \tag{4.14}$$

In denoting the candidate measure $\mathbb{Q}^{\alpha,\lambda,\kappa}$, we are conveying that the drift, arrival rate of MOs, and distribution of MOs have all been changed. The full class of new measures considered by the agent is

$$\mathcal{Q}^{\alpha,\lambda,\kappa} = \Bigg\{ \mathbb{Q}^{\alpha,\lambda,\kappa}(\eta, g) : \eta, g \text{ are } \mathcal{F}\text{-predictable}, \mathbb{E}^{\mathbb{P}}\left[\frac{\mathrm{d}\mathbb{Q}^{\alpha}(\eta)}{\mathrm{d}\mathbb{P}}\right] = 1,$$
$$\mathbb{E}^{\mathbb{Q}^{\alpha}(\eta)}\left[\frac{\mathrm{d}\mathbb{Q}^{\alpha,\lambda,\kappa}(\eta, g)}{\mathrm{d}\mathbb{Q}^{\alpha}(\eta)}\right] = 1, \text{ and } \mathbb{E}^{\mathbb{Q}^{\alpha,\lambda,\kappa}(\eta,g)}$$
$$\times \left[\int_0^T \int_0^\infty y^2 e^{g_t(y)} \nu_{\mathbb{P}}(\mathrm{d}y, \mathrm{d}t)\right] < \infty \Bigg\}. \tag{4.15}$$

The constraints imposed on the first two expectations ensure that the Radon–Nikodym derivatives in Equations 4.13 and 4.14 yield probability measures. The inequality constraint ensures that in the candidate measure, the profits earned by the agent have a finite variance. The set or candidate measures are parameterized by the process η and the random field g, and the dynamics of the midprice and MOs can be stated in terms of these quantities. In the candidate measure $\mathbb{Q}^{\alpha,\lambda,\kappa}(\eta, g)$, the drift of the midprice is no longer α, but is changed to η_t. Also in the candidate measure, the compensator of $\mu(\mathrm{d}y, \mathrm{d}t)$ becomes $\nu_{\mathbb{Q}}(\mathrm{d}y, \mathrm{d}t) = e^{g_t(y)}\, \nu_{\mathbb{P}}(\mathrm{d}y, \mathrm{d}t)$, see Jacod and Shiryaev (1987), Chapter III.3c, Theorem 3.17.

Before introducing the penalty function $\mathcal{H}$, we further decompose the measure change. First, note that according to the above discussion, if $g_t(y)$ does not depend on y, then the fill probability has not changed in the candidate measure, only the rate of MO arrival is different. Second, if $g_t(y)$ satisfies the equality

$$\int_0^\infty e^{g_t(y)} F(\mathrm{d}y) = 1, \tag{4.16}$$

for all $t \in [0, T]$, then only the fill probability has changed in the candidate measure and the intensity remains the constant λ. We are then able to break into two steps the measure change of Equation 4.14. Given a random field g, we define two new random fields by

$$g_t^{\lambda} = \log\left(\int_0^\infty e^{g_t(y)} F(\mathrm{d}y)\right), \tag{4.17}$$
$$g_t^{\kappa}(y) = g_t(y) - g_t^{\lambda}\,. \tag{4.18}$$

The random field g^{λ} does not depend on y, and it is easily shown that the random field g^{κ} satisfies Equation 4.16. Thus any measure $\mathbb{Q}^{\alpha,\lambda,\kappa}(\eta, g) \in \mathcal{Q}^{\alpha,\lambda\kappa}$ allows us to uniquely define two measures $\mathbb{Q}^{\alpha,\lambda}(\eta, g)$ and $\mathbb{Q}^{\alpha,\kappa}(\eta, g)$ via

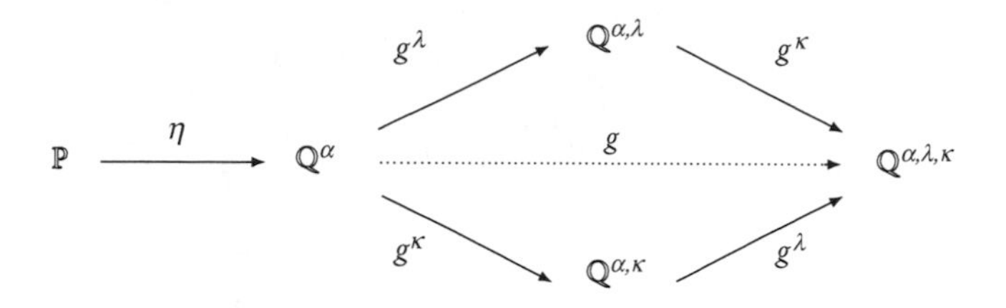

FIGURE 4.2: Three natural alternative routes from the reference measure $\mathbb{P}$ to a candidate measure $\mathbb{Q}^{\alpha,\lambda,\kappa}$ in which midprice drift, MO intensity, and execution price distribution of MOs have been altered.

Random–Nikodym derivatives:

$$\frac{d\mathbb{Q}^{\alpha,\lambda}(\eta,g)}{d\mathbb{Q}^{\alpha}(\eta)} = \exp\left\{-\int_0^T\int_0^\infty (e^{g_t^\lambda}-1)\,\nu_{\mathbb{P}}(dy,dt) + \int_0^T\int_0^\infty g_t^\lambda\,\mu(dy,dt)\right\},$$

$$\frac{d\mathbb{Q}^{\alpha,\kappa}(\eta,g)}{d\mathbb{Q}^{\alpha}(\eta)} = \exp\left\{-\int_0^T\int_0^\infty (e^{g_t^\kappa(y)}-1)\,\nu_{\mathbb{P}}(dy,dt) + \int_0^T\int_0^\infty g_t^\kappa(y)\,\mu(dy,dt)\right\}.$$

These measure changes represent decomposing the change from $\mathbb{Q}^{\alpha}(\eta)$ to $\mathbb{Q}^{\alpha,\lambda,\kappa}(\eta,g)$ into two subsequent measure changes, one in which only the intensity of orders changes and one in which only fill probability of orders changes. This is graphically represented in Figure 4.2.

To solve the robust optimization problem proposed in Equation 4.11, we consider g^λ and g^κ to be defined on their own, so that g^λ does not depend on y and g^κ satisfies Equation 4.16. Then $g_t(y) = g_t^\lambda + g_t^\kappa(y)$ parameterizes a full measure change. The penalty function we select is

$$\mathcal{H}\left(\mathbb{Q}^{\alpha,\lambda,\kappa}(\eta,g)\,|\,\mathbb{P}\right) = \begin{cases} \dfrac{1}{\varphi_\alpha}\log\left(\dfrac{d\mathbb{Q}^{\alpha}(\eta)}{d\mathbb{P}}\right) + \dfrac{1}{\varphi_\lambda}\log\left(\dfrac{d\mathbb{Q}^{\alpha,\lambda}(\eta,g)}{d\mathbb{Q}^{\alpha}(\eta)}\right) & \\ +\dfrac{1}{\varphi_\kappa}\log\left(\dfrac{d\mathbb{Q}^{\alpha,\lambda,\kappa}(\eta,g)}{d\mathbb{Q}^{\alpha,\lambda}(\eta,g)}\right), & \text{if } \varphi_\lambda \geq \varphi_\kappa, \\ & \\ \dfrac{1}{\varphi_\alpha}\log\left(\dfrac{d\mathbb{Q}^{\alpha}(\eta)}{d\mathbb{P}}\right) + \dfrac{1}{\varphi_\kappa}\log\left(\dfrac{d\mathbb{Q}^{\alpha,\kappa}(\eta,g)}{d\mathbb{Q}^{\alpha}(\eta)}\right) & \\ +\dfrac{1}{\varphi_\lambda}\log\left(\dfrac{d\mathbb{Q}^{\alpha,\lambda,\kappa}(\eta,g)}{d\mathbb{Q}^{\alpha,\kappa}(\eta,g)}\right), & \text{if } \varphi_\lambda \leq \varphi_\kappa. \end{cases} \tag{4.19}$$

A graphical representation of how this choice of penalty function places different weights on each step of the full measure change is given in Figure 4.3. Note from the definition 4.19 that when all three ambiguity weights are equal, the optimization problem 4.11 reduces to using relative entropy as the penalty.

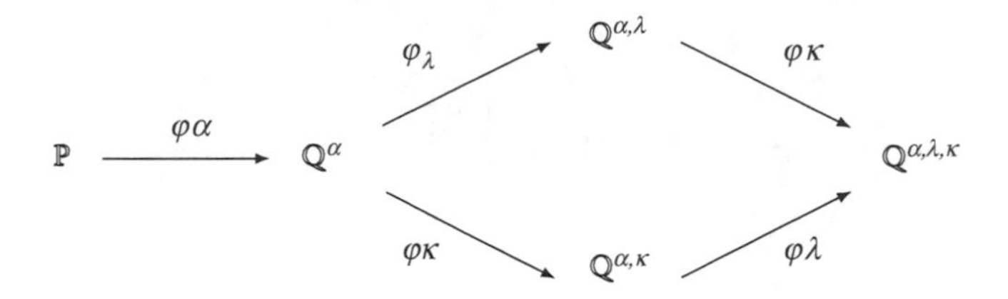

FIGURE 4.3: Ambiguity weights associated with each sequential step of the full measure change.

We return now to the optimization problem 4.11 which has an associated Hamilton–Jacobi-Bellman–Isaacs (HJBI) equation:

$$\begin{aligned} &\partial_t H + \frac{1}{2}\sigma^2 \partial_{SS} H + \inf_{\eta}\left\{ \eta\, \partial_S H + \frac{1}{2\varphi_\alpha}\left(\frac{\alpha-\eta}{\sigma}\right)^2 \right\} \\ &+ \sup_{\delta\geq 0} \inf_{g^\lambda} \inf_{g^\kappa\in\mathcal{G}} \left\{ \lambda \left[\int_\delta^\infty e^{g^\lambda + g^\kappa(y)} F(\mathrm{d}y) \right] \mathcal{D} H \mathbb{1}_{q\neq 0} \right. \\ &\left. + \mathcal{K}^{\varphi_\lambda,\varphi_\kappa}(g^\lambda, g^\kappa)\, \mathbb{1}_{\varphi_\lambda\geq\varphi_\kappa} + \mathcal{K}^{\varphi_\kappa,\varphi_\lambda}(g^\kappa, g^\lambda)\, \mathbb{1}_{\varphi_\lambda<\varphi_\kappa} \right\} = 0, \end{aligned} \tag{4.20}$$

with conditions

$$H(T, x, q, S) = x + q(S - \ell(q)) \tag{4.21}$$
$$H(t, x, 0, S) = x, \tag{4.22}$$

where the functional $\mathcal{K}$ is given by

$$\begin{aligned} \mathcal{K}^{c,d}(a,b) = &\frac{1}{c}\int_0^\infty \left[-(e^{a(y)} - 1) + a(y)\, e^{a(y)+b(y)} \right] \lambda\, F(\mathrm{d}y) \\ &+ \frac{1}{d}\int_0^\infty [-(e^{b(y)} - 1)\, e^{a(y)} + b(y)\, e^{a(y)+b(y)}]\, \lambda\, F(\mathrm{d}y), \end{aligned}$$

and the class of functions $\mathcal{G}$ is defined by an integral constraint:

$$\mathcal{G} = \left\{ g : \int_0^\infty e^{g(y)} F(\mathrm{d}y) = 1 \right\}. \tag{4.23}$$

In a similar fashion to Equation 4.2, this equation yields the ansatz $H(t, x, q, S) = x + q\,S + h_q(t)$, which when substituted into Equations 4.20

through 4.22 gives

$$\partial_t h_q + \inf_{\eta}\left\{\eta\, q + \frac{1}{2\varphi_\alpha}\left(\frac{\alpha-\eta}{\sigma}\right)^2\right\}$$
$$+\sup_{\delta\ge 0}\inf_{g^\lambda}\inf_{g^\kappa\in\mathcal{G}}\left\{\lambda\left[\int_\delta^\infty e^{g^\lambda+g^\kappa(y)}F(\mathrm{d}y)\right](\delta+h_{q-1}(t)-h_q(t))\mathbb{1}_{q\neq 0}\right. \tag{4.24}$$
$$\left.+\,\mathcal{K}^{\varphi_\lambda,\varphi_\kappa}(g^\lambda,g^\kappa)\,\mathbb{1}_{\varphi_\lambda\ge\varphi_\kappa}+\mathcal{K}^{\varphi_\kappa,\varphi_\lambda}(g^\kappa,g^\lambda)\,\mathbb{1}_{\varphi_\lambda<\varphi_\kappa}\right\}=0,$$
$$h(T,q) = -q\,\ell(q), \tag{4.25}$$
$$h(t,0) = 0\,. \tag{4.26}$$

Proposition 4.2 (Solution to HJBI equation). *Equation 4.24 along with the terminal and boundary conditions has a unique classical solution. Furthermore, the optimum in Equation 4.24 is achieved, where the optimizers are given by*

$$\delta_q^*(t) = \left(\tfrac{1}{\varphi_\kappa}\log\left(1+\tfrac{\varphi_\kappa}{\kappa}\right) - h_{q-1}(t)+h_q(t)\right)_+, \qquad q\neq 0,$$
$$\eta_q^*(t) = \alpha - \varphi_\alpha\,\sigma^2\, q,$$
$$g_q^{\lambda *}(t) = \left[\frac{\varphi_\lambda}{\varphi_\kappa}\log(1-e^{-\kappa\,\delta_q^*(t)}(1-e^{-\varphi_\kappa(\delta_q^*(t)+h_{q-1}(t)-h_q(t))}))\right]\mathbb{1}_{q\neq 0}, \tag{4.27}$$
$$g_q^{\kappa *}(t,y) = -\left[\log(1-e^{-\kappa\,\delta_q^*(t)}(1-e^{-\varphi_\kappa(\delta_q^*(t)+h_{q-1}(t)-h_q(t))}))\right.$$
$$\left.+\,\varphi_\kappa(\delta_q^*(t)+h_{q-1}(t)-h_q(t))\mathbb{1}_{y\ge\delta_q^*(t)}\right]\mathbb{1}_{q\neq 0}\,.$$

Proof. See Appendix. □

The feedback expressions above give the pointwise optimizer in the differential equation 4.6. Since we have a classical solution to this equation, the function H serves as a candidate value function. Also, by appropriately substituting the processes for the state variables in Equation 4.27, we get candidate optimal controls. The verification theorem below guarantees that these candidates are indeed the value function and optimal controls we seek.

Theorem 4.1 (Verification Theorem). *Let $h_q(t)$ be the solution to Equation 4.24 and let $H(t,x,q,S) = x+q\,S+h_q(t)$. Also let $\delta_t^\diamond = \delta_{q_t}^*(t)$, $\eta_t^\diamond = \eta_{q_t}^*(t)$, $g_t^{\lambda\diamond} = g_{q_t}^{\lambda *}(t)$, and $g_t^{\kappa\diamond}(y) = g_{q_t}^{\kappa *}(t,y)$ define processes. Then $\delta^\diamond$, $\eta^\diamond$, $g^{\lambda\diamond}$, and $g^{\kappa\diamond}$ are admissible controls. Further, H is the value function to the agent's control problem 4.11 and the optimum is achieved by these controls.*

Proof. See Appendix. □

4.4 Effects of Ambiguity on the Optimal Strategy

In this section, we investigate how the agent changes her optimal posting strategy based on her levels of ambiguity toward each source of uncertainty. For the three types of ambiguity, we compute the optimal strategy and compare it to the optimal strategy presented in Figure 4.1.

4.4.1 Arrival rate

When the agent is only ambiguous to the arrival rate of market orders, the expressions for δ^*, $g^{\lambda *}$, $g^{\kappa *}$ in Equation 4.27 are interpreted in terms of a limit as $\varphi_\kappa \to 0$ to arrive at

$$\begin{aligned}
\delta_q^*(t) &= \left(\frac{1}{\kappa} - h_{q-1}(t) + h_q(t)\right)_+, \qquad q \neq 0, \\
g_q^{\lambda *}(t) &= -\varphi_\lambda \, e^{-\kappa \, \delta_q^*(t)} (\delta_q^*(t) + h_{q-1}(t) - h_q(t)) \mathbb{1}_{q \neq 0}, \\
g_q^{\kappa *}(t, y) &= 0, \\
\eta_t^* &= \alpha \, .
\end{aligned}$$

When the agent is ambiguity averse only to the market order arrival rate, we expect $g^{\kappa *} = 0$ and $\eta^* = \alpha$. In other words, the agent is fully confident in her model of fill probabilities and drift of the midprice.

In Figure 4.4, we show a comparison of the optimal posting level when the agent is ambiguity averse to only market order arrival versus ambiguity neutral. Also shown is the market order intensity induced by the optimal candidate measure at various times and for various inventory levels ($\lambda \, e^{g_q^*(t)}$). Of particular interest is how the effective market order intensities differ as the strategy approaches maturity. For a fixed level of inventory, the effective market order intensity decreases as maturity approaches because that is the time in which such a change can impair the agent's performance the most. Also note that the largest change in the optimal posting occurs for the largest value of inventory. When the inventory is at its maximum, this is when the agent's fear of a misspecified arrival rate can have most significant impact on their trading performance.

4.4.2 Fill probability

When the agent is ambiguous only to fill probability, the quantities in Equation 4.27 are exactly as stated, with $g^{\lambda *} = 0$ and $\eta^* = \alpha$. This result can be explained along the same lines as the reasoning in the previous section.

The change in optimal posting due to ambiguity to the fill probability, shown in Figure 4.5, is qualitatively different from the case of ambiguity aversion specific to market order intensity. First, note that the change is smallest when the inventory position is largest. This is because the agent already has

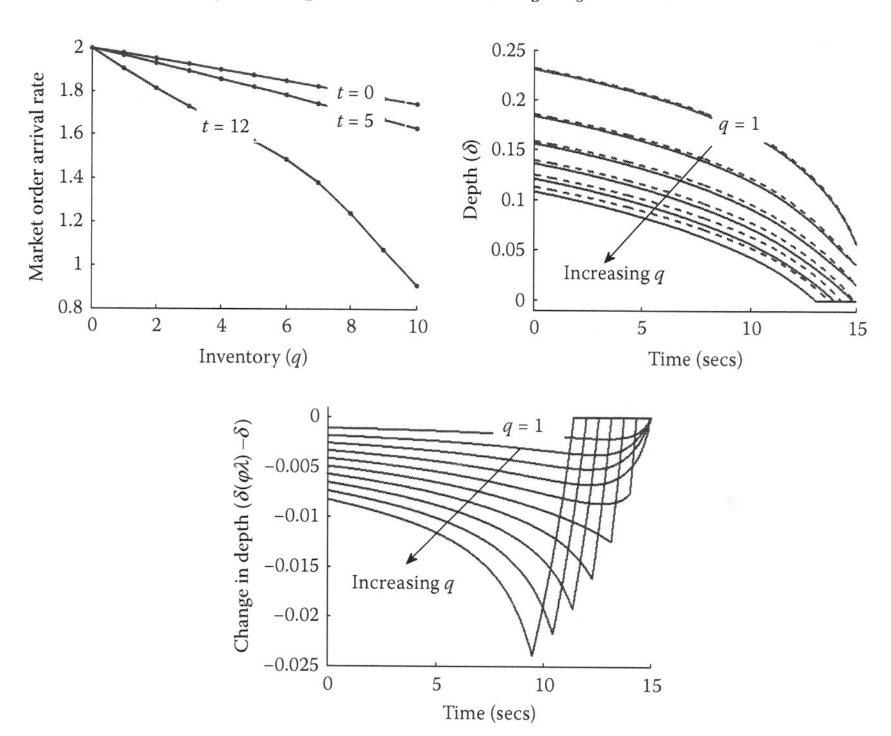

FIGURE 4.4: Market order arrival rate induced by optimal g_t, optimal depth, and change in depth due to ambiguity for an agent who is ambiguity averse to MO rate of arrival (dashed lines are ambiguity neutral depths). Parameter values are $\varphi_\lambda = 6$, $\kappa = 15$, $\lambda = 2$, $\sigma = 0.01$, $\alpha = 0$, $\ell(q) = \theta\, q$, $\theta = 0.01$, $Q = 10$, and $T = 15$.

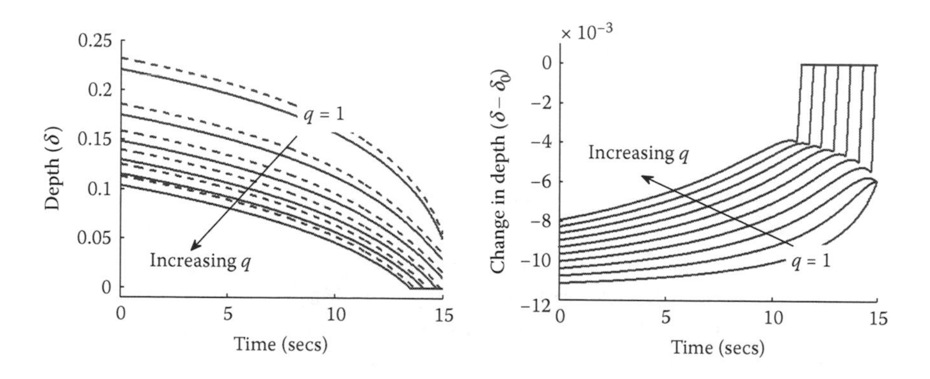

FIGURE 4.5: Optimal depth and change in depth due to ambiguity for an agent who is ambiguity averse to fill probability (dashed lines are ambiguity neutral depths). Parameter values are $\varphi_\kappa = 3$, $\kappa = 15$, $\lambda = 2$, $\sigma = 0.01$, $\alpha = 0$, $\ell(q) = \theta\, q$, $\theta = 0.01$, $Q = 10$, and $T = 15$.

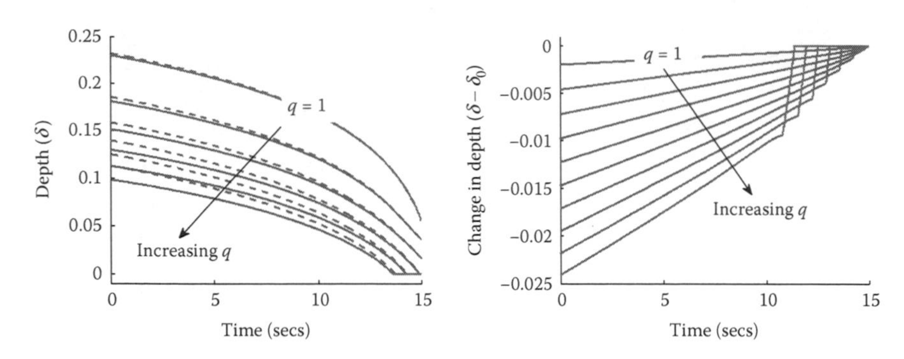

FIGURE 4.6: Optimal depth and change in depth due to ambiguity for an agent who is ambiguity averse to midprice drift (dashed lines are ambiguity neutral depths). Parameter values are $\varphi_\alpha = 5$, $\kappa = 15$, $\lambda = 2$, $\sigma = 0.01$, $\alpha = 0$, $\ell(q) = \theta\, q$, $\theta = 0.01$, $Q = 10$, and $T = 15$.

a natural level of protection to this type of ambiguity when inventory levels are large. When holding a large level of inventory, the agent will post lower prices in order to liquidate faster and avoid the terminal liquidation penalty. But when posting smaller prices, an equal change in the fill probability suffers a larger penalty (when compared to a similar change for a large price). The second difference to note is that the magnitude of the changes decreases as time approaches the trading horizon T rather than increases.

4.4.3 Midprice drift

When the agent is ambiguity averse to only midprice drift, market order dynamics follow that of the reference measure with constant intensity and exponential fill probabilities. However, the asset drift selected by the minimization aspect of the problem explicitly takes the form of a decreasing function of the agent's remaining inventory q.

4.4.3.1 Equivalence to inventory penalization

Here we briefly pose an equivalent interpretation of ambiguity with respect to midprice drift in terms of an inventory penalization. Once again, consider an agent with the goal of liquidating a number of shares before some time T, but instead of ambiguity aversion the agent enforces a quadratic inventory penalization so that her value function is

$$H^{\phi}(t,x,q,S) = \sup_{(\delta_u)_{t\le u\le T}\in\mathcal{A}} \mathbb{E}^{\mathbb{P}}_{t,x,q,S}\Bigg[X_{\tau\wedge T} + q_{\tau\wedge T}(S_{\tau\wedge T} - \ell(q_{\tau\wedge T})) - \frac{1}{2}\,\phi\,\sigma^2 \int_t^{\tau\wedge T} q_u^2\, \mathrm{d}u\Bigg].$$

After the usual ansatz, substitution into the corresponding HJB equation, and solving for the optimal controls δ, the resulting system of ODEs is

$$\partial_t h_q^\phi + \alpha\, q - \frac{1}{2}\phi\, \sigma^2\, q^2 + \sup_{\delta \geq 0}\left\{\lambda\, e^{-\kappa\,\delta}\left(\delta + h_{q-1}^\phi - h_q^\phi\right)\right\} \mathbb{1}_{q\neq 0} = 0\,.$$

This equation is equivalent to Equation 4.24 in the limits $\varphi_\lambda \to 0$ and $\varphi_\kappa \to 0$, and therefore we have

$$\begin{aligned} &\sup_{(\delta_u)_{t\leq u\leq T}\in\mathcal{A}} \mathbb{E}_{t,x,q,S}^{\mathbb{P}}\left[X_{\tau\wedge T} + q_{\tau\wedge T}(S_{\tau\wedge T} - \ell(q_{\tau\wedge T})) - \frac{1}{2}\,\phi\,\sigma^2 \int_t^{\tau\wedge T} q_u^2\,\mathrm{d}u\right] \\ =\; &\sup_{(\delta_u)_{t\leq u\leq T}\in\mathcal{A}}\; \inf_{\mathbb{Q}\in\mathcal{Q}^\alpha} \mathbb{E}_{t,x,q,S}^{\mathbb{Q}}\left[X_{\tau\wedge T} + q_{\tau\wedge T}(S_{\tau\wedge T} - \ell(q_{\tau\wedge T})) + \mathcal{H}_{t,\tau\wedge T}[\mathbb{Q}|\mathbb{P}]\right]. \end{aligned}$$

Since we are considering only ambiguity on the midprice drift, the infimum is taken over by equivalent measures where MO intensity and fill probability remain fixed to the reference measure. Thus a cumulative inventory penalization is equivalent to considering ambiguity on the drift of the midprice of the asset.

As shown in Figure 4.6, the change in the optimal depth is smallest when the inventory is smallest. However, the change in optimal depth consistently becomes more negative with time to maturity. Both of these behaviors make sense when midprice ambiguity is interpreted as a cumulative inventory penalization. The first characteristic is made clear by noting that the larger the agent's inventory position is, the larger the accumulation of the inventory penalty is, and the faster the agent desire to liquidate shares. The second characteristic is explained by noting that the impact on the strategy's performance due to a misspecified drift is approximately linear in time to maturity.

The effect of ambiguity on the drift of the midprice turns out to be more significant than ambiguity on the other two factors in the case of the liquidation problem. As shown in Section 4.5, specifically in Proposition 4.4, the optimal limit order price grows without bound as time to maturity increases when there is no ambiguity on drift. However, when ambiguity on drift is considered, all of the optimal posting levels become finite for all time.

4.5 Closed-Form Solutions

Suppose the agent has equal levels of ambiguity to MO arrival rate and fill probabilities so that $\varphi_\lambda = \varphi_\kappa = \varphi$. Suppose further that the optimal depths $\delta^*(t, q; \psi)$ given by Proposition 4.2 are positive for all t and q. Then substituting all of the optimal controls into Equation 4.24 gives

$$\partial_t h_q + \alpha\, q - \frac{1}{2}\varphi_\alpha\, \sigma^2\, q^2 + \frac{\xi}{\kappa}\, e^{-\kappa\,(-h_{q-1}+h_q)} = 0, \tag{4.28}$$

where

$$\xi = \left(1 + \frac{\varphi}{\kappa}\right)^{-(1+\frac{\kappa}{\varphi})} \lambda,$$

with terminal and boundary conditions $h_q(T) = -q\,\ell(q)$ and $h_0(t) = 0$.

Proposition 4.3 (Solving for $h_q(t)$ in Equation 4.28). *Let* $K_q = \alpha\,\kappa\,q - \frac{1}{2}\varphi_\alpha\,\sigma^2\,\kappa\,q^2$.

i. Suppose $K_q = 0$ *for all* q *(i.e.,* $\varphi_\alpha = 0$ *and* $\alpha = 0$*). Then*

$$h_q(t) = \frac{1}{\kappa}\log\left(\sum_{n=0}^{q} C_{q,n}\,(T-t)^n\right),$$

where

$$C_{q,n} = \frac{\xi^n}{n!}\,e^{-\kappa\,(q-n)\,\ell(q-n)}\,.$$

ii. Suppose $K_q = 0$ *only when* $q = 0$*. Then*

$$h_q(t) = \frac{1}{\kappa}\log\left(\sum_{n=0}^{q} C_{q,n}\,e^{K_n(T-t)}\right),$$

where

$$C_{n+j,n} = (-\xi)^j \prod_{p=1}^{j} \frac{1}{K_{n+p} - K_n}\,C_{n,n},$$

$$C_{q,q} = -\sum_{n=1}^{q-1} C_{q,n} + e^{-\kappa\,q\,\ell(q)},$$

$$C_{0,0} = 1\,.$$

Proof. See Appendix. □

The two cases considered in Proposition 4.3 preclude the case $K_q = 0$ for exactly two distinct values of q (the proposition only considers that either all K_q are zero or only $K_0 = 0$). Due to the quadratic dependence of K_q on q, this omitted case is the only one not considered. However, it is very unlikely that this case would arise in reality, for example, if the model was calibrated to market data. Even if it was the case, an arbitrarily small adjustment could be made to any of the parameters (the most reasonable choice would be φ_α) so that the ratio $\frac{\alpha}{\varphi_\alpha\,\sigma^2}$ is irrational.

Proposition 4.3 shows that the value function can have two different functional forms depending on the possible values of $K_q = \alpha\,\kappa\,q - \frac{1}{2}\,\varphi_\alpha\,\sigma^2\,\kappa\,q^2$. Thus how does this affect the optimal depth δ^*? The proposition below shows the difference in behavior of the optimal depths as time to maturity becomes arbitrarily large when (i) $K_q = 0, \forall q > 0$, and (ii) $K_q < 0, \forall q > 0$.

Proposition 4.4 (Behavior of optimal depths as $(T-t) \to \infty$). *Let $\tau = (T-t)$.*

In case (i) of Proposition 4.3, the optimal depths $\delta_q^*(t)$ grow as $\frac{1}{\kappa}\log\left(\frac{\xi}{q}\tau\right)$ as $\tau \to \infty$.

In case (ii) of Proposition 4.3, if $K_q < 0$, then the optimal depths $\delta_q^*(t)$ approach $\frac{1}{\varphi}\log\left(1+\frac{\varphi}{\kappa}\right)+\frac{1}{\kappa}\log\left(\frac{-\xi}{K_q}\right)$ as $\tau \to \infty$.

Proof. See Appendix. □

4.6 Inclusion of Market Orders

Up to this point, the agent has been restricted to trades using limit orders only. However, she may improve her trading performance if she also executes market orders. This improvement in performance would be due to submitting single trades which would assist in avoiding the large penalty invoked by submitting a large order at the end of the trading period. Mathematically, the inclusion of market orders corresponds to the additional use of an impulse control by the agent.

The effect of market orders on the agent's inventory and wealth dynamics is treated as follows: market sell orders executed by the agent are filled at a price $\frac{\Delta}{2}$ less than the midprice. Denote the number of market orders executed by the agent up to time t by J_t. Then the updated dynamics of the agent's inventory and wealth are

$$\begin{aligned} dq_t &= -dN_t - dJ_t, \\ dX_t &= (S_t + \delta_t)\, dN_t + \left(S_t - \frac{\Delta}{2}\right) dJ_t, \end{aligned}$$

respectively.

The agent's additional control consists of the set of times at which the process J increments.

4.6.1 Feedback controls

The agent must select a sequence of stopping times τ_k at which she executes a market order. The updated optimization problem for the ambiguity neutral agent is

$$H(t,x,q,S) = \sup_{\substack{(\delta_s)_{t\le s\le T} \in \mathcal{A} \\ (\tau_k)_{k=1,\dots,Q}}} \mathbb{E}^{\mathbb{P}}_{t,x,q,S}\left[X_{\tau\wedge T} + q_{\tau\wedge T}\left(S_{\tau\wedge T} - \ell(q_{\tau\wedge T})\right)\right].$$

The inclusion of market orders changes the equation satisfied by the value function $H(t,x,q,S)$. Rather than a standard HJB equation, H now satisfies a quasi-variational inequality of the following form:

$$\max\left\{\left(\partial_t H + \alpha\,\partial_S H + \tfrac{1}{2}\,\sigma^2\,\partial_{SS} H + \sup_{\delta\geq 0}\left\{\lambda\, e^{-\kappa\,\delta}\,\mathcal{D}H\right\}\mathbb{1}_{q\neq 0}\right);\right.$$
$$\left.\left(H\left(t,x+\left(S-\frac{\Delta}{2}\right),q-1,S\right)-H(t,x,q,S)\right)\right\} = 0. \quad (4.29)$$

From Equation 4.29, it is clear that one of the two terms must be equal to zero, and the other term must be less than or equal to zero. This allows the definition of a continuation region and execution region:

$$\mathcal{C} = \left\{(t,x,q,S) : \partial_t H + \alpha\partial_S H + \tfrac{1}{2}\sigma^2\partial_{SS}H + \sup_{\delta}\left\{\lambda\, e^{-\kappa\,\delta}\,\mathcal{D}H\right\}\mathbb{1}_{q\neq 0} = 0\right\};$$
$$\mathcal{E} = \left\{(t,x,q,S) : H(t,x+\left(S-\tfrac{\Delta}{2}\right),q-1,S) - H(t,x,q,S) = 0\right\}.$$

Whenever $(t,x,q,S)\in\mathcal{E}$, it is beneficial for the agent to execute a market order. When $(t,x,q,S)\in\mathcal{C}$, it is more beneficial for the agent to refrain from executing a market order and continue with the optimal placement of limit orders. In the case of using only limit orders, it was the optimal depths δ_t that were of interest. In this case, the boundary between the continuation and execution regions is also of interest. The ansatz $H(t,x,q,S) = x + q\,S + h_q(t)$ still applies, which simplifies the quasi-variational inequality to

$$\max\left\{\left(\partial_t h + \alpha\, q + \sup_{\delta\geq 0}\left\{\lambda\, e^{-\kappa\,\delta}\,(\delta + h_{q-1} - h_q)\right\}\mathbb{1}_{q\neq 0}\right);\right.$$
$$\left.\times\left(h_{q-1} - h_q - \tfrac{\Delta}{2}\right)\right\} = 0. \quad (4.30)$$

It is clear that the feedback expression for the optimal depths is of the same form as in Proposition 4.1. However, the numerical values of these depths will be different because the values of the functions $h_q(t)$ will be different. Also of importance is to note that after making the ansatz, the continuation and execution regions can be redefined in terms of $h_q(t)$ and therefore will not depend on the state variables x and S. The boundary between the two regions is a curve in the (t,q) plane. Figure 4.7 illustrates the optimal depths and market order execution boundary for an agent who liquidates a portfolio of assets with both limit and market orders. The notable difference in the optimal depths between this case and that of an agent who does not execute market orders is that presently, they are bounded below by $\max(0,\frac{1}{\kappa}-\frac{\Delta}{2})$, while without market orders they are bounded below by 0. This is easily seen from the feedback form of the depths, $\delta_q^*(t) = (\frac{1}{\kappa}-h_{q-1}(t)+h_q(t))_+$, combined with the inequality $h_{q-1}(t)-h_q(t)-\frac{\Delta}{2}\leq 0$.

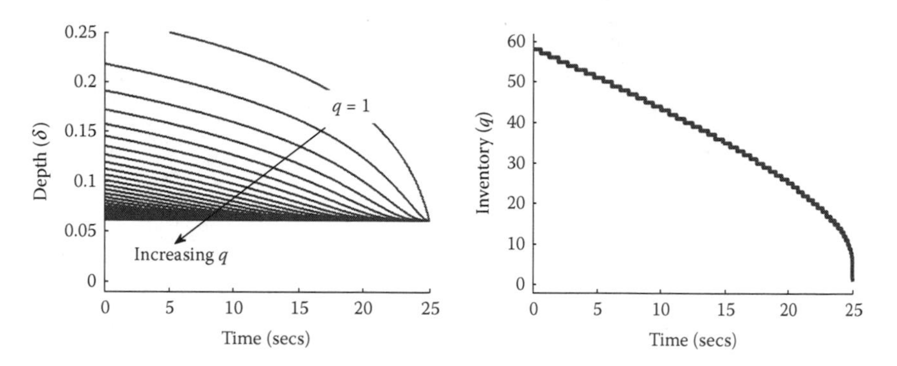

FIGURE 4.7: Optimal depth and market order execution boundary for ambiguity neutral agent. Parameter values are $\kappa = 15$, $\lambda = 2$, $\sigma = 0.01$, $\alpha = 0$, $\ell(q) = \theta\, q$, $\theta = 0.01$, $\Delta = 0.01$, $Q = 60$, and $T = 25$.

4.6.2 The effects of ambiguity aversion on market order execution

This section investigates the effects of ambiguity on the liquidating agent's use of market orders. These effects are observed as a modification to the market order exercise curve in the (t, q) plane depending on the levels of ambiguity. The appropriate quasi-variational inequality is obtained from Equation 4.24 in a similar manner as Equation 4.30 was obtained from Equation 4.6. The feedback form of the optimal controls is the same in each case, but they result in quantitatively different depths due to the change in the value function. The appropriate equation is

$$\max\left\{ \left(\partial_t h_q + \inf_{\eta}\left\{ \eta\, q + \frac{1}{2\,\varphi_\alpha}\left(\frac{\alpha-\eta}{\sigma}\right)^2 \right\} + \sup_{\delta\geq 0}\inf_{g^\lambda}\inf_{g^\kappa\in\mathcal{G}}\left\{ \lambda\left[\int_\delta^\infty e^{g^\lambda + g^\kappa(y)}\, F(\mathrm{d}y)\right](\delta + h_{q-1}(t) - h_q(t))\,\mathbb{1}_{q\neq 0}\right\}\right); \left(h_{q-1} - h_q - \frac{\Delta}{2}\right)\right\} = 0 \tag{4.31}$$

$$h(T,q) = -q\,\ell(q),$$
$$h(t,0) = 0.$$

The effect of ambiguity on the MO execution boundary can be explained with similar reasoning to the change in the optimal depths previously discussed. The most notable feature in Figure 4.8 is the magnitude of the change for an agent who is ambiguity averse to the midprice drift. The significant change in MO execution strategy can be intuitively understood again by interpreting this type of ambiguity as a cumulative inventory penalty. The agent

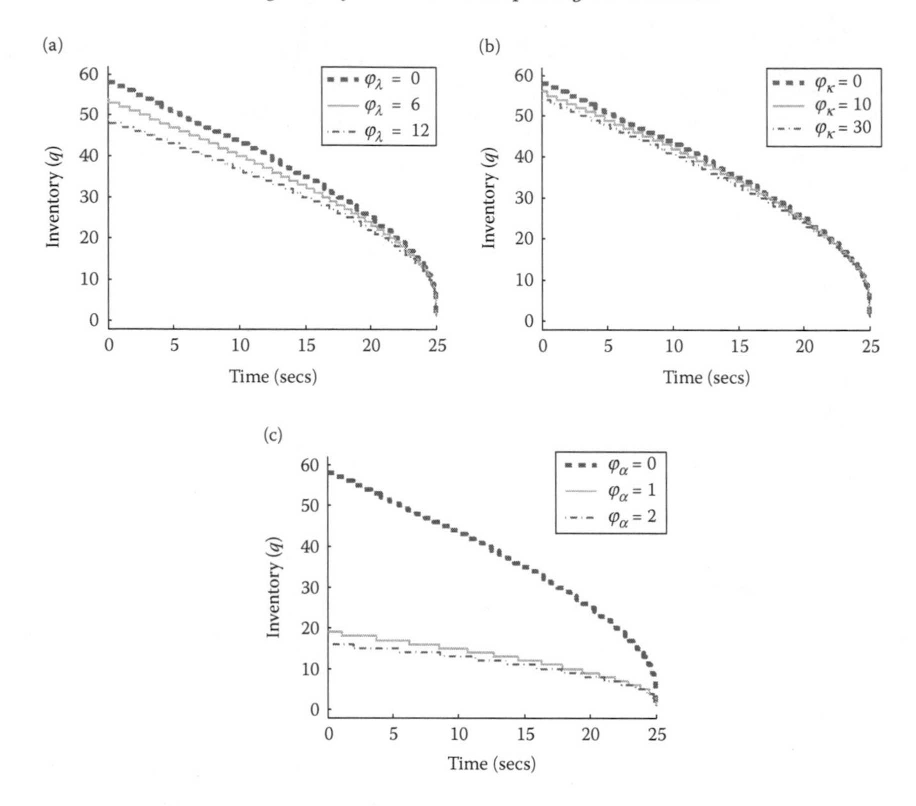

FIGURE 4.8: Effect on market order execution boundary for an agent who is ambiguity averse to different factors of the reference model. Parameter values are $\kappa = 15$, $\lambda = 2$, $\sigma = 0.01$, $\alpha = 0$, $\ell(q) = \theta\, q$, $\theta = 0.01$, $Q = 60$, and $T = 25$. (a) MO arrival rate, (b) fill probability, and (c) mid price drift.

is strongly encouraged to sell shares very quickly with this type of penalty, much more so than compared to only the terminal inventory penalty.

The relatively small change due to ambiguity on fill probability can be understood as follows: for large inventory positions, the natural inclination of the agent is to post small depths because she is in a hurry to liquidate the inventory position before maturity. But as the agent posts smaller depths, the change in the execution price distribution must be larger to have a significant impact on the fill probability and hence also the performance of the strategy. This type of large change is prevented by the entropic penalty. Essentially, by naturally wanting to post small depths, the agent has already gained a significant amount of protection against a misspecified fill probability.

Ambiguity with respect to market order arrival rate lies in a middle ground. On one hand, the significance of this type of ambiguity is less than that of the midprice drift because changing the arrival rate does not directly penalize the holding of inventory. On the other hand, posting smaller depths for larger inventory positions does not provide a natural protection against a change in the arrival rate as it did for a change in the distribution of trade executions.

4.7 Conclusion

We have shown how to incorporate ambiguity aversion into the context of an optimal liquidation problem and have investigated the impact of ambiguity with respect to different sources of randomness on the optimal trading strategy. The primary mathematical procedure which allows for the computation of the optimal strategy is in solving a PDE for the agent's value function. When the agent is only allowed to employ limit orders and when her ambiguity aversion levels satisfy a particular symmetry constraint, the solution to this PDE is known in closed form. This allows any application of the strategy to be implemented very efficiently by precomputing the optimal trading strategy.

When the ability to submit market orders is added to the model, we no longer have closed-form solutions for the optimal trading strategy. In this case, the PDE must be solved numerically, a task which can become computationally complex when the number of traded assets is increased or when additional features are added to the model. Other additional features which could also potentially cause a loss of closed-form solutions are the inclusion of a trade signal which indicates favorable or unfavorable market conditions, or the process of updating the agent's reference model in real time through a learning procedure.

Appendix 4A Proofs

Proof of Proposition 4.2. The minimization in η is independent of the optimization in δ, g^λ, and g^κ and so can be done directly. First-order conditions imply that $\eta^* = \alpha - \varphi_\alpha \sigma^2 q$, as desired. This value of η^* is easily seen to be unique as it is a quadratic optimization. For the optimization over δ, g^λ, and g^κ, first consider $\varphi_\lambda > \varphi_\kappa$. Then the term to be optimized is

$$\mathfrak{G}(\delta, g^\lambda, g^\kappa) = \lambda \left[\int_\delta^\infty e^{g^\lambda + g^\kappa(y)} F(\mathrm{d}y) \right] (\delta + h_{q-1} - h_q) + \mathcal{K}^{\varphi_\lambda, \varphi_\kappa}(g^\lambda, g^\kappa)$$

$$= \lambda \left[\int_\delta^\infty e^{g^\lambda + g^\kappa(y)} F(\mathrm{d}y) \right] (\delta + h_{q-1} - h_q) \ldots$$

$$+\frac{1}{\varphi_\lambda}[\lambda\int_0^\infty -(e^{g^\lambda}-1)+g^\lambda\, e^{g^\lambda+g^\kappa(y)}\,F(\mathrm{d}y)]\ldots$$

$$+\frac{1}{\varphi_\kappa}\left[\lambda\int_0^\infty -(e^{g^\kappa(y)}-1)\,e^{g^\lambda}+g^\kappa(y)\,e^{g^\lambda+g^\kappa(y)}\,F(\mathrm{d}y)\right]. \quad (A.1)$$

The remainder of the proof proceeds as follows:

1. Introduce a Lagrange multiplier γ corresponding to the constraint on $g^\kappa(y)$
2. Compute first-order conditions for the unconstrained $g^\kappa(y)$ which minimizes the Lagrange modified term
3. Compute the value of γ
4. Verify that the corresponding $g^{\kappa*}(y)$ provides a minimizer of $\mathfrak{G}(\delta, g^\lambda, g^\kappa)$ for all functions $g^\kappa \in \mathcal{G}$
5. Compute first-order conditions for g^λ
6. Verify that the corresponding $g^{\lambda*}$ provides a minimizer of $\mathfrak{G}(\delta, g^\lambda, g^{\kappa*})$
7. Compute first-order conditions for δ subject to the constraint $\delta \geq 0$
8. Verify that the corresponding δ^* provides a maximizer of $\mathfrak{G}(\delta, g^{\lambda*}, g^{\kappa*})$
9. Prove existence and uniqueness for the solution h

Parts 1 and 2: Solving for g^κ: The constraint $\int_0^\infty e^{g^\kappa(y)}\,F(\mathrm{d}y) = 1$ is handled by introducing a Lagrange multiplier γ and then minimizing over unconstrained g^κ. The optimization with respect to g^κ is handled in a pointwise fashion by minimizing the integrand with respect to $g^\kappa(y)$ for each value of $y \in [0, \infty)$. For $y \in (\delta, \infty)$, the quantity to be minimized is

$$\lambda\, e^{g^\lambda+g^\kappa(y)}(\delta+h_{q-1}-h_q)+\frac{\lambda}{\varphi_\lambda}(-(e^{g^\lambda}-1)+g^\lambda\, e^{g^\lambda+g^\kappa(y)})$$
$$+\frac{\lambda}{\varphi_\kappa}(-(e^{g^\kappa(y)}-1)\,e^{g^\lambda}+g^\kappa(y)\,e^{g^\lambda+g^\kappa(y)})+\gamma(e^{g^\kappa(y)}-1). \quad (A.2)$$

First-order conditions in $g^\kappa(y)$ give

$$g^\kappa(y) = -\frac{\varphi_\kappa}{\varphi_\lambda}g^\lambda-\frac{\gamma\varphi_\kappa}{\lambda}e^{-g^\lambda}-\varphi_\kappa(\delta+h_{q-1}-h_q)\,. \quad (A.3)$$

Similarly, first-order conditions in $g^\kappa(y)$ for $y \in [0, \delta]$ give

$$g^\kappa(y) = -\frac{\varphi_\kappa}{\varphi_\lambda}g^\lambda-\frac{\gamma\varphi_\kappa}{\lambda}e^{-g^\lambda}\,. \quad (A.4)$$

Combining Equations A.3 and A.4 gives

$$g^{\kappa *}(y) = -\frac{\varphi_\kappa}{\varphi_\lambda} g^\lambda - \frac{\gamma\varphi_\kappa}{\lambda} e^{-g^\lambda} - \varphi_\kappa(\delta + h_{q-1} - h_q)\mathbb{1}_{y>\delta}. \tag{A.5}$$

Part 3: Solving for γ: Substituting this expression into the integral constraint and performing some computations give an expression for γ:

$$\gamma = -\frac{g^\lambda\,\lambda}{\varphi_\lambda} e^{g^\lambda} + \frac{\lambda\, e^{g^\lambda}}{\varphi_\kappa} \log(1 - e^{-\kappa\,\delta} + e^{-\varphi_\kappa(\delta + h_{q-1} - h_q)}\, e^{-\kappa\,\delta}).$$

Substituting this into Equation A.5 gives

$$\begin{aligned} g^{\kappa *}(y) = &- \log(1 - e^{-\kappa\,\delta} + e^{-\varphi_\kappa(\delta + h_{q-1} - h_q)} e^{-\kappa\,\delta}) \\ &- \varphi_\kappa(\delta + h_{q-1} - h_q)\,\mathbb{1}_{y>\delta}. \end{aligned} \tag{A.6}$$

To prove that the expression for $g^{\kappa *}(y)$ is indeed a minimizer, it is convenient to introduce some shorthand notation:

$$\Delta h_q = h_{q-1} - h_q, \tag{A.7}$$

$$A = 1 - e^{-\kappa\,\delta} + e^{-\varphi_\kappa(\delta + \Delta h_q)}\, e^{-\kappa\,\delta}, \tag{A.8}$$

$$\underline{g} = - \log A, \tag{A.9}$$

$$\overline{g} = - \log A - \varphi_\kappa(\delta + \Delta h_q). \tag{A.10}$$

It is important to note that these quantities do not depend on g^λ. Also note that $\underline{g}$ and $\overline{g}$ are the two possible values that $g^{\kappa *}(y)$ can take depending on whether $y \le \delta$ or $y > \delta$. Let f be any other function in $\mathcal{G}$ and define $k(y) = e^{f(y)} - e^{g^{\kappa *}(y)}$. Then define $f_\epsilon(y) = \log(\epsilon\, k(y) + e^{g^{\kappa *}(y)})$. One can easily check that $f_\epsilon \in \mathcal{G}$ for all $\epsilon \in [0,1]$ and that $f_0 = g^{\kappa *}$ and $f_1 = f$. Let $m(\epsilon) = \mathfrak{G}(\delta, g^\lambda, f_\epsilon)$. We confirm that $g^{\kappa *}$ is the minimizer by showing that

$$\mathfrak{G}(\delta, g^\lambda, g^{\kappa *}) = m(0) \le m(1) = \mathfrak{G}(\delta, g^\lambda, f). \tag{A.11}$$

It is sufficient to show that m has a non-negative second derivative for all $\epsilon \in [0,1]$. Substituting expressions for f_ϵ and Equation A.6 into $\mathfrak{G}(\delta, g^\lambda, f_\epsilon)$ gives

$$\begin{aligned} m(\epsilon) = \lambda &\left[\int_\delta^\infty e^{g^\lambda + f_\epsilon(y)}\, F(\mathrm{d}y)\right](\delta + \Delta h_q) \ldots \\ &+ \frac{1}{\varphi_\lambda}\left[\lambda \int_0^\infty -(e^{g^\lambda} - 1) + g^\lambda e^{g^\lambda + f_\epsilon(y)}\, F(\mathrm{d}y)\right] \ldots \\ &+ \frac{1}{\varphi_\kappa}\left[\lambda \int_0^\infty -(e^{f_\epsilon(y)} - 1)e^{g^\lambda} + f_\epsilon(y) e^{g^\lambda + f_\epsilon(y)}\, F(\mathrm{d}y)\right] \end{aligned}$$

$$
\begin{aligned}
&= \lambda \Bigg[\int_{\delta}^{\infty} e^{g^{\lambda}} (e^{\overline{g}} + \epsilon k(y))\, F(\mathrm{d}y) \Bigg] (\delta + \Delta h_q) \ldots \\
&\quad + \frac{1}{\varphi_{\lambda}} \Bigg[\lambda \int_{0}^{\delta} -(e^{g^{\lambda}} - 1) + g^{\lambda} e^{g^{\lambda}} (e^{\underline{g}} + \epsilon k(y))\, F(\mathrm{d}y) \Bigg] \ldots \\
&\quad + \frac{1}{\varphi_{\lambda}} \Bigg[\lambda \int_{\delta}^{\infty} -(e^{g^{\lambda}} - 1) + g^{\lambda} e^{g^{\lambda}} (e^{\overline{g}} + \epsilon k(y))\, F(\mathrm{d}y) \Bigg] \ldots \\
&\quad + \frac{1}{\varphi_{\kappa}} \Bigg[\lambda \int_{0}^{\delta} -(e^{\underline{g}} + \epsilon k(y) - 1) e^{g^{\lambda}} + \log(e^{\underline{g}} + \epsilon k(y)) e^{g^{\lambda}} \\
&\qquad\qquad \times (e^{\underline{g}} + \epsilon k(y))\, F(\mathrm{d}y) \Bigg] \ldots \\
&\quad + \frac{1}{\varphi_{\kappa}} \Bigg[\lambda \int_{\delta}^{\infty} -(e^{\overline{g}} + \epsilon k(y) - 1) e^{g^{\lambda}} + \log(e^{\overline{g}} + \epsilon k(y)) e^{g^{\lambda}} \\
&\qquad\qquad \times (e^{\overline{g}} + \epsilon k(y))\, F(\mathrm{d}y) \Bigg],
\end{aligned}
$$

and taking a derivative with respect to ϵ gives

$$
\begin{aligned}
m'(\epsilon) &= \lambda \Bigg[\int_{\delta}^{\infty} e^{g^{\lambda}} k(y)\, F(\mathrm{d}y) \Bigg] \Big(\delta + \Delta h_q \Big) + \frac{1}{\varphi_{\lambda}} \Bigg[\lambda \int_{0}^{\infty} g^{\lambda} e^{g^{\lambda}} k(y)\, F(\mathrm{d}y) \Bigg] \ldots \\
&\quad + \frac{1}{\varphi_{\kappa}} \Bigg[\lambda \int_{0}^{\delta} -k(y) e^{g^{\lambda}} + k(y) e^{g^{\lambda}} + k(y) \log(e^{\underline{g}} + \epsilon k(y)) e^{g^{\lambda}}\, F(\mathrm{d}y) \Bigg] \ldots \\
&\quad + \frac{1}{\varphi_{\kappa}} \Bigg[\lambda \int_{\delta}^{\infty} -k(y) e^{g^{\lambda}} + k(y) e^{g^{\lambda}} + k(y) \log(e^{\overline{g}} + \epsilon k(y)) e^{g^{\lambda}}\, F(\mathrm{d}y) \Bigg] \\
&= \lambda \Bigg[\int_{\delta}^{\infty} e^{g^{\lambda}} k(y)\, F(\mathrm{d}y) \Bigg] \Big(\delta + \Delta h_q \Big) \ldots \\
&\quad + \frac{1}{\varphi_{\kappa}} \Bigg[\lambda \int_{0}^{\delta} k(y) \log(e^{\underline{g}} + \epsilon k(y)) e^{g^{\lambda}}\, F(\mathrm{d}y) \Bigg] \ldots \\
&\quad + \frac{1}{\varphi_{\kappa}} \Bigg[\lambda \int_{\delta}^{\infty} k(y) \log(e^{\overline{g}} + \epsilon k(y)) e^{g^{\lambda}}\, F(\mathrm{d}y) \Bigg].
\end{aligned}
$$

Evaluating this expression at $\epsilon = 0$ gives

$$
\begin{aligned}
m'(0) &= \lambda\left[\int_{\delta}^{\infty} e^{g^{\lambda}} k(y)\, F(\mathrm{d}y)\right]\left(\delta + \Delta h_q\right) \dots \\
&\quad + \frac{1}{\varphi_\kappa}\left[\lambda \int_0^{\delta} k(y) \underline{g} e^{g^{\lambda}}\, F(\mathrm{d}y)\right] + \frac{1}{\varphi_\kappa}\left[\lambda \int_{\delta}^{\infty} k(y) \overline{g} e^{g^{\lambda}}\, F(\mathrm{d}y)\right] \\
&= \lambda\left[\int_{\delta}^{\infty} e^{g^{\lambda}} k(y)\, F(\mathrm{d}y)\right]\left(\delta + \Delta h_q\right) - \frac{1}{\varphi_\kappa}\lambda \int_0^{\delta} k(y)\log(A) e^{g^{\lambda}}\, F(\mathrm{d}y) \dots \\
&\quad - \frac{1}{\varphi_\kappa}\lambda \int_0^{\delta} k(y)\left(\log(A) + \varphi_\kappa(\delta + \Delta h_q)\right) e^{g^{\lambda}}\, F(\mathrm{d}y) \\
&= 0
\end{aligned}
$$

as expected. Continuing by taking a second derivative with respect to ϵ:

$$
\begin{aligned}
m''(\epsilon) &= \frac{1}{\varphi_\kappa}\lambda \int_0^{\delta} \frac{e^{g^{\lambda}} k^2(y)}{e^{\underline{g}} + \epsilon k(y)}\, F(\mathrm{d}y) + \frac{1}{\varphi_\kappa}\lambda \int_{\delta}^{\infty} \frac{e^{g^{\lambda}} k^2(y)}{e^{\overline{g}} + \epsilon k(y)}\, F(\mathrm{d}y) \\
&= \frac{1}{\varphi_\kappa}\lambda \int_0^{\infty} \frac{e^{g^{\lambda}} k^2(y)}{e^{g^{\kappa *}(y)} + \epsilon k(y)}\, F(\mathrm{d}y) \\
&= \frac{1}{\varphi_\kappa}\lambda \int_0^{\infty} \frac{e^{g^{\lambda}} k^2(y)}{e^{f_\epsilon(y)}}\, F(\mathrm{d}y).
\end{aligned}
$$

This expression is non-negative for all $\epsilon \in [0, 1]$, showing that indeed the expression for $g^{\kappa *}(y)$ in Equation A.6 is a minimizer. This expression is strictly positive unless $k \equiv 0$, showing that the inequality in Equation A.11 is strict unless $f = g^{\kappa *}$, therefore $g^{\kappa *}$ is the unique minimizer.

Part 5: First-order conditions for g^{λ}: After substituting the expression A.6 into the term to be minimized (see Equation A.1) and performing some tedious computations, we must minimize the following with respect to g^{λ}:

$$
\begin{aligned}
&\lambda\, e^{g^{\lambda}} e^{\overline{g}}(\delta + \Delta h_q) e^{-\kappa\,\delta} \\
&+ \frac{\lambda}{\varphi_\lambda}(-(e^{g^{\lambda}} - 1) + g^{\lambda} e^{g^{\lambda}} e^{\overline{g}}) e^{-\kappa\,\delta} + \frac{\lambda}{\varphi_\kappa}(-(e^{\overline{g}} - 1) e^{g^{\lambda}} + \overline{g} e^{g^{\lambda}} e^{\overline{g}}) e^{-\kappa\,\delta} \\
&+ \frac{\lambda}{\varphi_\lambda}(-(e^{g^{\lambda}} - 1) + g^{\lambda} e^{g^{\lambda}} e^{\underline{g}})(1 - e^{-\kappa\,\delta}) \\
&+ \frac{\lambda}{\varphi_\kappa}(-(e^{\underline{g}} - 1) e^{g^{\lambda}} + \underline{g} e^{g^{\lambda}} e^{\underline{g}})(1 - e^{-\kappa\,\delta}).
\end{aligned}
\tag{A.12}
$$

Applying first-order conditions in g^λ and carrying out some tedious computations gives the candidate minimizer:

$$g^{\lambda *} = \frac{\varphi_\lambda}{\varphi_\kappa} \log A = \frac{\varphi_\lambda}{\varphi_\kappa} \log(1 - e^{-\kappa\,\delta} + e^{-\varphi_\kappa(\delta + h_{q-1} - h_q)} e^{-\kappa\,\delta}). \qquad \text{(A.13)}$$

This is the unique root corresponding to the first-order conditions.

Part 6: Verify that $g^{\lambda *}$ is a minimizer: Taking two derivatives of Equation A.12 with respect to g^λ and cancelling terms gives

$$\frac{g^\lambda e^{g^\lambda}}{\varphi_\lambda} + \frac{e^{g^\lambda}}{\varphi_\lambda} - \frac{\log(A) e^{g^\lambda}}{\varphi_\kappa}.$$

When the expression A.13 is substituted above, this becomes $\frac{A^{(\varphi_\lambda/\varphi_\kappa)}}{\varphi_\lambda}$, which is always positive because $A > 0$. Thus this value of $g^{\lambda *}$ provides a minimizer. Uniqueness of the root corresponding to first-order conditions (and the fact that it is the only critical value) implies that $g^{\lambda *}$ is the unique minimizer.

Part 7: Solving for δ: Substituting expressions A.7 through A.10 and A.13 into A.12, after some tedious computations we must maximize the following expression over δ:

$$\frac{\lambda}{\varphi_\lambda}\left(1 - \exp\left\{\frac{\varphi_\lambda}{\varphi_\kappa} \log(1 - e^{-\kappa\,\delta} + e^{-\varphi_\kappa(\delta + h_{q-1} - h_q)} e^{-\kappa\,\delta})\right\}\right).$$

Maximizing this term is equivalent to minimizing

$$\exp\left\{\frac{\varphi_\lambda}{\varphi_\kappa} \log(1 - e^{-\kappa\,\delta} + e^{-\varphi_\kappa(\delta + h_{q-1} - h_q)} e^{-\kappa\,\delta})\right\},$$

which is equivalent to minimizing

$$1 - e^{-\kappa\,\delta} + e^{-\varphi_\kappa(\delta + h_{q-1} - h_q)} e^{-\kappa\,\delta}. \qquad \text{(A.14)}$$

Computing first-order conditions for δ gives

$$\delta^* = \frac{1}{\varphi_\kappa} \log\left(1 + \frac{\varphi_\kappa}{\kappa}\right) - h_{q-1} + h_q. \qquad \text{(A.15)}$$

If this value is positive, we check that it is a minimizer of Equation A.14 by taking a second derivative. If it is non-negative, we show that the first derivative of Equation A.14 is positive for all $\delta > 0$, meaning that the desired value of δ^* is 0.

Part 8: Verify that δ^* is a minimizer of Equation A.14: Suppose the value given by Equation A.15 is positive. Taking two derivatives of Equation A.14 with respect to δ gives

$$-\kappa^2 e^{-\kappa\,\delta} + (\varphi_\kappa + \kappa)^2 e^{-\varphi_\kappa(\delta + h_{q-1} - h_q)} e^{-\kappa\,\delta}.$$

Substituting Equation A.15 into this expression gives

$$\kappa \varphi_\kappa e^{-\kappa\,\delta^*} > 0,$$

and so the value in Equation A.15 minimizes Equation A.14. Now suppose the value in Equation A.15 is nonpositive. This means the following inequality holds:

$$e^{-\varphi_\kappa(h_{q-1}-h_q)} \le \frac{\kappa}{\kappa+\varphi_\kappa}.$$

The first derivative of Equation A.14 with respect to δ is

$$(\kappa - (\kappa+\varphi_\kappa)e^{-\varphi_\kappa(\delta+h_{q-1}-h_q)})e^{-\kappa\,\delta},$$

and the preceding inequality implies that this is non-negative for all $\delta \ge 0$, implying that $\delta^* = 0$ is the minimizer of Equation A.14. Thus the value of δ which maximizes the original term of interest is

$$\delta^* = \left(\frac{1}{\varphi_\kappa}\log(1+\frac{\varphi_\kappa}{\kappa}) - h_{q-1} + h_q\right)_+ ,$$

as desired. The case of $\varphi_\kappa > \varphi_\lambda$ is essentially identical.

Part 9: Existence and uniqueness of h: Begin by substituting the optimal feedback controls, η^*, $g^{\lambda*}$, and $g^{\kappa*}(y)$ into Equation 4.24. This results in

$$\begin{aligned}\partial_t h_q + \alpha\, q - \tfrac{1}{2}\varphi_\alpha\sigma^2 q^2 + \sup_{\delta\ge 0}\Bigg\{\tfrac{\lambda}{\varphi_\lambda}\Bigg(1-\exp\Bigg\{\tfrac{\varphi_\lambda}{\varphi_\kappa}\log(1-e^{-\kappa\,\delta}\\ +e^{-\kappa\,\delta-\varphi_\kappa(\delta+h_{q-1}-h_q)})\Bigg\}\Bigg)\Bigg\}\mathbb{1}_{q\ne 0} = 0,\\ h_q(T) = -q\,\ell(q).\end{aligned} \tag{A.16}$$

This is a system of ODEs of the form $\partial_t\mathbf{h} = \mathbf{F}(\mathbf{h})$. To show existence and uniqueness of the solution to this equation, the function $\mathbf{F}$ is shown to be bounded and globally Lipschitz. It suffices to show that the function f is bounded and globally Lipschitz, where f is given by

$$f(x,y) = \sup_{\delta\ge 0}\left\{\frac{\lambda}{\varphi_\lambda}\left(1-\exp\left\{\frac{\varphi_\lambda}{\varphi_\kappa}\log(1-e^{-\kappa\,\delta}+e^{-\kappa\,\delta-\varphi_\kappa(\delta+x-y)})\right\}\right)\right\}.$$

Boundedness and the global Lipschitz property of f implies the same for $\mathbf{F}$, and so existence and uniqueness follows from the Picard–Lindelöf theorem. The global Lipschitz property is a result of showing that all directional derivatives of f exist and are bounded for all $(x,y)\in\mathbb{R}^2$.

The supremum is attained at $\delta^* = \left(\frac{1}{\varphi_\kappa}\log(1+\frac{\varphi_\kappa}{\kappa}) - x + y\right)_+$. Thus two separate domains for f must be considered: $\frac{1}{\varphi_\kappa}\log(1+\frac{\varphi_\kappa}{\kappa}) > x-y$ and $\frac{1}{\varphi_\kappa}\log(1+\frac{\varphi_\kappa}{\kappa}) \le x-y$. First consider $\frac{1}{\varphi_\kappa}\log(1+\frac{\varphi_\kappa}{\kappa}) > x-y$ so that $\delta^* = \frac{1}{\varphi_\kappa}\log(1+\frac{\varphi_\kappa}{\kappa}) - x + y$. Substituting this into the expression for f yields:

$$\begin{aligned} f(x,y) &= \frac{\lambda}{\varphi_\lambda}\left(1-\exp\left\{\frac{\varphi_\lambda}{\varphi_\kappa}\log\left(1-e^{-\frac{\kappa}{\varphi_\kappa}\log(1+\frac{\varphi_\kappa}{\kappa})+\kappa(x-y)}\right.\right.\right.\\ &\quad\left.\left.\left.\times(1-e^{-\log(1+\frac{\varphi_\kappa}{\kappa})})\right)\right\}\right)\\ &= \frac{\lambda}{\varphi_\lambda}\left(1-\exp\left\{\frac{\varphi_\lambda}{\varphi_\kappa}\log\left(1-Be^{\kappa(x-y)}\right)\right\}\right), \end{aligned}$$

where $B = \left(\frac{\kappa}{\varphi_\kappa+\kappa}\right)^{\frac{\kappa}{\varphi_\kappa}} \frac{\varphi_\kappa}{\varphi_\kappa+\kappa} > 0$. Letting $z = B\, e^{\kappa\,(x-y)}$, the inequality $\frac{1}{\varphi_\kappa}\log(1+\frac{\varphi_\kappa}{\kappa}) > x-y$ implies

$$0 < z < \left(\frac{\kappa}{\varphi_\kappa+\kappa}\right)^{\frac{\kappa}{\varphi_\kappa}} \frac{\varphi_\kappa}{\varphi_\kappa+\kappa} e^{\frac{\kappa}{\varphi_\kappa}\log(1+\frac{\varphi_\kappa}{\kappa})}$$
$$= \left(\frac{\kappa}{\varphi_\kappa+\kappa}\right)^{\frac{\kappa}{\varphi_\kappa}} \frac{\varphi_\kappa}{\varphi_\kappa+\kappa} (\frac{\varphi_\kappa+\kappa}{\kappa})^{\frac{\kappa}{\varphi_\kappa}} = \frac{\varphi_\kappa}{\varphi_\kappa+\kappa} < 1.$$

Since z is positive we have

$$f(x,y) = \frac{\lambda}{\varphi_\lambda}\left(1 - \exp\left\{\frac{\varphi_\lambda}{\varphi_\kappa}\log\left(1-z\right)\right\}\right) < \frac{\lambda}{\varphi_\lambda}.$$

Taking partial derivatives of f in this domain gives

$$\partial_x f(x,y) = -\partial_y f(x,y) = \frac{\lambda}{\varphi_\kappa} e^{\frac{\varphi_\lambda}{\varphi_\kappa}\log\left(1-Be^{\kappa(x-y)}\right)} \frac{B\kappa e^{\kappa(x-y)}}{1-Be^{\kappa(x-y)}}$$
$$= \frac{\lambda}{\varphi_\kappa} e^{\frac{\varphi_\lambda}{\varphi_\kappa}\log\left(1-z\right)} \frac{z}{1-z}. \tag{A.17}$$

This expression is non-negative and continuous for $0 \le z \le \frac{\varphi_\kappa}{\varphi_\kappa+\kappa}$, and therefore achieves a finite maximum somewhere on that interval. Thus $\partial_x f$ and $\partial_y f$ are bounded in this domain, and so directional derivatives exist and are also bounded everywhere in the interior of the domain. On the boundary, directional derivatives exist and are bounded if the direction is toward the interior of the domain.

Now consider $\frac{1}{\varphi_\kappa}\log(1+\frac{\varphi_\kappa}{\kappa}) \le x-y$, which implies $\delta^* = 0$. The expression for $f(x,y)$ in this domain is

$$f(x,y) = \frac{\lambda}{\varphi_\lambda}(1 - e^{-\varphi_\lambda(x-y)}),$$

which is bounded by $\frac{\lambda}{\varphi_\lambda}\left(1 - e^{-\frac{\varphi_\lambda}{\varphi_\kappa}\log(1+\frac{\varphi_\kappa}{\kappa})}\right)$. Partial derivatives of f are given by

$$\partial_x f(x,y) = -\partial_y f(x,y) = \lambda\, e^{-\varphi_\lambda(x-y)}. \tag{A.18}$$

In this domain, the derivatives $\partial_x f$ and $\partial_y f$ are bounded by $\lambda\, e^{-\frac{\varphi_\lambda}{\varphi_\kappa}\log(1+\frac{\varphi_\kappa}{\kappa})}$. So similarly to the first domain, directional derivatives exist and are bounded in the interior. On the boundary, they exist and are bounded in the direction toward the interior of the domain. Thus we have existence and boundedness on the boundary toward either of the two domains. The directional derivative on the boundary is zero when the direction is parallel to the boundary. Existence and boundedness of directional derivatives for all $(x, y) \in \mathbb{R}^2$ allow us to show the Lipschitz condition easily:

$$
\begin{aligned}
|f(x_2, y_2) - f(x_1, y_1)| &= \left| \int\limits_C \nabla f(x, y) \cdot \mathrm{d}\vec{r} \right| \le \int\limits_C |\nabla f(x, y)| \mathrm{d}s \le \int\limits_C A \mathrm{d}s \\
&= A |(x_2, y_2) - (x_1, y_1)|,
\end{aligned}
$$

where C is the curve that connects (x_1, y_1) to (x_2, y_2) in a straight line and A is a uniform bound on the gradient of f. This proves that there exists a unique solution h to Equation 4.24.

□

Proof of Theorem 4.1. Let h be the solution to Equation 4.24 with terminal conditions $h_q(T) = -q\,\ell(q)$, and define a candidate value function by $\hat{H}(t, x, q, S) = x + q\,S + h_q(t)$. From Ito's lemma we have

$$
\begin{aligned}
\hat{H}(T, X^\delta_{T-}, S_{T-}, q^\delta_{T-}) &= \hat{H}(t, x, S, q) + \int\limits_t^T \partial_t h_{q_s}(s) \mathrm{d}s + \int_t^T \alpha\, q_s \mathrm{d}s + \sigma \int\limits_t^T q_s \mathrm{d}W_s \\
&+ \int\limits_t^T \int\limits_{\delta_s}^{\infty} (\delta_s + h_{q_{s-}-1}(s) - h_{q_{s-}}(s)) \mu(\mathrm{d}y, \mathrm{d}s).
\end{aligned}
$$

Note that for any admissible measure $\mathbb{Q}(\eta, g)$ and admissible control δ, we have

$$
\begin{aligned}
\mathbb{E}^{\mathbb{Q}(\eta,g)}\left[\int\limits_0^T \int\limits_{\delta_t}^{\infty} (\delta_t)^2\, \nu_{\mathbb{Q}(\eta,g)}(\mathrm{d}y, \mathrm{d}t) \right] &= \mathbb{E}^{\mathbb{Q}(\eta,g)}\left[\int\limits_0^T \int\limits_{\delta_t}^{\infty} (\delta_t)^2 e^{g_t(y)}\, \nu_{\mathbb{P}}(\mathrm{d}y, \mathrm{d}t) \right] \\
&\le \mathbb{E}^{\mathbb{Q}(\eta,g)}\left[\int\limits_0^T \int\limits_{\delta_t}^{\infty} y^2 e^{g_t(y)}\, \nu_{\mathbb{P}}(\mathrm{d}y, \mathrm{d}t) \right] \\
&\le \mathbb{E}^{\mathbb{Q}(\eta,g)}\left[\int\limits_0^T \int\limits_0^{\infty} y^2 e^{g_t(y)}\, \nu_{\mathbb{P}}(\mathrm{d}y, \mathrm{d}t) \right] < \infty.
\end{aligned}
$$

The remainder of the proof proceeds as follows:

1. We show that the feedback forms of $\delta^\diamond$, $\eta^\diamond$, $g^{\lambda\diamond}$, and $g^{\kappa\diamond}$ are admissible.
2. For an arbitrary admissible $\delta = (\delta_t)_{0 \le t \le T}$, we define an admissible response measure indexed by M which is denoted as $\mathbb{Q}^{\alpha,\lambda,\kappa}(\eta(\delta), g_M(\delta))$.
3. We show that M can be taken sufficiently large, independent of t and δ, such that the response measure is pointwise (in t) ϵ-optimal.

4. We show that the candidate function $\hat{H}$ satisfies $\hat{H}(t,x,S,q) \geq H(t,x,S,q)$.

5. We show that the candidate function $\hat{H}$ satisfies $\hat{H}(t,x,S,q) \leq H(t,x,S,q)$.

Step 1: $\delta^\diamond$, $\eta^\diamond$, $g^{\lambda\diamond}$, and $g^{\kappa\diamond}$ are admissible: Since q is bounded between 0 and Q, $\eta^\diamond$ is bounded and therefore admissible. The existence and uniqueness of a classical solution for h means that it achieves a finite maximum and minimum for some $q \in \{0,\ldots,Q\}$ and $t \in [0,T]$. Thus, from the feedback expressions for $g^{\lambda\diamond}$ and $g^{\kappa\diamond}$, we see that they are also bounded and therefore admissible. Admissibility of $\delta^\diamond$ is clear.

Step 2: Defining admissible response measure: Let $\delta = (\delta_t)_{0\leq t\leq T}$ be an arbitrary admissible control and define pointwise minimizing response controls by

$$\begin{aligned}
\eta_t(\delta) &= \alpha - \varphi_\alpha \sigma^2 q_t, \\
g_t^\lambda(\delta) &= \frac{\varphi_\lambda}{\varphi_\kappa} \log(1 - e^{-\kappa\,\delta_t}(1 - e^{-\varphi_\kappa(\delta_t + h_{q_t-1}(t) - h_{q_t}(t))})), \\
g_t^\kappa(y;\delta) &= -\log(1 - e^{-\kappa\,\delta_t}(1 - e^{-\varphi_\kappa(\delta_t + h_{q_t-1}(t) - h_{q_t}(t))})) \\
&\quad - \varphi_\kappa(\delta_t + h_{q_t-1}(t) - h_{q_t}(t))\mathbb{1}_{y\geq\delta_t}.
\end{aligned}$$

These processes each have the same form as the pointwise minimizers found in Proposition 4.2, and so for a given $\delta = (\delta_t)_{0\leq t\leq T}$, these controls achieve the pointwise infimum in Equation 4.24. Since h is a classical solution to Equation 4.24, it is bounded for $t \in [0,T]$ and $0 \leq q \leq Q$. Using the boundedness of h, we see that $g_t^\lambda(0)$ is finite and bounded with respect to t, and $\lim_{\delta\to\infty} g_t^\lambda(\delta) = 0$, therefore $g_t^\lambda(\delta)$ is bounded. It is also clear that $\eta_t(\delta)$ is bounded. However, $g_t^\kappa(y;\delta)$ is only bounded from above, so it is possible that the pair $(\eta_t(\delta), g_t(\delta))$ does not define an admissible measure as per the definition in Equation 4.15. In order to proceed, we use a modification of g_t^κ:

$$\begin{aligned}
g_{t,M}^\kappa(y;\delta) &= -\log(1 - e^{-\kappa\,\delta_t}(1 - e^{-\varphi_\kappa(\delta_t + h_{q_t-1}(t) - h_{q_t}(t))})) \\
&\quad - \varphi_\kappa \min(\delta_t + h_{q_t-1}(t) - h_{q_t}(t), M)\mathbb{1}_{y\geq\delta_t}.
\end{aligned}$$

Since $g_{t,M}^\kappa$ is bounded, letting $g_{t,M}(y;\delta) = g_t^\lambda(\delta) + g_{t,M}^\kappa(y;\delta)$, the pair $(\eta_t(\delta), g_M(\delta))$ does define an admissible measure $\mathbb{Q}^{\alpha,\lambda,\kappa}(\eta(\delta), g_M(\delta))$. Note that for a fixed t and δ_t, $g_{t,M}^\kappa(y;\delta) \to g_t^\kappa(y;\delta)$ as $M \to \infty$ pointwise in y and in $L^1(F(\mathrm{d}y))$.

Step 3: Showing pointwise ϵ-optimality: As in the proof of Proposition 4.2, consider the functional

$$\begin{aligned}
\mathfrak{G}(t,\delta,g^\lambda,g^\kappa) &= \lambda\left[\int_\delta^\infty e^{g^\lambda + g^\kappa(y)} F(y)\mathrm{d}y\right](\delta + h_{q-1}(t) - h_q(t)) \\
&\quad + \mathcal{K}^{\varphi_\lambda,\varphi_\kappa}(g^\lambda,g^\kappa)\mathbb{1}_{\varphi_\lambda>\varphi_\kappa} + \mathcal{K}^{\varphi_\kappa,\varphi_\lambda}(g^\kappa,g^\lambda)\mathbb{1}_{\varphi_\kappa>\varphi_\lambda}.
\end{aligned}$$

We now show

$$\lim_{M\to\infty} \mathfrak{G}(t,\delta_t,g_t^\lambda(\delta),g_{t,M}^\kappa(\cdot;\delta)) = \mathfrak{G}(t,\delta_t,g_t^\lambda(\delta),g_t^\kappa(\cdot;\delta))$$

uniformly in t and δ. Consider the first term only, and compute the difference when evaluated at both $g_{t,M}^\kappa(\cdot;\delta_t)$ and $g_t^\kappa(\cdot;\delta_t)$, which we denote by

$$\begin{aligned}
&\mathfrak{J}(t,\delta,M)\\
&= \lambda\, e^{g_t^\lambda(\delta)} \left|\delta_t + h_{q-1}(t) - h_q(t)\right| \left| \int_{\delta_t}^{\infty} e^{g_{t,M}^\kappa(y;\delta)}\, F(\mathrm{d}y) - \int_{\delta_t}^{\infty} e^{g_t^\kappa(y;\delta)}\, F(\mathrm{d}y)\right|\\
&= \lambda\, e^{(1-\frac{\varphi_\kappa}{\varphi_\lambda})g_t^\lambda(\delta)} \left|\delta_t + h_{q-1}(t) - h_q(t)\right| e^{-\kappa\,\delta_t} \left| e^{-\varphi_\kappa \min(\delta_t+h_{q-1}(t)-h_q(t),M)}\right.\\
&\qquad \left. - e^{-\varphi_\kappa(\delta_t+h_{q-1}(t)-h_q(t))}\right|\\
&= \lambda\, e^{(1-\frac{\varphi_\kappa}{\varphi_\lambda})g_t^\lambda(\delta)} \left|\delta_t + h_{q-1}(t) - h_q(t)\right| e^{-\kappa\,\delta_t} e^{-\varphi_\kappa M}\\
&\qquad \times \left|1 - e^{-\varphi_\kappa(\delta_t+h_{q-1}(t)-h_q(t)-M)}\right| \mathbb{1}_{\delta_t+h_{q-1}(t)-h_q(t)\ge M}\\
&\le \lambda\, e^{(1-\frac{\varphi_\kappa}{\varphi_\lambda})g_t^\lambda(\delta)} \left|\delta_t + h_{q-1}(t) - h_q(t)\right| e^{-\kappa\,\delta_t} e^{-\varphi_\kappa M}.
\end{aligned}$$

As previously noted, both h and g^λ are uniformly bounded, say by C and D respectively, so clearly $\mathfrak{J}(t,\delta,M)$ is bounded. For an arbitrary $\epsilon' > 0$, we may choose M sufficiently large such that

$$\mathfrak{J}(t,\delta,M) \le \lambda\, e^{|1-\frac{\varphi_\kappa}{\varphi_\lambda}|D}(\delta_t + 2C)e^{-\kappa\,\delta_t}e^{-\varphi_\kappa M} < \epsilon' \quad \text{for all} \quad \delta_t \ge 0.$$

Showing uniform convergence of $\mathcal{K}^{\varphi_\lambda,\varphi_\kappa}(g_t^\lambda(\delta), g_{t,M}^\kappa(\cdot;\delta))\mathbb{1}_{\varphi_\lambda>\varphi_\kappa}$, and $\mathcal{K}^{\varphi_\kappa,\varphi_\lambda}$ $(g_{t,M}^\kappa(\cdot,\delta), g^\lambda(\delta))\mathbb{1}_{\varphi_\kappa>\varphi_\lambda}$ is essentially the same and so the details are omitted.

Let $\epsilon > 0$ be arbitrary and let M be sufficiently large (chosen independently of t and δ) so that

$$0 < \mathfrak{G}(t,\delta_t,g_t^\lambda(\delta),g_{t,M}^\kappa(\cdot;\delta)) - \mathfrak{G}(t,\delta_t,g_t^\lambda(\delta),g_t^\kappa(\cdot;\delta)) < \epsilon.$$

Then since δ is arbitrary and h satisfies Equation 4.24, the following inequality holds almost surely for every t:

$$\begin{aligned}
&\partial_t h_{q_t} + \eta_t(\delta) q_t + \frac{1}{2\varphi_\alpha}\left(\frac{\alpha - \eta_t(\delta)}{\sigma}\right)^2\\
&\quad + \lambda\left[\int_{\delta_t}^{\infty} e^{g_t^\lambda(\delta)+g_{t,M}^\kappa(y;\delta)}\, F(\mathrm{d}y)\right](\delta_t + h_{q_t-1}(t) - h_{q_t}(t))\\
&\quad + \mathcal{K}^{\varphi_\lambda,\varphi_\kappa}(g_t^\lambda(\delta), g_{t,M}^\kappa(\cdot;\delta))\,\mathbb{1}_{\varphi_\lambda\ge\varphi_\kappa} + \mathcal{K}^{\varphi_\kappa,\varphi_\lambda}(g_{t,M}^\kappa(\cdot;\delta), g_t^\lambda(\delta))\,\mathbb{1}_{\varphi_\lambda<\varphi_\kappa} < \epsilon.
\end{aligned} \tag{A.19}$$

Thus the measure $\mathbb{Q}^{\alpha,\lambda,\kappa}(\eta(\delta), g_M(\delta))$ is pointwise (in t) ϵ-optimal, uniformly in δ.

Step 4: Showing $\hat{H}(t,x,S,q) \geq H(t,x,S,q)$: Taking an expectation of $\hat{H}(T, X^\delta_{T-}, S_{T-}, q^\delta_{T-})$ in the measure $\mathbb{Q}^{\alpha,\lambda,\kappa}(\eta(\delta), g_M(\delta))$, and using Equation A.19, gives

$$
\begin{aligned}
&\mathbb{E}_{t,x,q,S}^{\mathbb{Q}^{\alpha,\lambda,\kappa}(\eta(\delta),g_M(\delta))}\left[\hat{H}(T, X^\delta_{T-}, S_{T-}, q^\delta_{T-})\right] \\
&\quad = \hat{H}(t,x,S,q) + \mathbb{E}_{t,x,q,S}^{\mathbb{Q}^{\alpha,\lambda,\kappa}(\eta(\delta),g_M(\delta))}\left[\int_t^T \partial_t h_{q_s}(s)\mathrm{d}s + \int_t^T \alpha\, q_s \mathrm{d}s + \sigma \int_t^T q_s \mathrm{d}W_s \right.\\
&\qquad \left. + \int_t^T \int_{\delta_s}^\infty (\delta_s + h_{q_s-1}(s) - h_{q_s}(s))\, \nu_{\mathbb{Q}^{\alpha,\lambda,\kappa}(\eta(\delta),g_M(\delta))}(\mathrm{d}y, \mathrm{d}s)\right]\\
&\quad \leq \hat{H}(t,x,S,q) + \epsilon(T-t) + \mathbb{E}_{t,x,q,S}^{\mathbb{Q}^{\alpha,\lambda,\kappa}(\eta(\delta),g_M(\delta))}\left[-\frac{1}{2\varphi_\alpha}\int_t^T \left(\frac{\alpha - \eta_s(\delta)}{\sigma}\right)^2 \mathrm{d}s \right.\\
&\qquad - \int_t^T \Big(\mathcal{K}^{\varphi_\lambda,\varphi_\kappa}(g^\lambda_s(\delta), g^\kappa_{s,M}(\cdot;\delta))\, \mathbb{1}_{\varphi_\lambda \geq \varphi_\kappa} \\
&\qquad\qquad \left. + \mathcal{K}^{\varphi_\kappa,\varphi_\lambda}(g^\kappa_{s,M}(\cdot;\delta), g^\lambda_s(\delta))\, \mathbb{1}_{\varphi_\lambda < \varphi_\kappa}\Big)\mathrm{d}s \right].
\end{aligned}
$$

Therefore, the candidate function satisfies

$$
\begin{aligned}
&\hat{H}(t,x,S,q) + \epsilon(T-t) \\
&\quad \geq \mathbb{E}_{t,x,q,S}^{\mathbb{Q}^{\alpha,\lambda,\kappa}(\eta(\delta),g_M(\delta))}\left[\hat{H}(T, X^\delta_{T-}, S_{T-}, q^\delta_{T-}) + \frac{1}{2\varphi_\alpha}\int_t^T \left(\frac{\alpha - \eta_s(\delta)}{\sigma}\right)^2 \mathrm{d}s \right.\\
&\qquad + \int_t^T \Big(\mathcal{K}^{\varphi_\lambda,\varphi_\kappa}(g^\lambda_s(\delta), g^\kappa_{s,M}(\cdot;\delta))\, \mathbb{1}_{\varphi_\lambda \geq \varphi_\kappa} \\
&\qquad\qquad \left. + \mathcal{K}^{\varphi_\kappa,\varphi_\lambda}(g^\kappa_{s,M}(\cdot;\delta), g^\lambda_s(\delta))\, \mathbb{1}_{\varphi_\lambda < \varphi_\kappa}\Big)\mathrm{d}s\right] \\
&\quad = \mathbb{E}_{t,x,q,S}^{\mathbb{Q}^{\alpha,\lambda,\kappa}(\eta(\delta),g_M(\delta))}\left[\hat{H}(T, X^\delta_T, S_T, q^\delta_T) + \mathcal{H}_{t,T}\left(\mathbb{Q}^{\alpha,\lambda,\kappa}(\eta(\delta), g_M(\delta))|\mathbb{P}\right)\right] \\
&\quad = \mathbb{E}_{t,x,q,S}^{\mathbb{Q}^{\alpha,\lambda,\kappa}(\eta(\delta),g_M(\delta))}\left[X^\delta_T + q^\delta_T(S_T - \ell(q^\delta_T)) + \mathcal{H}_{t,T}\left(\mathbb{Q}^{\alpha,\lambda,\kappa}(\eta(\delta), g_M(\delta))|\mathbb{P}\right)\right].
\end{aligned}
$$

Since this holds for one particular choice of admissible measure $\mathbb{Q}^{\alpha,\lambda,\kappa}(\eta(\delta), g_M(\delta))$, we have

$$
\hat{H}(t,x,S,q) + \epsilon(T-t) \geq \inf_{\mathbb{Q}\in\mathcal{Q}} \mathbb{E}^{\mathbb{Q}}_{t,x,q,S}\left[X^\delta_T + q^\delta_T(S_T - \ell(q^\delta_T)) + \mathcal{H}_{t,T}(\mathbb{Q}|\mathbb{P})\right].
$$

This inequality holds for the arbitrarily chosen control δ, therefore

$$\hat{H}(t,x,S,q) + \epsilon(T-t) \geq \sup_{(\delta_s)_{t\leq s\leq T}\in\mathcal{A}} \inf_{\mathbb{Q}\in\mathcal{Q}} \mathbb{E}^{\mathbb{Q}}_{t,x,q,S} \times \left[X^{\delta}_T + q^{\delta}_T(S_T - \ell(q^{\delta}_T)) + \mathcal{H}_{t,T}(\mathbb{Q}|\mathbb{P})\right] = H(t,x,S,q),$$

and letting $\epsilon \to 0$ we finally obtain

$$\hat{H}(t,x,S,q) \geq H(t,x,S,q)\,. \tag{A.20}$$

Step 5: Showing $\hat{H}(t,x,S,q) \leq H(t,x,S,q)$: Now, let $\delta^{\diamond} = (\delta^{\diamond}_t)_{0\leq t\leq T}$ be the control process defined in the statement of the theorem, and let η_t, g^{λ}_t, and $g^{\kappa}_t(y)$ be arbitrary such that they induce an admissible measure $\mathbb{Q}^{\alpha,\lambda,\kappa}(\eta,g) \in \mathcal{Q}^{\alpha,\lambda,\kappa}$. Then from Ito's lemma and the fact that h satisfies Equation 4.24

$$\begin{aligned}
&\mathbb{E}^{\mathbb{Q}^{\alpha,\lambda,\kappa}(\eta,g)}_{t,x,q,S}\left[\hat{H}(T, X^{\delta^{\diamond}}_{T-}, S_{T-}, q^{\delta^{\diamond}}_{T-})\right] \\
&= \hat{H}(t,x,S,q) + \mathbb{E}^{\mathbb{Q}^{\alpha,\lambda,\kappa}(\eta,g)}_{t,x,q,S}\Bigg[\int_t^T \partial_t h_{q_s}(s)\mathrm{d}s + \int_t^T \alpha\, q^{\delta^{\diamond}}_s\,\mathrm{d}s + \sigma\int_t^T q^{\delta^{\diamond}}_s\,\mathrm{d}W_s \\
&\qquad + \int_t^T\int_{\delta^{\diamond}_s}^{\infty}\Big(\delta^{\diamond}_s + h_{q_s-1}(s) - h_{q_s}(s)\Big) \\
&\qquad \times \nu_{\mathbb{Q}^{\alpha,\lambda,\kappa}(\eta,g)}(\mathrm{d}y,\mathrm{d}s)\Bigg] \\
&\geq \hat{H}(t,x,S,q) + \mathbb{E}^{\mathbb{Q}^{\alpha,\lambda,\kappa}(\eta,g)}_{t,x,q,S}\Bigg[-\frac{1}{2\varphi_\alpha}\int_t^T\left(\frac{\alpha-\eta_s}{\sigma}\right)^2\mathrm{d}s \\
&\qquad - \int_t^T\Big(\mathcal{K}^{\varphi_\lambda,\varphi_\kappa}(g^{\lambda}_s, g^{\kappa}_s)\,\mathbb{1}_{\varphi_\lambda\geq\varphi_\kappa} \\
&\qquad + \mathcal{K}^{\varphi_\kappa,\varphi_\lambda}(g^{\kappa}_s, g^{\lambda}_s)\,\mathbb{1}_{\varphi_\lambda<\varphi_\kappa}\Big)\mathrm{d}s\Bigg].
\end{aligned}$$

And so the candidate function satisfies

$$\begin{aligned}
\hat{H}(t,x,S,q) \leq \mathbb{E}^{\mathbb{Q}^{\alpha,\lambda,\kappa}(\eta,g)}_{t,x,q,S}\Bigg[&\hat{H}(T, X^{\delta^{\diamond}}_{T-}, S_{T-}, q^{\delta^{\diamond}}_{T-}) + \frac{1}{2\varphi_\alpha}\int_t^T\left(\frac{\alpha-\eta_s}{\sigma}\right)^2\mathrm{d}s \\
&+ \int_t^T\Big(\mathcal{K}^{\varphi_\lambda,\varphi_\kappa}(g^{\lambda}_s, g^{\kappa}_s)\,\mathbb{1}_{\varphi_\lambda\geq\varphi_\kappa} \\
&+ \mathcal{K}^{\varphi_\kappa,\varphi_\lambda}(g^{\kappa}_s, g^{\lambda}_s)\,\mathbb{1}_{\varphi_\lambda<\varphi_\kappa}\Big)\mathrm{d}s\Bigg]
\end{aligned}$$

$$= \mathbb{E}_{t,x,q,S}^{\mathbb{Q}^{\alpha,\lambda,\kappa}(\eta,g)}\left[\hat{H}(T, X_T^{\delta^\diamond}, S_T, q_T^{\delta^\diamond}) + \mathcal{H}_{t,T}(\mathbb{Q}^{\alpha,\lambda,\kappa}(\eta,g)|\mathbb{P})\right]$$
$$= \mathbb{E}_{t,x,q,S}^{\mathbb{Q}^{\alpha,\lambda,\kappa}(\eta,g)}\left[X_T^{\delta^\diamond} + q_T^{\delta^\diamond}(S_T - \ell(q_T^{\delta^\diamond})) + \mathcal{H}_{t,T}(\mathbb{Q}^{\alpha,\lambda,\kappa}(\eta,g)|\mathbb{P})\right].$$

Since this holds for any arbitrary admissible measure $\mathbb{Q}^{\alpha,\lambda,\kappa}(\eta,g)$, we have

$$\hat{H}(t,x,S,q) \leq \inf_{\mathbb{Q}\in\mathcal{Q}^{\alpha,\lambda,\kappa}} \mathbb{E}_{t,x,q,S}^{\mathbb{Q}}\left[X_T^{\delta^\diamond} + q_T^{\delta^\diamond}(S_T - \ell(q_T^{\delta^\diamond})) + \mathcal{H}_{t,T}(\mathbb{Q}|\mathbb{P})\right].$$

Therefore,

$$\hat{H}(t,x,S,q) \leq \sup_{(\delta_s)_{t\leq s\leq T}\in\mathcal{A}} \inf_{\mathbb{Q}\in\mathcal{Q}^{\alpha,\lambda,\kappa}} \mathbb{E}_{t,x,q,S}^{\mathbb{Q}}\left[X_T^{\delta} + q_T^{\delta}(S_T - \ell(q_T^{\delta})) + \mathcal{H}_{t,T}(\mathbb{Q}|\mathbb{P})\right]$$
$$= H(t,x,S,q)\,. \tag{A.21}$$

Combining Equations A.20 and A.21 gives

$$\hat{H}(t,x,q,S) = H(t,x,q,S),$$

as desired. □

Proof of Proposition 4.3. Let $\omega_q(t) = e^{\kappa h_q(t)}$, or equivalently, $h_q(t) = \frac{1}{\kappa}\omega_q(t)$. Substituting this into Equation 4.28 gives

$$\frac{\partial_t\omega_q}{\kappa\omega_q} + \alpha\,q - \frac{1}{2}\varphi_\alpha\sigma^2 q^2 + \frac{\xi}{\kappa}\frac{\omega_{q-1}}{\omega_q}\mathbb{1}_{q\neq 0} = 0$$
$$\partial_t\omega_q + \alpha\kappa q\omega_q - \frac{1}{2}\varphi_\alpha\kappa\sigma^2 q^2\omega_q + \xi\omega_{q-1}\mathbb{1}_{q\neq 0} = 0$$
$$\partial_t\omega_q + K_q\omega_q + \xi\omega_{q-1}\mathbb{1}_{q\neq 0} = 0, \tag{A.22}$$

together with terminal conditions $\omega_q(T) = e^{-\kappa q\,\ell(q)}$. To prove Part (i), Equation A.22 becomes

$$\partial_t\omega_q = -\xi\omega_{q-1}\mathbb{1}_{q\neq 0}\,.$$

For $q = 0$, this results in $\omega_0(t) = 1$. For $q > 0$, integrating both sides yields

$$\omega_q(T) - \omega_q(t) = -\xi\int_t^T \omega_{q-1}$$
$$\omega_q(t) = \xi\int_t^T \omega_{q-1}(u)\mathrm{d}u + \omega_q(T)\,. \tag{A.23}$$

Since $\omega_0(t)$ is a constant and each ω_q results from the integral of ω_{q-1}, $\omega_q(t)$ can be written as

$$\omega_q(t) = \sum_{n=0}^{q} C_{q,n}(T-t)^n, \tag{A.24}$$

where each $C_{q,n}$ must be computed. Substituting Equation A.24 into Equation A.23 gives

$$\begin{aligned}
\omega_q(t) &= \xi \int_t^T \sum_{n=0}^{q-1} C_{q-1,n}(T-u)^n + \omega_q(T) \\
&= \xi \sum_{n=0}^{q-1} C_{q-1,n} \int_t^T (T-u)^n + \omega_q(T) \\
&= \xi \sum_{n=0}^{q-1} C_{q-1,n} \frac{(T-t)^{n+1}}{n+1} + \omega_q(T) \\
&= \xi \sum_{n=1}^{q} C_{q-1,n-1} \frac{(T-t)^{n}}{n} + \omega_q(T).
\end{aligned}$$

From here it is easy to see that

$$C_{q,n} = \frac{\xi}{n} C_{q-1,n-1}$$

and

$$C_{q,0} = \omega_q(T)\,.$$

Together, these give the desired result

$$C_{q,n} = \frac{\xi^n}{n!}\omega_{q-n}(T) = \frac{\xi^n}{n!} e^{-\kappa(q-n)\ell(q-n)}.$$

To prove Part (ii), return to Equation A.22. Since $K_0 = 0$, it is easily seen that $\omega_0(t) = 1$. For $q > 0$, a recursive solution to Equation A.22 can be written as

$$\begin{aligned}
\partial_t\omega_q + K_q\omega_q + \xi\omega_{q-1} &= 0 \\
\partial_t(e^{K_q t}\omega_q(t)) &= -\xi e^{K_q t}\omega_{q-1} \\
e^{K_q T}\omega_q(T) - e^{K_q t}\omega_q(t) &= -\xi \int_t^T e^{K_q u}\omega_{q-1}(u)\mathrm{d}u \\
\omega_q(t) &= \xi e^{-K_q t} \int_t^T e^{K_q u}\omega_{q-1}(u)\mathrm{d}u + \omega_q(T)e^{K_q(T-t)}.
\end{aligned} \tag{A.25}$$

With $\omega_q(t) = 1$ and each $\omega_{q-1}(t)$ being integrated against $e^{K_q t}$, the general form of $\omega_q(t)$ can be written as

$$\omega_q(t) = \sum_{n=0}^{q} C_{q,n} e^{K_n(T-t)}, \tag{A.26}$$

where each $C_{q,n}$ must be computed. Substituting Equation A.26 for ω_{q-1} into Equation A.25 gives

$$\begin{aligned}
\omega_q(t) &= \xi e^{-K_q t} \int_t^T e^{K_q u} \sum_{n=0}^{q-1} C_{q-1,n} e^{K_n(T-u)} \mathrm{d}u + \omega_q(T) e^{K_q(T-t)} \\
&= \xi e^{-K_q t} \int_t^T \sum_{n=0}^{q-1} C_{q-1,n} e^{K_n T} e^{(K_q - K_n)u} \mathrm{d}u + \omega_q(T) e^{K_q(T-t)} \\
&= \xi e^{-K_q t} \sum_{n=0}^{q-1} C_{q-1,n} e^{K_n T} \int_t^T e^{(K_q - K_n)u} \mathrm{d}u + \omega_q(T) e^{K_q(T-t)} \\
&= \xi e^{-K_q t} \sum_{n=0}^{q-1} C_{q-1,n} e^{K_n T} \frac{e^{(K_q - K_n)T} - e^{(K_q - K_n)t}}{K_q - K_n} + \omega_q(T) e^{K_q(T-t)} \\
&= \xi \sum_{n=0}^{q-1} C_{q-1,n} \frac{e^{K_q(T-t)} - e^{K_n(T-t)}}{K_q - K_n} + \omega_q(T) e^{K_q(T-t)} \\
&= -\xi \sum_{n=0}^{q-1} \frac{C_{q-1,n}}{K_q - K_n} e^{K_n(T-t)} + \left(\xi \sum_{n=0}^{q-1} \frac{C_{q-1,n}}{K_q - K_n} + \omega_q(T) \right) e^{K_q(T-t)}.
\end{aligned} \tag{A.27}$$

From the first summation term above, it is deduced that $C_{q,n} = \frac{-\xi C_{q-1,n}}{K_q - K_n}$ for $q > n$. This recursive relation leads to

$$C_{n+j,n} = (-\xi)^j \prod_{p=1}^{j} \frac{1}{K_{n+p} - K_n} C_{n,n}.$$

The second summation term in Equation A.27 leads to $C_{q,q} = \xi \sum_{n=0}^{q-1} \frac{C_{q-1,n}}{K_q - K_n} + \omega_q(T)$, which from the previous recursion is the same as

$$C_{q,q} = -\sum_{n=0}^{q-1} C_{q,n} + \omega_q(T).$$

These equations together with $C_{0,0} = 1$ yield all of the coefficients. □

Proof of Proposition 4.4. **Case (i):** For case (i) with $K_q = 0$ for all q, the value function is of the form:

$$h_q(t) = \frac{1}{\kappa}\log\left(\sum_{n=0}^{q} C_{q,n}(T-t)^n\right),$$

where each $C_{q,n} > 0$, and the feedback form of the optimal depth is

$$\delta_q^*(t) = \frac{1}{\varphi}\log\left(1+\frac{\varphi}{\kappa}\right) - h_{q-1}(t) + h_q(t).$$

Substituting the value function into the feedback expression gives

$$\delta_q^*(t) = \frac{1}{\varphi}\log\left(1+\frac{\varphi}{\kappa}\right) + \frac{1}{\kappa}\log\left(\frac{\sum_{n=0}^{q} C_{q,n}(T-t)^n}{\sum_{n=0}^{q-1} C_{q-1,n}(T-t)^n}\right).$$

As $T-t \to \infty$, the argument of the logarithm approaches ∞ since the numerator is a polynomial of higher degree than the denominator. The argument of the logarithm clearly grows as $\frac{C_{q,q}}{C_{q-1,q-1}}(T-t)$. With the expression for $C_{q,n}$ from Proposition 4.3, this means the depth grows as

$$\delta_q^*(t) \approx \frac{1}{\kappa}\log\left(\frac{\xi}{q}(T-t)\right).$$

Case (ii): For case (ii) with $K_q < 0$ for all $q > 0$, the value function is of the form:

$$h_q(t) = \frac{1}{\kappa}\log\left(\sum_{n=0}^{q} C_{q,n}e^{K_n(T-t)}\right),$$

and so the optimal depth is equal to

$$\delta_q^*(t) = \frac{1}{\varphi}\log\left(1+\frac{\varphi}{\kappa}\right) + \frac{1}{\kappa}\log\left(\frac{\sum_{n=0}^{q} C_{q,n}e^{K_n(T-t)}}{\sum_{n=0}^{q-1} C_{q-1,n}e^{K_n(T-t)}}\right).$$

Since each K_n is negative (except for K_0 which is equal to 0), this clearly converges as $(T-t) \to \infty$ to

$$\begin{aligned}\delta_q^*(t) &\to \frac{1}{\varphi}\log\left(1+\frac{\varphi}{\kappa}\right) + \frac{1}{\kappa}\log\left(\frac{C_{q,0}}{C_{q-1,0}}\right)\\ &= \frac{1}{\varphi}\log\left(1+\frac{\varphi}{\kappa}\right) + \frac{1}{\kappa}\log\left(\frac{-\xi}{K_q}\right).\end{aligned}$$

□

References

Almgren, R. and Chriss, N., 2001. Optimal execution of portfolio transactions. *Journal of Risk*, 3, 5–40.

Avellaneda, M. and Stoikov, S., 2008. High-frequency trading in a limit order book. *Quantitative Finance*, 8(3), 217–224.

Cartea, Á., Donnelly, R., and Jaimungal, S., 2017. Algorithmic trading with model uncertainty. *SIAM Journal on Financial Mathematics*, 8(1), 635–671.

Cartea, Á. and Jaimungal, S., 2015a. Optimal execution with limit and market orders. *Quantitative Finance*, 15(8), 1279–1291.

Cartea, Á. and Jaimungal, S., 2015b. Risk metrics and fine tuning of high-frequency trading strategies. *Mathematical Finance*, 25(3), 576–611.

Cartea, Á., Jaimungal, S., and Penalva, J., 2015. *Algorithmic and High-Frequency Trading*. Cambridge University Press, Cambridge, United Kingdom.

Guéant, O. and Lehalle, C.-A., 2015. General intensity shapes in optimal liquidation. *Mathematical Finance*, 25(3), 457–495.

Guéant, O., Lehalle, C.-A., and Fernandez-Tapia, J., 2012. Optimal portfolio liquidation with limit orders. *SIAM Journal on Financial Mathematics*, 3(1), 740–764.

Guéant, O., Lehalle, C.-A., and Fernandez-Tapia, J., 2013. Dealing with the inventory risk: a solution to the market making problem. *Mathematics and Financial Economics*, 7(4), 477–507.

Jacod, J. and Shiryaev, A. N., 1987. *Limit Theorems for Stochastic Processes*. Grundlehren der mathematischen Wissenschaften. Springer-Verlag, Berlin, Germany.

Kharroubi, I. and Pham, H., 2010. Optimal portfolio liquidation with execution cost and risk, *SIAM Journal on Financial Mathematics*, 1(1), 897–931.

Obizhaeva, A. A. and Wang, J., 2013. Optimal trading strategy and supply/demand dynamics. *Journal of Financial Markets*, 16(1), 1–32.

Chapter 5

Challenges in Scenario Generation: Modeling Market and Non-Market Risks in Insurance

Douglas McLean

CONTENTS

5.1 Introduction

5.1.1 The challenge of negative nominal interest rates

In August 2014, the euro-denominated German 1 year bond yield dropped below zero and stubbornly stayed there until the time of writing (late November 2017). Like many other rates, including more recently, the German 10-year rate in June 2016, this heralded an unprecedented era of low interest rates. Symptomatic of the policy of quantitative easing from the European Central Bank, the bond market has been reacting to a surfeit of cash. Now overnight bank deposits incur charges and borrowers would *receive* interest rather than pay it. Such a policy was designed to stimulate lending and punish those who might otherwise save. For economic scenario generation, this has exposed a challenge: *model risk*. In what will be discussed in the sequel, this has led to a quandary in terms of the log-normal basis in which the so-called market implied volatilities are typically quoted. It was now no longer possible to *receive* market quotes from data vendors as the log-normal implied volatilities had ceased to exist. The Libor Market Model (e.g., Brigo and Mercurio 2006; Wu and Zhang 2006) could not be calibrated as implied volatilities had ceased to exist. The ramification for economic scenario generators (ESG) with a knock-on effect to insurance companies' asset and liability modeling systems (ALM for short) was that they were no longer going to be able to make stochastic projections of the most fundamental quantity in the market—the yield curve. Something needed to be done. The error had been in a lack of foresight: the assumption that nominal interest rates could go negative had been inconceivable yet it had happened. The obvious solution would be to switch to a different model that could cope with negative rates. But which one? Fortunately, the investment banking industry had already foreseen a negative nominal rate scenario and had been using Bachelier's arithmetic Brownian motion as a model (1901) for the underlying interest rates for some time. Bachelier's model allowed negative nominal rates where Black's model did not. Furthermore, the data providers had been offering *absolute* implied volatilities to match. The switch was made and, for the time at least, another ESG challenge had been resolved.

5.1.2 Objectives

I purposely romanticized the situation of negative nominal interest rates to emphasize the scale of the challenge that faces anyone who sets out to construct an ESG. It is not easy. For one thing, it involves modeling each of the major asset classes in the financial markets and then "hanging" them together within a coherent co-dependency structure. The structure of choice is almost always the Gaussian copula and this is almost never appropriate within a risk-management system. It is not by accident that this is referred to as the *formula that killed Wall Street* (e.g., Mackenzie and Spears 2012). Its misuse certainly contributed to the 2007/2008 financial crisis. In simply setting a correlation matrix (whether it is a Pearson or a ranks-based Spearman correlation matrix) and correlating asset shocks by a Cholesky matrix factorization (e.g., Noble and Daniel 1977), one makes a de facto choice of the Gaussian copula. In Section 5.3 I will discuss alternative copulas which do not suffer from the detraction of permitting full portfolio diversification in times of financial crisis. One must also be mindful that any ESG must function in a multi-economy setting: there will almost certainly be more than one economy, perhaps 30 or 40, and one will want to forecast and price both assets and liabilities within each consistently. One final introductory note is to the question of model calibration which arises perpetually. Equivalent to non-convex optimization, model calibration is one of the most challenging areas of mathematics and must be tackled if one is to realistically parametrize the models within an ESG.

The aim of this chapter is therefore to illustrate some of the challenges that need to be overcome and to set them in the context of the regulatory world of insurance that they inhabit. It must be said that the Solvency 2 directive (and other rather onerous regulatory regimes around the world) provide the demand for ESGs. To begin with, therefore, I will set the scene by focusing largely on Pillar 1 of the Solvency 2 regulation of insurance companies in the European Union.

5.1.3 ESG and solvency 2

Insurance and reinsurance companies rely on ALM systems such as SunGard's *Prophet Asset Liability Strategy*[1]; Willis Towers Watson's *Replica*, *Igloo*, and *MoSes*[2] solutions; and Aon Benfield's *ReMetrica*[3]; to model both the asset and the liability sides of their balance sheets. In the European Union, the Solvency 1 regime was the first attempt to assess solvency of pan-European insurance companies in a coherent way. I do not discuss this regime in any detail except to say that it was hard to compare insurers precisely on this measure and that it did not allow for the time value of derivative assets to be fairly reported on insurers' balance sheets. The Solvency 2 directive, issued by the European Insurance and Occupational Pensions Authority (EIOPA)[4] was introduced to improve upon the Solvency 1 directive. Perhaps the greatest change in moving to Solvency 2 is the enhanced use of stochastic scenarios for valuation and forecasting of insurers' balance sheets. The transition to Solvency 2 has been challenging for many continental European insurers who have not been used to operating under this paradigm but less onerous for those in the United Kingdom where there has been a regulatory requirement for such reporting with systems already in place: for example, in the calculation of derivative styled market consistent embedded value. Now, however,

[1] http://www.prophet-web.com
[2] http://www.towerswatson.com
[3] http://www.aon.com/reinsurance/analytics/remetrica.jsp
[4] https://eiopa.europa.eu

the Solvency 2 directive instructs all European insurers to work on this basis. It is worthy of note that other regimes like Solvency 2 are in force elsewhere in the world. The Swiss Solvency Test (SST) has been granted full equivalence to Solvency 2 while regulatory regimes in Australia, Bermuda, Brazil, Canada, Mexico, and the United States have been granted Solvency 2 equivalence for a period of 10 years under Pillar 1. This is largely driven out of necessity for non-EU domiciled insurers who participate in EU markets.

This chapter focuses on the use of an ESG to generate stochastic scenario sets for use within the context of the European Union's Solvency 2 directive under Pillar 1 (and potentially under Pillar 2). Note that by *scenario* I mean the market dynamics that can be simulated from a given multi-variate probability distribution via coupled systems of stochastic differential or econometric equations. Monte-Carlo simulation is the vehicle which permits any practical number of simulations each representing a *state of the world*. Each scenario unfolds differently given the same, current market conditions. Except to say that they may be used within the context of ORSA (Pillar 2), I do not discuss the careful crafting of specific scenarios which are aligned with *stress testing* in the Basel II banking regulations. These scenario-set sizes are usually small and are aimed at testing whether a business can withstand a highly stylized specific shock such as a worldwide pandemic, financial crisis or terrorist attack. Rather the scenarios which emerge from an ESG are randomly generated from a well-defined stochastic process and are usually generated on the order of thousands if not tens or even hundreds of thousands.

The Solvency 2 regulatory regime came into vigor on January 1, 2016 following a period of extended industry consultation lasting more than a decade. It replaces 13 individual regulatory regimes over 27 jurisdictions in Europe (including the United Kingdom). There are three *Pillars* of Solvency 2 which each assess fundamental components of good insurance practices. Pillar 1 is by far the most prescriptive of the three and requires each insurance company to report its solvency capital requirement (SCR). This is a quantitative measure of its ability to withstand a severe 1-in-200-year event over a 1-year horizon: that is, it is the 1-year value-at-risk (VaR) at the 99.5th percentile.

The local regulator (such as the Prudential Regulatory Authority, PRA, in the United Kingdom) is then at liberty to add a *capital buffer* if it judges the SCR to be too low given the specific nature of an insurer's business. An additional metric is computed: the *minimum capital requirement* or MCR; which the PRA sets and lies between 25% and 45% of the SCR (including any capital buffer). If the regulatory capital held by an insurer falls below this MCR then it is deemed to be *insolvent* and its business is put into administration. If, however, the regulatory capital held is more than the MCR yet less than the SCR then the regulator issues a requirement to the insurer to detail how it will increase its regulatory capital above the level required by the SCR. This plan must be returned within one month after which the regulator may directly intervene in the management of the insurance company if it is not satisfied. The t-year VaR at level α is defined on the (stochastic) loss variable L_t:

$$\mathrm{VaR}_t^\alpha := \inf_{x\in\mathbb{R}}\{x : \Pr(L_t \leq x) \geq \alpha\} \tag{5.1}$$

It is worthy to note the choice and the stability of the metric that is being used to define the SCR: the 1-year VaR at 99.5%. The rationale for the 1-year VaR measure is that it predicts the amount of extra capital that would have to be raised now and invested at the risk-free rate so that in one year's time an insurer would be solvent (McNeil et al., 2015). This argument can be used to show that the SCR is a quantile of the loss distribution. However, McNeil et al. (2015) detail the properties a good risk measure should have and go on to show that the VaR risk measure does not satisfy each of these. The four properties of a coherent risk measure $\tilde{\rho}(L)$ as a function of the loss L (considered a random variable over some linear space $\mathcal{M}$ of losses) are:

1. Monotonicity: for losses $L_1 < L_2, \tilde{\rho}_1(L_1) \leq \tilde{\rho}_2(L_2)$
2. Translational invariance: $\forall \ell \in \mathbb{R}$ deterministic, $\tilde{\rho}(L + \ell) = \tilde{\rho}(L) + \ell$
3. Sub-additivity: any L_1, L_2, $\tilde{\rho}(L_1 + L_2) \leq \tilde{\rho}(L_1) + \tilde{\rho}(L_2)$
4. Positive homogeneity: any $\lambda > 0, \tilde{\rho}(\lambda L) = \lambda\tilde{\rho}(L)$.

The interpretation of these axioms is as follows. Monotonicity shows that positions that incur higher losses require higher capital. Adding a constant to a loss means simply adding a constant to the risk measure. Sub-additivity and positive homogeneity both show that the risk measure of any two combined losses cannot lead to a greater loss than when they are measured separately: diversification is always worthwhile. McNeil et al. (2015) then detail how VaR is not sub-additive as a risk measure. Since an insurance company is typically composed of several *business units* and under Solvency 2 each business unit must calculate its own VaR with an *aggregated* VaR measure over the whole business finally being submitted to the regulator then breaking into further business units may reduce the VaR measure further. Other risk measures satisfying all four constraints are termed *coherent*. For example, there is the coherent expected shortfall measure:

$$\mathrm{ES}_\alpha := \frac{1}{1-\alpha}\int_\alpha^1 \mathrm{VaR}_x(L)\mathrm{d}x; \tag{5.2}$$

which is arguably a better choice than the simple VaR measure as it averages each VaR value above the α-quantile of the loss distribution. In any event, a risk measure is simply a gross summary statistic from an insurer's multivariate risk-driver distribution. Given the large dimensionality directly implicated in this multivariate density (see Sections 5.2 and 5.3), taking such a gross summary is a severe marginalization of a high-dimensional object. Since some aspects of the density will be better estimated than others and some will be based on very little information at all (if only by *expert judgement*, e.g., losses incurred by operational risks have little available data), then the sensitivity of the VaR measure to small perturbations ought to be given scrutiny. A small change in assumptions leading to a slightly different model parametrization, for example, has the potential to lead to an unpredictable change in an insurer's capital requirements, be it for better or for worse.

Compared to Pillar 1, Pillar 2 is a broad instruction to insurers to make their *own risk and solvency assessment* (ORSA). The wording is deliberately vague and is intended to instill a holistic risk management ethos. Insurers must demonstrate that they have considered, understood, and quantified all possible risks their businesses face over a long time horizon (beyond the 1-year horizon of Pillar 1). It is also a chance for an insurer to use its own in-house modeling approach that accurately represents the risks on its balance sheet. One approach is to use an ESG in a multi-period projection of market and non-market risks. This can be rather onerous on the necessary scenario budget, however. Given the vague wording of Pillar 2, an insurer may opt for a small number of *stylized* scenarios. These are like the Basel II stressed scenarios used in banking regulation. Typically, they test the robustness of an institution to specific outcomes in the future such as a second financial crisis. In any case, insurers must demonstrate that they have implemented an appropriate risk management strategy to mitigate their risks. Ultimately, the regulator reviews their approach and can choose whether to accept or reject it. In the latter case, an insurer would be required to update their risk management procedures to a new acceptable level and resubmit for approval: a task which is certainly costly.

Pillar 3 focuses on disclosure and transparency, and sets out regulatory reporting standards. Under Pillar 3, insurers are required to submit two reports annually: A *Solvency and Financial Condition Report* and a *Regulator Supervisory Report*; the former is made public and the latter is private. Since scenario generation is primarily concerned with Pillar 1 (and increasingly with Pillar 2) I refer the interested reader to the website of the Bank of England[5] for more information on Pillar 3 and continue with my discussion of Pillar 1.

Computation of the SCR may follow one of two paths: either insurance companies use a prescribed *standard formula* in their calculations or they may define and use their own *internal model*. Being formulaic, the standard formula model is easier to implement but has the downside of being relatively insensitive to the specifics of an insurer's business. For example, if a book of business contains mainly *plain vanilla* products, that is, such as standard annuities or self-invested personal pension schemes, then the standard formula may be a sufficient and cost-effective way of evaluating the SCR. On the other hand, if the book of business contains defined benefit schemes or guaranteed products with derivative styled payoffs, it will certainly be advantageous to

[5]http://www.bankofengland.co.uk/pra/Pages/regulatorydata/insurance/reporting.aspx

incur the often substantial cost of building one's own internal model that will give a more accurate (and hopefully less punitive) SCR.

The regulatory supervisor recognizes the use of stochastic models of market dynamics and goes on further to say that these should be supplied by an ESG. Market dynamics may be simulated under one of the two paradigms: the real world $\mathbb{P}$-measure or the risk-neutral/market consistent $\mathbb{Q}$-measure. Valuation of derivative styled optionality embedded in the insurer's balance sheet is then computed under the $\mathbb{Q}$-measure, which (typically) guarantees the absence of arbitrage.[6] I say *typically* since it is not always possible to find such a $\mathbb{Q}$-measure. For example, when pricing long-dated bonds (or anything derived from them such as long-dated interest rate hedges), the market may be quite illiquid with only a few players on the buy and sell sides. Reliable prices may not exist and so the arbitrage-free $\mathbb{Q}$ measure may be unreliable. This represents a significant challenge in economic scenario generation and is the subject of current research (Salahnejhad and Pelsser 2015). Forecasting an insurer's assets and liabilities over, for example, a 1-year horizon to establish a loss distribution must be done under the real-world $\mathbb{P}$-measure. This is the best forward-looking measure given the state of the world today and where economic expectations may reasonably be expected to fall over time. Under specific technical conditions (see, e.g., Baxter and Rennie 1996), the real-world $\mathbb{P}$- and risk-neutral/market-consistent $\mathbb{Q}$-measures may be related via Girsanov's theorem and the Radon-Nikodym derivative. One may simply suffice to postulate the existence of a *market price of risk* as compensation for bearing the risk in a world that is not arbitrage-free. This introduces a *drift correction* to the growth rates to compensate. Having established the need for two measures one must choose the most appropriate stochastic models accordingly and calibrate their parameters. This represents a challenging problem in scenario generation: *parameter calibration*; a challenge I will return to discuss in Section 5.2. Typically, one calibrates models under the risk-neutral/market-consistent $\mathbb{Q}$-measure to a snapshot of the current market. Models under the real-world $\mathbb{P}$-measure are often fitted to historical data. The obvious difficulty is then in reconciling the financial theory of measure changes between the risk-neutral and the real world via a drift correction with the statistical goodness-of-fit in models under each measure. If the fit is good under one measure yet poor under the other one may opt for different models under each measure which aren't related by a measure change. There is no easy way to reconcile this dichotomy but in practice one needs to be pragmatic.

ALM systems model, at the most granular level, the behavior of the policyholder schemes and the assets backing them. An essential ingredient of each ALM system is its market model which is supplied by an ESG. Dependent on the precise flavor of the insurance fund under consideration, these are invariably sensitive to equity, foreign exchange and credit risk but are always

[6]For an introduction to the idea of risk-neutral pricing, see Hull (2005) or Baxter and Rennie (1996).

sensitive to interest rate and inflation risk. When an ALM is coupled to scenarios emerging from an ESG, it becomes a Monte-Carlo simulation engine that is either used for pricing or forecasting. When pricing, any liability with derivative styled payoffs may be valued consistently to the market by approximating the expectation of the discounted payoff under the $\mathbb{Q}$-measure, and by taking the sample mean of Monte-Carlo-simulated market scenarios. It is imperative that the simulations are run under the risk-neutral measure for this approach to be appropriate. Indeed, market-consistent valuation occurs on both sides of the balance sheet: assets as well as liabilities. The real-world $\mathbb{P}$ measure is required under Pillar 1 of the Solvency 2 SCR calculation when projecting assets and liabilities onto the 1-year horizon. This develops the loss distribution from which the SCR's 99.5% 1-year VaR may be taken. However, the loss distribution will only be completely defined if the variables in the balance sheet appear at the 1-year horizon. This is not the case for derivatives such as equity options and interest rate swaptions. Importantly, if an insurer's business incorporates policyholder guarantees then these, having derivative styled characteristics, will not yet be valued either. For example, a minimum money-back guarantee in a with-profits fund hides latent value or *moneyness* at any point in time if the fund has policies *in-force*. One must make a further projection of the fund but this time under the market-consistent $\mathbb{Q}$-measure and discount/deflate this value back to the 1-year horizon. This allows a value to be established that considers the *time value of guarantees* (TVOG) or the *market-consistent embedded value* (MCEV) of the guarantees.

The need for a second projection under the market risk-neutral measure beyond the 1-year horizon leads to what is known as the *nested stochastic problem* and is a significant challenge in economic scenario generation. For example, if an insurer chooses to build the loss distribution using a thousand real-world forecasts at 1 year then, for each of these, a similar number (say another one thousand[7]) of market-consistent scenarios are needed to capture any latent value inherent in the derivative styled behavior of an insurer's assets and liabilities. In some instances, derived securities, such as equity options, a Black–Scholes model could theoretically replace the need for the second (inner) set of market risk-neutral simulations. However, such formulas are not available for more complicated equity models, or for certain interest rate and credit models. Whenever derivative securities are on an insurer's balance sheet then they will need to be valued and one simple solution is to use the Monte-Carlo simulation to a given time horizon, discount and take a sample mean. For liabilities, there are certainly no closed-form solutions or numerical methods for assessing their embedded value, and the Monte-Carlo simulation is the *only* way they may be estimated.[8] An open question is whether such estimators are

[7]Even here, this number may not be enough and is highly problem specific.

[8]With the possible exception of using a *replicating portfolio* of matching assets. However, getting such a portfolio to match well enough is a somewhat intractable problem.

biased or if they have an unacceptably high variance. One may, for example, settle for an estimator with a much lower variance if its bias is controllably small somehow: for example, by using a weighted estimator where the weights are chosen optimally by minimizing some out-of-sample metric. Opting for a better estimator may reduce the number of scenarios required which, in the naive case of the arithmetic sample mean estimator, is of the order of a million. Such large numbers of scenarios produce a serious performance bottleneck: ALM systems are currently not sufficiently fast enough to process this number of scenarios in a reasonable amount of time. Indeed, typically scenarios from an ESG are *written out* to a csv (or equivalent) file and then *read into* an ALM system. The I/O burden of processing such a large scenario set file using, say, a monthly time step over multiple risk drivers in multiple economies is, at least currently, prohibitive.

Mercifully, some solutions to the nested stochastic bottleneck exist. They include the use of replicating portfolios, curve-fitting and least-squares Monte Carlo (see Cathcart 2012). The latter technique approximates the asset and/or liability values with regression functions that have been fitted to noisy versions of the nested stochastic output. To be more precise, instead of running one thousand market-consistent simulations for every real-world scenario, a very small number of market-consistent simulations is run per real-world scenario. This brings the overall scenario budget down from a million to only a few thousand: something which is much more manageable for an ALM system. The challenge here is now to produce a few thousand market-consistent calibrations for each of the few thousand states of the world emerging at the 1-year horizon: a non-trivial exercise if this is to be done in any reasonable time. Functional approximations are typically made to produce calibrations quickly rather than force them through time-consuming numerical optimization algorithms. One should exercise caution as, yet again, the parameter calibration problem appears and care must be taken to ensure it is realistic (see Section 5.2). This scenario set is run through the ALM system and a regression function (such as a multiple polynomial regression function, or other, in the underlying variables) is fitted to the (discounted) desired asset or liability payoffs. The idea behind the least-squares Monte-Carlo technique is that the expectation of the discounted payoff may be approximated by the linear predictor coming from the polynomial regression function. If one can show that the conditions of the Gauss–Markov theorem for best linear unbiased estimators (e.g., Zyskind and Martin 1202; DeGroot and Schervish 2013) is satisfied, then this approximation is asymptotically rigorous. It is clear that there will be assets and/or liabilities where this is not the case or where one is not in the asymptotic limit and so one must use these estimators after they have been validated for bias and variance. How easy or otherwise this is to validate is problem specific but it seems that in many cases the simple linear predictor from an ordinary least-squares regression is adequate. In problem cases, for example, where the data display heteroskedasticity, simultaneously modeling the mean and the variance with regression functions can help (see the generalized additive models for location, shape, and scale[9]). For Pillar 2 ORSA, regression functions could be generated at each time period of interest.

Finally, truly high-performance solutions, such as that offered by Willis Towers Watson's *MoSes HPC* or Aon Benfield's PathWise$^{(\mathrm{TM})}$, Phillips (2016), have only very recently become available. It may take some time before these solutions can be fully absorbed and accepted by the insurance industry and so, for the moment at least, approximations are *de rigeur*.

5.1.4 Layout

This chapter is therefore laid out as follows. In Section 5.2, I discuss the make-up of a typical ESG including the major asset classes that it models. At a high level, I will illustrate some of the challenges that face the modeling exercise such as parameter calibration and discuss some high-performance computing techniques that can be deployed to accelerate scenario set generation.[10] In Section 5.3, I will discuss copula co-dependency structures introduced by Sklar (1959) and go on to introduce the risk scenario generator (RSG). This is a natural extension of an ESG to allow for non-market risks such as policyholder lapse and operational risk. In Section 5.4, I will illustrate in some detail two problems that are typical of the degree of complexity faced by ESGs. There I give a new method to represent the marginal distribution functions of composite variables. Given this representation, one can simulate correlated instances of these variables through the copula-marginal factorization. I conclude with a discussion in Section 5.5.

5.2 Economic Scenario Generators

Several commercial and free ESGs exist. Examples of commercially available ESGs are: Moody's Analytics' *RiskIntegrity Suite ESG*[11] (formerly the Barrie and Hibbert ESG), Willis Towers Watson's *Star ESG*,[12] Numerix' *Oneview ESG*,[13] Conning's *GEMS ESG*,[14] Ortec's *Dynamic Scenario Generator*,[15] and Deloitte's *XSG*.[16] A free ESG (Moudiki and Planchet 2016) is available from the **R** statistical software package's **CRAN** website[17] of which there are two relevant packages: **ESG** and **ESGtoolkit**.[18] Compared

[9] http://www.gamlss.org

[10] To which special thanks is due to Colin Carson for many helpful discussions on HPC.

[11] http://www.barrhibb.com/products_and_services/detail/scenario_generation

[12] http://www.towerswatson.com

[13] http://www.numerix.com

[14] http://www.conning.com

[15] http://www.ortec-finance.co.uk

[16] http://www2.deloitte.com

[17] http://www.r-project.org

[18] https://cran.r-project.org/package=ESG and ESGtoolkit

to its commercially available competitors, the **rESG** is somewhat limited in the range of stochastic models it offers and, at the time of writing, it supports only market-consistent modeling. A real-world ESG could theoretically be obtained by estimating and inserting a drift correction into the underlying stochastic differential equations (SDEs) but this often leads to unsatisfactory real-world probability distributions. It is, however, a genuine achievement to create and distribute a free scenario generator and credit is due to its creator, Thierry Moudiki.

A significant risk in scenario generation is that of parameter calibration of the underlying stochastic models. Not all scenario generators, commercial or free, offer model calibration toolkits, which is certainly a serious limitation any insurer should consider. As I will describe in Section 5.2.5, good model calibrations are essential for an ESG but they often come at a cost of significant analyst intervention. Moody's Analytics, for example, offers a calibration toolkit for the technically adept users but also a calibration service (both standard and bespoke) which addresses this challenge. Indeed, many insurance companies have dedicated teams that provide internal model calibrations. Together with the problem of model calibrations, significant challenges to scenario generation also come from discretization and model error.

Discretization error arises from the choice of time-step size in a simulation. Processes which are manifestly Gaussian are easier to treat (cf. a Vasicek model of interest rates to a Libor market model with stochastic volatility, Wu and Zhang 2006). However, one should be aware that discretization error comes not only from the stochastic shock terms but also from the deterministic drift. The error in the drift is on the order of the time-step size[19] and this is cumulative in the number of time steps taken. If the drift is subject to a systematic bias coming from discretization error, all probability distributions will end up located in the wrong place. Smaller time steps certainly alleviate this issue but are computationally more onerous, leading to increased run-time and infeasibly large scenario sets. The associated I/O bottleneck as they are uploaded to an ALM system can be punitive and the temptation to use fewer Monte-Carlo trials can have a deleterious effect on the standard errors in the estimates of TVOG, MCEV, and ultimately to the SCR.

Difficulties incurred by model error are much subtler: if the given model of choice does not fit the data very well one would wish to be able to detect this. Good practice would be to fit more than one model and compare model diagnostic plots. For instance, and continuing within the context of the example of interest rates again, perhaps a Libor market model with stochastic volatility is the preferred choice of interest rate model. It has many desirable properties: it can model variations in implied volatility by maturity, tenor, and strike. It also models market observable forward rates. If rates are expected to be positive, it may be a very good choice indeed. However, it has a rather large number of parameters that need adjusting to match to the market data (e.g., to caps or floors, or to swaptions). The Vasicek model is a short rate model (it does

[19]Under a Euler–Maruyama discretization.

not model a market observable quantity) and more limited in its ability to describe market features such as implied volatility skew. It also predicts negative rates which, until recently, had been an undesirable property for nominal interest rates. The Vasicek model does have far fewer parameters and as such is a much simpler model than a Libor market model with stochastic volatility. If after optimizing over the mean-squared error each model produces similar qualities of fit for similar optimal objective values and diagnostic plots, or if the simpler model provides a *better* fit then this calls into question the use of the more complicated model. It is possible that the data simply doesn't support the more detailed model even if that is the true generation process. A simple analogy might be when one tries to regress noisy univariate response data Y on a predictor X. The true process may be quadratic in X, or perhaps even cubic or quartic in X; but if the data are insufficient to pin down the model's parameters, only noise will be represented by the more complicated model.

These risks serve to illustrate the complexity and challenges involved in creating an effective scenario generator. The *Moody's Analytics* ESG won the 2015 *Risk.Net* award for the best ESG based on a holistic approach servicing the insurance sector. Its *RiskIntegrity* suite included, in addition to the base ESG, an *Automation Module* permitting an automatic calibration and scenario set generation ability to any number of economies and models. This alleviates a significant burden from insurance companies. In addition, the Enterprise-level *Proxy Generator* allows clients to directly address the nested-stochastic problem inherent in the SCR calculation under Pillar 1 of Solvency 2.

Before I move on to motivate the basic ESG engine, its asset classes and challenges, I refer the interested actuarial reader to Varnell (2011) who discusses scenario generators in the context of Solvency 2 (and to the references contained therein).

5.2.1 What does an ESG do?

At a high level, an ESG provides a coherent way of evolving the possible states of a financial market (or markets) through simulated time using the Monte-Carlo simulation technique (e.g., Glasserman 2003). It will likely be used to evolve the financial variables from several national economies simultaneously under an appropriate co-dependency structure. Care therefore needs to be taken in how one sets both the intra- and inter-economy correlations. From an insurance company's point of view, the role of the ESG is to create scenario sets that can be uploaded to its ALM system. The scenario sets generated from an ESG are the missing piece of the jigsaw which allows insurers to make forecasts under the real-world $\mathbb{P}$-measure or to price their assets and liabilities under the market-consistent/risk-neutral $\mathbb{Q}$-measure. As such, an insurer will select the measure and variables so that an output scenario set reflects the risks it faces on its balance sheet.

Each variable is modeled as a random process in time either using a stochastic differential equation (SDE) or by an econometric/time-series model. Typically, SDEs are used in a discretized form to model market risks (e.g., interest rate, credit, or equity risk) as there is financial theory underpinning their dynamics that SDEs have been formulated to reflect. For non-market risks, such as gross domestic product or an insurer's mortality risk, an econometric model may be preferred. This is because the economic theory for non-market risks is much less well developed and more open to statistical modeling although this partitioning is not strict. One may use an econometric model for real-world interest rates for example. I will concentrate on SDE models in this chapter.

A simple example of an SDE is the model of the *lognormal* asset-price process sometimes referred to as *geometric Brownian motion*. Let S_t be the price of a stock (or index) at time $t \geq 0$. Then its return may be modeled as

$$\frac{\mathrm{d}S_t}{S_t} = \mu(S_t, t)\,\mathrm{d}t + \sigma(S_t, t)\,\mathrm{d}W_t, \quad S_0 \text{ given} \tag{5.3}$$

Here μ is the *drift* which may depend on the price S_t and time t, and $\mathrm{d}t$ is the deterministic differential in t. The second term on the right-hand side is the purely random term. The *volatility* σ may depend on the price S_t as well as on time t like in the drift term μ. The random *Brownian* component W_t is written in terms of its stochastic differential $\mathrm{d}W_t$. Mathematically, Equation 5.3 is shorthand for a given integral representation that I omit here (but see, e.g., Baxter and Rennie 1996). The properties of W_t are:

1. Initially, the process is zero: $W_0 = 0$ a.s.
2. Increments are independent: $W_{t+\Delta t} - W_t$ is independent of the sigma algebra generated by W_t each t, $\Delta t \geq 0$
3. The difference $W_{t+\Delta t} - W_t \sim N(0, \Delta t)$
4. W_t is continuous with probability 1.

The Euler–Maruyama discretized version (Glasserman 2003) at step $i\Delta t$ of Equation 5.3 is

$$S_{i+1} = S_i[1 + \mu(S_i, i\Delta t)]\Delta t + S_i\sigma(S_i, i\Delta t)\Delta t^{1/2} Z_i, \quad i \in \mathbb{Z}^+, \quad S_0 \text{ given} \tag{5.4}$$

Writing, in an abuse of notation S_i for $S_{i\Delta t}$ and where $Z_i \sim N(0,\, 1)$. To evolve S_t one must be able to generate standard normal variables quickly. The Box–Muller (see Glasserman 2003) method generates two standard normal variables from two standard uniform variables in a computationally inexpensive transformation. The generation of other random variables, as shall be seen in Section 5.4, can be much more onerous. This is particularly true since *correlation* of the risk variables is sought. By necessity, therefore, an

ESG designer is obliged to use some random number generator and several of these exist: for example, Wichmann and Hill (1982, 1984) and the Mersenne Twister (Matsumoto and Nishimura 1998). These are industry standard methods for random number generation and they work reasonably well. However, one should always be cautious of random number generation: if an insufficiently elegant generator is used, the numbers emerging can eventually show deterministic patterns.

Many introductory textbooks on financial engineering (e.g., Hull 2005; Baxter and Rennie 1996) show, by means of the Girsanov theorem, how a simple change-of-measure may relate the Brownian shocks $dW_t^{\mathbb{P}}$ under the $\mathbb{P}$-forecasting measure to those under the pricing measure $\mathbb{Q}$:

$$W_t^{\mathbb{Q}} = W_t^{\mathbb{P}} + \lambda t \tag{5.5}$$

Here λ is the all-important *market-price-of-risk* for the lognormal process (Equation 5.3) modeling the compensation per unit of volatility an investor should receive for bearing risk in the real world. The change-of-measure often translates simply into a deterministic correction to an asset's growth rate although in more complicated models this correction may itself be stochastic (e.g., the CIR process Hull 2005). An important consequence of *arbitrage-free-pricing* is that under the risk-neutral $\mathbb{Q}$-measure, all assets grow with the growth rate of the *numeraire* or *reference* asset of that class. To see how this sets the market price of risk, observe that in the case of a constant interest rate r, drift μ and volatility σ then Equation 5.3 has the solution (using Ito's formula):

$$\begin{aligned} S_t &= S_0 \exp\left[\left(\mu + \frac{1}{2}\sigma^2\right)t + \sigma W_t^{\mathbb{P}}\right] \\ &= S_0 \exp\left[\left(\mu - \lambda\sigma + \frac{1}{2}\sigma^2\right)t + \sigma W_t^{\mathbb{Q}}\right] \qquad (5.6) \\ &= S_0 \exp\left[\left(r + \frac{1}{2}\sigma^2\right)t + \sigma W_t^{\mathbb{Q}}\right] \qquad (5.7) \end{aligned}$$

where λ is necessarily $(\mu - r)/\sigma$. This highlights the duality that exists between the $\mathbb{P}$ and $\mathbb{Q}$ measures and illustrate the two paradigms under which an ESG must be able to simulate, here in a simple setting. One *observes* the stock-price process Equation 5.6 and is able, at least in principle, to find an estimate for μ. For pricing purposes, however, one must use the growth rate r from the numeraire asset and simulate using Equation 5.7. If the price today of a 3-month equity put option is sought, then the numeraire asset is the 3-month nominal interest rate government bond. I say it is possible to estimate μ "in principle" because this can be a challenging exercise in robust statistical estimation. One must appeal to data to find a suitable historical period and then use the method of moments, maximum likelihood or use a Bayesian approach to estimate μ. The question of precisely which historical period becomes primordial: too short a period and sampling error will dominate an estimate, too long and the lack of stationarity may bias an estimate. A more robust approach might be to make an *economic assumption* about the long-term trend in the growth rates that can take a holistic approach in view of the entire market. One may then observe r and compute a risk premium $\mu^* = \mu - r = \lambda\sigma$. A calibration approach in the real world is therefore to produce a consistent set of long-term calibration *targets* in terms of assets' risk premia. For more detailed assets, such as real or nominal interest rates, or credit, the calculation of risk-premia becomes more challenging as a term structure of targets emerges: one is liable to make functional assumptions about risk-premia. When making assumptions on risk premia, one must assess their sensitivity to economic assumptions. This is a major challenge in economic scenario generation made more difficult owing to the amount of analyst intervention that is needed to estimate the risk-premia. Unlike a market-consistent calibration where clearly available market data are available, the situation in real-world modeling is much more fluid and open to interpretation.

The requirement of assets' growth to be fixed at the numeraire rate in the market-consistent setting affords us a particularly useful yet simple validation technique that is essential for any well-functioning ESG. Specifically, one may use a simple charting functionality to demonstrate how well a market-consistent ESG is functioning. As another example, consider a 10-into-20-year receiver swaption struck at par. This is a derivative contract giving the holder the right but not the obligation to enter into a 10-into-20-year swap contract in 10 years' time. If market swap rates fall below the current par-yield 10 years' hence then the holder may exercise his or her right to enter into the more valuable swap contract at its option maturity date. Otherwise, the swaption matures worthless as swap rates are more valuable in the market at maturity. The *reference* asset for this swaption is a 20-year fixed term deferred annuity whose first payment is in 10 years' time. Arbitrage theory coupled with the Girsanov theorem tells us that unless the growth rates of the swaption and the annuity are the same then a risk-free profit may be locked in by constructing a portfolio which is short one of the assets and long the other in a way that is known in advance. One way to *validate* the output from a risk-neutral ESG is to check that such derivative securities are *martingales*. Recall that for a process M_t to be a martingale with respect to a numeraire B_t:

$$\mathbb{E}_{\mathbb{Q}}\left[\frac{M_t}{B_t}\middle|\mathcal{F}_s\right] = \frac{M_s}{B_s} \quad 0 \leq s \leq t; \tag{5.8}$$

in the filtered probability space $(\{\Omega, \mathcal{F}, \{\mathcal{F}_t\}_{t\geq 0}, \mathbb{Q})$. So, one may reasonably compute the value of the 10-into-20-year swaption M_t over simulated time and plot the ratio of it with the deferred annuity B_t. If the resulting time-series plot shows an absence of a trend (i.e., an absence of growth) and wanders randomly around a value of 1 then one may have confidence that the pricing measure is sound. This may be repeated for the term structure of swaptions struck at par (or other) and overlay error bars for a compact validation plot.

Several challenges face this approach to validation. First, a typical ESG works in discrete-time over time steps, which are perhaps monthly or annual. It does not work in the continuous-time framework required by the financial theory and so one may fail to validate M_t as a martingale when it is indeed a martingale in discrete time. Indeed, one needs to make a choice regarding the equivalent measure to be used as a reference: often the discretely compounded rolled-up cash account is used. A Girsanov transform is then employed whenever this is not the appropriate numeraire. However, an *additional* drift correction will be needed to correct for the discrete-time approximation to continuous-time that is being made. Fortunately, this does not present any burden, simply something to be mindful of. Another practical issue for ESG users is the precise number of ESG sample paths needed: too many and an ALM system becomes prohibitively slow, too few and sampling error will dominate to the extent that illusory trends in the drift may be seen. A smaller number of paths is obviously preferred, but how small? Given a regulator may require an insurer to offer some evidence that a pricing measure is indeed arbitrage-free at a given significance level $\alpha = 1\%$ (say) there is no way around the hurdle of having to use a sufficient sample size to demonstrate the martingale property. One possible work-around may be to generate more scenarios than can be handled by the ALM system in a reasonable time, verify its martingale property and then optimally select the best scenario subset of desired size that minimally disturbs the distribution while constraining the drift to be the drift of the full set. At the very least, this disturbance would only be a drift correction from the noisy (incorrect) drift of the small sample to the more accurate drift of the full sample.

Although scenario generation under the pricing measure $\mathbb{Q}$ relies on concepts from theoretical financial engineering, insurance users will wish to extend the interest rate models over periods of time that naturally cover their liabilities (and assets). This becomes problematic when market data are sparse or illiquid. To cover their policyholder liabilities which tend to be long-dated, insurers naturally hold long-dated bonds and contingent claims on these. However, no market exists for these in any great depth: it is largely illiquid and insurance companies are, in effect, *making* the market. Although illiquidity is not a problem for insurers per se[20] the difficulty that emerges is more a regulatory one. An insurer is required to find an appropriate fair value for their assets, illiquid or otherwise. In the absence, therefore, of a viable risk-neutral measure economic theory can be useful. It can be used to establish *unconditional forward rates* but these are assumptive in nature and should be justified (e.g., by a sensitivity analysis). Salahnejhad and Pelsser (2015) give a theoretical basis for pricing in illiquid markets. In any event, an ESG makes certain assumptions and insurance companies must be aware of them.

[20]Note that it does expose them to policyholder lapse and surrender risk should they need access to cash quickly in times of market distress.

In the sequel, I will continue my discussion of SDE models by introducing a taxonomy of asset classes beginning with what is potentially the most fundamental construct in an ESG—the yield curve. Several techniques exist for extracting the current yield curve from market data. Under Solvency 2, the method of choice is due to Smith and Wilson (see EIOPA-BoS-15/035 2015). I offer an alternative due to Antonio and Roseburgh (2010): a cubic spline method where there is market data available and an extrapolation beyond the last market data using the Nelson and Siegel (1987) functional form as is implemented by Moody's Analytics. The cubic spline is *latent* meaning that the market data itself is not interpolated but rather for an *a priori* fixed set of knot points, the space of cubic splines is minimized with respect to a given criteria satisfying given smoothness conditions. This is often a characteristic of ESG modeling: the challenge is to extract a quantity of interest that isn't directly observable: the instantaneous forward rates representation of the yield curve is an example.

Following interest rate modeling, I will briefly discuss credit and then equities before concluding the section with a note on high-performance computing in insurance and the important considerations of calibrating models to market data. Although of obvious importance to ESG, I will delay discussion of co-dependency structures until Section 5.3 on risk scenario generators (RSGs): co-dependency being of equivalent importance to both economic and market risks. In the interests of space, I therefore omit discussion of the important asset classes of foreign exchange and inflation, and focus on market consistent modeling. I leave the interested reader to follow up these topics with the reference: McNeil et al. (2015), Brigo and Mercurio (2006); and those contained therein.

5.2.2 The yield curve

The yield curve is the most fundamental ingredient to an ESG. It underpins most of the other asset classes. For instance, an equity option price is the expectation under the risk-neutral measure of the discounted option payoff. Discount factors naturally come from the nominal yield curve of a given economy where the option is traded and are expressed using the yield curve spot rate. Calibration of a yield curve model is the first step in the introduction of a full interest rate model that I will discuss in the next section. For the moment, I will describe a method that can be used for modeling the yield curve on a given date using bond and swap data that is due to Antonio and Roseburgh (2010). It is relatively robust, allows a mixture of bonds and swaps to be used where they are liquid and matches to a specialized functional form beyond the last available maturity. Although my discussion is aimed at capturing *nominal* yield curves, there is no reason why this method cannot be applied to model the *real* yield curve. One would simply replace the government treasury bonds and swap data with inflation proofed index-linked gilts and savings certificates (in the United Kingdom).

This is an alternative approach to that of Smith and Wilson (EIOPA-BoS-15/035, EIOPA 2015) which is the method of choice under Solvency 2. Like Smith and Wilson, calibration may be to either bonds or swaps or both but the yield curve data is not interpolated but, rather, is represented as a cubic spline with a Nelson–Siegel functional extrapolation. The instantaneous forward rate curve $f(t)$ is modeled, being considered latent, that is, it is not directly observed, rather the parameters of the model are extracted by minimizing a functional. Assume the relationship between the zero-coupon bond prices and $f(t)$ is, for bonds with coupons $c \geq 0$ and frequency ω per annum:

$$P(T;c) = \sum_{i=1}^{n} (\delta_{in} + c)\exp\left(-\int_0^{t_i} f(t)\mathrm{d}t\right), \quad n = \omega T. \tag{5.9}$$

The forward rates are represented by a latent cubic spline modeling the forward rates curve at the $K + 1$ *a priori* specified knot points t_k for $k = 0, 1, \ldots, K$ ($t_K = T_M$, the longest bond maturity) with a Nelson and Siegel (1987) extrapolation in $t_K < t \leq t_{\max}$ (typically $t_{\max} = 120$ years):

$$f_{\mathrm{ns}}(t|\boldsymbol{\beta}) = \beta_1 + [\beta_2 + \beta_3(t - t_{\max})]\mathrm{e}^{-\lambda(t-t_{\max})}, \quad \boldsymbol{\beta} = (\beta_1, \beta_2, \beta_3)' \in \mathbb{R}^3; \tag{5.10}$$

as

$$f(t|\boldsymbol{\theta}_{1:K}) = \sum_{k=1}^{K} \mathbb{I}(t_{k-1} \leq t < t_k) s(t|\boldsymbol{\theta}_k) + \mathbb{I}(t_K \leq t \leq t_{\max}) f_{\mathrm{ns}}(t|\boldsymbol{\beta}); \tag{5.11}$$

where each of the K cubic splines are parametrized by a parameter vector $\boldsymbol{\theta} = (a, b, c, d)' \in \mathbb{R}^4$ with $s(t|\boldsymbol{\theta}) = a + bt + ct^2 + dt^3$.

The Nelson and Siegel rate parameter λ is set to an economically plausible $O(0.01)$ value and Antonio and Roseburgh (2010) set $\beta_1 = f_{t\max} - \beta_2$ to fix the unconditional forward rate[21] at $f_{t\max}$. To ensure a smooth transition to the extrapolated forward curve from the last available market data point at T_M, a first derivative "gradient" smoothing penalty of magnitude w_1 is applied over the final 20% of the available market data: $[T_2, T_M]$. To guard against overfit, a second derivative curvature penalty of magnitude w_2 is applied from the first 20% of the available market data till T_M: $[T_1, T_M]$.

The bond price may then be written succinctly as

$$P(T|\boldsymbol{\Theta}) = \exp(-\boldsymbol{\tau}_T' \boldsymbol{\Theta}) \tag{5.12}$$

where $\boldsymbol{\tau}_T \in \mathbb{R}^{4K+3}$ contains the parameter-free knot point detail from the cubic spline and $\boldsymbol{\Theta} \doteq \mathrm{vec}(\boldsymbol{\theta}_{1:K}, \boldsymbol{\beta})' \in \mathbb{R}^{4K+3}$ contains the parameters from each spline component and the extrapolant. If the modeled forward rates given in Equation 5.11 evaluated at the knot points $t_{0:K}$ are: $\mathbf{f} = (f_0, f_1, \ldots, f_K)' \in \mathbb{R}^{K+1}$; then the relationship between the spline's parameter vector $\boldsymbol{\Theta}$ and the

[21]The forward rate at the longest modeled maturity t_{max}.

discrete forward rates vector $\mathbf{f}$ is found through the matrix–vector relationship arising from the cubic spline:

$$A\mathbf{\Theta} = B\mathbf{f} + \mathbf{u}, \quad A \in \mathbb{R}_{4K+3,4K+3}, \quad B \in \mathbb{R}_{4K+3,K+1}, \quad \mathbf{u} \in \mathbb{R}^{4K+3}. \quad (5.13)$$

Here A is a known invertible matrix and B is a known rectangular matrix. The vector $\mathbf{u} := f_{t_{\max}} \hat{\mathbf{e}}_{4K+3}$ where $\hat{\mathbf{e}}_k$ is the unit basis vector along the kth axis, $k \in [1, \ldots, 4K+3]$. Most importantly, the cubic spline matrix–vector relationship Equation 5.13 shows that the parameter vector $\mathbf{\Theta}$ can be eliminated in favor of the forward rates vector $\mathbf{f}$. If the set of bond price/duration data is denoted by $\mathcal{P} = \{(P_j, D_j) \in \mathbb{R}^2 : j = 1, \ldots, M\}$ and modeled bond prices by $\hat{P}_j(\mathbf{f})$ using Equation 5.12, then the objective criteria are

$$\begin{aligned} g(\mathbf{f}|\mathcal{P}) = \sum_{j=1}^{M} \left(\frac{P_j - \hat{P}_j(\mathbf{f})}{D_j} \right)^2 + w_1 \int_{T_1}^{T_M} \left(\frac{\partial f}{\partial t}[t|\mathbf{\Theta}(\mathbf{f})] \right)^2 dt \\ + w_2 \int_{T_2}^{T_M} \left(\frac{\partial^2 f}{\partial t^2}[t|\mathbf{\Theta}(\mathbf{f})] \right)^2 dt \end{aligned} \quad (5.14)$$

As a direct consequence of (5.13) g may be directly minimized over $\mathbf{f}$ rather than over the spline-coefficient vector $\mathbf{\Theta}$. Applying a nonlinear optimization technique such as Levenberg–Marquardt (Transtrum and Sethna 2012; Press et al. 2007) leads to the solution $\hat{\mathbf{f}}$.

In the case that swap rates are also required to aid in the calibration of the yield curve, note that the T-year forward swap rate at time $t = 0$ is

$$S(T) = \frac{1 - \prod_{j=1}^{n} [1 + f(T_j)/\omega]^{-1}}{\sum_{i=1}^{n} \omega^{-1} \prod_{j=1}^{i} [1 + f(T_j)/\omega]^{-1}} \quad (5.15)$$

and it is noted that the swap rate is a function of the forward rates at the required knot points: $S[T_j|\mathbf{\Theta}(\mathbf{f})]$. The objective Equation 5.14 may be updated by adding a term:

$$w_0 \sum_{j=1}^{M} \left(S_j - \hat{S}[T_j|\mathbf{\Theta}(\mathbf{f})] \right)^2 \quad (5.16)$$

Some adjustment of the relative proportions of the objective function hyper-parameters is needed: w_0, w_1, and w_2; and this can be handled by 10-fold cross-validation (Hastie et al. 2009) or other out-of-sample validation technique.

5.2.3 Nominal interest-rate models

Dynamic models of the interest rate typically come in one of two flavors: short rate and forward rate models. Models of the short rate r_t are invariably the simplest attempting only to model the interest rate manifesting itself over an infinitesimal period of time. This is not a market observable quantity.

However, it can be integrated to produce zero coupon bond prices:

$$P(t,T) = \mathbb{E}_{\mathbb{Q}}\left[\exp\left(-\int_t^T r_s \mathrm{d}s\right)\middle|\mathcal{F}_t\right]$$

where $P(t, T)$ is the maturity-T zero coupon bond viewed at time $t \geq 0$, $\mathbb{Q}$ is the risk-neutral measure and $\mathcal{F}_t$ records the market history up until time t. From these, market observable coupon bearing bond prices may be derived. Derivative quantities are valued in a similar way. Once one has specified dynamics for the short rate, the pricing problem comes down to one of integration. For a certain class of interest rate models there can be a significant simplification in the integration, specifically for the *affine term structure* models. Here the (exponentially) affine form for the zero-coupon bond is sought:

$$P(t,T) = \exp[A(t,T) - B(t,T)r_t]$$

The challenge is then to find functions A and B consistent with the dynamics for r_t:

$$\mathrm{d}r_t = \mu(t, r_t)\mathrm{d}t + \sigma(t, r_t)\mathrm{d}W_t$$

If one seeks affine transforms of the short rate for the drift and (squared) volatility, then a coupled system of ordinary differential equations emerges (maturity T is considered a parameter):

$$\begin{aligned}
\frac{\mathrm{d}A}{\mathrm{d}t} - \beta(t)B + \frac{1}{2}\delta(t)B^2 &= 0, \quad A(T,T) = 0 \\
\frac{\mathrm{d}B}{\mathrm{d}t} + \alpha(t)B - \frac{1}{2}\gamma(t)B^2 &= -1, \quad B(T,T) = 0
\end{aligned}$$

For the choice of $\mu = b - ar$ and σ constant then $\alpha = -a, \beta = b, \gamma = 0$ and $\delta = \sigma^2$ then the Vasicek dynamic model emerges:

$$\mathrm{d}r_t = (b - ar_t)\mathrm{d}t + \sigma \mathrm{d}W_t$$

with A and B functions as

$$\begin{aligned}
A(t,T) &= a^{-2}[B(t,T) - T + t]\left(ab - \frac{1}{2}\sigma^2\right) - \frac{1}{4}a^{-1}\sigma^2[B(t,T)]^2 \\
B(t,T) &= a^{-1}[1 - \exp(-a(T-t))]
\end{aligned}$$

The conditional distribution for the short rate given any time horizon is normal and the benefit of such simple dynamics is in its analytically tractability: a model such as the Vasicek model is easily implemented in an ESG. However, a detraction of this model is precisely that its conditional distribution is normal: this implies negative short rates and, until recently, nominal rates have not been negative. The Vasicek model still has its place as a real interest rate model where rates, measuring the purchasing power of a basket of goods, can realistically be negative. Still, a single-factor model may not produce enough dispersion and for this one may reasonably take a two-factor Vasicek model and this is done by introducing a second SDE to model its long-term mean. Typically, one may work with

$$\begin{aligned} \mathrm{d}r_t &= \alpha_1(m_t - r_t)\mathrm{d}t + \Sigma_1 \mathrm{d}W_t^r \\ \mathrm{d}m_t &= \alpha_2(\mu - m_t)\mathrm{d}t + \Sigma_2 \mathrm{d}W_t^m \\ \langle \mathrm{d}W_t^r, \mathrm{d}W_t^m \rangle &= 0 \end{aligned}$$

While this leads to a more flexible affine term-structure model, it does not yet allow for the *initial* yield curve to be modeled. Rather, at time zero some *Vasicek implied* yield curve emerges over which there is no control. To obviate this challenge, one moves to the more flexible framework of the Hull and White model (see Hull and White 1990; Hull 2005). The two-factor Hull and White model is

$$\begin{aligned} r_t &= \phi(t) + x_t^1 + x_t^2 \\ \mathrm{d}x_t^i &= -\alpha_i^h x_t^i \mathrm{d}t + \sigma_i \mathrm{d}W_t^{(i)}, \quad i = 1, 2, \\ \langle \mathrm{d}W_t^1, \mathrm{d}W_t^2 \rangle &= \rho\, \mathrm{d}t \end{aligned}$$

where, crucially, introduction of the function $\phi(t)$ allows the initial yield curve to be matched exactly. If it is chosen to be the *Vasicek implied* initial yield curve, then $\phi(t) \equiv \mu$ and the Vasicek and Hull and White models may then be related by the simple change of state variables:

$$\begin{aligned} r_t &= \mu + x_t^1 + x_t^2 \\ m_t &= \mu - \frac{\alpha_2 - \alpha_1}{\alpha_1} x_t^2 \end{aligned}$$

which *precludes* $\alpha_2 = \alpha_1$ (but this is easily remedied by considering the special case in isolation). The parameters are related as follows:

$$\begin{aligned} \alpha_{1,2}^h &= \alpha_{1,2}, \quad \sigma_1 = \sqrt{\Sigma_1^2 \left(\frac{\alpha_1}{\alpha_2 - \alpha_1} \right)^2 \Sigma_2^2}, \\ \sigma_2 &= \left| \frac{\alpha_1}{\alpha_2 - \alpha_1} \right| \Sigma_2, \quad \rho = -\frac{\Sigma_2}{\sigma_1} \left| \frac{\alpha_1}{\alpha_2 - \alpha_1} \right| \end{aligned}$$

A consequence, therefore, is that under the Hull and White representation, the stochastic factors now have a non-zero correlation coefficient. In fact, it is manifestly negative. The main detraction with the Vasicek and Hull and White models is that they are fundamentally normal implying that the short rate may become negative. Until recently, it was not anticipated that nominal interest rates could become negative. One final affine model is available: the multi-factor Cox–Ingersoll–Ross (CIR) model. Here, the interest rates are guaranteed to be positive since the solution of the CIR SDE is a scaled non-central χ^2-variable. I defer discussion of the CIR process until the next section on credit models. The final short rate model I will describe is the two-factor Black–Karasinski model. Its dynamics are

$$\begin{aligned} \mathrm{d}\ln r_t &= \alpha_1[\ln m_t - \ln r_t]\mathrm{d}t + \sigma_1 \mathrm{d}W_1 \\ \mathrm{d}\ln m_t &= \alpha_2[\mu' - \ln m_t]\mathrm{d}t + \sigma_2 \mathrm{d}W_2 \end{aligned}$$

The conditional and unconditional distribution of r_t is log-normal the benefit of which is that it can never be negative. This desirable feature is rather offset by the analytical intractability of this model. Not being of the affine class, its calibration is hampered by the lack of any analytically tractable formulae for its bond prices nor swaptions. Indeed, a two-dimensional recombinant binomial tree can be used for both calibration and simulation. The challenge here is to store the tree sensibly so that its time step is sufficiently small to obviate discretization error of the SDEs.

Heath, Jarrow, and Morton (1990) described a framework for the evolution of the continuous forward rates term structure. If $f(t, T)$ is the continuous compounding T-year forward rate at time $t > 0$ then under the HJM framework its evolution is simply described by

$$\mathrm{d}f(t,T) = \mu(t,T)\mathrm{d}t + \Sigma(t,T)'\mathrm{d}W_t$$

Here, Σ is a d-dimensional vector of factor loadings for the d-dimensional vector of Brownian increments $\mathrm{d}W_t$. In a market-consistent risk-neutral setting, the drift term is completely specified in order that $f(t, T)$ is driftless with respect to its numeraire. Note that it is typically non-zero as W_t is measured with respect to a common numeraire for simulation purposes: the cash account (say). Non-zero drift terms appear in the risk-neutral setting owing to an application of the Girsanov transform.

Practical use of the HJM framework is hampered by the fact that the state variables: $f(t, s)$; are not market observable quantities. Brace, Gatarek, and Musiela (1997) mitigated this problem by introducing a model that would become something of an industry standard: the *Libor Market Model* (also referred to as the BGM model). This models the forward rates that are traded in the market. In London, these are the Libor rates but the model applies widely to any market (Euribors, for example). If K distinct Libor forward rates each with maturity T_k $(k = 1, \ldots, K)F_k(t,T)$ are being modeled, the BGM model is a log-normal model in $F_{k,t}$:

$$\frac{\mathrm{d}F_{k,t}}{F_{k,t}} = \sigma'\mathrm{d}W_t$$

for vectors of length p of local volatilities σ and Brownian shocks W_t (each specified under their own numeraire). Another departure from the HJM framework is that the BGM model uses discrete compounding rather than continuous compounding. The relationship between the kth rate and the zero-coupon bond prices is

$$F_{k,t} = \frac{1}{T_{k+1} - T_k}\left(\frac{P(t, T_k)}{P(t, T_{k+1})} - 1\right)$$

A first criticism of the basic BGM model is in its insensitivity to *skew* in the market implied volatility data: for example, European styled payer or receiver swaptions show variations in their implied volatility relative to those struck at par. Market implied volatility data depend on three parameters: the maturity of the swaption, the tenor of the underlying swap contract *and* the strike relative to the par yield curve at contract inception. An insurer whose liabilities are exposed to a fall in interest rates may purchase a receiver swaptions portfolio to hedge this risk away. The insurer may like to strike the swaptions in the hedge at the guaranteed rate of interest it has promised to its policyholders. This is unlikely to be at the current par yield and so a correct price is sought *away-from-the-money*. The BGM model is insensitive to this strike; put another way, the BGM implied volatility (hyper-) surface is constant across strike. This problem may be alleviated by the introduction of a *stochastic volatility* process V_t modeled by, perhaps, a CIR process. The downside here is that the analytical formulae for the European swaption is lost and one must appeal to semi-analytical formulae such as that found by Wu and Zhang (2006).

A second criticism of the BGM model is in its fundamental assumption of log-normality. This precludes rates ever becoming negative but as illustrated in the introduction nominal rates *can* become negative. Market-implied volatilities in the log-normal BGM model require the initial yield curve to be everywhere positive, however. Moreover, whenever the yield-curve approaches zero it becomes computationally onerous to compute log-normal implied volatilities from swaption price data: they become unstable. When the yield curve is negative it is no longer possible to compute swaption prices using the standard Black's formula. Log-normal implied volatilities cease to exist and the consequences for parameter calibration are obvious. A solution is to use displaced forward rates: that is, a constant term is added to the forward rates. The problem then is how to reliably set this displacement term? Furthermore, the BGM model is quite likely to achieve exponentially large forward rates in finite time. While this is not an issue for market-consistent pricing per se, it does mean that scenario sets created by an ESG will have values in it that are too large to be represented by computational precision: there will be Nan's. This is unacceptable for an insurer's ALM system. A solution is to switch from the log-normal BGM model to a normal equivalent model with a stochastic CIR process V_t:

$$\mathrm{d}F_{k,t} = \sqrt{V_t}\sigma'\mathrm{d}W_t$$

One may calibrate this model using absolute implied volatility data that can be obtained from market data providers such as Markit and SuperDerivatives. Exponential *blow-up* is largely mitigated and the rate may become negative, but nowadays, this is something that is very real.

Other dynamic models of interest rates exist such as the SABR-LMM model. I refer the interested reader to the text Brigo and Mercurio (2006) for more information.

5.2.4 Credit models

These are split into two categories: the structural and the reduced form models. Structural models attempt to model the credit risky nature at the enterprise level by considering their equity as a call option on their assets. This type of model directly estimates their risk of downgrade or default. The Moody's KMV model Berndt et al. (2004) is a commercial example of this type of model. Structural models, such as the Jarrow–Landau–Turnbull model, use market credit spreads of the broad ratings classes: AAA, AA, ..., CCC, et cetera; to measure the risk-neutral default probabilities (with assumptions about recovery rates) and relate these through coupled systems of SDEs to model the spreads themselves. An ESG model either models the credit spread through default-only models or through downgrade models where a Markov-chain styled transition matrix of probabilities is specified. When any drivers are used to describe the credit spreads, the greatest challenge in credit modeling is in the reliable calibration and estimation of models, which can become quickly over-parametrized. The output from an ESG is in terms of spreads split by credit class and maturity. The complexity of the calculation is similar to that of the forward interest rates models. The interested reader is referred to Lando (2004).

5.2.5 Equity models

An important asset class synonymous with financial markets is the equity market. Models of equity assets reach as far back as Bachelier (1901) who described the first SDE of an equity. Arithmetic Brownian motion is described thus

$$\mathrm{d}S_t = \mu \mathrm{d}t + \sigma \mathrm{d}W_t, \quad S_0 \text{ given} \tag{5.17}$$

Contrast this for a moment with the SDE for geometric Brownian motion, simplified from Equation 5.3:

$$\frac{\mathrm{d}S_t}{S_t} = \mu \mathrm{d}t + \sigma \mathrm{d}W_t, \quad S_0 \text{ given} \tag{5.18}$$

The single most-important observation is that arithmetic Brownian motion is unconditionally normally distributed and S_t can be either positive or negative. This is unrealistic for an equity process and so Bachelier's model put an end to the development of financial engineering for the better part of half a century. The development and success of Black and Scholes (1973) equity option pricing theory heralded a new era in financial engineering leading to such developments as the Libor market model (as discussed earlier). The field has even turned full circle and embraced once again Bachelier's arithmetic model for the purposes of solving the issue of negative nominal interest rates. For its part, the geometric model Equation 5.18 is lognormally distributed and non-negative. Non-negativity is certainly sensible for an index or stock price, but is log-normality always sensible? In the long term, one might envisage a different unconditional distribution with more or less skew and kurtosis.

For market-consistent pricing of equity derivatives, the lognormal equity asset model is a practical solution. Its main detraction, however, is that its price predictions are at odds with the markets. Striking in- and out-of-the money produces different prices not predicted by the lognormal model. The volatility surface (Gatheral 2006) is sensitive to strike as well as option maturity. Alternatives to the lognormal model are Merton's jump diffusion model (Merton 1976):

$$\begin{aligned} \frac{\mathrm{d}S_t}{S_t} &= -\lambda\bar{\mu}\mathrm{d}t + (\eta_t - 1)\mathrm{d}N_t \\ \eta_t &\sim \log - N(\bar{\mu}, \sigma^2) \end{aligned} \tag{5.19}$$

and $\mathrm{d}N_t \sim \lim_{\delta t \to 0^+} \mathrm{Po}(\lambda \delta t)$; and Heston's stochastic volatility model (Heston 1993):

$$\begin{aligned} \frac{\mathrm{d}S_t}{S_t} &= \mu\mathrm{d}t + \sqrt{v_t}\mathrm{d}W_t^{(1)} \\ \mathrm{d}v_t &= \alpha(\theta - v_t)\mathrm{d}t + \xi\sqrt{v_t}\mathrm{d}W_t^{(2)} \end{aligned} \tag{5.20}$$

Jump-diffusions perform well when pricing derivatives of short maturities and can produce volatility skew there but at longer maturities they do not perform so well. Stochastic volatility models longer maturities well but suffers on short maturities. The combined stochastic volatility jump diffusion Bates (1996) works well:

$$\frac{\mathrm{d}S_t}{S_t} = (\mu - \lambda\bar{\mu})\,\mathrm{d}t + \sqrt{v_t}\,\mathrm{d}W_t^{(1)} + (\eta_t - 1)\,\mathrm{d}N_t \tag{5.21}$$

$$\begin{aligned} \mathrm{d}v_t &= \alpha(\theta - v_t)\mathrm{d}t + \xi\sqrt{v_t}\mathrm{d}W_t^{(2)} \\ \eta_t &\sim \log - N(\bar{\mu}, \sigma^2) \end{aligned}$$

The drift terms above are each carefully set to ensure the zero drift of any derived quantities of interest under the appropriate numeraire measure. The remaining difficulty with the Bates model is in a relative inability to correlate Bates processes with other processes within an ESG. Although one can correlate the Brownian shock by means of a Cholesky factorization of the correlation matrix, it is less than evident how to correlate the combined equity asset process which involves, in addition to the Brownian shock, the effect of stochastic volatility and a compound Poisson process. I address how to do this in Section 5.4.

Other popular models that involve stochastic volatility are those of Duffie et al. (2000), Levy processes (e.g., Carr et al. 2002) and the SABR mixed local volatility and stochastic volatility models by Hagan et al. (2002).

5.2.6 Calibration issues

It is now an opportune moment to pause and reflect on the sheer complexity of the calibration problem having introduced two very detailed models for interest rates and equities. In the market-consistent nominal interest rate setting, one popular modeled is to calibrate to market swaption prices (or to their implied volatilities). One produces modeled prices for each market available instrument and builds an objective criterion such as the mean-squared error. This metric may be minimized using a numerical optimization algorithm such as the Levenberg–Marquardt algorithm (e.g., Transtrum and Sethna 2012; Press et al. 2007) or other. Since the resulting optimization problem is nonconvex without a directly accessible functional form for the Jacobian nor Hessian, optimization is challenging and potentially onerous.

If pricing derivatives given a set of parameters is the forward problem, then optimizing a metric given market data to find the *best* parameter set is the *inverse* problem. In such nonlinear optimizations one is not guaranteed to find the best solution and usually ends up settling for something which is satisfactory in some sense (if indeed one can find a solution at all). Harder still, finding the unique global optimum is *the* fundamental challenge. Indeed, there may be many local optima and the iterative scheme created by the Levenberg–Marquardt algorithm is not guaranteed to converge to the global minimum. Global optimization schemes do exist, notably Nelder–Meade or differential evolution (Press et al. 2007). However, they are all computationally much more onerous than local optimization schemes. Ultimately, any iterative scheme reduces to a multi-dimensional non-autonomous discrete dynamical system. Such systems, even when the dynamics are known, are subject to multiple critical points, limit cycles and, most interesting of all, *strange attractors* (Hale and Kocak 1991). It is worthy of note that strange attractors may be found in the most innocuous of difference equation systems. Their key defining characteristic is that any solution *tends* to the trajectory defined by the strange attractor yet this trajectory never repeats itself nor intersects its past trajectory. In a very real sense, it is space-filling in some intermediate sub-space that lies *between* integer dimensions and every point along this path is optimal. In continuous time, strange attractors are precluded from coupled systems of two differential equations by the Poincare–Bendixon theorem but no result holds for coupled systems of two or more *difference* equations. Seeking an optimal calibration is potentially equivalent to tracing the trajectory laid by a strange attractor as the critical phenomena of a system of nonlinear difference equations. Care must be taken to ensure that of the critical phenomena listed, a critical point has indeed been obtained: any of the phenomena can show good validation to vanilla option prices yet still leave the analyst unaware of the precise nature of the phenomena realized. By a naive stopping rule, one may have exited the dynamical difference system at some point on a limit cycle, or at some large but finite number of steps along the trajectory of a strange attractor thinking that one has found the unique critical point.

Luckily there are some remedies to this problem. First, start the optimization scheme repeatedly from as many different points in the parameter space and record the ultimate destinations reached. Examine the index plots of the optimal metric and if a large proportion of the solutions with the smallest indices have roughly the same metric, one can be assured that a near global solution has been reached. Now examine the ultimate parameter values using the same ranking: do the most optimal calibrations correspond to similar values of the parameters? If so, this is the best outcome. On the other hand, if the index plot of the parameters shows a high degree of variability, one may deduce that the market data does not drive a unique solution. This may be due to the manifestation of a limit cycle or to a strange attractor. Calculation of the Lyapunov exponent[22] for the system may serve to characterize the convergent phenomena. If one does discover either of these phenomena, the current market snapshot does not define a unique model. A solution is to complement the snapshot with more information. If one naively inserts restrictive parameter limits, one runs the risk of finding solutions along the boundary. This is problematic as a calibration found on the boundary is equivalent to reducing the parameter count by one and grossly simplifies the calibration problem in an arbitrary way that is difficult to control. This introduces a redundancy to the difference equation system and one ends up solving an alternative optimization program that was not intended. One needs to add more information into the system *sensibly*. One way to achieve this is to weigh the current solution set with the empirical density function generated from the most recent well-behaved solution set. This makes it possible to select a solution from the current data in such a way that is *close* to the previous data (Hastie et al. 2009). At the very least, when one comes to value one's liabilities, the change in parameters will not be unusually profound and the value of the liabilities today will be consistent with the last time a price was sought.

Having a robust calibration technique that does not change significantly across consecutive time periods except by an amount that is, in some sense "reasonable," is certainly desirable. However, a model calibrated to one type of market data need not necessarily lead to a similar calibration of the same model under another. For example, receiver swaption interest rate data is better defined where it is in-the-money for low interest rates. Payer swaption prices tend to be smaller where receiver swaption prices are larger and consequently, owing to a disparity in scales, there is less information available for calibration under payers in this instance. One tail of the underlying density may be less well defined when comparing across calibrations and hence lead to different parametrizations. One may calibrate to market-implied volatilities to obviate this risk, but while receiver and payer swaption generated implied volatilities are theoretically equal, they aren't in practice and one finds oneself in the same quandary as with prices. If one replaces swaptions

[22]The Lyapunov exponent measures the extent to which two solutions to a dynamical system will tend to diverge after some period (Hale and Kocak 1991) *given* they both have the same initial conditions.

with calibrations to interest rate derivatives such as caps and/or floors, then the calibration may change yet again despite the theory implying that the same risk-neutral density should pervade. One must be pragmatic, therefore, and test the sensitivity of any calibration to data that is *out-of-sample*. This is a rather serious issue as I note that the sensitivity to MCEV and TVOG, and ultimately to the SCR under Solvency 2, depends upon it.

Finally, it is noted that the calibration of an equity SVJD model in eight parameters necessarily requires the interest rate model to be calibrated first. The equity model is coupled to the interest rate model through the stochastic interest rate discount factor. Some approximations can usually be made but this serves to illustrate the depth and difficulty of the general calibration problem to market data.

5.2.7 High-performance computing in ESGs

The main use of high-performance computing in an ESG is to distribute many Monte-Carlo trials across multiple servers so that they may be executed in parallel. The data are then collated by the client to create the complete dataset. This principal of tasks decomposition and distribution can be applied to many related problems such as RSG, or to scenario generation for least-squares Monte-Carlo proxy model fitting. It is, however, unclear how one should apply high-performance computing to best aid in model parameter calibration. Essentially, anything that may be linearly decomposed into a repetitive sequence of like tasks may make use of high-performance computing. Parameter calibrations are almost certainly iterative and nonlinear in nature. Since reinitializing a calibration to a new point helps obviate the local nature of some optimizers, then this is one use of HPC. A second use of HPC is in the calibration of models to exotic derivatives (e.g., callable swaps needed to hedge portfolios of fixed rate mortgages) where no analytics are available or in the calibration of a more complicated interest rate model, again for whom there are no available option pricing formulae. Here, option payoffs may be simulated using the Monte-Carlo technique. While this is computationally prohibitive for one quad-core desktop PC, a multi-core HPC cluster may speed up the calculation sufficiently to make it a viable calibration alternative. One must overcome the HPC communication bottleneck of collating results by averaging trials from different cores. Some empirical tuning of the right balance of trials-to-cores is needed.

The use of Microsoft's Azure servers in the cloud can speed up scenario set generation and locally, Digipede Grids[23] offer a robust local solution. These provide an array of servers which can process individual fragments of an ESG simulation. For the least-squares Monte-Carlo Solvency 2 SCR calculation, the use of bespoke in-house HPC grids to parallelize ESG execution and regression calculations is necessary, however. The same Digipede Grids are not useful

[23] http://digipede.net

because the market-consistent inner component of the least-squares Monte-Carlo technique has a very low trial budget. A more efficient approach is to run whole simulations in parallel rather than breaking jobs up trial wise. Other third-party HPC platforms could be used to achieve similar results.

As mentioned, the *cloud* is most synonymous with HPC and there are two way of leveraging such platforms. Products may work on an *Infrastructure as a Service* (IaaS) paradigm. Here, more servers are added to increase core and memory capacity on any environment. The move toward *Platform as a Service* (PaaS) is new. The main difference between Iaas and PaaS is the removal of the notion of "a server" and one works with a pool of resources: cores and memory; dynamically deploying services to perform calculations. See *the NIST Definition of Cloud Computing* from the National Institute of Science and Technology or Chang et al. (2010). The main players in service-based computation are *Apache Spark*, Microsoft's *Service Fabric*, and Google's *App Engine*. They offer similar capabilities in the form of a scalable platform that will allow applications to dynamically burst upwards to a high number of parallel service instances as demand grows.

In terms of local HPC solutions, NVidia's GPU computation using *OpenCL* or *CUDA* can yield impressive results. GPU appliances are now available in *Amazon Web Services* (AWS) and on Microsoft's *Azure* permitting software engineers access to considerable computing power: see NVidia's new Pascal-based servers.[24] Maxeler Technologies' *Field Programmable Gate Arrays*[25] (FPGA) are programmable compute nodes that can be configured to perform a specific set calculation very quickly. At each clock cycle, more calculations can be loaded leading to an impressive throughput. Finally, quantum computation, if a little speculative at present, works in a similar manner to FPGAs. They are not classic general processors but, like FPGAs, are devices that can be configured to solve a particular problem very quickly.

5.3 Risk-Scenario Generators

Generally speaking, non-market risks are any risks faced by an insurer that are not economic and are typified by the fact that deep and liquid markets do not exist for trading in them. Historical data in non-market risks is difficult to obtain and maintain. Solvency 2's Pillar I and II recognize the importance of these through the potential impact on an insurer's balance sheet. A non-exhaustive list of such risks includes industry-wide systemic risks (e.g., the financial crisis of 2008/2009), mortality and longevity risks, policyholder lapse and surrender risk, and operational risk. Being characterized by a general lack of data is the challenge an insurer has in modeling them.

[24] http://www.nvidia.co.uk/object/cuda-parallel-computing-uk.html
[25] http://www.maxeler.com

Codependencies and tail risk emerging from, for example, a systemic risk shock, lead to a plethora of correlation structures. The copula-marginal distribution factorization (Sklar, 1959) is a key ingredient to any RSG. In the multivariate risk setting it extends the idea of simple Gaussian correlation of Brownian shocks via a Cholesky factorization to all risk factors. In what is clearly unsatisfactory modeling behavior, the Gaussian correlation approach still allows diversification of risks in times of crisis. During periods of relatively quiescent and stable economic behavior, diversification seems reasonable but when concerted action is happening *together* such as a run on the banks, it is not. Changing co-dependency structure from a Gaussian to a T-copula on a low degree of freedom can model the effects of herding in times of crisis yet still allow for diversifiable behavior during periods of relative calm.

A RSG, therefore, needs to incorporate an ESG but must also provide a coherent framework for modeling the non-market risks *alongside* the market risks. Since the theory for non-market risks is not as well developed as for market risks, one generally resorts to statistical models for these. RSGs need to provide an insurer with a statistical toolbox of marginal distribution functions and copulas to model the joint distribution of market *and* non-market risks. At some point downstream of the RSG, the results will be aggregated to produce capital charges which monetize the impact of these risks on the business. This section provides some detail on co-dependency structures and non-market risks.

5.3.1 Co-dependency structures and simulation

McNeil et al. (2015) discuss in detail the simulation of variables from multivariate distributions. For many of the processes discussed in the previous section, the Brownian shock is the key ingredient and over a timestep Δt one may conveniently approximate the stochastic differential $\mathrm{d}W_t$ by $N(0, \Delta t)$. In the multivariate setting of the ESG one seeks to correlate many such variables ($n \gg 1$, say) to a user-specified positive semi-definite correlation matrix.

Generating a correlated set of normally distributed variables is relatively easy. One begins by using an industry standard pseudo-random number generator such as the Mersenne twister (Matsumoto and Nishimura 1998) to create uniform random variables on the interval $[0, 1]$. These are grouped into pairs and, using the algorithm of Box and Muller (1958), are converted into pairs of independent and identically distributed pseudo-random normal deviates, which are then used to populate a p-dimensional vector z. Performing the Cholesky decomposition of the positive semi-definite correlation matrix P into LL' (where L is a lower triangular matrix) and constructing the vector $x := Lz$ produces a realization from a multivariate normal distribution with correlation matrix P and standard normal marginals. The challenge comes when one requires to correlate non-Gaussian variables.

Sklar's theorem (Sklar 1959), or rather its converse gives a general and flexible framework for the factorization of multivariate distribution functions.

Sklar showed that for every multivariate distribution function F there exists a unique (in the case of distribution functions comprised of continuous variables with certain restrictions to those comprising discrete variables) function C (the copula) such that for each of the distribution functions marginals $F_{1:p}$:

$$F(x_1, \ldots, x_p) = C(F_1(x_1), \ldots, F_p(x_p))$$

Furthermore, if the density function is f with marginals $f_{1:p}$ (and copula density c) then

$$f(x_1, \ldots, x_p) = c(F_1(x_1), \ldots, F_p(x_p)) \prod_{n=1}^{p} f_n(x_n)$$

If one specifies the copula, $C(u_1, \ldots, u_p)$, then one may combine it with *any* desired marginal distributions and the resulting multivariate distribution function is unique (see McNeil et al. 2015). The strength in this result for scenario generators is that to simulate from a multivariate distribution, one need only simulate a multivariate instance from the copula: $(u_1, \ldots, u_p)'$; and then evaluate the quantile functions of each of the marginal variables at these copula values: $(q_1(u_1), \ldots, q_p(u_p))'$; to obtain a desired (ranks-based) instance of the joint distribution function. Since it is often easy to simulate from a copula (at least from many of the most popular multivariate distributions: e.g., the normal, t-copula, or the grouped t-copula) then one may use whatever marginal distributions desired: *they need not be normal* nor relate in any way to the copula.

I note that it is not always possible to generate random variables with any prescribed correlation under the Pearson product moment coefficient. However, since the copular approach uses a theoretical ranks-based correlation between variables then practically any correlation *is* possible although it may not be practically possible to obtain a good sample correlation estimate in the presence of degenerate variables containing multiple repeated instances of the same value (e.g., the PEDCP case to be discussed in Section 5.4).

Since my motivation was to introduce more realistic co-dependency structures than the simple Gaussian copula, I will now illustrate the benefits of alternatives to it in modeling tail risk. For illustration, I will compare the Gaussian and T-copula on low degrees of freedom in the bivariate case.

The procedure for representing a bivariate copula underlying given bivariate data is easily done by plotting the normalized ranks of the first variable against the normalized ranks of the second variable. In the theoretical setting of a known normal or T-copula then this is achieved by applying the marginal distribution functions to the realized bivariate data. McNeil et al. (2015) detail how this is achieved. In this example, $n = 2.5 \times 10^4$ pairs of illustrative data from a bivariate normal distribution were generated with a given zero mean and covariance matrix:

$$\Sigma = \begin{bmatrix} 1 & \rho \\ \rho & 1 \end{bmatrix}$$

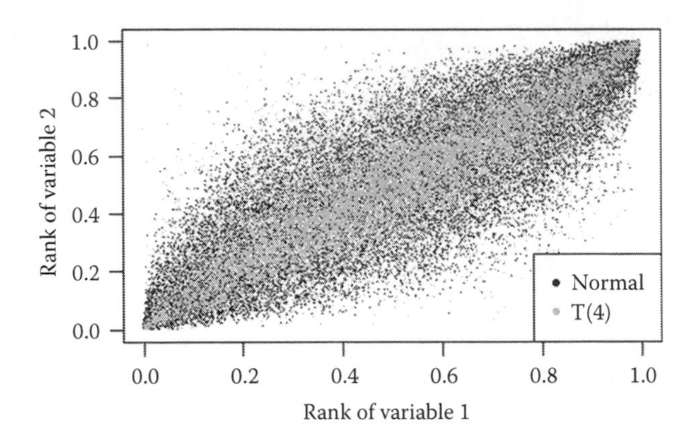

FIGURE 5.1: Normal and T(4) copulas represented as scatterplots.

where $\rho = 0.8$ (Σ is also the correlation matrix P and both marginals are standard normal). A bivariate T-distribution is created using the same parameters and an additional degrees-of-freedom parameter ν here set to 4. The T-distribution is created from the bivariate normal data by generating a vector v of length 2.5×10^4 containing $\chi^2(4)$-variables. Scaling the i-th pair of normal data by ν/v_i generates bivariate $T(\nu)$-data which is easily accomplished in any ESG. Since the marginal distributions of a multivariate $T(\nu)$ distribution are univariate t_ν distributions, then $T(\nu)$-copula data is easily obtained by applying the t_ν distribution function to each element of the pair.

Comparing scatterplots of the simulated copula data sample serves to illustrate the point. Focusing in the upper right-hand corner of the scatterplot corresponds to large extremes of both variables: for example, times of high financial stress. As variable pairs move further into this corner they are coerced to move closer and closer together under the $T(\nu)$ copula than under the more dispersed normal copula. This goes some way in modeling the so-called *herding behavior* of the markets in times of financial stress. The $T(\nu)$-copula can be extended to the more versatile grouped T-copula in which sets of *like* variables in multivariate data are grouped and allocated a group degrees-of-freedom parameter. Other copulas have been used in finance but have not, to the best of my knowledge, been applied within the insurance context such as the Archimedean and Gumbel–Hougaard copulas (see McNeil et al. 2015); see Figure 5.1.

In the context of ESG and RSG simulations, the copula approach is often essential in effectively simulating appropriately correlated risk drivers. Finally, I note that if one intends to simulate the effect of a $T(\nu)$-copula (or other) over more than one time step, say n steps, then one must decompose the copula such that after the application of n steps the $T(\nu)$-copula emerges. If this decomposition is not made and a $T(\nu)$-copula is simulated at *each* time step, then a dilution effect occurs which is sometimes referred to as a *copula central limit theorem* and a Gaussian copula emerges after n steps for large enough n. This leads on to the discussion of α-stable distributions and the reader is referred to McNeil et al. (2015) for more details.

In Section 5.4 where I illustrate challenges in scenario generation, it is as a direct result of the copula/marginal factorization and Sklar's converse that I can simulate correlated instances of some rather challenging variables. In the remainder of this section I cover some of the major non-market risk drivers. As a consequence of the converse to Sklar's theorem, they may be represented by any univariate distribution function as long as it is possible to simulate these using the probability integral transform. This is possible for variables such as the normal, t, chi-squared, and Gamma whose densities are readily available. For more challenging variables, one solution is to represent them in some way such as that given for the examples in Section 5.4.

5.3.2 Mortality and longevity risk

Mortality risk is felt by an insurer during its accrual phase of a life assurance policy. Essentially, if mortality is seen to worsen, then fewer policyholders will be able to pay insurance premiums and the business can lose a core proportion of its value. Models of mortality are of the Age-Period-Cohort type and their generalizations. The model of Lee and Carter is a good starting point (e.g., Haberman and Renshaw 2008).

The Institute and Faculty of Actuaries maintains a *Continuous Mortality Investigation* or (CMI[26]) and cites the following as critical to the study of mortality risk:

1. Annuitant mortality
2. Critical illness and mortality
3. Income protection
4. Self-administered pension scheme (SAPS) mortality.

On the flip-side to mortality risk there is longevity risk. Potentially much more serious than mortality risk, it measures the risk during the payout phase of an insurance policy. Since life expectancy has been increasing since the latter half of the twentieth century it implies that insurers are going to have to pay more people in their retirement for longer. The potential shortfall in funds could be very severe depending on the composition of an insurer's business. Defined benefit schemes have been all but replaced by defined contribution schemes for this reason. To mitigate longevity risk, insurers have the option of entering into a longevity swap agreement with an investment bank, say. Inevitably this will be costly, however.

[26]https://www.actuaries.org.uk/learn-and-develop/continuous-mortality-investigation

5.3.3 Lapse and surrender risk

New government legislation makes it easier for pensioners to take their pension as a lump sum upon retirement. In bullish markets people surrender their policies because they can achieve better returns elsewhere (and vice versa). In bear markets, surrender risk is minimized but an insurer will generally find it difficult to honor guarantees on policies during this time. In times of market distress, lapse may become an important feature and may lead to a *contraction* of the market.

5.3.4 Operational risk

These are often seen as a *catch-all* category for *other* risks on an insurer's balance sheet. An operational risk is any risk that can be caused either by human error (e.g., logging a trade in millions rather than tens of millions) to failing to comply with regulatory tax requirements, for example. Essentially, anything that arises as an error incurred during the *normal* operation of a business.

More and more prominence is being attributed to operational risk/loss as there is a very real chance it could contain the one *killer* risk rendering an insurer insolvent. Under Solvency 2's Pillar I, the standard formula SCRop is volume driven and based on an insurer's premiums. This may lead to excessive capital charges and, in turn, drive an insurer toward a partial internal model. It is true, however, that under Pillar II, insurers must develop full and independent risk-monitoring frameworks including operation risk and justify them to the regulator anyway.

Where an internal model is used, it tends to follow the bespoke scenario stress testing and loss distribution approach familiar in Basel II operational risk analyses in the banking sector. The procedure as outlined by Neil et al. (2012) is as follows:

1. Use Pillar II/ORSA risk control infrastructure

2. Create operational loss scenarios

3. Identify the most potent loss making scenarios

4. Derive theoretical loss distributions for each of these

5. Assess correlations between these scenarios

6. Use a copula co-dependency to aggregate losses and obtain a capital charge

Despite this framework, operational loss capital charges are notoriously hard to estimate as so few data exist either within companies or in the public domain. Given the requirements of Solvency 2 Pillars I and II, insurers are beginning to gather and monitor their own data. Targeted at insurance companies under Solvency 2, the Operational Risk Consortium[27] (ORIC) was founded in 2005 to provide a focal point for insurers to share their operational risk data anonymously. It also provides industry benchmarks and thought leadership.

Blending an insurer's own data with the online ORIC resource is certainly one approach. However, calibrations of operational risk models can still be problematic: risk drivers may flip in and out of models frequently dependent on the precise historical period used. Insurer's may prefer resort to independent expert or panel judgement. Neil et al. (2012), for example, suggest an alternative approach based on Bayesian networks that specifically model the links between different risk drivers.

In Section 5.4, I provide a detailed specification for a marginal risk-driver that can be used to measure operational losses: the truncated Pareto event-driven compound Poisson distribution. Since the events themselves are of power-law Pareto type I class, no analytical formula exists for their distribution function (unlike in the case of a Gamma event-driven variable, for example). However, I derive a new technique that enables a representation of the distribution functions of this and other compound variables in terms of polynomial bases. The method is possible when one may simulate a specific variable without knowledge of its distribution function. The truncated Pareto event-driven compound Poisson variable can, by its nature, lead to very peculiar distributions but the most challenging aspect of it from a risk-scenario generator's perspective is in how to correlate it with other risk drivers. Since I offer a means to simulate it from its marginal, this can be achieved using a copula-marginal factorization and the probability integral transform.

5.4 Examples of Challenges in Scenario Generation

This section considers how to derive the distribution function of composite variables where, although it is possible to simulate these variables, access to their distribution function is unavailable by analytical means. This is problematical for an economic or RSG because it is not at all obvious how to correlate such variables. If access to some *representation* of their distribution functions were possible, a scenario generator *would* be able to correlate them via the probability integral transform. Two variables are considered: first, the truncated Pareto-event-driven compound Poisson (PEDCP) variable and second, the conditional equity asset shock distribution of a stochastic volatility and jump diffusion (SVJD) model. The former is represented by a finite-element basis and the latter by an orthogonal series of Chebyshev polynomials. As I will show, if one can simulate instances of the variables in isolation then

[27] https://www.oricinternational.com

the coefficients in their representations may be estimated. The key result is that one need only hold a relatively small number of coefficients to represent a variable's distribution function well by a function that is at least continuous and often differentiable. This is in stark contrast to using an empirical cumulative distribution function (ECDF) representation where one is obliged to hold a large amount of simulated data. Whenever the ECDF is to be evaluated, one must sift the ranked data until the appropriate quantile has been bounded by two adjacent data values and interpolate. Besides being computationally demanding it is also memory intensive compared to accessing a good distribution function representation. The methods are new.

5.4.1 PEDCP representation

Let the random variable measuring operational loss be X and let us assume that it is an event-driven compound Poisson variable and write $X \sim \text{PEDCP}(x_m, x^*, \alpha, \lambda)$ where the count N is Poisson distributed with rate parameter $\lambda > 0$. The event distribution is taken to be the *truncated* Pareto Type I distribution. The events are simulated as Pareto Type I events: they have a minimum value of $x_m > 0$; but these have an artificial ceiling applied to them at level $x^* > x_m$. Truncation of variables is something that insurers may wish to model as it gives an extra degree of freedom to the event variable. However, it does lead to a more complicated distribution function.

Recall the Pareto Type I density and distribution function:

$$f_{\text{Pa}}(x) = \frac{\alpha x_m^\alpha}{x^{\alpha+1}}, \quad F_{\text{Pa}}(x) = 1 - \left(\frac{x_m}{x}\right)^\alpha \quad \forall x \geq x_m$$

The Pareto Type I power-law decay is denoted $\alpha > 0$. The effect of truncating such a variable above a certain value is to stack the remaining probability at that ceiling. The truncated density is

$$\begin{aligned} f_{\text{t.-Pa}}(x) &= (1 - F_{\text{Pa}}(x^\star))\, f_{\text{Pa}}(x) + F_{\text{Pa}}(x^\star)\delta(x - x^\star), \\ F_{\text{t.-Pa}}(x) &= \mathbb{I}_{x \leq x^\star} F_{\text{Pa}}(x) + \mathbb{I}_{x \geq x^\star} \end{aligned}$$

Writing $X_n \sim t. - Pa(x_m,\, x^*,\, \alpha) \forall n \in \text{N}$ and $X_0 = 0$, the conditional compound Poisson variable is $X|N = \Sigma_{n=0}^N X_n$. By the total law of probability, the density of the unconditional PEDCP variable is

$$f_X(x) = f_{\text{Po}}(0; \lambda) f_{X|N=0}(x) + [1 - f_{\text{Po}}(0; \lambda)] f_{X|N>0}(x)$$

Since the distribution of $f_{X|N=0}(x)$ is degenerate being equal to $\delta(x)$ then from here on the distribution of X *given* $N > 0$ is sought and to simplify notation I simply write X in place of $X|N > 0$.

The first step is to simulate instances of this variable in isolation for use in estimating the parameters in a representation of the variable's distribution (and latterly, the density) function. To motivate an interesting example, the following parameters are taken: $\lambda = 1$, $\alpha = 3$, $x_m = 0.1$, and $x^* = 0.15$. First,

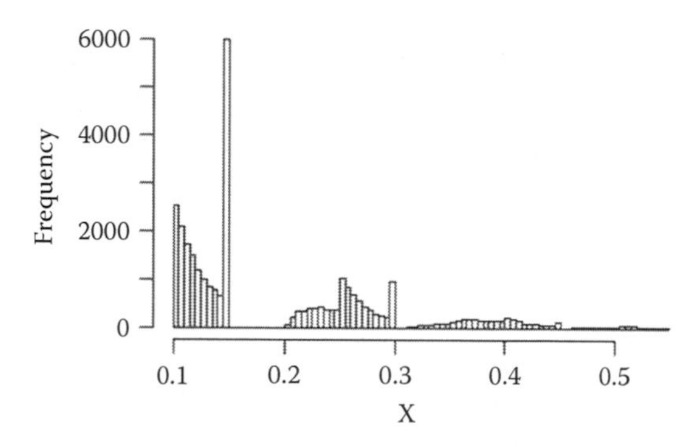

FIGURE 5.2: PEDCP histogram given $N > 0$ events.

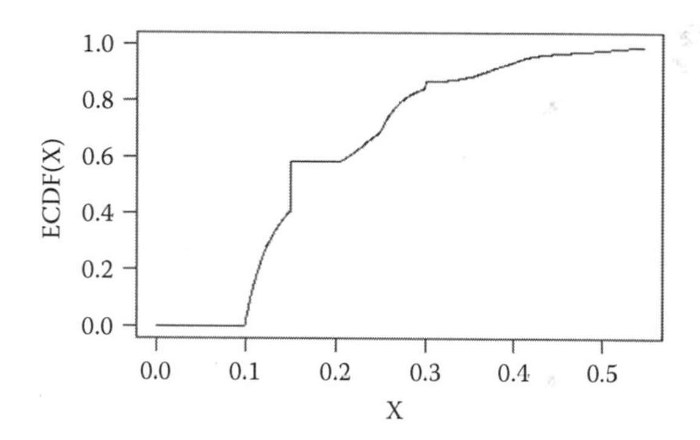

FIGURE 5.3: Compound Poisson empirical distribution function given $N > 0$ events.

5×10^4 Poisson variables Po(1) are simulated using **R**'s **rpois()** function giving the following summary:

```
##      0      1     2     3    4    5   6  7
##  18560  18271  9024  3156  818  139  24  8
```

Note that, of the $N_{reps} = 5 \times 10^4$ Poisson variates simulated, some 31440 were non-zero, that is, 62.88% of the variates which approximates well the theoretical value $1 - e^{-\lambda} \approx 63.21\%$. Then, enough Pareto variables were simulated by generating standard uniform variables u using **R**'s **runif()** function and inserting them into the Pareto quantile function: $q(u) = x_m(1-u)^{-1/\alpha}$. After truncating the Pareto variables at $x^* = 0.15$ the requisite numbers given the Poisson counts N were summed to give X.

A histogram of the non-zero results (and truncating all variates simulated beyond the 99%-ile for exposition only) and its empirical distribution function, as computed by **R**'s **ecdf()** function, are given, respectively, in Figures 5.2 and 5.3.

Before beginning, the support of the PEDCP variable (denoted here by $X \in [x_m, \infty) \subset \mathbb{R}$) is transformed into the finite interval [0, 1). Thus, let the transformed variable be Y and the transformation be $y = \varphi(x;\kappa) := 1 - e^{-\kappa(x-xm)}$ some $\kappa \in \mathbb{R}^+$ (here fixed at $\kappa = 0.8$). The distribution function $F_X(x)$ is then:

$$\begin{aligned}
F_X(x) &= \int_{x_m}^{x} f_X(x')\mathrm{d}x' \\
&= \int_0^{\varphi(x;\kappa)} f_X(\varphi^{-1}(y';\kappa))\mathrm{d}\varphi^{-1}(y';\kappa) \\
&= \int_0^{\varphi(x;\kappa)} \frac{f_X(\varphi^{-1}(y';\kappa))}{\kappa(1-y')}\mathrm{d}y' \\
&= \int_0^{y} g_Y(y';\kappa)\mathrm{d}y' =: G_Y(y;\kappa) \quad \text{where} \quad g_Y(y;\kappa) := \frac{f_X(\varphi^{-1}(y;\kappa))}{\kappa(1-y)}
\end{aligned}$$

Note that the steps above are possible since the inverse transformation of φ is found via

$$x = x_m + \frac{1}{\kappa}\log\left(\frac{1}{1-y}\right) =: \varphi^{-1}(y;\kappa)$$

and so

$$\frac{\mathrm{d}}{\mathrm{d}y}\varphi^{-1}(y;\kappa) = \frac{1}{\kappa}\frac{\mathrm{d}}{\mathrm{d}y}\log\left(\frac{1}{1-y}\right) = \frac{1}{\kappa(1-y)}$$

In what follows, finite-element representations of both the PEDCP distribution and density functions are given. As will be seen, the latter leads to a more satisfactory representation.

5.4.1.1 Distribution function

Let the (transformed) distribution function G_Y be given a representation in terms of the linear finite-element basis $\mathcal{B}_K$ over a given number $K \in \mathbb{N}$ of elements:

$$\begin{aligned}
\mathcal{B}_K = \{&\phi_k \in C^0([0,1]) : \phi_k(y) = K\max(\min(y - y_{k-1}, y_{k+1} - y), 0)/2 \\
&y_k = k/K,\ 0 \leq k \leq K\}
\end{aligned}$$

Each element of the basis is of the form of a "witches hat": they are zero everywhere except over three consecutive nodes: y_{k-1}, y_k and y_{k+1}; where they interpolate the values 0, 1, and 0, respectively. The representation is

$$G_Y(y) = \sum_{k=0}^{K} \tilde{G}_k\phi_k(y), \quad y \in [0,1]$$

and is equivalent to a continuous linear spline function. Note that $G_Y(y_k) = \tilde{G}_k$ by construction as $\phi_k(y_k) = 1$. To determine the coefficients, observe that for each $j \in [0, K]$:

$$\sum_{k=0}^{K} \tilde{G}_k \int_0^1 \phi_j(y)\phi_k(y)\mathrm{d}y = \int_0^1 G_Y(y)\phi_j(y)\mathrm{d}y$$

$$\text{iff} \quad \sum_{k=0}^{K} A_{jk}\tilde{G}_k = G_Y(y)\Phi_j(y)|_0^1 - \int_0^1 g_Y(y)\Phi_j(y)\mathrm{d}y$$

$$= \Phi_j(1) - \mathbb{E}_Y[\Phi_j(Y)]$$

This is of the form $A\tilde{\mathbf{G}} = \mathbf{\Phi}(1) - \mathbb{E}-[\mathbf{\Phi}(Y)]$. The elements of A are, for $0 < j = k < K$:

$$A_{jj} = \int_0^1 \phi_j(y)^2 \mathrm{d}y = \frac{2}{3K}$$

with $A_{00} = A_{KK} = (1/3K)$. Elsewhere, for $\leq j, k \leq K$ and $|j - k| = 1$:

$$A_{jk} = \int_0^1 \phi_j(y)\phi_k(y)\mathrm{d}y = \frac{1}{6K}$$

while for $|j - k| > 1 A_{jk} = 0$. Finally, observe that Φ_j is an anti-derivative of ϕ_j. Thus, for $0 < j < K$:

$$\Phi_j(y) = \begin{cases} 0 & \text{if } y < y_{j-1} \\ K(y - y_{j-1})^2/2 & \text{if } y_{j-1} \leq y < y_j \\ 1/K - K(y_{j+1} - y)^2/2 & \text{if } y_j \leq y < y_{j+1} \\ 1/K & \text{if } y \geq y_{j+1} \end{cases}$$

while for $j = 0$:

$$\Phi_0(y) = \begin{cases} 1/(2K) - K(y_1 - y)^2/2 & \text{if } y_0 \leq y < y_1 \\ 1/(2K) & \text{if } y \geq y_1 \end{cases}$$

and for $j = K$:

$$\Phi_K(y) = \begin{cases} 0 & \text{if } y < y_{K-1} \\ K(y - y_{K-1})^2/2 & \text{if } y_{K-1} \leq y < y_K \end{cases}$$

Here, some $K = 1000$ finite elements were used to represent the information carried in 31440 non-zero instances of the PEDCP variable that was simulated. This is the number of doubles that would otherwise be held in memory if the empirical distribution function were to be used for simulation. The estimated distribution function is shown in Figure 5.4.

Unfortunately, the distribution function is non-monotone and so this representation is unsatisfactory: it may not be used well with the probability integral transform for simulating instances of the PEDCP variable. A representation which is monotone is required. A good finite-element representation of the *density* function, that is, one which was everywhere non-negative, would ensure that the resulting distribution function (now composed of monotone increasing quadratic segments) would be everywhere increasing. This is the subject of the next section.

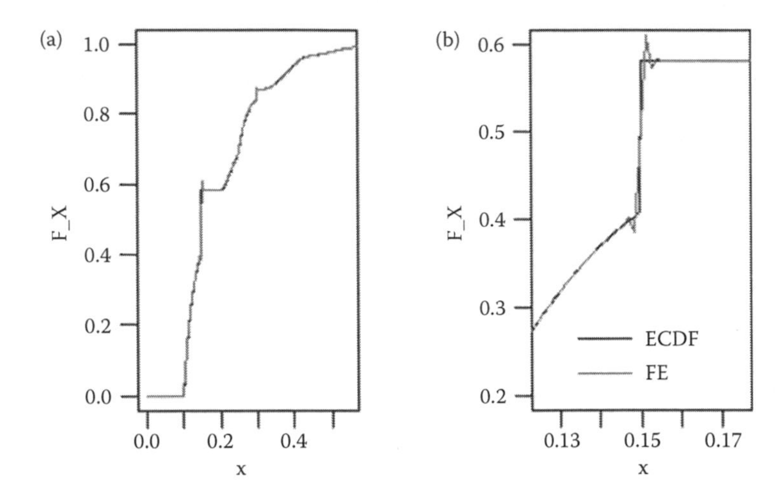

FIGURE 5.4: Finite-element representation of the PEDCP distribution function. In (a) the whole distribution modeled is shown and in (b) a zoomed view around the "step" illustrates undesirable non-monotone behavior.

5.4.1.2 Density function

Give the (transformed) density function a representation in terms of finite elements:

$$g_Y(y) = \sum_{k=0}^{K} \tilde{g}_k \phi_k(y), \quad y \in [0,1]$$

and note that $g_Y(y_k) = \tilde{g}_k$ since $\phi_k(y_k) = 1$ (the values in between nodes y_k are linearly interpolated). Note also that $\int_{-1}^{1} g_Y(y)\,\mathrm{d}y = 1$ and this can be maintained by scaling the coefficients $\tilde{g}_k$ accordingly. If the integral's value is currently A then:

$$a = \sum_{k=0}^{K} \tilde{g}_k \int_{-1}^{1} \phi_k(y)\mathrm{d}y = \frac{1}{K}\left(\frac{1}{2}\tilde{g}_0 + \sum_{\substack{K-1\\k=1}} \tilde{g}_k + \frac{1}{2}\tilde{g}_K\right)$$

One may then scale each coefficient by a to obtain a proper density function. It is henceforth assumed that this is done. For $j \in [0, K]$:

$$\sum_{k=0}^{K} \tilde{g}_k \int_0^1 \phi_j(y)\phi_k(y)\mathrm{d}y = \int_0^1 g_Y(y)\phi_j(y)\mathrm{d}y$$

$$\text{iff} \quad \sum_{k=0}^{K} A_{jk}\tilde{g}_k = \int_0^1 g_Y(y)\phi_j(y)\mathrm{d}y = \mathbb{E}_Y[\phi_j(Y)]$$

This is of the form $A\mathbf{g} = \mathbf{e}$ where $\boldsymbol{e} = \mathbb{E}_Y[\phi(Y)]$ like the distribution function representation. The estimated density function is shown in Figure 5.5.

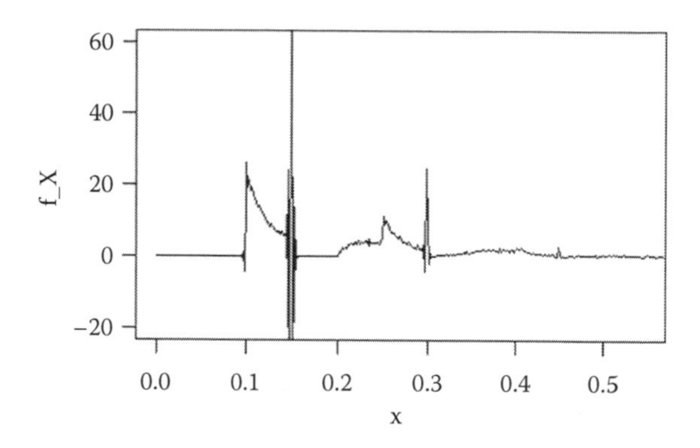

FIGURE 5.5: Finite-element representation of the density function (given $N > 0$ events).

This solution is noisy and is negative valued for some values of the variable. To mitigate these problems, seek instead a viscosity solution employing a regularization technique. Note that the solution $\mathbf{x} = \mathbf{g}$ of the linear system described above and denoted by $A\mathbf{x} = \mathbf{e}$ is also the solution to a quadratic programming problem. Let $||\boldsymbol{v}||_2 = \sqrt{\Sigma_{i=1}^n v_i^2}$ be the familiar 2-norm over all vectors $\boldsymbol{v} \in \mathbb{R}^n$. If $h(\mathbf{x};\lambda) = ||A\mathbf{x} - \mathbf{e}||_2^2 + \lambda||\mathbf{x}||_2^2$ ($\lambda \in \mathbb{R}^+$) then the equivalent quadratic program is

$$\mathbf{g} = \text{argmin}_{\mathbf{x}\in\mathbb{R}^{K+1}} h(\mathbf{x}; 0)$$

The *viscosity* solution requires $\lambda > 0$ and is

$$\mathbf{g}^\lambda = \text{argmin}_{\mathbf{x}\in\mathbb{R}^{K+1},\lambda>0} h(\mathbf{x}; \lambda)$$

Note that the objective may be written as

$$\begin{aligned} h(\mathbf{x};\lambda) &= \mathbf{x}'A'A\mathbf{x} - 2\mathbf{e}'A\mathbf{x} + \mathbf{e}'\mathbf{e} + \lambda\mathbf{x}'\mathbf{x} \\ &= \mathbf{x}'(A'A + \lambda I)\mathbf{x} - 2\mathbf{e}'A\mathbf{x} + \mathbf{e}'\mathbf{e} \end{aligned}$$

Let the matrix $A'A + \lambda I$ have the Cholesky factorization LL' for the lower triangular matrix $L \in \mathbb{R}_{n\times n}$. Then defining $\mathbf{e}^* := L^{-1}A'\mathbf{e}$ (or equivalently, $\mathbf{e} = A^{-T}L\mathbf{e}^*$) gives

$$\begin{aligned} h(\mathbf{x};\lambda) &= \mathbf{x}'LL'\mathbf{x} - 2(A^{-T}L\mathbf{e}^\star)'A\mathbf{x} + (A^{-T}L\mathbf{e}^\star)'(A^{-T}L\mathbf{e}^\star) \\ &= ||L'\mathbf{x} - L^{-1}A'\mathbf{e}||_2^2 + ||\mathbf{e}||_2^2 - ||L^{-1}A'\mathbf{e}||_2^2 \end{aligned}$$

To find the λ-viscosity solution one may not simply proceed by simultaneously minimizing h over $\mathbf{x}$ and λ as the solution $\lambda \to 0$ will be sought. Instead, one observes that for fixed $\lambda > 0$ any solution satisfies:

$$\mathbf{x} = L^{-T}L^{-1}A'\mathbf{e} = (LL')^{-1}A'\mathbf{e} = (A'A + \lambda I)^{-1}A'\mathbf{e}$$

The λ hyper-parameter is estimated by an out-of-sample-method using 10-fold cross-validation (Figure 5.6).

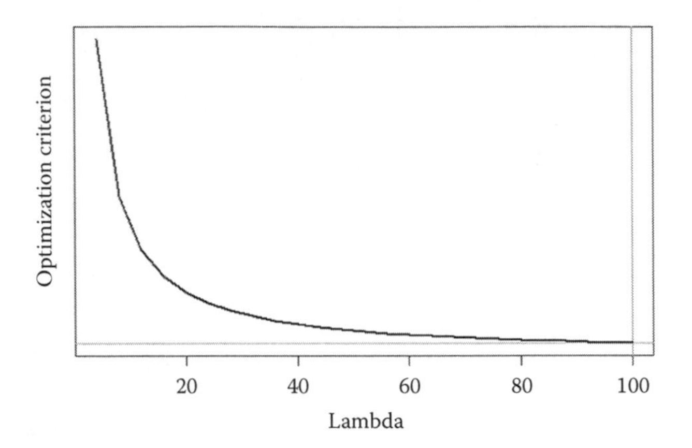

FIGURE 5.6: Hyper-parameter λ estimation using 10-fold cross-validation. The y-axis scale is suppressed as the difference between the smallest and largest objective criteria was 2.29×10^{-9} with the lower grey horizontal line measuring 0.03583.

The normalization of the density, in order such that it should integrate to one over its domain, prevents further reduction in the objective after around a value of $\lambda \approx 100$ where the solution asymptotes to a constant yet non-zero value: 0.0358326. The regularized solution may therefore be taken as the solution for large λ and is therefore:

$$\mathbf{g}^{\infty} = \frac{A'\mathbf{e}}{a}$$

where a is the normalization constant:

$$a = \frac{1}{K}\left(\frac{1}{2}g_0^{\infty} + \sum_{i=1}^{K-1} g_i^{\infty} + \frac{1}{2}g_K^{\infty}\right)$$

Therefore, there is no need to solve the system $(A'A + \lambda I)\mathbf{x} = A'\mathbf{e}$ and hence no need to hold the (often prohibitively large) matrix $A'A + \lambda I$. The results are shown in Figure 5.7a where the viscosity solution is shown in red superimposed on top of the non-regularized solution.

Prior to seeking a viscosity solution ($\lambda = 0$) the range of values for g_Y was $(-66.67, 272.7)$. Having taken the regularized solution (i.e., where λ is assumed sufficiently large) the range of g_Y is (0, 111.3). There is no need to artificially floor values at zero to ensure a valid density function *while* the associated distribution function will necessarily be increasing on its domain.

Given the density function representation for g_Y a distribution function G_Y is sought. This can be obtained by directly integrating the finite-element

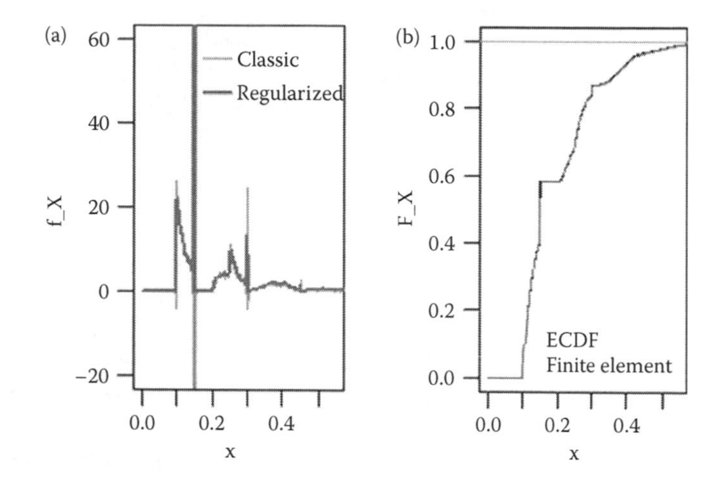

FIGURE 5.7: Regularized finite-element representation of the PEDCP (a) density and (b) distribution function.

representation. This reduces to integrating each element ϕ_k previously derived as Φ_k:

$$G_Y(y) = \sum_{i=0}^{K} \tilde{g}_k \Phi_k(y)$$

The results are shown in Figure 5.7b. One may simulate from this distribution function using the probability integral transform assured that it is continuous, monotone increasing and smooth (being everywhere a quadratic function). This is unlike the case of the representation of the PEDCP distribution function which was non-monotone and only continuous. This is made even more remarkable given that it was not necessary to solve a large linear system in achieving this result.

5.4.2 Stochastic volatility and jump diffusion representation

A detailed equity model within an ESG might be the *stochastic volatility and jump diffusion* (SVJD) model. As previously discussed in Section 5.2.4 it is comprised of two parts: the first is Heston's stochastic volatility model and the second is Merton's jump diffusion model. The aim of this section is to enable the equity model to hit the target correlations that have been set between its returns and any other variables in the ESG (including possibly other instances of SVJD models). A difficulty occurs if one correlates the equity asset shock distribution *conditioned upon* the size of the stochastic variance and the jump shock with other asset-shocks. Although incorrect, this may not seem an unreasonable approach since most asset shocks are simple Brownian shocks. Rather, it is the detailed nature of the SVJD's asset's unconditional shock that causes the oversight as it is composed of five random components: the equity asset's Brownian shock, the Merton jump frequency and size, the current level of the stochastic variance and the change in the level of the stochastic variance. In correlating only the Brownian shock, one discovers the returns correlations emerging from a scenario generator simulation file are subject to a systematic bias compared to their targets. The solution is to correlate the *unconditional* equity asset shock with any other asset shock but this has no known distributional form and is, hence, unavailable analytically.

In this section I will solve the SVJD asset returns correlation problem by giving a representation of the unconditional equity asset shock distribution at each modeled period. I will then be able to correlate the equity returns with any other assets using the probability integral transform. As in Section 5.4.1, this is done by first simulating instances of the process in isolation. The difference between this and the previous section is that an entire process is simulated in isolation, rather than a single random variable, which leads to a consistent sequence of conditional marginals indexed by time. Quasi-random number generation is used to improve the standard error in the representations' coefficients. This approach is new.

5.4.2.1 Distributional representation

Suppose the distribution function of the SVJD equity asset shock (for the moment, denote this by X) is to be given a representation in terms of the orthogonal Chebyshev polynomial basis of the first kind:

$$\mathcal{B}' = \{T_n \in C^{\infty}(\mathbb{R}) : T_n(x) = \cos[n\cos^{-1}x], n \in \mathbb{Z}^+\}$$

Observe that this basis satisfies the orthogonality condition:

$$\int_{-1}^{1} T_m(x)T_n(x)w(x)\mathrm{d}x = \frac{1}{2}\pi(1+\delta_{m0})\delta_{mn}, \quad w(x) = \frac{1}{\sqrt{1-x^2}}, \quad m,n \in \mathbb{Z}^+$$

The support of X is the entire real line $\mathbb{R}$ and must be mapped into the finite interval $(-1, 1)$. Let the transformed variable be Y and the transformation: $y = \varphi(x;\kappa) := \tanh(\kappa x)$, some $\kappa \in \mathbb{R}^+$ (to be chosen presently). The distribution function $F_X(x)$ is then:

$$\begin{aligned}
F_X(x) &= \int_{x_m}^{x} f_X(x')\mathrm{d}x' \\
&= \int_{-1}^{\varphi(x;\kappa)} f_X(\varphi^{-1}(y';\kappa))\mathrm{d}\varphi^{-1}(y';\kappa) \\
&= \int_{-1}^{\varphi(x;\kappa)} \frac{f_X(\varphi^{-1}(y';\kappa))}{\kappa(1-(y')^2)}\mathrm{d}y' \\
&= \int_{-1}^{y} g_Y(y';\kappa)\mathrm{d}y' =: G_Y(y;\kappa) \quad \text{where} \quad g_Y(y;\kappa) := \frac{f_X(\varphi^{-1}(y;\kappa))}{\kappa(1-y^2)}
\end{aligned}$$

Note that the steps above are possible since the inverse transformation of φ is found via:

$$x = x_m + \frac{1}{2\kappa}\log\left(\frac{1+y}{1-y}\right) =: \varphi^{-1}(y;\kappa) \quad \text{iff} \quad \frac{\mathrm{d}}{\mathrm{d}y}\varphi^{-1}(y;\kappa)$$
$$= \frac{1}{\kappa(1-y^2)}$$

Let the (transformed) distribution function G_Y be given by the representation in terms of the orthogonal Chebyshev polynomial basis $\mathcal{B}'$ (and truncated after $K \in \mathbb{N}$ terms):

$$G_Y(y) = \sum_{k=0}^{\infty} \tilde{G}_k T_k(y) \approx \sum_{k=0}^{K-1} \tilde{G}_k T_k(y)$$

I now proceed directly to derive expressions for the coefficients $\tilde{G}_k$. First, observe the following intermediate result which holds for Chebyshev polynomials. Since $y \in [-1,1]$ then let $y = \cos u$ so $\mathrm{d}y = -\sin u\mathrm{d}u$, $\forall k > 0$:

$$\int T_k(y)w(y)\mathrm{d}y = \int \frac{\cos[k\cos^{-1}y]}{\sqrt{1-y^2}}\mathrm{d}y = -\frac{1}{k}\sin[k\cos^{-1}y] + c$$
$$= -\frac{1}{k}T_k^{\#}(y) + c$$

having defined $T_k^{\#}(y) = \sin[k\cos^{-1}y]$. Note that $T_k^{\#}(1) = \sin(0) = 0$ and $T_k^{\#}(-1) = \sin(k\pi) = 0$. If $k = 0$ then the above integral is $-\cos^{-1}y + c$. The coefficients $\tilde{G}_k$ may be obtained, for $k = 0$ as

$$a_{00}\tilde{G}_0 = \int_{-1}^{1} G_Y(y)w(y)\,\mathrm{d}y$$
$$= -G_Y(y)\cos^{-1}(y)|_{-1}^{1} + \int_{-1}^{1} g_Y(y)\cos^{-1}y\,\mathrm{d}y$$
$$\text{iff } \tilde{G}_0 = \frac{1}{a_{00}}\mathbb{E}_Y[\cos^{-1}Y]$$

and $\forall k = 0,\ldots,K-1$, as:

$$a_{kk}\tilde{G}_k = \int_{-1}^{1} G_Y(y)T_k(y)w(y)\,\mathrm{d}y$$
$$= -\frac{1}{k}G_Y(y)T_k^{\#}(y)\Big|_{-1}^{1} + \frac{1}{k}\int_{-1}^{1} g_Y(y)T_k^{\#}(y)\,\mathrm{d}y$$
$$= \frac{1}{k}\int_{-1}^{1} g_Y(y)T_k^{\#}(y)\,\mathrm{d}y \quad \text{iff} \quad \tilde{G}_k = \frac{1}{a_{kk}k}\mathbb{E}_Y[T_k^{\#}(Y)]$$

In practice, one can now empirically estimate the coefficients $\tilde{G}_k$ by generating enough samples from the unconditional distribution of X, transforming to $Y = \varphi(X;\kappa)$, and approximating the expectations above by their sample means. The objective is now to simulate the correct variable X that will enable, ultimately, recovery of F_X.

5.4.2.2 The SVJD model and the combined equity asset shock

The SVJD model has the continuous time SDE representation:

$$\begin{aligned}\frac{\mathrm{d}S_t}{S_t} &= (\mu - \lambda\bar{\mu})\mathrm{d}t + \sqrt{v_t}\mathrm{d}W_t^{(1)} + (\eta_t - 1)\mathrm{d}N_t \\ \mathrm{d}v_t &= \alpha(\theta - v_t)\mathrm{d}t + \xi\sqrt{v_t}\mathrm{d}W_t^{(2)}\end{aligned}$$

where $W_t^{(i)} (i = 1, 2)$ are correlated Brownian motions: $\mathrm{Cor}[\delta W_t^{(1)}, \delta W_t^{(2)}] = \rho$ in the limit[28] as $\delta t \to 0$. The stochastic variable N_t is the standard Poisson counting process with $\delta N_t \sim \mathrm{Po}(\lambda\delta t)$, again in the limit as $\delta t \to 0$. The jump sizes are controlled by the stochastic variable $\eta_t \sim \log - N(\bar{\mu}, \sigma^2)$.

Consider now the equivalent form:

$$\begin{aligned}\frac{\mathrm{d}S_t}{S_t} &= (\mu - \lambda\bar{\mu})\mathrm{d}t + \mathrm{d}X_t \\ \mathrm{d}X_t &= \sqrt{v_t}\mathrm{d}W_t^{(1)} + (\eta_t - 1)\mathrm{d}N_t \\ \mathrm{d}v_t &= \alpha(\theta - v_t)\mathrm{d}t + \xi\sqrt{v_t}\mathrm{d}W_t^{(2)}\end{aligned}$$

and it is about the distribution function of the variable $\delta X_t | \mathscr{F}_t$ that I seek a representation in terms of orthogonal polynomials. For the equity S_t and equity shock term δX_t, consider the Euler–Maruyama discretization from continuous time to discrete time[29] $t = i\Delta t$, $i \in \mathbb{N}$, is (S_0 given):

$$\begin{aligned}\frac{\Delta S_i}{S_{i-1}} &= (\mu - \lambda\bar{\mu})\Delta t + \Delta X_i \\ \Delta X_i &= \sqrt{v_{i-1}\Delta t}\left(\sqrt{1-\rho^2}Z_i^{(1)} + \rho Z_i^{(2)}\right) + \left(\mathrm{e}^{\bar{\mu}+\sigma Z_i^{(3)}} - 1\right)\Delta N_i \\ \Delta v_i &= \alpha(\theta - v_{i-1})\Delta t + \xi\sqrt{v_{i-1}\Delta t}Z_i^{(2)}\end{aligned}$$

Here, $Z_i^{(1,3)} \sim^{iid} N(0,1)$ and $\Delta N_i \sim \mathrm{Po}(\lambda\Delta t)$. Since the equity shock depends on the stochastic variance at time index $i - 1$ and upon current level of the stochastic variance's normal shock $Z_i^{(2)}$ then the equity shock depends upon

[28] The notation δX_t, for some stochastic variable X_t indexed by continuous time $t > 0$, is a short-hand for $X_t - X_{t-\delta t}$ where $\delta t \in \mathbb{R}$ s.t. $0 < \delta t \ll 1$.

[29] Again, but in discrete time I write $\Delta X_i = X_i - X_{i-1}$.

both v_{i-1} and Δv_i as follows:

$$\begin{aligned}\Delta X|(v,\Delta v) &:= \sqrt{v\Delta t}\left(\sqrt{1-\rho^2}Z^{(1)}+\rho Z^{(2)}\right)+\left(e^{\bar{\mu}+\sigma Z^{(3)}}-1\right)\Delta N,\\ &= \sqrt{v\Delta t}\left\{\sqrt{1-\rho^2}Z^{(1)}+\rho\left(\frac{\Delta v-\alpha(\theta-v)\Delta t}{\xi\sqrt{v\Delta t}}\right)\right\}\\ &\quad+\left(e^{\bar{\mu}+\sigma Z^{(3)}}-1\right)\Delta N\\ &= \frac{\rho}{\xi}\left(\Delta v-\alpha(\theta-v)\Delta t\right)+\sqrt{(1-\rho^2)v\Delta t}Z^{(1)}\\ &\quad+\left(e^{\bar{\mu}+\sigma Z^{(3)}}-1\right)\Delta N\end{aligned}$$

However, in order to be able to correlate two SVJD processes (or an SVJD process and another asset), a representation is needed for the {unconditional} distribution of the variable ΔX_i *marginalizing out* the values v for the variance (corresponding to its value at time index $i-1$), to its step change Δv over the time interval $t_{i-1}=(i-1)\Delta t$ to $t_i=i\Delta t$, to the equity asset's Brownian shock, to its Poisson jump intensity and log-normal jump size. Since this is to be achieved through simulation, the process of marginalization is trivial: one simulates each of the quantities at a given time step and once the equity asset shock has been formed, they are simply discarded. One must be mindful to observe the correct conditioning on the step change in the variance, but this is simple to arrange.

For the purposes of exposition, the following parametrization for the SVJD model was considered: $\mu=0, \bar{\mu}=-0.4, \lambda=0.1, \sigma=0.2767, \alpha=0.02462, \xi=0.1088, \theta=0.25, v_0=0.02206$ and $\rho=-0.9462$. In simulating SVJD paths over 30 years, a time step of $\Delta t=1/12$ was used leading to 360 time steps in total. A Spearman ranks-based shock correlation of 0.8 with a second, identically parametrized yet independent, SVJD process was set. Since the support of the unconditional equity asset shock variable ΔX is the whole real line, a value of $\kappa=3.664$ was found to be appropriate[30] for the transformation $y=\varphi(\Delta x;\kappa)=\tanh(\kappa\Delta x)$. A representation in Chebyshev polynomials for the distribution function of ΔX is now sought: $F_{\Delta X}(x)=F_{\Delta X}(\varphi^{-1}(y;\kappa))=: G_Y(y;\kappa)$.

5.4.2.3 Unconditional equity asset shock distribution

To simulate the unconditional ΔX shock (which will depend on time horizon T because the scaled non-central chi-squared solution to the CIR process depends on time horizon T), the conditional equity asset shock representation of the preceding section was used to marginalize the joint

[30] By taking $\kappa=log(2/\varepsilon-1)$, with $0<\varepsilon\ll 1$ (practically taking $\varepsilon=0.05$), this guarantees the transformation $y=\varphi(x;\kappa)=\tanh(\kappa x)$ will have a value of $-1+\varepsilon$ at $x=-1$. This ensures that the lower tail of the equity asset shock distribution, here dominated by the lognormal variable, is adequately modeled by part of the transform before the asymptotic behavior ensues.

distribution of (ΔX, v, Δv, $Z^{(1)}$, $Z^{(3)}$, ΔN) over the five-tuple (v, Δv, $Z^{(1)}$, $Z^{(3)}$, ΔN):

$$\Delta X = \frac{\rho}{\xi}\left(\Delta v - \alpha(\theta - v)\Delta t\right) + \sqrt{(1-\rho^2)v\Delta t}Z^{(1)} + \left(\mathrm{e}^{\bar{\mu}+\sigma Z^{(3)}} - 1\right)\Delta N$$

for every time horizon t of interest, that is, $t_i = i\Delta t$ each $i = 1, 2, 3, ..., T_{\max} = 360$. At times $t > s \geq 0$ the continuous-time CIR process conditioned on information up until time s is

$$v_t | v_s \sim k(t-s)\chi^2(\nu, \lambda(t-s, v_s))$$

where

$$k(t) = \frac{\xi^2(1-\mathrm{e}^{-\alpha t})}{4\alpha}, \quad \nu = \frac{4\alpha\theta}{\xi^2}, \quad \lambda(t,v) = \frac{4\alpha v}{\xi^2(\mathrm{e}^{\alpha t}-1)}$$

Using these results, variables $v_i \sim k(t_i)\chi^2(\nu, \lambda(t_i, v_0)), v_i|v_{i-1} \sim k(\Delta t)\chi^2(\Delta t, v_i)$, $Z^{(1:3)} \overset{iid}{\sim} N(0,1)$ were generated and the variance shock set to $\Delta v_i = (v_i|v_{i-1}) - v_{i-1}$. At each time-index i, the unconditional equity asset shock orthogonal polynomial representations were created for each variable ΔX_i.

To understand precisely how each ΔX_i was represented, consider the following. First, a $5 \times 10^4 \times 5$ array composed of 5×10^4 instances of a five-dimensional quasi-random uniform sequence was created using *Rmetrics*' **R** package **fOptions**.[31] Quasi-random variables were used rather than pseudo-random variables to avoid *clumping* or *clustering* of variables. Their effect is to span more homogeneously the support of the equity asset shocks with a smaller number of variables than would otherwise have been possible had pseudo-random numbers been used. I note that quasi-random rather than pseudo-random numbers could have been used in the previous section on PEDCP variables but this was not found to be necessary. It was advantageous when modeling the equity asset process since an entire process of 360 distributions were sought with much larger memory implications. The function **runif.sobol()** was used with a seed value of 1983 and Owen and Faure-Tezuka type scrambling. The following steps in generating the data were: (i) to generate two 5×10^4-vectors of standard normal deviates and one 5×10^4-vector of Poisson data using the third, fourth and fifth columns of the Sobol array and the appropriate quantile functions; (ii) to loop over the 360 time-periods constructing the scaled non-central chi-squared level and step-change data using the first and second columns of the Sobol array and appropriate quantile functions, and then marginalizing to determine the equity asset shock data; and (iii) to determine the coefficients in the Chebyshev representations using $K = 100$ terms.

It is important to emphasize that the derivation of the equity asset shock representations is a one-off cost that was done upfront in relatively little time.

[31] A package for "Pricing and Evaluating Basic Options" maintained on *CRAN* by the *Rmetrics Association Zurich*, https://www.rmetrics.org.

Knowledge of these distributions permits the evolution of the equity asset from S_{i-1} to S_i through application of the discrete equity asset dynamics for any number of Monte-Carlo trials. It is also not necessary to propagate the stochastic variance process v_i and this can be ignored. However, it can be done if desired in the following way. Knowledge of it is available at time zero: v_0; and so, computation of the value $v_i|(\Delta X_i, v_{i-1})$ is necessary to the required conditional distribution function. This is done by appealing to and rearranging the expression for the conditional equity asset shock given in the preceding section.

$$v_i|(\Delta X_i, v_{i-1}) = \underbrace{\frac{\xi}{\rho}\Delta X_i + \alpha\theta + v_{i-1}(1-\alpha)\Delta t}_{=:\chi(\Delta X_i, v_{i-1}) \text{ deterministic}}$$
$$\underbrace{-\frac{\xi}{\rho}\left[\sqrt{(1-\rho^2)v_{i-1}\Delta t}Z_i^{(1)} + \left(e^{\bar{\mu}+\sigma Z_i^{(3)}} - 1\right)\Delta N_i\right]}_{=:u_i|v_{i-1} \text{ stochastic}}$$

Recovery of v_i is now only attainable conditioned upon ΔX_i and v_{i-1}. Using the quasi-random five-dimensional Sobol set, one must represent the conditional distribution of v_i as a function of the tuple $(\Delta X_i,\ v_{i-1})$ in a bivariate Chebyshev expansion *at each step of the simulation.* Only in this way can *the* stochastic variance process v_i be consistently realized alongside its equity asset shock process ΔX_i at the time of simulation. Given a relatively small number of terms was necessary for an accurate representation of the equity asset shock process (around 15 terms were found to be sufficient, see Figure 5.9), one anticipates this number squared of terms to represent the variance process. It is stressed that this step is *not necessary* in the projection of correlated equity assets. However, it must be done in this way to produce a consistent stochastic variance process *if that is desired.*

The empirical distribution functions of the equity asset shock distributions ΔX_i at different time horizons are shown in Figure 5.8. The representations in terms of Chebyshev polynomials are omitted from the plot as they are so similar the differences are immaterial.

Bar plots of the coefficients at different timesteps: 1 month, 90 months (7.5 years), 180 months (15 years), and 360 months (30 years) are shown in Figure 5.9. Out of the $K = 100$ coefficients, only the first 15 have non-negligible coefficients.

Two equity asset processes $S_t^{(1)}$ and $S_t^{(2)}$ with the same underlying parametrization (as given above) are now simulated where the correlation[32] between their marginal shocks is set to $\tilde{\rho} = 0.8$. This is done by modifying the standard uniform shock pair $\mathbf{u} = (u_1, u_2)'$ to $\mathbf{v} = (v_1, v_2)'$ that seeds the equity asset shock quantile function. If the correlation matrix between the two

[32]Note: this is different from the correlation between the equity asset and stochastic variance shocks ρ.

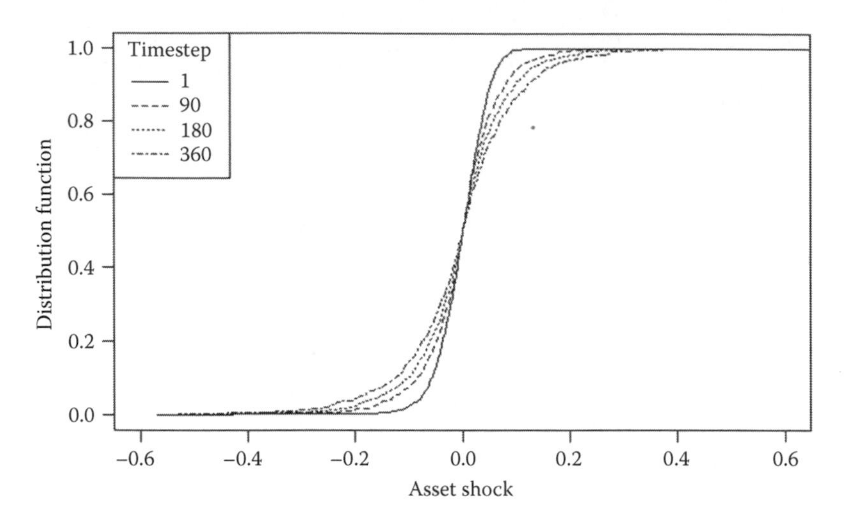

FIGURE 5.8: SVJD equity asset shock distributions.

equity asset shocks is (somewhat trivially):

$$R = \begin{bmatrix} 1 & \tilde{\rho} \\ \tilde{\rho} & 1 \end{bmatrix}$$

then the Cholesky factorization of R is LL^T where:

$$L = \begin{bmatrix} 1 & 0 \\ \tilde{\rho} & \eta \end{bmatrix}, \quad \eta := \sqrt{1 - \tilde{\rho}^2}$$

Let $\mathbf{\Phi} = (\Phi, \Phi)'$, where Φ is the standard normal distribution function, and $\mathbf{q} = (\Phi^{-1}, \Phi^{-1})'$, where q is the standard normal quantile function, be the independent (bivariate) standard normal distribution and quantile functions, respectively. The modified shock pair $\mathbf{v} = (v_1, v_2)'$ becomes $\mathbf{\Phi}(L\mathbf{q}(\mathbf{u}))$:

$$\begin{pmatrix} v_1 \\ v_2 \end{pmatrix} = \mathbf{\Phi}\left(\begin{bmatrix} 1 & 0 \\ \rho & \eta \end{bmatrix} \boldsymbol{q}\begin{pmatrix} u_1 \\ u_2 \end{pmatrix}\right)$$
$$= \begin{pmatrix} u_1 \\ \Phi\left(\rho\Phi^{-1}(u_1) + \eta\Phi^{-1}(u_2)\right) \end{pmatrix}$$

A plot of the equity asset paths is given in Figure 5.10, which also shows a scatterplot of the simulated equity asset shocks. The Spearman's correlation between the two equity asset processes $S_i^{(1)}$ and $S_i^{(2)}$ is: 0.2422. The correlation between the equity asset *shock* processes $\Delta X_i^{(1)}$ and $\Delta X_i^{(2)}$ that is targeted with a value of $\tilde{\rho} = 0.8$ was 0.8026 demonstrating the success of this exercise. To further verify the veracity of this modeling, one could compare the behavior of the Chebyshev asset shock representation with the standard coupled SDE representations when pricing path-dependent options. This is the subject of a future work.

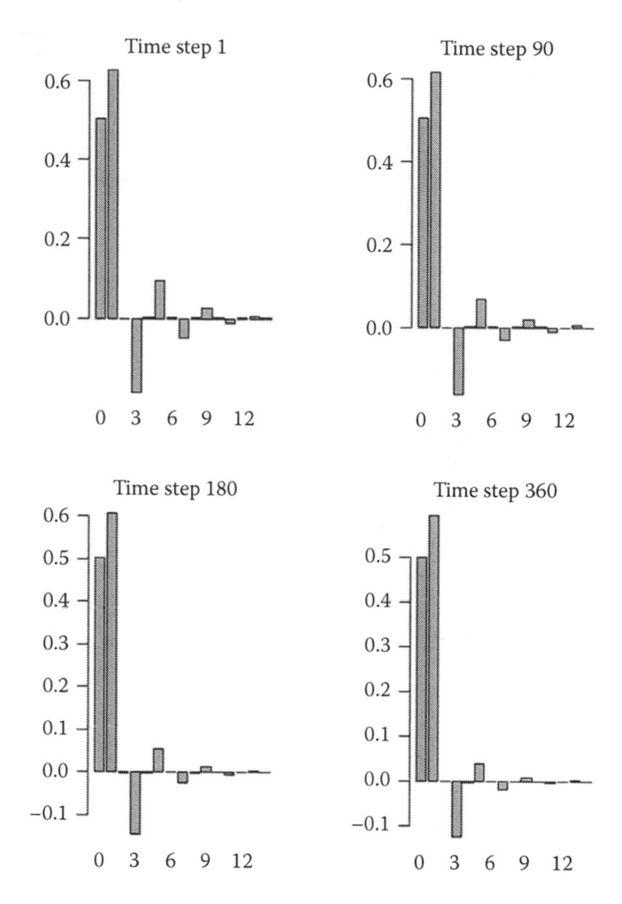

FIGURE 5.9: Bar plots of the coefficients from the Chebyshev polynomial representation of the SVJD equity asset shock distribution.

5.5 Discussion

In this chapter I have described the role of scenario generators within the context of insurance and have outlined some challenges that they present. Motivating the chapter with a high-level overview of the Solvency 2 European directive, I illustrated the use of scenario generators particularly when an insurer chooses to build its own internal model. This approach is costly but the cost may be offset against the potentially punitive regulatory capital laid down by the simpler yet prescriptive *standard model*. I described how Pillar 1 required a quantitative measure of downside risk: the 1 year VaR from an insurer's loss distribution. A scenario generator is then required to upload economic and risk scenarios to its asset and liability system to enable modeling of its market and non-market risks. Moreover, under Pillar 2, an insurer must

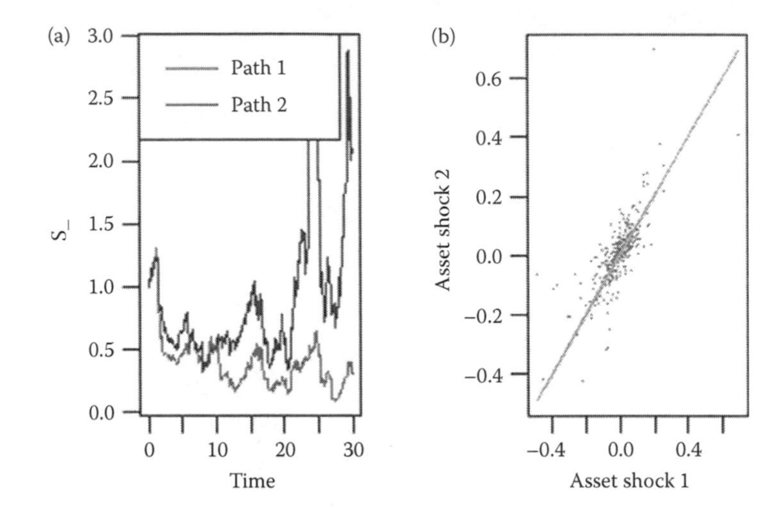

FIGURE 5.10: (a) Two correlated SVJD paths and (b) scatterplot of the SVJD equity asset shocks (correlation = 0.8026).

go beyond the 1-year horizon and a multi-year scenario generator is potentially required. The precise wording of Pillar 2 is deliberately vague, however, and a small set of *stylized* scenarios could be used instead. Indeed, a diligent insurer who uses Monte-Carlo trials to forecast its balance sheet must be able to price under the risk-neutral measure *at every time step* of the forecast. Rather than enduring a complexity of $O(N^2)$ under Pillar 1, an insurer would necessarily endure a cost of $O(p \times N^2)$ where $p > 2$ is the maximum number of periods projected: a truly daunting task if $p \gg 2$. This is perhaps the single most-challenging problem facing economic scenario generation: it is not that scenarios can't be generated fast enough it is that they face an I/O bottleneck when they are uploaded to an ALM system.

The bottleneck begs the question of whether *in-built* scenario generators within ALM systems could alleviate this problem? Unfortunately, a paradigm shift in the way insurers model their assets and liabilities would be required. One approach is to simplify the problem to one that is still amenable to current ALM systems. The *Least-Squares Monte Carlo* method has been pioneered by Cathcart (2012) and implemented as a commercial software solution by Moody's Analytics complementing its flagship *Scenario Generator* product. It leverages the use of statistical regression functions to proxy market-consistent Monte-Carlo valuations allowing calculations in $O(N)$-time. An adjacent approach is that of *curve fitting* (see also Cathcart (2012)). Truly high-performance computing solutions, for their part, are beginning to enter the sector. Aon Benfield Securities have implemented *PathWise*(™): a large-scale hedging tool for portfolios of variable annuities. Using a farm of GPUs, gamma losses are hedged effectively in real time with rebalancing happening in fractions of a second, Phillips (2016).

In Section 5.2, I outlined the modeling challenges faced by scenario generators. Each of the major asset classes requires a model before scenario generation can begin. Indeed, there is a rich literature in financial and quantitative engineering that scenario generators may rely upon. For example, I described the taxonomy of equity models that led from the simplest arithmetic Brownian motion model to the most complicated stochastic volatility jump diffusion (SVJD) model. The great statistician George Box is alleged to have said that *all models are wrong but some are useful* and I am inclined to agree with this sentiment. I would go further and say that: *a model is useful particularly when it is robust and parsimonious.* Given the difficulties of nonlinear and non-convex optimization that one naturally encounters in model calibration, sometimes less is more: difficult calibrations may serve as a warning signal that a little bias is better than large over-fit (or no fit at all when the optimizer ranges over the trajectory of a strange attractor). The thorny issue of nonlinear and non-convex optimization in multiple dimensions is a classic unsolved problem of modern mathematics. It must be treated with great care and often, simpler is not just better, it is the *only* option for coherent fits.

In Section 5.3, I introduced the concept of the *risk scenario generator* as a generalization of the *ESG* to include non-market risks such as policyholder lapse and surrender risk, or operational risk. RSGs typically appeal to statistical distributions to describe these risks in the absence of a robust financial theory such as that for nominal interest rate modeling. The introduction of statistical risk drivers tied in to a brief discussion on co-dependency and to the challenging problem of variable correlation. The converse of Sklar's theorem (1959) shows how to represent a multivariate risk distribution by its *copula* and *marginal distributions*: a flexible solution to the specification of full multivariate distributions. Having access to a risk driver's marginal distribution becomes important and problematic when no such analytical form exists for it. In an approach that is new, I showed how to develop representations of two challenging marginal risk-driver distributions in the absence of any analytical formulae. This type and level of challenge is typical of those faced in ESG and RSG development. First, in a statistical model for operational risk and second for the SVJD equity asset shock distribution. Correlation is once again possible and the copula-marginal factorization may continue to be used.

I purposely avoided discussion of the estimation of the term premia in equity, fixed interest, credit, and other models when working in the real world. Estimation of these is demanding and if using historical data or noisy data, one must take care to perform sensitivity analyses to check for robust parameter estimates. For example, in one approach to corporate credit spread modeling at a very high level, it may not even be clear from the data to which credit class a given set of spread data belongs. One may be forced to *bucket* the data in a manner that is arbitrary. How would estimates change if the buckets were modified? It is here, rather in the fitting of models to market-consistent data, that one should be mindful to produce standard error analyses. Economic justification for assumptions such as the unconditional forward interest rate (i.e., the interest rate applying to very long dated bonds) must be given. Justification of all assumptions is core to Pillar 3 of the Solvency 2 directive.

Despite the challenges faced by scenario generators, and I have by no means covered them all in this short chapter (and any views being my own, of course and not necessarily representative of the views held by Moody's Analytics), they will continue to remain core to an insurer's ability to set its regulatory and other capital requirements. As to the future, developments in high-performance computing have been slow to enter the arena. However, Solvency 2 has come into force only relatively recently. When insurers realize that they can manage their businesses more cost-effectively and more profitably by making more use of new technologies such as SaaS, HPC, and the cloud, they will drive the pace of change and Solvency 2 will have been the catalyst.

References

Antonio, D. and Roseburgh, D. *Fitting the Yield Curve: Cubic Spline Interpolation and Smooth Extrapolation.* Barrie & Hibbert Knowledge Base Article, Edinburgh, UK, 2010.

Bachelier, L. Théorie mathématique du jeu. *Annales Scientifiques de l'Ecole Normale Supérieure.*, Vol 18, pp 143–210, 1901.

Bates, D. S. Jumps and stochastic volatility: Exchange rate processes implicit in deutsche mark options. *Rev. Fin. Stud.*, Vol 9(1), pp 69–107, 1996.

Baxter, M. and Rennie, A. *Financial Calculus: An Introduction to Derivative Pricing.* Cambridge University Press. Cambridge, 1996.

Berndt, A. R., Douglas, R., Duffie, D., Ferguson, F., and Schranz, D. *Measuring Default Risk in Premia from Default Swap Rates and EDFs.* Preprint, Stanford University. 2004.

Black, F. and Scholes, M. The Pricing of Options and Corporate Liabilities. *J. Political Econ.*, Vol 81(3), pp 637–654, 1973.

Box, G. E. P. and Muller, E. A note on the generation of random normal deviates. *Annals Math. Stat.*, Vol 29(2), pp 610–611, 1958.

Brace, A., Gatarek, D., and Musiela, M. The market model of interest rate dynamics. *Math. Fin.*, Vol 7(2), pp 127–154, 1997.

Brigo, D. and Mercurio, F. *Interest Rates Models: Theory and Practice.* Springer, Berlin, 2006.

Carr, P., Geman, H., Madan, D. and Yor, M. The fine structure of asset returns: An empirical investigation. *J. Busin.*, Vol 75(2), pp 305–332, 2002.

Cathcart, M. J. Monte-Carlo simulation approaches to the valuation and risk management of unit-linked insurance products with guarantees. Doctoral thesis, School of Mathematical and Computer Sciences, Heriot-Watt University, UK. 2012.

Chang, W. Y., Abu-Amara, H., and Sanford, J. F. *Transforming Enterprise Cloud Services.* Springer, London, 2010.

DeGroot, M. H. and Schervish, M. *Probability and Statistics.* (4th ed.), Pearson, Cambridge, 2013.

Duffie, D., Pan, J., and Singleton, K. Transform analysis and asset for affine jump-diffusions. *Econometric*, Vol 68(6), pp 1343–1376, 2000.

EIOPA. *Technical Documentation of the Methodology to Derive EIOPA's Risk-free Interest Rate Term Structures.* EIOPA technical documentation, EIOPA-BoS-15/035. https://eiopa.europa.eu/Publications/Standards/. 2015.

Gatheral, J. *The Volatility Surface: A Practitioner's Guide.* Wiley Finance. Hoboken, NJ.2006.

Glasserman, P. *Monte-Carlo Methods in Financial Engineering.* Springer. New York, NY.2003.

Haberman, S. and Renshaw, A. 2008. Mortality, longevity and experiments with the Lee-Carter model. *Lifetime Data Anal.*, Vol 14, pp 286, doi: 10.1007/s10985-008-9084-2

Hagan, P., Kumar, D., Lesniewski, A. E., and Woodward, D. Managing smile risk. *Wilmott Magazine*, Vol 1, pp 84–108, 2002.

Hale, J. K. and Kocak, H. Dynamics and bifurcations. *Texts Appl. Math.*, Vol 3, pp 444–494, Springer 1991.

Hastie, T., Tibshirani, R., and Friedman, J. *The Elements of Statistical Learning: Data Mining, Inference and Prediction.* (2nd ed.), Springer, New York, NY, 2009.

Heath, D., Jarrow, R., and Morton, A. Bond pricing and the term structure of interest rates: A discrete time approximation. *J. Fin. Quant. An.*, Vol 25, pp 419–440, 1990.

Heston, S. L. A closed-form solution for options with stochastic volatility with applications to bond and currency options. *Rev. Fin. Studies.*, Vol 6(2), pp 327–343, 1993.

Hull, J. C. *Options, Futures and Other Derivatives.* (6th ed.), Prentice-Hall, Upper Saddle River, NJ, 2005.

Hull, J. and White, A. Pricing interest-rate derivative securities. *Rev. Fin. Stud.*, Vol 3(4), pp 573–592, 1990.

Lando, D. *Credit Risk Modelling: Theory and Applications.* Princeton University Press, Princeton, 2004.

Mackenzie, D. and Spears, T. *The Formula that Killed Wall Street?*, The Gaussian Copula and the Material Cultures of Modelling. Working Paper, 2012.

Matsumoto, M. and Nishimura, T. Mersenne twister: A 623-dimensionally equidistributed uniform pseudo-random number generator. *ACM Trans. Mod. Comp. Sim.*, Vol 8(1), pp 3–30, 1998.

McNeil, A. J., Frey, R., and Embrechts, P. *Quantitative Risk Management: Concepts, Techniques and Tools.* (2nd ed.), Princeton University Press. Woodstock, UK; New Jersey, USA, 2015.

Merton, R. C. Option pricing when underlying stock returns are discontinuous. *J. Fin. Econ.*, Vol 3, pp 125–144, 1976.

Moudiki, T. and Planchet, F. Economic scenario generators. Book chapter in Laurent, J. P., Ragnar, N., Planchet, F. (eds) *Modelling in Life Insurance—A Management Perspective. EAA Series*, Springer, 2016. doi: 10.1007/978-3-319-29776-7.

Nelsen, C. R. and Siegel, A. F. Parsimonious modeling of yield curves. *J. Bus.*, Vol 60(4), pp 4733–4489, 1987.

Neil, C., Clark, D., Kent, J., and Verheugen, H. *A Brief Overview of Current Approaches to Operational Risk under Solvency II.* Milliman White Paper series, 2012.

Noble, B. and Daniel, J. W. *Applied Linear Algebra.* Prentice-Hall, Englewood Cliffs, NJ, 1977.

Phillips, P. *PathWise(TM) High Productivity Computing Platform.* Aon Securities Inc., Aon Benfield. http://www.aon.com/reinsurance/aon-securities/pathwise-solutions-group.jsp, 2016.

Press, W. H., Teukolsky, S. A., Vetterling, W. T., and Flannery, B. P. *Numerical Recipes: The Art of Scientific Computation* (3rd ed.), Cambridge University Press, New York, NY, 2007.

Salahnejhad Ghalehjooghi, A. and Pelsser, A. *Time-Consistent Actuarial Valuations.* Available at SSRN: http://ssrn.com/abstract=2614002 or http://dx.doi.org/10.2139/ssrn.2614002, 2015.

Sklar, A. *Fonctions de repartition a n dimensions et leurs marges.* Publ. Inst. Statist. Univ. Paris, Vol 8, pp 229–231. 1959.

Transtrum, M. K. and Sethna, J. P. *Improvements to the Levenberg-Marquardt Algorithm for Nonlinear Least-Squares Minimization.* arXiv preprint, arXiv:1201.5885, https://arxiv.org/abs/1201.5885. 2012.

Varnell, E. M. Economic scenario generators and solvency II. *BAJ*, Vol 16, pp 121–159, 2011. doi: 10.1017/S1357321711000079.

Wichmann, B. and Hill, D. Algorithm AS 183: An efficient and portable Pseudo-random number generator. *J. Roy. Stat. Soc. C (Appl. Stat.)*, Vol 31(2), pp 188–190, 1982.

Wichmann, B. and Hill, D. Correction: Algorithm AS 183: An efficient and portable pseudo-random number generator. *J. Roy. Stat. Soc. C (Appl. Stat.)*, Vol 33(1), pp 123, 1984.

Wu, L. and Zhang, F. Libor market model with stochastic volatility. *J. Indust. Mgmt. Opt.*, Vol 2(2), pp 199–227, 2006.

Zyskind, G. and Martin, F. B. On best linear estimation and general Gauss-Markov theorem in linear models with arbitrary nonnegative covariance structure. *SIAM J. App. Math.*, Vol 17(6), pp 1190–1202.

Part II

Numerical Methods in Financial High-Performance Computing (HPC)

Chapter 6

Finite Difference Methods for Medium- and High-Dimensional Derivative Pricing PDEs

Christoph Reisinger and Rasmus Wissmann

CONTENTS

6.1 Introduction

Many models in financial mathematics and financial engineering, particularly in derivative pricing, can be formulated as partial differential equations (PDEs). Specifically, for the most commonly used continuous-time models of asset prices the value function of a derivative security, that is the option value as a function of the underlying asset price, is given by a PDE. This opens up the possibility to use accurate approximation schemes for PDEs for the numerical computation of derivative prices.

As the computational domain is normally a box, or can be restricted to one by truncation, the construction of tensor product meshes and spatial finite difference stencils is straightforward [1]. Accurate and stable splitting methods have become standard for efficient time integration [2].

Notwithstanding this, the most common approach in the financial industry appears to be Monte Carlo methods. This is partly a result of the perception that PDE schemes, although highly efficient for simple contracts, are less flexible and harder to adapt to more exotic features. In particular, the widespread belief is that PDE schemes become too slow for practical use if the number of underlying variables exceeds 3.

Indeed, the increase in computational time and memory requirements of standard mesh-based methods with the dimension is exponential and has become known as the "curse of dimensionality." Various methods, such as sparse grids [3,4], radial basis functions [5], and tensor approaches ([6] for an application to finance and [7] for a literature survey), have been proposed to break this curse. These methods can perform remarkably well for special cases, but have not been demonstrated to give accurate enough solutions for truly high dimensions in applications (larger than, say, 5).

In conversations about numerical methods for high-dimensional PDEs inevitably the question comes up: "How high can you go?" This is a meaningful question if one considers a specific type of PDE with closely defined characteristics. But even within the fairly narrow class of linear second-order parabolic PDEs which are most common in finance, the difficulty of solving them varies vastly and depends on a number of factors: the input data (such as volatilities and correlations), the boundary data (payoff), and the quantity of interest (usually the solution of the PDE at a single point).

It is inherent to the methods presented in this chapter that it is not the nominal dimension of a PDE which matters. A PDE which appears inaccessible to numerical methods in its raw form may be very easily approximated if a more adapted coordinate system is chosen. This can be either because the solution is already adequately described by a low number of principal components (it has low "truncation dimension"), or because it can be accurately represented as the sum of functions of a low number of variables (it has low "superposition dimension").

To exploit such features, we borrow ideas from data analysis to represent the solutions by sums of functions which can be approximated by PDEs with low effective dimension. More specifically, the method is a "dynamic" version of the anchored-ANOVA decompositions which were applied to integration problems in finance in Reference 8. A version which is equivalent in special cases has been independently derived via PDE expansions in Reference 3; a detailed error analysis is found in Reference 9 and also in Reference 10; an efficient parallelization strategy is proposed in Reference 11; and the method is extended to complex derivatives in Reference 12 and to CVA computations in Reference 13. The link of these methods to anchored-ANOVA is already observed in References 14 and 15. We present here a systematic approach which extends [9] from Black–Scholes to more general models, and analyze the accuracy of the approximations by way of carefully chosen numerical tests.

In the remainder of this section, we describe the mathematical framework. Then, in Section 6.2, we describe the standard approximation schemes. In Section 6.3, we define and explain in detail a dimension-wise decomposition. Section 6.4 summarizes known theoretical results for the constant coefficient case and offers a heuristic argument for the accuracy of a variable coefficient extension. Section 6.5 gives numerical results for test cases. We draw a conclusion in Section 6.6.

Throughout this chapter, we study asset price processes of the form

$$dS_t^i = \mu_i(S_t, t)\, dt + \sigma_i(S_t, t)\, dW_t^i, \quad i = 1, \ldots, N, \quad t > 0, \tag{6.1}$$

$$S_0^i = s_i, \quad i = 1, \ldots, N, \tag{6.2}$$

where W is an N-dimensional standard Brownian motion, $s \in \mathbb{R}^N$ is a given initial state, the drift μ_i and local volatility σ_i are functions $\mathbb{R}^N \times [0, T] \to \mathbb{R}$, and we will allow the correlation between the Brownian drivers also to be "local," that is, given S_t at time t the instantaneous correlation matrix is $(\rho_{ij}(S_t, t))_{1 \le i,j \le N}$. We consider European-style financial derivatives on S_T with maturity $T > 0$ and payoff function $h : \mathbb{R}^N \to \mathbb{R}$, whose value function $V : \mathbb{R}^N \times [0, T] \to \mathbb{R}$ can be written as

$$V(s, t) = \mathbb{E}[\exp(-\textstyle\int_t^T \alpha(S_u, u)\, du) h(S_T) | S_t = s],$$

where α is a discount factor, possibly stochastic through its dependence on S, and V satisfies the Kolmogorov backward PDE [16]

$$\frac{\partial V}{\partial t} + \sum_{i=1}^{N} \mu_i \frac{\partial V}{\partial s_i} + \frac{1}{2} \sum_{i,j=1}^{N} \sigma_i \sigma_j \rho_{ij} \frac{\partial^2 V}{\partial s_i \partial s_j} - \alpha V = 0,$$

$$V(s, T) = h(s).$$

For simplicity, we consider functions defined on the whole of $\mathbb{R}^N$, but it will become clear how to deal with bounded domains.

Let $p(y, t; s, 0)$ be the transition density function of S_t at y given state s at $t = 0$. Then if α does not depend on S, we can write

$$V(s, 0) = \exp(-\textstyle\int_0^T \alpha(u)\, du) \displaystyle\int_{\mathbb{R}^N} p(y, T; s, 0) h(y)\, dy.$$

Here, p satisfies the Kolmogorov forward equation

$$-\frac{\partial p}{\partial t} - \sum_{i=1}^{N} \frac{\partial}{\partial y_i} (\mu_i p) + \frac{1}{2} \sum_{i,j=1}^{N} \frac{\partial^2}{\partial y_i \partial y_j} (\sigma_i \sigma_j \rho_{ij} p) = 0,$$

$$p(y, 0; s, 0) = \delta(y - s),$$

where δ is the Dirac distribution centered at 0.

Most commonly, one is interested in approximating the value of $V(s_0, 0)$ for a given, fixed $s_0 \in \mathbb{R}^N$, and derivatives of V with respect to s_0.

As a first step, we change the time direction to *time-to-maturity*, $t \to T-t$, to obtain

$$\frac{\partial V}{\partial t} = \sum_{i=1}^{N} \mu_i \frac{\partial V}{\partial s_i} + \frac{1}{2} \sum_{i,j=1}^{N} \sigma_i \sigma_j \rho_{ij} \frac{\partial^2 V}{\partial s_i \partial s_j} - \alpha V, \tag{6.3}$$

$$V(s, 0) = h(s), \tag{6.4}$$

where we keep the symbols t and V for simplicity. We now transform the PDE into a standard form by using a rotation and subsequent translation of the spatial coordinates. For a given orthogonal matrix $Q \in \mathbb{R}^{N \times N}$, define $\beta : \mathbb{R}^N \times [0, T] \to \mathbb{R}^N$ componentwise by

$$\beta_i(x, t) \equiv \sum_{j=1}^{N} Q_{ji} \int_0^t \mu_j(x, T-u)\,\mathrm{d}u \tag{6.5}$$

for $1 \le i \le N$. We then introduce new spatial coordinates x via

$$x(s, t) = Q^{\mathrm{T}} s + \beta(s_0, t) \tag{6.6}$$

and set

$$a = Q^{\mathrm{T}} s_0 + \beta(s_0, T). \tag{6.7}$$

We write $s(x, t) = Q(x - \beta(s_0, t))$ for the inverse transform.

A simple calculation shows that the PDEs (6.3–6.4) transform into

$$\frac{\partial V}{\partial t} = \mathcal{L}V \quad := \sum_{k,l=1}^{N} \lambda_{kl} \frac{\partial^2 V}{\partial x_k \partial x_l} + \sum_{k=1}^{N} \kappa_k \frac{\partial V}{\partial x_k} - \alpha V, \tag{6.8}$$

$$V(x, 0) = g(x) \; := h(s(x, 0)), \tag{6.9}$$

for a function $V : \mathbb{R}^N \times [0, T] \to \mathbb{R}$, $T > 0$, where we still call the transformed function V by slight abuse of notation, and

$$\begin{aligned} \lambda_{kl}(x, t) &= \frac{1}{2} \sum_{i,j=1}^{N} Q_{ik} Q_{jl} \sigma_i \sigma_j \rho_{ij}, \\ \kappa_k(x, t) &= \sum_{i=1}^{N} Q_{ik} \left[\mu_i - \mu_i(s_0, T-t)\right], \end{aligned} \tag{6.10}$$

where σ_i and ρ_{ij} are functions of $(s(x, t), T-t)$.

For a constant (i.e., independent of time and the spatial coordinates), positive semidefinite coefficient matrix $\Sigma = (\Sigma_{ij})_{1 \le i,j \le N} = (\sigma_i \sigma_j \rho_{ij})_{1 \le i,j \le N}$, we can choose Q to be the matrix of eigenvectors of Σ sorted by eigenvalue size,[1] that is,

$$Q = (q_1, \ldots, q_N), \quad \frac{1}{2}\Sigma q_i = \lambda_i q_i, \ \ \lambda_1 \geq \cdots \geq \lambda_N \geq 0, \tag{6.11}$$

and get $(\lambda_{kl})_{1\leq k,l,\leq N} = \mathrm{diag}(\lambda_1, \ldots, \lambda_N)$ as a constant diagonal matrix.

If μ does not depend on the spatial coordinates x but only on t, then the difference under the sum in Equation 6.10 vanishes identically and thus $\kappa(x,t) \equiv 0$.

Moreover, if α is also only a function of t, the zero order term can be eliminated from Equation 6.8 by considering $\exp\left(\int_0^t \alpha(T-u)\,\mathrm{d}u\right) V$.

If all this is satisfied, then $\mathcal{L}$ simplifies to the N-dimensional heat operator in Equation 6.12. Keeping the symbol V for the transformed value function and $\mathcal{L}$ for the operator for simplicity, we obtain

$$\frac{\partial V}{\partial t} = \mathcal{L}V = \sum_{k=1}^{N} \lambda_k \frac{\partial^2 V}{\partial x_k^2}, \tag{6.12}$$

$$V(x,0) = g(x), \tag{6.13}$$

for $x \in \mathbb{R}^N$, $t \in (0,T)$, $\lambda = (\lambda_1, \ldots, \lambda_N) \in \mathbb{R}_+^N$.

In all other cases, that is, if Σ is not constant and μ depends on s, a transformation to a diagonal diffusion without drift is generally not possible. By translation to $s = s_0 + \int_0^T \mu(s_0,u)\,\mathrm{d}u$ and choosing Q as the eigenvectors of $\Sigma(s,T)$, one obtains $\lambda_{kl}(a,0) = 0$ for $k \neq l$ and $\kappa_k(a,0) = 0$, but these coefficients are nonzero for other (x,t).

6.2 Finite Difference Schemes

In this section, we describe the finite difference schemes used for the one- and two-dimensional versions of Equations 6.12 and 6.8 which we will need to construct the dimension-wise splitting introduced in Section 6.3. We choose the Crank–Nicolson scheme for the one-dimensional equations, Brian's scheme [17] for multidimensional PDEs without cross-derivatives, and the Hundsdorfer–Verwer (HV) scheme [18] for PDEs with cross-derivative terms. These are established techniques from the literature which are routinely used in financial institutions for derivative pricing, and can be replaced by a method of choice. As such, this section can be skipped without loss of continuity.

We follow standard procedure [1] to define a finite difference approximation V_h to V, where $h = (\Delta t, \Delta x_1, \ldots, \Delta x_d)$ contains both the time step size

[1] If Σ has eigenvectors with multiplicity larger than 1, then this decomposition is not unique. In that case, we can simply choose any such matrix Q.

$\Delta t > 0$ and the spatial mesh sizes $\Delta x_i > 0$, $i = 1, \ldots, d$, where d is the dimension of the PDE. We first define basic finite difference operators

$$\begin{aligned}
\delta_t V_h(\cdot, t) &= \frac{V_h(\cdot, t + \Delta t) - V_h(\cdot, t)}{\Delta t}, \\
\delta_x^i V_h(\cdot, t) &= \frac{V_h(\cdot + \Delta x_i, t) - V_h(\cdot - \Delta x_i, t)}{2\Delta x_i}, \\
\delta_x^{i,i} V_h(\cdot, t) &= \frac{V_h(\cdot + \Delta x_i, t) - 2V_h(\cdot, t) + V_h(\cdot - \Delta x_i, t)}{\Delta x_i^2}, \\
\delta_x^{i,j} V_h &= \delta_x^i \delta_x^j V, \quad i \neq j,
\end{aligned}$$

and then an approximation to $\mathcal{L}$ by

$$L(t) = \sum_{i=1}^{d} \kappa_i(\cdot, t)\, \delta_x^i + \sum_{i,j=1}^{d} \lambda_{ij}(\cdot, t)\, \delta_x^{i,j} - \alpha(\cdot, t),$$

where the operator $\kappa_i(\cdot, t_n)\delta_x^i$, applied to V_h, at a point $x = (x_{j_1}, \ldots, x_{j_d})$ is

$$((\kappa_i(\cdot, t_n)\delta_x^i)V_h)_{j_1, \ldots, j_d} = \kappa_i(x, t_n) \frac{V_h(x + \Delta x_i e_i, t_n) - V_h(x - \Delta x_i e_i, t_n)}{2\Delta x_i},$$

where e_i, the ith unit vector, and similar for the σ and α terms.

Ignoring spatial boundaries for the time being, V_h is defined for all $(x, t) \in \mathbb{R}^d \times \{0, \Delta t, \ldots, T\}$ by the scheme

$$\begin{aligned}
\delta_t V_h &= \theta L(t + \Delta t) V_h(t + \Delta t) + (1 - \theta) L(t) V_h(t), \qquad (6.14) \\
V_h(x, T) &= \phi(x),
\end{aligned}$$

where $\theta \in [0, 1]$. Here, $\Delta t = T/N_t$, where N_t is the number of timesteps.

In practice, the scheme and solution need to be restricted to a bounded domain, and for simplicity we restrict ourselves here to a box where $x_{i,\min} \leq x_i \leq x_{i,\max}$. These may be given naturally, for example, $x_{\min} = 0$ if x is a positive stock price, or by truncation of an infinite interval at suitably large values, for example, a certain number of standard deviations away from the spot. Then with N_i the number of mesh intervals in coordinate direction x_i, $\Delta x_i = (x_{i,\max} - x_{i,\min})/N_i$, the mesh points are $x_{i,j} = x_{i,\min} + j\Delta x_i$ for $j = 0, \ldots, N_i$, $i = 1, \ldots, d$. We denote the numerical solution on this mesh by U_n, this being the vector $(V_h((x_{i,j_i})_{i=1,\ldots,N}, t_n))_{j_i=0,\ldots,N_i}$.

Let $L^n \equiv L(t_n)$ be the discretization matrix at time step t_n, then this matrix is first decomposed into

$$L^n = L_0^n + L_1^n + \cdots + L_d^n,$$

where the individual $L_i^n, 1 \leq i \leq d$, contain the contribution to L stemming from the first- and second-order derivatives in the ith dimension,

$$L_i^n = \kappa_i(\cdot, t_n)\delta_x^i + \lambda_{ii}(\cdot, t_n)\delta_x^{i,i} - \frac{1}{d}\alpha(\cdot, t_n),$$

and, following Reference 2, we define one matrix F_0 which accounts for the mixed derivative terms,

$$L_0^n = \sum_{i \neq j} \lambda_{ij}(\cdot, t_n)\delta_x^{i,j}.$$

For $L_0^n = 0$, which contains the discretization of Equation 6.12 as a special case, a simple splitting scheme is given by the Douglas scheme [19],

$$\begin{aligned} Y_0 &= U_{n-1} + \Delta t L^{n-1} U_{n-1}, \\ (I - \theta \Delta t L_j^n) Y_j &= Y_{j-1} - \theta \Delta t L_j^{n-1} U_{n-1}, \quad j = 1, \ldots, d, \\ U_n &= Y_d. \end{aligned} \tag{6.15}$$

The scheme is unconditionally stable for all $\theta \geq 1/2$ and of second order in time for $\theta = 1/2$ (otherwise of first order, see Reference 20).

A second-order modification of the above scheme was proposed by Brian [17], where the first two steps are as above with $\theta = 1$ and step size $\Delta t/2$, and the last step (6.15) is replaced by a Crank–Nicholson-type step

$$\frac{U_n - U_{n-1}}{\Delta t} = \sum_{j=1}^{d} \frac{1}{2}(L_j^n + L_j^{n-1}) Y_j.$$

For $L_0^n \neq 0$, that is, with cross-derivative terms present as in the general case of Equation 6.14, second order gets lost and an iteration of the idea is needed. The HV scheme [18],

$$\begin{aligned} Y_0 &= U_{n-1} + \Delta t L^{n-1} U_{n-1}, \\ (I - \theta \Delta t L_j^n) Y_j &= Y_{j-1} - \theta \Delta t L_j^{n-1} U_{n-1}, \quad j = 1, 2, 3, \\ \widetilde{Y}_0 &= Y_0 + \frac{1}{2} \Delta t \left[L^n Y_3 - L^{n-1} U_{n-1} \right], \\ (I - \theta \Delta t L_j^n) \widetilde{Y}_j &= Y_{j-1} - \theta \Delta t L_j^n Y_j, \quad j = 1, 2, 3, \\ U_n &= \widetilde{Y}_3, \end{aligned}$$

defines a second-order consistent ADI splitting for all θ, and can be shown to be *von Neumann* stable for $\theta \in \left[\frac{1}{2} + \frac{1}{6}\sqrt{3}, 1\right]$ [21]. We use $\theta = \frac{1}{2} + \frac{1}{6}\sqrt{3} \approx 0.789$ in the computations.

A severe computational difficulty arises for d larger than approximately 3, as the total number of operations is proportional to $N_t N_1 \ldots N_d$, that is, grows exponentially in the dimension. In the numerical tests, we will use $N_1 = N_2 = 800$ and $N_t = 1000$ for the two-dimensional equations. These involve 6.4×10^8 unknowns. In Reference 9, for a second-order extension, $N_1 = N_2 = N_3 = 500$ and $N_t = 50$ are used for the three-dimensional equations involved, that is, 6.25×10^9 unknowns. It is clear that within this framework a further increase in the dimension will only be practically feasible by reducing the number of mesh points in each direction and consequently sacrificing accuracy.

6.3 Decomposition Methods

In order to accurately approximate derivative prices with $N > 3$ factors, we define an approximate dimension-wise decomposition, in the spirit of anchored-ANOVA decompositions. Here, the starting point a of the transformed process, from Equation 6.7, serves as an "anchor." We show the basic concept in a static setting in Section 6.3.1, and its application to constant and variable coefficient stochastic processes and PDEs in the subsequent sections.

We assume in this section that a suitable rotation and translation (see the end of Section 6.1) has taken place, so that

$$\lambda_{ij}(a, 0) = 0, \quad i \neq j, \tag{6.16}$$

$$\kappa_i(a, 0) = 0. \tag{6.17}$$

We then denote for simplicity

$$\lambda_i(x, t) \equiv \lambda_{ii}(x, t).$$

For brevity, we set $\alpha = 0$ in this section, but the extension to $\alpha \neq 0$ is straightforward.

6.3.1 Anchored-ANOVA decomposition

We follow here Reference 8 to define the anchored-ANOVA decomposition of a function $g : \mathbb{R}^N \to \mathbb{R}$, with a given "anchor" $a \in \mathbb{R}^N$. For a given index set $u \subset \mathcal{N} = \{i : 1 \leq i \leq N\}$, denote by $a \backslash x_u$ the N-vector

$$(a \backslash x_u)_i = \begin{cases} x_i, & i \in u, \\ a_i, & i \notin u. \end{cases}$$

Then $g_u(a; \cdot)$ defined for all $x \in \mathbb{R}^N$ by $g_u(a; x) = g(a \backslash x_u)$ is a projection of g, where we make the dependence of g_u on the anchor a explicit in the notation.

We proceed to define a difference operator Δ recursively through $\Delta g_\emptyset = g_\emptyset$ and, for $u \neq \emptyset$,

$$\Delta g_u = g_u - \sum_{w \subset u} \Delta g_w = \sum_{w \subseteq u} (-1)^{|w|-|u|} g_w.$$

An exact decomposition of g is then given by the identity

$$g = \sum_{u \subseteq \mathcal{N}} \Delta g_u = \sum_{k=0}^{N} \sum_{|u|=k} \Delta g_u. \tag{6.18}$$

This enables the definition in Reference 8 of successive dimension-wise approximations to the integral of g by truncation of the series.

6.3.2 Constant coefficient PDEs

We start by considering the N-dimensional heat equation

$$\frac{\partial V}{\partial t} = \mathcal{L}V = \sum_{k=1}^{N} \lambda_k \frac{\partial^2 V}{\partial x_k^2}, \tag{6.19}$$

$$V(\cdot, 0) = g, \tag{6.20}$$

with constant λ.

Given an initial-value problem of the form 6.19 and 6.20, and an index set $u \subseteq \mathcal{N}$, define a differential operator

$$\mathcal{L}_u = \sum_{k \in u} \lambda_k \frac{\partial^2}{\partial x_k^2},$$

and an approximation V_u of V as the solution to

$$\frac{\partial V_u}{\partial t} = \mathcal{L}_u V_u, \tag{6.21}$$

$$V_u(\cdot, 0) = g. \tag{6.22}$$

The definition in Equation 6.21 is equivalent to saying

$$\frac{\partial V_u}{\partial t} = \mathcal{L}V_u,$$

$$V_u(x, 0) = g(a \backslash x_u),$$

that is, projecting the initial condition, but it is not normally true that V_u from Equation 6.21 is the projection of the solution V of Equation 6.19 in the sense of Section 6.3.1.

From here on, we can proceed as in Section 6.3.1 to set

$$\Delta V_u == \sum_{w \subseteq u} (-1)^{|w|-|u|} V_w.$$

To approximate V by lower dimensional functions, we truncate the series in Equation 6.18 and define

$$V_{0,s} = \sum_{k=0}^{s} \sum_{|u|=k} \Delta V_u = \sum_{k=0}^{s} c_k \sum_{|u|=k} V_u, \tag{6.23}$$

where c_k are integer constants that depend on the dimensions N and s. The point to note is that V_u is essentially a $|u|$-dimensional function as it only depends on the fixed anchor and $|u|$ components of x.

In situations where one or several coordinates play a dominant role, it will be useful to consider a generalization of Equation 6.23 to

$$V_{r,s} = \sum_{k=0}^{s} c_k \sum_{|u|=k} V_{u \cup \{1,\ldots,r\}}, \quad r + s \leq N. \tag{6.24}$$

Here, all components $V_{u \cup \{1,\ldots,r\}}$ depend on all the $x_1, \ldots, x_r$.

6.3.3 Variable coefficients: Full freezing

The simplest way to deal with variable coefficients is to "freeze" them at a constant value and then apply the methodology from Section 6.3.2. As we are interested in the PDE solution at the anchor point a, the obvious choice is to approximate κ_i and λ_{ij} by $\kappa_i(a, 0)$ and $\lambda_{ij}(a, 0)$.

For a given subset $u \subseteq \mathcal{N}$, we then define (note that in this case $\kappa_i(a, 0) = 0$ and $\lambda_{ij} = 0$, $i \neq j$)

$$\frac{\partial V_u}{\partial t} = \sum_{i \in u} \lambda_{ii}(a, 0) \frac{\partial^2 V_u}{\partial x_i^2},$$
$$V_u(x, 0) = g(x).$$

6.3.4 Partial freezing

The full freezing approximation in Section 6.3.3 throws away more information than needed. In the following extension, we keep as much as possible of the original dynamics of the process in the low-dimensional cross-section the process is restricted to.

For given subset $u \subseteq \mathcal{N}$, we now define

$$\frac{\partial V_u}{\partial t} = \sum_{i \in u} \kappa_i(a \backslash x_u, t) \frac{\partial V_u}{\partial x_i} + \sum_{i,j \in u} \lambda_{ij}(a \backslash x_u, t) \frac{\partial^2 V_u}{\partial x_i \partial x_j},$$
$$V_u(x, 0) = g(x).$$

Given the variability of the coefficients, there is generally no static coordinate transformation that reduces the PDE to the heat equation. The difference to the localized problem in the previous section is that since the PDE coefficients $\lambda(x, t)$ and $\kappa(x, t)$ change with spatial and time coordinates, the PDE will in general contain first-order and nondiagonal second-order terms.

6.3.5 Partial freezing and zero-correlation approximation

Here, motivated by $\lambda_{ij}(a, 0) = 0$ for all $i \neq j$, we make the additional approximation that this holds for all x and t. So we define now

$$\frac{\partial V_u}{\partial t} = \sum_{i \in u} \kappa_i(a \backslash x_u, t) \frac{\partial V_u}{\partial x_i} + \sum_{i \in u} \lambda_{ii}(a \backslash x_u, t) \frac{\partial^2 V_u}{\partial x_i^2},$$
$$V_u(x, 0) = g(x).$$

This extra approximation in addition to Section 6.3.4 does not give any further dimension reduction, but simplifies the PDEs somewhat, that is, no cross-derivative terms are present, which simplifies the construction of numerical schemes.

6.4 Theoretical Results

In this section, we review the rigorous error analysis from Reference 9 for the constant coefficient case in Section 6.3.2, and give a novel, more heuristic extension of this analysis to the variable coefficient setting of Section 6.3.4.

What is essential in the analysis is clearly the size of the diffusion and drift coefficients in the various directions, as well as the variability of the initial data jointly with respect to different sets of variables. The relevant measure of variability is defined in the following.

Definition 6.1. *Let*

$$C^{j,k,\text{mix}} = \left\{ g \in C^b : \partial^j_{i_1} \dots \partial^j_{i_k} g \in C^b, \; \forall 1 \le i_1 < \dots < i_k \le N \right\},$$
$$C^b = \left\{ g : \mathbb{R}^N \to \mathbb{R} \; \textit{continuous} : |g(x)| \le c \, \textit{for all } x \textit{ for some } c > 0 \right\}.$$

The spaces of functions in Definition 6.1 allow us to measure whether a function is truly multidimensional by its cross-derivative with respect to sets of variables. The growth condition ensures well-posedness of the PDE.

6.4.1 Constant coefficients

We follow here Reference 9. Let $\widehat{V}_{r,s} = V_{r,s} - V$ be the approximation error of $V_{r,s}$ from Equation 6.24. Then the following holds.

Theorem 6.1 (Theorems 5 and 14 in Reference 9).

1. *Assume* $g \in C^{2,2,\text{mixed}}$ *in Equations 6.19 and 6.20. Then the expansion error* $\widehat{V}_{r,1}$ *satisfies*

$$\left\| \widehat{V}_{r,1}(\cdot, t) \right\|_\infty \le t^2 \sum_{r<i<j\le N} \lambda_k \lambda_i \left\| \frac{\partial^4 g}{\partial x_i^2 \partial x_j^2} \right\|_\infty . \tag{6.25}$$

2. *Assume* $g \in C^{2,3,\text{mix}}$ *in Equations 6.19–6.20. Then the expansion error* $\widehat{V}_{r,2}$ *satisfies*

$$\left\| \widehat{V}_{r,2}(\cdot, t) \right\|_\infty \le t^3 \sum_{r<i<j<k\le N} \lambda_i \lambda_j \lambda_k \left\| \frac{\partial^6 g}{\partial x_i^2 \partial x_j^2 \partial x_k^2} \right\|_\infty . \tag{6.26}$$

The analysis in Reference 9 derives PDEs for the error itself, and then makes use of standard maximum principle-type arguments to estimate the size of the error.

For instance, by using the PDEs satisfied by V and $V_{\{1,\ldots,r,i\}}$ for different i, it can be shown that

$$\begin{aligned}\frac{\partial}{\partial t}\widehat{V}_{r,1} &= \mathcal{L}_{\{1,\ldots,r\}}\widehat{V}_{r,1} \\ &\quad + \sum_{i=r+1}^{N}\left[\mathcal{L}_{\{1,\ldots,r,i\}} - \mathcal{L}_{\{1,\ldots,r\}}\right]V_{\{1,\ldots,r,i\}} + \left[\mathcal{L}_{\{1,\ldots,r\}} - \mathcal{L}\right]V \\ &= \sum_{k=1}^{r}\lambda_k\frac{\partial^2}{\partial x_k^2}\widehat{V}_{r,1} + \sum_{k=r+1}^{N}\lambda_k\frac{\partial^2}{\partial x_k^2}\left[V_{\{1,\ldots,r,k\}} - V\right]. \end{aligned} \tag{6.27}$$

This is an inhomogeneous heat equation for $\widehat{V}_{r,1}$ with zero initial data and a right-hand side which can be shown to be small. As a consequence, the solution itself is small. Informally, the terms on the right-hand side $V^{\{1,\ldots,r,k\}} - V$ are of order $O(\lambda_{r+1} + \cdots + \lambda_N - \lambda_k)$, and hence the right-hand side is of order $O(\sum_{r<i<j\leq N}\lambda_i\lambda_j)$. A slightly more careful argument gives the precise bound (Equation 6.25), and a similar but lengthier argument for $V_{r,2}$ gives Equation 6.26.

A number of comments are in order regarding the smoothness requirements dictated by the error bounds. First, most option payoffs are nonsmooth, have kinks and discontinuities. This would appear to render Equation 6.25 and its higher order versions meaningless. A reworking of the derivation shows that g can actually be replaced by $V_{r,0}$, which is the solution to

$$\frac{\partial V_{r,0}}{\partial t} = \sum_{k=1}^{r}\lambda_k\frac{\partial^2 V_{r,0}}{\partial x_k^2},$$
$$V_{r,0}(x,0) = g(x).$$

So even if g itself is not smooth, $V_{r,0}$ will be smooth except in degenerate situations which are analyzed in detail in Reference 9. Roughly speaking, as long as the locations of kinks and discontinuities are not parallel to all of the first r coordinate axes, $V_{r,0}$ is smooth enough for the expansion error to be well defined.

The second important point is that as Equation 6.25 contains only mixed derivative terms, for any payoffs which depend only on, say, x_1 and x_k for some $k > 1$, the decomposition of the option price is exact. Moreover, the value of any derivative that can be statically replicated by options with such simple payoffs is found exactly. Again, a more detailed discussion is found in Reference 9.

6.4.2 Variable coefficients

The transformation 6.6 with appropriate Q (see the discussion at the end of Section 6.1) ensures Equations 6.16 and 6.17 but this is only true at $t = 0$ and $x = a$. However, using arguments similar to Reference 9 and Section 6.4.1, we can still derive a PDE for the expansion error even for nonconstant coefficients. Straightforward calculus yields an expression similar to Equation 6.27, namely

$$\frac{\partial}{\partial t}\widehat{V}_{r,1} = \sum_{k,l=1}^{r} \lambda_{kl}(z,t)\frac{\partial^2}{\partial x_k x_l}\widehat{V}_{r,1}$$

$$+ \sum_{k=r+1}^{N} \left[\lambda_{kk}\frac{\partial^2}{\partial x_k^2} + \sum_{l=1}^{r}\lambda_{kl}\frac{\partial^2}{\partial x_k x_l}\right]\left[V_{\{1,\ldots,r,k\}} - V\right] \quad (6.28)$$

$$- \sum_{k,l=r+1,k\neq l}^{N} \lambda_{kl}\frac{\partial^2}{\partial x_k x_l}V \quad (6.29)$$

$$+ \sum_{k=r+1}^{N} \kappa_k \frac{\partial}{\partial x_k}\left[V_{\{1,\ldots,r,k\}} - V\right] \quad (6.30)$$

This equation contains three source terms, which determine the error size:

- The first term, (see Equation 6.28), is similar to the source term appearing in the constant coefficient case. It is essentially a restricted differential operator applied to the difference between full and partial solutions.
- The second term, (see Equation 6.29), consists of the nondiagonal terms not captured at all in the expansion applied to the full solution. It contains the full solution rather than the difference between full and partial ones, but the λ_{kl} involved are zero for $t = 0$ and $x = a$.
- The third term, (see Equation 6.30), where $\kappa_k(a,0) = 0$, captures the changes in κ and again acts on the differences between partial and full solutions.

At $t = 0$ and $x = a$, all three source terms are zeros, because

$$V_{\{1,\ldots,r,k\}}(x,0) - V(x,0) = 0 \quad \forall x \in \mathbb{R}^N \text{ and } \lambda_{kl}(a,0) = 0, k \neq l.$$

Away from these initial coordinates, the terms grow slowly and drive a nonzero error.

Instead of investigating this further theoretically, we give quantitative examples in the following section.

6.5 Numerical Examples

In this section, we analyze the numerical accuracy of the decomposition from Section 6.3 for the approximation of European basket options, where the

TABLE 6.1: Different base cases with nonconstant parameters

Nonconstant component	Parameter	Example
Time-dependent drift	$\mu = \mu(t)$	Exactly described by Equations 6.5 and 6.6
Time-dependent volatilities	$\sigma = \sigma(t)$	Sections 6.5.3 and 6.5.4
Time-dependent correlation	$\rho = \rho(t)$	Sections 6.5.1 and 6.5.2
Asset-dependent drift	$\mu = \mu(S)$	LIBOR market model in Reference 12
Asset-dependent volatilities	$\sigma = \sigma(S)$	Local vol—not considered
Asset-dependent correlation	$\rho = \rho(S)$	Section 6.5.5

model for the underlying stock has variable coefficients. We list six "base" cases of how the PDE coefficients can be varied in Table 6.1.

Consider assets whose dynamics for the prices of $S_t^1, \ldots, S_t^N$ are given by

$$\mathrm{d}(\log S_t^i) = -\frac{1}{2}\sigma_i^2(S_t, t)\,\mathrm{d}t + \sigma_i(S_t, t)\,\mathrm{d}W_t^i, \quad 1 \le i \le N,$$

under the risk-neutral measure with zero interest rates. By considering log prices as primitive variable in Equation 6.1, in a Black–Scholes setting, that is, if σ and ρ are constant, the PDE coefficients are constant. Generally, the Brownian motions W^i are correlated according to the correlation matrix

$$(\rho_{ij}(S,t))_{1 \le i,j \le N}.$$

We consider two possible correlation structures:

$$\rho_{\text{simple}}(\gamma) = \begin{pmatrix} 1 & \gamma & \gamma & \cdots & \gamma \\ \gamma & 1 & \gamma & \cdots & \gamma \\ \vdots & & \ddots & \vdots & \\ \gamma & \gamma & \gamma & \cdots & 1 \end{pmatrix}$$

for $\gamma \in (-1, 1)$ and

$$\rho_{\exp,ij}(\gamma) = \exp(-\gamma|i-j|)$$

for $\gamma > 0$, where we replace γ by a function $\gamma : \mathbb{R}^N \times [0, T] \to \mathbb{R}$, possibly being asset- and time dependent. The covariance matrix $\Sigma(S, t)$ is then fully characterized via $\Sigma_{ij}(S,t) = \sigma_i(S,t)\sigma_j(S,t)\rho_{ij}(S,t)$. Due to the asset- and time dependency of correlations and volatilities, the asset distributions are no longer log-normal and hence a transformation of the pricing PDE to the standard heat equation is generally not possible.

As a test case, we choose a European arithmetic basket option with $N = 10$. The payout at maturity $T = 1$ is

$$h(S) = \max\left(\sum_{i=1}^{N} \omega_i S_i - K, 0\right),$$

with strike $K = 100$ and weights $\omega_i \in \mathbb{R}$, $i = 1, \ldots, N$. We will examine the value at the point $S_{0,i} = 100$ for all i. As payout weight vectors ω, we consider

$$\begin{aligned}
\omega^1 &= (1/10, 1/10, 1/10, 1/10, 1/10, 1/10, 1/10, 1/10, 1/10, 1/10),\\
\omega^2 &= (4/30, 4/30, 4/30, 4/30, 4/30, 2/30, 2/30, 2/30, 2/30, 2/30),\\
\omega^3 &= (1/4, 1/4, 1/4, 1/4, 1/4, 1/4, 1/4, -1/4, -1/4, -1/4).
\end{aligned}$$

Using $V_{1,1}$ as approximation to V, we expect that the accuracy will be best for ω^1 and worst for ω^3, because ω^1 is parallel to the principal component of Σ and ω^3 closer to orthogonal.

The numerical parameters chosen were $N_1 = N_2 = 800$ and $N_t = 1000$, corresponding to a time step of size $\Delta t = 0.001$. For the reference Monte Carlo estimator V_{MC}, we used 10^8 paths. This setup reduces the discretization and simulation errors sufficiently for us to determine a good estimate of the expansion method's accuracy.

We implemented and tested two numerical algorithms for the solution of the PDE problems. One algorithm is the diagonal ADI method from Section 6.3.5 (with results denoted by $V_{\text{PDE}}^{\text{diagADI}}$), where we updated the diffusion coefficient values at every time step, and the PDE is solved numerically by Brian's scheme. The second method from Section 6.3.4 does incorporate the off-diagonal terms in the lower dimensional problems (denoted $V_{\text{PDE}}^{\text{HV}}$), where the numerical PDE solution is based on the HV scheme.

We also compute the results for the fully frozen model from Section 6.3.3, that is, with covariance matrix fixed at $\Sigma(s_0, T)$, both for the expansion ($V_{\text{PDE}}^{\text{loc}}$) and a full Monte Carlo estimator ($V_{\text{MC}}^{\text{loc}}$). This allows us to understand what contribution to the error comes from the variability of the coefficients, compared to the decomposition error already present for constant coefficients.

Our primary intention here is to give a proof of concept, rather than an in-depth study of the performance and convergence. We want to demonstrate that and how expansion methods can be used for variable coefficients.

6.5.1 Time-dependent simple correlation

For time-dependent simple correlation $\rho(t) = \rho_{\text{simple}}(t)$, the eigenvalues change over time. However, the lower $N - 1$ eigenvalues are identical and the subspace spanned by their eigenvectors does not change.

Table 6.2 shows results for $\sigma_i = 0.2$ and

$$\rho(t) = \rho_{\text{simple}}(0.8 - 0.8 \cdot (t/T - 0.5)^2) \in [\rho_{\text{simple}}(0.6), \rho_{\text{simple}}(0.8)].$$

TABLE 6.2: Time-dependent simple correlation

		V_{MC}	$V_{\text{PDE}}^{\text{diagADI}}$	$V_{\text{PDE}}^{\text{HV}}$	$V_{\text{MC}}^{\text{loc}}$	$V_{\text{PDE}}^{\text{loc}}$
ω^1		6.9463	6.9451	6.9451	6.3784	6.3715
	σ_{MC}	0.0011			0.0010	
	Δ_{abs}		−0.0012	−0.0012		−0.0069
	Δ_{rel}		−0.02%	−0.02%		−0.11%
	$\Delta_{\text{abs}}/\sigma_{\text{MC}}$		−1.06	−1.06		−6.73
ω^2		6.9602	6.9584	6.9584	6.3991	6.3932
	σ_{MC}	0.0011			0.0010	
	Δ_{abs}		−0.0018	−0.0018		−0.0059
	Δ_{rel}		−0.03%	−0.03%		−0.09%
	$\Delta_{\text{abs}}/\sigma_{\text{MC}}$		−1.57	−1.57		−5.75
ω^3		7.5631	7.5816	7.5816	7.3585	7.4069
	σ_{MC}	0.0012			0.0012	
	Δ_{abs}		0.0185	0.0185		0.0484
	Δ_{rel}		0.24%	0.24%		−0.66%
	$\Delta_{\text{abs}}/\sigma_{\text{MC}}$		14.96	14.96		−40.58

PDE/ADI and PDE/HV results were almost identical and very close to the MC results. Only in the third case of ω^3 did they even differ in a statistically significant way, that is, relative to the standard error σ_{MC}, from the MC computation. It is worth noting that the errors are even slightly larger in the fully frozen case, implying that the variable coefficients present no particular problem in this model.

6.5.2 Time-dependent exponential correlation

For a time-dependent exponential correlation $\rho(t) = \rho_{\text{exp}}(t)$, the eigenvalues and eigenvectors change substantially over time, resulting in a significant contribution from nonzero off-diagonal elements in $\lambda(t)$.

Table 6.3 shows results for $\sigma_i = 0.2$ and

$$\rho(t) = \rho_{\text{exp}}(0.25 - 0.6 \cdot (t/T - 0.5)^2) \in [\rho_{\text{exp}}(0.1), \rho_{\text{exp}}(0.25)].$$

PDE/ADI results are again close to the MC results. The PDE/HV results differ somewhat more, against the expectation, but note that both solutions are significantly more accurate than the constant coefficient approximation. The third case, ω^3, is again the most challenging one for the dimension-wise method.

6.5.3 Time-dependent volatilities, simple correlation

For time-dependent $\sigma_i = \sigma(t)$, that is, the case where all volatilities are time dependent but equal, the eigenvalues $\lambda_1, \ldots, \lambda_N$ of Σ are simply scaled up or down over time and the matrix of eigenvectors stays constant. This means that all nondiagonal terms of λ vanish and the transformation to the

TABLE 6.3: Time-dependent exponential correlation

		V_{MC}	$V_{\text{PDE}}^{\text{diagADI}}$	$V_{\text{PDE}}^{\text{HV}}$	$V_{\text{MC}}^{\text{loc}}$	$V_{\text{PDE}}^{\text{loc}}$
ω^1		6.0662	6.0738	6.0590	6.8534	6.8477
	σ_{MC}	0.0010			0.0011	
	Δ_{abs}		0.0076	0.0885		−0.0057
	Δ_{rel}		0.13%	1.46%		−0.08%
	$\Delta_{\text{abs}}/\sigma_{\text{MC}}$		7.82	90.88		−5.11
ω^2		6.1646	6.1695	6.1547	6.9109	6.9085
	σ_{MC}	0.0010			0.0011	
	Δ_{abs}		0.0049	−0.0099		−0.0024
	Δ_{rel}		0.08%	−0.16%		−0.03%
	$\Delta_{\text{abs}}/\sigma_{\text{MC}}$		4.92	−10.00		−2.15
ω^3		9.6062	9.5346	9.7786	9.2907	9.3279
	σ_{MC}	0.0015			0.0015	
	Δ_{abs}		−0.0716	0.1724		0.0372
	Δ_{rel}		−0.75%	1.80%		−0.40%
	$\Delta_{\text{abs}}/\sigma_{\text{MC}}$		−46.34	111.54		24.44

heat equation is exact. This case is simple: it merely requires the solution of a heat equation with time-dependent diffusion coefficients.

For time-dependent $\sigma_i = \sigma_i(t)$, that is, the case where the volatilities vary differently over time, the eigenvectors change with t. This in general leads to the appearance of nonzero off-diagonal terms. With no dependency on the asset values S, the initial PDE transformation means that those terms vanish at time $t = 0$ and then grow over time for $t > 0$.

Table 6.4 shows results for $\rho = \rho_{\text{simple}}(0.7)$ and

$$\sigma_i(t) = 0.1(1 + t/T)\left(1 + \frac{i-1}{N-1}\right) \in [0.1, 0.2]\left(1 + \frac{i-1}{N-1}\right).$$

Both the PDE/diagonal ADI and PDE/HV results are fairly accurate for the first two test cases. They both struggle with the third one, producing errors of 2.42% and 2.66%. Given that a similar error is present in the fully localized case, that is, for the model with constant coefficients, we conclude that this error is primarily due to the expansion method being applied to the challenging payout direction ω^3, rather than the nonconstant nature of the coefficients.

6.5.4 Time-dependent volatilities, exponential correlation

Table 6.5 shows results for

$$\sigma_i(t) = 0.1(1 + t/T)\left(1 + \frac{i-1}{N-1}\right) \in [0.1, 0.2]\left(1 + \frac{i-1}{N-1}\right)$$

and

$$\rho(t) = \rho_{\text{exp}}(0.25 - 0.6 \cdot (t/T - 0.5)^2) \in [\rho_{\text{exp}}(0.1), \rho_{\text{exp}}(0.25)].$$

TABLE 6.4: Time-dependent volatilities, simple correlation

		$V_{\mathbf{MC}}$	$V_{\mathbf{PDE}}^{\mathbf{diagADI}}$	$V_{\mathbf{PDE}}^{\mathbf{HV}}$	$V_{\mathbf{MC}}^{\mathbf{loc}}$	$V_{\mathbf{PDE}}^{\mathbf{loc}}$
ω^1		7.7987	7.7947	7.8234	5.1128	5.1123
	σ_{MC}	0.0013			0.0008	
	Δ_{abs}		−0.0040	0.0248		−0.0005
	Δ_{rel}		−0.05%	0.32%		−0.01%
	$\Delta_{\text{abs}}/\sigma_{\text{MC}}$		−3.10	19.19		−0.57
ω^2		7.3183	7.3151	7.3416	4.7972	4.7961
	σ_{MC}	0.0012			0.0008	
	Δ_{abs}		−0.0032	0.0233		−0.0011
	Δ_{rel}		−0.04%	0.32%		−0.02%
	$\Delta_{\text{abs}}/\sigma_{\text{MC}}$		−2.67	19.39		−1.41
ω^3		6.2074	6.3579	6.3723	4.0555	4.1658
	σ_{MC}	0.0010			0.0006	
	Δ_{abs}		0.1504	0.1649		0.1103
	Δ_{rel}		2.42%	2.66%		−2.72%
	$\Delta_{\text{abs}}/\sigma_{\text{MC}}$		150.26	164.72		174.38

TABLE 6.5: Time-dependent volatilities, exponential correlation

		$V_{\mathbf{MC}}$	$V_{\mathbf{PDE}}^{\mathbf{diagADI}}$	$V_{\mathbf{PDE}}^{\mathbf{HV}}$	$V_{\mathbf{MC}}^{\mathbf{loc}}$	$V_{\mathbf{PDE}}^{\mathbf{loc}}$
ω^1		6.9951	7.0905	7.1454	5.1602	5.1595
	σ_{MC}	0.0012			0.0008	
	Δ_{abs}		0.0955	0.1503		−0.0007
	Δ_{rel}		1.36%	2.15%		−0.01%
	$\Delta_{\text{abs}}/\sigma_{\text{MC}}$		83.12	130.87		−0.86
ω^2		6.5570	6.7953	6.7047	4.8380	4.8382
	σ_{MC}	0.0011			0.0008	
	Δ_{abs}		0.2383	0.1477		0.0002
	Δ_{rel}		3.63%	2.25%		0.00%
	$\Delta_{\text{abs}}/\sigma_{\text{MC}}$		223.82	138.71		0.27
ω^3		9.6494	10.0537	9.8383	5.6868	5.7252
	σ_{MC}	0.0015			0.0009	
	Δ_{abs}		0.4042	0.1889		0.0384
	Δ_{rel}		4.19%	1.96%		0.67%
	$\Delta_{\text{abs}}/\sigma_{\text{MC}}$		265.62	124.13		43.29

By combining time-dependent volatilities with time-dependent correlation, we have created a challenging scenario for our method. The PDE/diagonal ADI approach starts to be insufficient for the more complicated cases, differing by more than 4% for ω^3. The PDE/HV algorithm produces a relatively constant error of about 2% in all three test cases.

Contrasting with the fully frozen approximation, it is evident that this is the first scenario in which the variability of the coefficients creates a major contribution to the overall error.

TABLE 6.6: Asset-dependent correlation

		$V_{\mathbf{MC}}$	$V_{\mathbf{PDE}}^{\mathbf{diagADI}}$	$V_{\mathbf{PDE}}^{\mathbf{HV}}$	$V_{\mathbf{MC}}^{\mathbf{loc}}$	$V_{\mathbf{PDE}}^{\mathbf{loc}}$
ω^1		6.7937	1.4032	6.7393	7.2138	7.2147
	σ_{MC}	0.0108			0.0010	
	Δ_{abs}		−5.3905	−0.0544		−0.0009
	Δ_{rel}		−79.35%	−0.80%		−0.01%
	$\Delta_{\mathrm{abs}}/\sigma_{\mathrm{MC}}$		−497.65	−5.02		−0.75
ω^2		6.7910	6.7910	6.7534	7.2232	7.2239
	σ_{MC}	0.0109			0.0010	
	Δ_{abs}		−5.3660	−0.0376		−0.0008
	Δ_{rel}		−79.02%	−0.55%		−0.01%
	$\Delta_{\mathrm{abs}}/\sigma_{\mathrm{MC}}$		−494.23	−3.47		−0.64
ω^3		7.4977	2.4238	7.3838	7.6708	7.6663
	σ_{MC}	0.0122			0.0010	
	Δ_{abs}		−5.0739	−0.1139		0.0045
	Δ_{rel}		−67.67%	−1.52%		0.06%
	$\Delta_{\mathrm{abs}}/\sigma_{\mathrm{MC}}$		−416.90	−9.36		3.56

6.5.5 Asset-dependent correlation

Table 6.6 shows results for $\sigma_i = 0.2$ and

$$\rho(S) = \rho_{\mathrm{simple}}\left(0.6 + 0.2\exp\left(-\frac{1}{N}\sum_i^N \frac{|S_i - 100|}{10}\right)\right) \\ \in \left[\rho_{\mathrm{simple}}(0.6), \rho_{\mathrm{simple}}(0.8)\right].$$

Because of the added computational complexity of having to calculate the correlation for every vector of asset values encountered, these calculations were done with 10^6 Monte Carlo paths, $J = 400$ grid points and $M = 400$ time steps.

Clearly, the PDE/diagonal ADI approach is insufficient and the nondiagonal PDE terms are necessary for the solution. The PDE/HV approach, which incorporates them, correspondingly gives fairly accurate results for ω^1 and ω^2, relative to the MC variance. As before, the accuracy decreases for the ω^3 case, which coincidentally depends only weakly on the chosen correlation dynamics.

6.6 Conclusion

This chapter describes a systematic approach to approximating medium- to high-dimensional PDEs in derivative pricing by a sequence of lower dimensional PDEs, which are then accessible to state-of-the-art finite difference methods. The splitting is accurate especially in situations where the dynamics of the underlying stochastic processes can be described well by a lower number of components. In such situations, the decomposition can loosely be interpreted as a Taylor expansion with respect to small perturbations in the other directions.

To complement the theoretical analysis of the method in the constant parameter setting in earlier work, we describe here various extensions to variable parameters and analyze their accuracy through extensive numerical tests. Although the examples are necessarily specific, they are chosen to cover a spectrum of effects which occur in derivative pricing applications. As the approximation errors are determined locally by the variability of the solution and the parameters with respect to the different coordinates and time, the examples are to some extent representative of a wider class of situations.

Specifically, we designed test cases where different parameters varied with respect to spatial coordinates and time, and where the payoff varied most rapidly in different directions relative to the principle component of the covariance matrix. Across all cases, the ω^1 case, where the payout vector is parallel to the first eigenvector of Σ, showed the best accuracy, while the ω^3 case showed the worst. This was expected from the theoretical analysis and the results for constant coefficients, see Section 6.4.1.

Overall, our computations demonstrate that expansion methods can in principle be applied in this fashion to some variable coefficient asset models. Higher order methods or other extensions might be necessary to reduce the error sufficiently for real-world financial applications.

References

1. Tavella, D. and Randall, C. *Pricing Financial Instruments: The Finite Difference Method.* Wiley, New York, 2000.

2. in 't Hout, K. J. and Foulon, S. ADI finite difference schemes for option pricing. *Int. J. Numer. Anal. Mod.*, 7(2):303–320, 2010.

3. Reisinger, C. and Wittum, G. Efficient hierarchical approximation of high-dimensional option pricing problems. *SIAM J. Sci. Comp.*, 29(1):440–458, 2007.

4. Leentvaar, C. C. W. and Oosterlee, C. W. On coordinate transformation and grid stretching for sparse grid pricing of basket options. *J. Comput. Appl. Math.*, 222:193–209, 2008.

5. Pettersson, U., Larsson, E., Marcusson, G., and Persson, J. Improved radial basis function methods for multi-dimensional option pricing. *J. Comput. Appl. Math.*, 222(1):82–93, 2008.

6. Kazeev, V., Reichmann, O., and Schwab, C. Low-rank tensor structure of linear diffusion operators in the TT and QTT formats. *Linear Algebra Appl.*, 438(11):4204–4221, 2013.

7. Grasedyck, L., Kressner, D., and Tobler, C. A literature survey of low-rank tensor approximation techniques. *GAMM-Mitteilungen*, 36(1):53–78, 2013.

8. Griebel, M. and Holtz, M. Dimension-wise integration of high-dimensional functions with applications to finance. *J. Complex.*, 26(5):455–489, 2010.

9. Reisinger, C. and Wissmann, R. Error analysis of truncated expansion solutions to high-dimensional parabolic PDEs. *arXiv preprint arXiv:1505.04639*, 2015.

10. Hilber, N., Kehtari, S., Schwab, C., and Winter, C. Wavelet finite element method for option pricing in high-dimensional diffusion market models. *Technical Report* 2010–01, SAM, ETH Zürich, 2010.

11. Schröder, P., Schober, P., and Wittum, G. Dimension-wise decompositions and their efficient parallelization. *Electronic version of an article published in Recent Developments in Computational Finance, Interdisciplinary Mathematical Sciences*, 14:445–472, 2013.

12. Reisinger, C. and Wissmann, R. Numerical valuation of derivatives in high-dimensional settings via PDE expansions. *J. Comp. Fin.*, 18(4):95–127, 2015.

13. de Graaf, C. S. L., Kandhai, D., and Reisinger, C. Efficient exposure computation by risk factor decomposition. *arXiv preprint arXiv:1608.01197*, 2016.

14. Reisinger, C. Asymptotic expansion around principal components and the complexity of dimension adaptive algorithms. In Garcke, J. and Griebel, M., editors, *Sparse Grids and Applications*, number 88 in Springer Lectures Notes in Computational Science and Engineering, Springer-Verlag, Berlin, Heidelberg, pages 263–276, 2012.

15. Schröder, P., Gerstner, T., and Wittum, G. Taylor-like ANOVA-expansion for high dimensional problems in finance. Working paper, 2012.

16. Musiela, M. and Rutkowski, M. *Martingal Methods in Financial Modelling*. Springer, Berlin, 2nd edition, 2005.

17. Brian, P. L. T. A finite-difference method of higher-order accuracy for the solution of three-dimensional transient heat conduction problems. *AIChE J.*, 7:367–370, 1961.

18. Hundsdorfer, W. and Verwer, J. *Numerical Solution of Time-Dependent Advection–Diffusion Reaction Equations*, volume 33. Springer-Verlag, Berlin, Heidelberg, 2013.

19. Douglas, J. Alternating direction methods for three space variables. *Numer. Math.*, 4(1):41–63, 1962.

20. in 't Hout, K. J. and Mishra, C. Stability of ADI schemes for multidimensional diffusion equations with mixed derivative terms. *Appl. Numer. Math.*, 74:83–94, 2013.

21. Haentjens, T. and in 't Hout, K. J. ADI finite difference schemes for the Heston–Hull–White PDE. *Journal of Computational Finance*, 16(1):83–110, 2012.

Chapter 7

Multilevel Monte Carlo Methods for Applications in Finance*

Michael B. Giles and Lukasz Szpruch

CONTENTS

*Previously published in *Recent Developments in Computational Finance: Foundations, Algorithms and Applications*, Thomas Gerstner and Peter Kloeden, editors,

7.1 Introduction

In 2001, Heinrich [1] developed a multilevel Monte Carlo method for parametric integration, in which one is interested in estimating the value of $\mathbb{E}[f(x,\lambda)]$, where x is a finite-dimensional random variable and λ is a parameter. In the simplest case in which λ is a real variable in the range $[0,1]$, having estimated the value of $\mathbb{E}[f(x,0)]$ and $\mathbb{E}[f(x,1)]$, one can use $\frac{1}{2}(f(x,0)+f(x,1))$ as a control variate when estimating the value of $\mathbb{E}[f(x,\frac{1}{2})]$, since the variance of $f(x,\frac{1}{2})-\frac{1}{2}(f(x,0)+f(x,1))$ will usually be less than the variance of $f(x,\frac{1}{2})$. This approach can then be applied recursively for other intermediate values of λ, yielding large savings if $f(x,\lambda)$ is sufficiently smooth with respect to λ.

Giles' multilevel Monte Carlo path simulation [2] is both similar and different. There is no parametric integration, and the random variable is infinite-dimensional, corresponding to a Brownian path in the original paper. However, the control variate viewpoint is very similar. A coarse path simulation is used as a control variate for a more refined fine path simulation, but since the exact expectation for the coarse path is not known, this is in turn estimated recursively using even coarser path simulation as control variates. The coarsest path in the multilevel hierarchy may have only one timestep for the entire interval of interest.

A similar two-level strategy was developed slightly earlier by Kebaier [3], and a similar multilevel approach was under development at the same time by Speight [4,5].

In this review article, we start by introducing the central ideas in multilevel Monte Carlo (MLMC) simulation, and the key theorem from Reference 2 which gives the greatly improved computational cost if a number of conditions are satisfied. The challenge then is to construct numerical methods which satisfy these conditions, and we consider this for a range of computational finance applications.

7.2 Multilevel Monte Carlo

7.2.1 Monte Carlo

Monte Carlo simulations have become an essential tool in the pricing of derivatives security and in risk management. In the abstract setting, our goal is to numerically approximate the expected value $\mathbb{E}[Y]$, where $Y = P(X)$ is a functional of a random variable X. In most financial applications, we are not able to sample X directly and hence, in order to perform Monte Carlo simulations, we approximate X with $X_{\Delta t}$ such that $\mathbb{E}[P(X_{\Delta t})] \to \mathbb{E}[P(X)]$, when $\Delta t \to 0$. Using $X_{\Delta t}$ to compute N approximation samples produces the standard Monte Carlo estimate

$$\hat{Y} = \frac{1}{N}\sum_{i=1}^{N} P(X^i_{\Delta t}),$$

where $X^i_{\Delta t}$ is the numerical approximation to X on the ith sample path and N is the number of independent simulations of X. By standard Monte Carlo results $\hat{Y} \to \mathbb{E}[Y]$, when $\Delta t \to 0$ and $N \to \infty$. In practice, we perform a Monte Carlo simulation with given $\Delta t > 0$ and finite N producing an error to the approximation of $\mathbb{E}[Y]$. Here we are interested in the mean square error that is

$$MSE \equiv \mathbb{E}\left[(\hat{Y} - \mathbb{E}[Y])^2\right].$$

Our goal in the design of the Monte Carlo algorithm is to estimate Y with accuracy root-mean-square error ε ($MSE \leq \varepsilon^2$), as efficiently as possible. That is to minimize the computational complexity required to achieve the desired mean square error. For standard Monte Carlo simulations, the mean square error can be expressed as

$$\begin{aligned}\mathbb{E}\left[(\hat{Y} - \mathbb{E}[Y])^2\right] &= \mathbb{E}\left[(\hat{Y} - \mathbb{E}[\hat{Y}] + \mathbb{E}[\hat{Y}] - \mathbb{E}[Y])^2\right] \\ &= \underbrace{\mathbb{E}\left[(\hat{Y} - \mathbb{E}[\hat{Y}])^2\right]}_{\text{Monte Carlo variance}} + \underbrace{\left(\mathbb{E}[\hat{Y}] - \mathbb{E}[Y])^2\right)}_{\text{bias of the approximation}}.\end{aligned}$$

The Monte Carlo variance is proportional to $(1/N)$

$$\mathbb{V}[\hat{Y}] = \frac{1}{N^2}\mathbb{V}\left[\sum_{i=1}^{N} P(X^i_{\Delta t})\right] = \frac{1}{N}\mathbb{V}[P(X_{\Delta t})].$$

For both Euler–Maruyama (EM) and Milstein approximations, $|\mathbb{E}[\hat{Y}] - \mathbb{E}[Y]| = \mathcal{O}(\Delta t)$, typically. Hence the mean square error for standard Monte Carlo simulations is given by

$$\mathbb{E}\left[(\hat{Y} - \mathbb{E}[Y])^2\right] = \mathcal{O}\left(\frac{1}{N}\right) + \mathcal{O}(\Delta t^2).$$

To ensure the root-mean-square error is proportional to ε, we must have $MSE = \mathcal{O}(\varepsilon^2)$ and therefore $1/N = \mathcal{O}(\varepsilon^2)$ and $\Delta t^2 = \mathcal{O}(\varepsilon^2)$, which means $N = \mathcal{O}(\varepsilon^{-2})$ and $\Delta t = \mathcal{O}(\varepsilon)$. The computational cost of a standard Monte Carlo simulation is proportional to the number of paths N multiplied by the cost of generating a path, that is, the number of timesteps in each sample path. Therefore, the cost is $C = \mathcal{O}(\varepsilon^{-3})$. In the following section, we will show that using MLMC we can reduce the complexity of achieving root mean square error ε to $\mathcal{O}(\varepsilon^{-2})$.

7.2.2 Multilevel Monte Carlo theorem

In its most general form, MLMC simulation uses a number of levels of resolution, $\ell = 0, 1, \ldots, L$, with $\ell = 0$ being the coarsest, and $\ell = L$ being the finest. In the context of an SDE simulation, level 0 may have just one timestep for the whole time interval $[0, T]$, whereas level L might have 2^L uniform timesteps $\Delta t_L = 2^{-L}T$.

If P denotes the payoff (or other output functional of interest), and P_ℓ denotes its approximation on level l, then the expected value $\mathbb{E}[P_L]$ on the finest level is equal to the expected value $\mathbb{E}[P_0]$ on the coarsest level plus a sum of corrections which give the difference in expectation between simulations on successive levels,

$$\mathbb{E}[P_L] = \mathbb{E}[P_0] + \sum_{\ell=1}^{L} \mathbb{E}[P_\ell - P_{\ell-1}]. \tag{7.1}$$

The idea behind MLMC is to independently estimate each of the expectations on the right-hand side of Equation 7.1 in a way which minimizes the overall variance for a given computational cost. Let Y_0 be an estimator for $\mathbb{E}[P_0]$ using N_0 samples, and let Y_ℓ, $\ell > 0$ be an estimator for $\mathbb{E}[P_\ell - P_{\ell-1}]$ using N_ℓ samples. The simplest estimator is a mean of N_ℓ independent samples, which for $\ell > 0$ is

$$Y_\ell = N_\ell^{-1} \sum_{i=1}^{N_\ell} (P_\ell^i - P_{\ell-1}^i). \tag{7.2}$$

The key point here is that $P_\ell^i - P_{\ell-1}^i$ should come from two discrete approximations for the same underlying stochastic sample [6], so that on finer levels of resolution the difference is small (due to strong convergence) and so its variance is also small. Hence very few samples will be required on finer levels to accurately estimate the expected value.

The combined MLMC estimator $\hat{Y}$ is

$$\hat{Y} = \sum_{\ell=0}^{L} Y_\ell.$$

We can observe that

$$\mathbb{E}[Y_\ell] = N_\ell^{-1} \sum_{i=1}^{N_\ell} \mathbb{E}[P_\ell^i - P_{\ell-1}^i] = \mathbb{E}[P_\ell^i - P_{\ell-1}^i],$$

and

$$\mathbb{E}[\hat{Y}] = \sum_{\ell=0}^{L} \mathbb{E}[Y_\ell] = \mathbb{E}[P_0] + \sum_{\ell=1}^{L} \mathbb{E}[P_\ell - P_{\ell-1}] = \mathbb{E}[P_L].$$

Although we are using different levels with different discretization errors to estimate $\mathbb{E}[P]$, the final accuracy depends on the accuracy of the finest level L.

Here we recall the Theorem from Reference 2 (which is a slight generalization of the original theorem in Reference 2) which gives the complexity of MLMC estimation.

Theorem 7.1. *Let P denote a functional of the solution of a stochastic differential equation (SDE), and let P_ℓ denote the corresponding level ℓ numerical approximation. If there exist independent estimators Y_ℓ based on N_ℓ Monte Carlo samples, and positive constants $\alpha, \beta, \gamma, c_1, c_2, c_3$ such that $\alpha \geq \frac{1}{2} \min(\beta, \gamma)$ and*

(i) $|\mathbb{E}[P_\ell - P]| \leq c_1 \, 2^{-\alpha \ell}$

(ii) $\mathbb{E}[Y_\ell] = \begin{cases} \mathbb{E}[P_0], & \ell = 0 \\ \mathbb{E}[P_\ell - P_{\ell-1}], & \ell > 0 \end{cases}$

(iii) $\mathbb{V}[Y_\ell] \leq c_2 \, N_\ell^{-1} 2^{-\beta \ell}$

(iv) $C_\ell \leq c_3 \, N_\ell \, 2^{\gamma \ell}$*, where C_ℓ is the computational complexity of Y_ℓ*

then there exists a positive constant c_4 such that for any $\epsilon < e^{-1}$ there are values L and N_ℓ for which the multilevel estimator

$$Y = \sum_{\ell=0}^{L} Y_\ell$$

has a mean-square-error with bound

$$MSE \equiv \mathbb{E}\left[(Y - \mathbb{E}[P])^2\right] < \epsilon^2$$

with a computational complexity C with bound

$$C \leq \begin{cases} c_4 \, \epsilon^{-2}, & \beta > \gamma, \\ c_4 \, \epsilon^{-2} (\log \epsilon)^2, & \beta = \gamma, \\ c_4 \, \epsilon^{-2-(\gamma-\beta)/\alpha}, & 0 < \beta < \gamma. \end{cases}$$

7.2.3 Improved multilevel Monte Carlo

In the previous section, we showed that the key step in MLMC analysis is the estimation of variance $\mathbb{V}[P_\ell^i - P_{\ell-1}^i]$. As it will become more clear in the next section, this is related to the strong convergence results on approximations of SDEs, which differentiates MLMC from standard MC, where we only require a weak error bound for approximations of SDEs.

We will demonstrate that in fact the classical strong convergence may not be necessary for a good MLMC variance. In Equation 7.2, we have used the same estimator for the payoff P_ℓ on every level ℓ, and therefore Equation 7.1 is a trivial identity due to the telescoping summation. However, in Reference 7 Giles demonstrated that it can be better to use different estimators for the finer and coarser of the two levels being considered, P_ℓ^{f} when level ℓ is the finer level, and P_ℓ^{c} when level ℓ is the coarser level. In this case, we require that

$$\mathbb{E}[P_\ell^{\mathrm{f}}] = \mathbb{E}[P_\ell^{\mathrm{c}}] \quad \text{for } \ell = 1, \ldots, L, \tag{7.3}$$

so that

$$E[P_L^{\mathrm{f}}] = \mathbb{E}[P_0^{\mathrm{f}}] + \sum_{\ell=1}^{L} \mathbb{E}[P_\ell^{\mathrm{f}} - P_{\ell-1}^{\mathrm{c}}].$$

The MLMC Theorem is still applicable to this modified estimator. The advantage is that it gives the flexibility to construct approximations for which $P_\ell^{\mathrm{f}} - P_{\ell-1}^{\mathrm{c}}$ is much smaller than the original $P_\ell - P_{\ell-1}$, giving a larger value for β, the rate of variance convergence in condition (iii) in the theorem. In the following sections, we demonstrate how suitable choices of P_ℓ^{f} and P_ℓ^{c} can dramatically increase the convergence of the variance of the MLMC estimator.

The good choice of estimators, as we shall see, often follows from analysis of the problem under consideration from the distributional point of view. We will demonstrate that methods that had been used previously to improve the weak order of convergence can also improve the order of convergence of the MLMC variance.

7.2.4 Stochastic differential equations

First, we consider a general class of d-dimensional SDEs driven by Brownian motion. These are the primary object of studies in mathematical finance. In subsequent sections, we demonstrate extensions of MLMC beyond the Brownian setting.

Let $(\Omega, \mathcal{F}, \{\mathcal{F}_t\}_{t\geq 0}, \mathbb{P})$ be a complete probability space with a filtration $\{\mathcal{F}_t\}_{t\geq 0}$ satisfying the usual conditions, and let $w(t)$ be a m-dimensional Brownian motion defined on the probability space. We consider the numerical approximation of SDEs of the form

$$\mathrm{d}x(t) = f(x(t))\,\mathrm{d}t + g(x(t))\,\mathrm{d}w(t), \tag{7.4}$$

where $x(t) \in \mathbb{R}^d$ for each $t \geq 0$, $f \in C^2(\mathbb{R}^d, \mathbb{R}^d)$, $g \in C^2(\mathbb{R}^d, \mathbb{R}^{d\times m})$, and for simplicity we assume a fixed initial value $x_0 \in \mathbb{R}^d$. The most prominent example of SDEs in finance is a geometric Brownian motion

$$dx(t) = \alpha x(t)\,dt + \beta x(t)\,dw(t),$$

where $\alpha, \beta > 0$. Although, we can solve this equation explicitly it is still worthwhile to approximate its solution numerically in order to judge the performance of the numerical procedure we wish to apply to more complex problems. Another interesting example is the famous Heston stochastic volatility model

$$\begin{cases} ds(t) = rs(t)\,dt + s(t)\sqrt{v(t)}\,dw_1(t) \\ dv(t) = \kappa(\theta - v(t))\,dt + \sigma\sqrt{v(t)}\,dw_2(t) \\ dw_1\,dw_2 = \rho\,d\,t, \end{cases} \tag{7.5}$$

where $r, \kappa, \theta, \sigma > 0$. In this case, we do not know the explicit form of the solution and therefore numerical integration is essential in order to price certain financial derivatives using the Monte Carlo method. At this point, we would like to point out that the Heston model 7.5 does not satisfy standard conditions required for numerical approximations to converge. Nevertheless, in this paper we always assume that coefficients of SDEs 7.4 are sufficiently smooth. We refer to References 8–10 for an overview of the methods that can be applied when the global Lipschitz condition does not hold. We also refer the reader to Reference 11 for an application of MLMC to the SDEs with additive fractional noise.

7.2.5 Euler and Milstein discretizations

The simplest approximation of SDEs 7.4 is an EM scheme. Given any step size Δt_ℓ, we define the partition $\mathcal{P}_{\Delta t_\ell} := \{n\Delta t_\ell : n = 0, 1, 2, \ldots, 2^\ell\}$ of the time interval $[0, T]$, $2^\ell \Delta t = T > 0$. The EM approximation $X_n^\ell \approx x(n\,\Delta t_\ell)$ has the form [12]

$$X_{n+1}^\ell = X_n^\ell + f(X_n^\ell)\,\Delta t_\ell + g(X_n^\ell)\,\Delta w_{n+1}^\ell, \tag{7.6}$$

where $\Delta w_{n+1}^\ell = w((n+1)\Delta t_\ell) - w(n\Delta t_\ell)$ and $X_0 = x_0$. Equation 7.6 is written in a vector form and its ith component reads as

$$X_{i,n+1}^\ell = X_{i,n}^\ell + f_i(X_n^\ell)\,\Delta t_\ell + \sum_{j=1}^m g_{ij}(X_n^\ell)\,\Delta w_{j,n+1}^\ell.$$

In the classical Monte Carlo setting, we are mainly interested in the weak approximation of SDEs 7.4. Given a smooth payoff $P : \mathbb{R}^d \to \mathbb{R}$, we say that $X_{2^\ell}^\ell$ converges to $x(T)$ in a weak sense with order α if

$$|\mathbb{E}[P(x(T))] - \mathbb{E}[P(X_T^\ell)]| = \mathcal{O}(\Delta t_\ell^\alpha).$$

Rate α is required in condition (i) of Theorem 7.1. However, for MLMC condition (iii) of Theorem 7.1 is crucial. We have

$$\mathbb{V}_\ell \equiv \text{Var}\,(P_\ell - P_{\ell-1}) \leq \mathbb{E}\left[(P_\ell - P_{\ell-1})^2\right]$$

and

$$\mathbb{E}\left[(P_\ell - P_{\ell-1})^2\right] \leq 2\,\mathbb{E}\left[(P_\ell - P)^2\right] + 2\,\mathbb{E}\left[(P - P_{\ell-1})^2\right].$$

For Lipschitz continuous payoffs, $(P(x) - P(y))^2 \leq L\left\|x - y\right\|^2$, we then have

$$\mathbb{E}\left[(P_\ell - P)^2\right] \leq L\,\mathbb{E}\left[\left\|x(T) - X_T^\ell\right\|^2\right].$$

It is clear now that in order to estimate the variance of the MLMC, we need to examine strong convergence property. The classical strong convergence on the finite time interval $[0, T]$ is defined as

$$\left(\mathbb{E}\left[\left\|x(T) - X_T^\ell\right\|^p\right]\right)^{1/p} = \mathcal{O}(\Delta t_\ell^\xi), \quad \text{for } p \geq 2.$$

For the EM scheme, $\xi = 0.5$. In order to deal with path-dependent options, we often require measuring the error in the supremum norm:

$$\left(\mathbb{E}\big[\sup_{0\leq n\leq 2^\ell} \left\|x(n\Delta t_\ell) - X_n^\ell\right\|^p\big]\right)^{1/p} = \mathcal{O}(\Delta t^\xi) \quad \text{for } p \geq 2.$$

Even in the case of globally Lipschitz continuous payoff P, the EM does not achieve $\beta = 2\xi > 1$ which is optimal in Theorem 7.1. In order to improve the convergence of the MLMC variance, the Milstein approximation $X_n \approx x(n\,\Delta t_\ell)$ is considered, with ith component of the form [12]

$$\begin{aligned} X_{i,n+1}^\ell = X_{i,n}^\ell + f_i(X_n^\ell)\,\Delta t_\ell + \sum_{j=1}^m g_{ij}(X_n^\ell)\,\Delta w_{j,n+1}^\ell \\ + \sum_{j,k=1}^m h_{ijk}(X_n^\ell)\left(\Delta w_{j,n}^\ell \Delta w_{k,n}^\ell - \Omega_{jk}\,\Delta t_\ell - A_{jk,n}^\ell\right) \end{aligned} \tag{7.7}$$

where Ω is the correlation matrix for the driving Brownian paths, and $A_{jk,n}^\ell$ is the Lévy area defined as

$$A_{jk,n}^\ell = \int_{n\Delta t_\ell}^{(n+1)\Delta t_\ell} \left(\Big(w_j(t) - w_j(n\Delta t_\ell)\Big)\,\mathrm{d}w_k(t) - \Big(w_k(t) - w_k(n\Delta t_\ell)\Big)\,\mathrm{d}w_j(t)\right).$$

The rate of strong convergence ξ for the Milstein scheme is double the value we have for the EM scheme and therefore the MLMC variance for Lipschitz payoffs converges twice as fast. However, this gain does not come without a price. There is no efficient method to simulate Lévy areas, apart from dimension 2 [13–15]. In some applications, the diffusion coefficient $g(x)$ satisfies a commutativity property which gives

$$h_{ijk}(x) = h_{ikj}(x) \quad \text{for all } i, j, k.$$

In that case, because the Lévy areas are antisymmetric (i.e., $A_{jk,n}^l = -A_{kj,n}^l$), it follows that $h_{ijk}(X_n^\ell)\,A_{jk,n}^l + h_{ikj}(X_n^\ell)\,A_{kj,n}^l = 0$ and therefore the terms involving the Lévy areas cancel and so it is not necessary to simulate them. However, this only happens in special cases. Clark and Cameron [16] proved for a particular SDE that it is impossible to achieve a better order of strong convergence than the EM discretization when using just the discrete increments of the underlying Brownian motion. The analysis was extended by Müller-Gronbach [17] to general SDEs. As a consequence if we use the standard MLMC method with the Milstein scheme without simulating the Lévy areas, the complexity will remain the same as for EM. Nevertheless, Giles and Szpruch showed in Reference 18 that by constructing a suitable antithetic estimator one can neglect the Lévy areas and still obtain a multilevel correction estimator with a variance which decays at the same rate as the scalar Milstein estimator.

7.2.6 Multilevel Monte Carlo algorithm

Here we explain how to implement the Monte Carlo algorithm. Let us recall that the MLMC estimator Y is given by

$$\hat{Y} = \sum_{\ell=0}^{L} Y_\ell.$$

We aim to minimize the computational cost necessary to achieve desirable accuracy ε. As for standard Monte Carlo, we have

$$\mathbb{E}\left[(Y - \mathbb{E}[P(X)])^2\right] = \underbrace{\mathbb{E}\left[(Y - \mathbb{E}[\hat{Y}])^2\right]}_{\text{Monte Carlo variance}} + \underbrace{\left(\mathbb{E}[P_L] - \mathbb{E}[P(X)]\right)^2}_{\text{bias of the approximation}}.$$

The variance is given by

$$\mathbb{V}[Y] = \sum_{\ell=0}^{L} \mathbb{V}[Y_\ell] = \sum_{\ell=0}^{L} \frac{1}{N_\ell} V_\ell,$$

where $V_\ell = \mathbb{V}[P_\ell - P_{\ell-1}]$. To minimize the variance of Y for fixed computational cost $C = \sum_{\ell=0}^{L} N_\ell \Delta t_\ell^{-1}$, we can treat N_ℓ as continuous variable and use the Lagrange function to find the minimum of

$$L = \sum_{\ell=0}^{L} \frac{1}{N_\ell} V_\ell + \lambda \left(\sum_{\ell=0}^{L} N_\ell \Delta t_\ell^{-1} - C \right).$$

First-order conditions show that $N_\ell = \lambda^{-\frac{1}{2}} \sqrt{V_\ell \Delta t_\ell}$, therefore

$$\mathbb{V}[Y] = \sum_{\ell=0}^{L} \frac{V_\ell}{N_\ell} = \sum_{\ell=0}^{L} \frac{\sqrt{\lambda}}{\sqrt{V_\ell \Delta t_\ell}} V_\ell.$$

Since we want $\mathbb{V}[Y] \leq \frac{\varepsilon^2}{2}$, we can show that

$$\lambda^{-\frac{1}{2}} \geq 2\varepsilon^{-2} \sum_{\ell=0}^{L} \sqrt{V_\ell / \Delta t_\ell},$$

thus the optimal number of samples for level ℓ is

$$N_\ell = \left\lceil 2\varepsilon^{-2} \sqrt{V_\ell \Delta t_\ell} \sum_{\ell=0}^{L} \sqrt{V_\ell / \Delta t_\ell} \right\rceil. \tag{7.8}$$

Assuming $\mathcal{O}(\Delta t_\ell)$ weak convergence, the bias of the overall method is equal to $c\Delta t_L = cT\,2^{-L}$. If we want the bias to be proportional to $\frac{\varepsilon}{\sqrt{2}}$, we set

$$L_{\max} = \frac{\log\left(\varepsilon/(cT\sqrt{2})\right)^{-1}}{\log 2}.$$

From here we can calculate the overall complexity. We can now outline the algorithm

1. Begin with $L = 0$
2. Calculate the initial estimate of V_L using 100 samples
3. Determine optimal N_ℓ using Equation 7.8
4. Generate additional samples as needed for new N_ℓ
5. If $L < L_{\max}$, set $L := L + 1$ and go to 2

Most numerical tests suggest that $L_{\max}$ is not optimal and we can substantially improve MLMC by determining optimal L by looking at bias. For more details, see Reference 2.

7.3 Pricing with Multilevel Monte Carlo

A key application of MLMC is to compute the expected payoff of financial options. We have demonstrated that for globally Lipschitz European payoffs, convergence of the MLMC variance is determined by the strong rate of convergence of the corresponding numerical scheme. However, in many financial applications payoffs are not smooth or are path dependent. The aim of this section is to overview results on mean square convergence rates for EM and Milstein approximations with more complex payoffs. In the case of EM, the majority of payoffs encountered in practice have been analyzed by Giles et al. [19]. Extension of this analysis to the Milstein scheme is far from obvious.

This is due to the fact that the Milstein scheme gives an improved rate of convergence on the grid points, but this is insufficient for path-dependent options. In many applications, the behavior of the numerical approximation between grid points is crucial. The analysis of the Milstein scheme for complex payoffs was carried out in Reference 20. To understand this problem better, we recall a few facts from the theory of strong convergence of numerical approximations. We can define a piecewise linear interpolation of a numerical approximation within the time interval $[n\Delta t_\ell, (n+1)\Delta t_\ell)$ as

$$X^\ell(t) = X_n^\ell + l_\ell(X_{n+1}^\ell - X_n^\ell), \quad \text{for } t \in [n\Delta t_\ell, (n+1)\Delta t_\ell), \tag{7.9}$$

where $l_\ell \equiv (t - n\Delta t_\ell)/\Delta t_\ell$. Müller-Gronbach [21] has shown that for the Milstein scheme 7.9, we have

$$\mathbb{E}\left[\sup_{0\leq t\leq T} \left\|x(t) - X^\ell(t)\right\|^p\right] = \mathcal{O}(|\Delta t_\ell \log(\Delta t_\ell)|^{p/2}), \quad p \geq 2, \tag{7.10}$$

that is the same as for the EM scheme. In order to maintain the strong order of convergence, we use Brownian Bridge interpolation rather than basic piecewise linear interpolation:

$$\tilde{X}^\ell(t) = X_n^\ell + \lambda_\ell\,(X_{n+1}^\ell - X_n^\ell) + g(X_n^\ell)\,\left(w(t) - w(n\Delta t_\ell) - \lambda\,\Delta w_{n+1}^l\right), \tag{7.11}$$

for $t \in [n\Delta t_\ell, (n+1)\Delta t_\ell)$. For the Milstein scheme interpolated with Brownian bridges, we have [21]

$$\mathbb{E}\left[\sup_{0\leq t\leq T} \left\|x(t) - \tilde{X}^\ell(t)\right\|^p\right] = \mathcal{O}(|\Delta t_\ell \log(\Delta t_\ell)|^{p}).$$

Clearly $\tilde{X}^\ell(t)$ is not implementable, since in order to construct it, the knowledge of the whole trajectory $(w(t))_{0\leq t\leq T}$ is required. However, we will demonstrate that combining $\tilde{X}^\ell(t)$ with conditional Monte Carlo techniques can dramatically improve the convergence of the variance of the MLMC estimator. This is due to the fact that for suitable MLMC estimators only distributional knowledge of certain functionals of $(w(t))_{0\leq t\leq T}$ will be required.

7.3.1 Euler–Maruyama scheme

In this section, we demonstrate how to approximate the most common payoffs using the EM scheme 7.6.

The Asian option we consider has the payoff

$$P = \left(T^{-1}\int_0^T x(t)\,\mathrm{d}t - K\right)^+.$$

Using the piecewise linear interpolation 7.9, one can obtain the following approximation:

$$P_l \equiv T^{-1} \int_0^T X^\ell(t)\,\mathrm{d}t = T^{-1} \sum_{n=0}^{2^\ell-1} \tfrac{1}{2}\,\Delta t_\ell\,(X_n^\ell + X_{n+1}^\ell).$$

Lookback options have payoffs of the form:

$$P = x(T) - \inf_{0\le t\le T} x(t).$$

A numerical approximation to this payoff is

$$P_\ell \equiv X_T^\ell - \inf_{0\le n\le 2^\ell} X_n^\ell.$$

For both of these payoffs, it can be proved that $V_\ell = \mathcal{O}(\Delta t_\ell)$ [19].

We now consider a digital option, which pays one unit if the asset at the final time exceeds the fixed strike price K and pays zero otherwise. Thus the discontinuous payoff function has the form:

$$P = \mathbf{1}_{\{x(T)>K\}},$$

with the corresponding EM value

$$P_\ell \equiv \mathbf{1}_{\{X_T^\ell>K\}}.$$

Assuming boundedness of the density of the solution to Equation 7.4 in the neighborhood of the strike K, it has been proved in Reference 19 that $V_\ell = o(\Delta t_\ell^{1/2-\delta})$, for any $\delta > 0$. This result has been tightened by Avikainen [22] who proved that $V_\ell = \mathcal{O}(\Delta t_\ell^{1/2} \log \Delta t_\ell)$.

An up-and-out call gives a European payoff if the asset never exceeds the barrier, B, otherwise it pays zero. So, for the exact solution we have

$$P = (x(T) - K)^+ \mathbf{1}_{\{\sup_{0\le t\le T} x(t)\le B\}},$$

and for the EM approximation

$$P_\ell \equiv (X_T^\ell - K)^+ \mathbf{1}_{\{\sup_{0\le n\le 2^\ell} X_n^\ell\le B\}}.$$

A down-and-in call knocks in when the minimum asset price dips below the barrier B, so that

$$P = (x(T) - K)^+ \mathbf{1}_{\{\inf_{0\le t\le T} x(t)\le B\}},$$

and, accordingly,

$$P_l \equiv (X_T^\ell - K)^+ \mathbf{1}_{\{\inf_{0\le n\le 2^\ell} X_n^\ell\le B\}}.$$

For both of these barrier options, we have $\mathbb{V}_\ell = o(\Delta t_\ell^{1/2-\delta})$, for any $\delta > 0$, assuming that $\inf_{0\le t\le T} x(t)$ and $\sup_{0\le t\le T} x(t)$ have bounded density in the neighborhood of B [19].

As summarized in Table 7.1, numerical results taken from Reference 7 suggest that all of these results are near-optimal.

TABLE 7.1: Orders of convergence for V_ℓ as observed numerically and proved analytically for both Euler discretizations; δ can be any strictly positive constant

	Euler	
Option	**Numerical**	**Analysis**
Lipschitz	$\mathcal{O}(\Delta t_\ell)$	$\mathcal{O}(\Delta t_\ell)$
Asian	$\mathcal{O}(\Delta t_\ell)$	$\mathcal{O}(\Delta t_\ell)$
Lookback	$\mathcal{O}(\Delta t_\ell)$	$\mathcal{O}(\Delta t_\ell)$
Barrier	$\mathcal{O}(\Delta t_\ell^{1/2})$	$\mathcal{O}(\Delta t_\ell^{1/2-\delta})$
Digital	$\mathcal{O}(\Delta t_\ell^{1/2})$	$\mathcal{O}(\Delta t_\ell^{1/2} \log \Delta t_\ell)$

7.3.2 Milstein scheme

In the scalar case of SDEs 7.4 (i.e. with $d = m = 1$), the Milstein scheme has the form:

$$X_{n+1}^\ell = X_n^\ell + f(X_n^{2^l})\Delta t_\ell + g(X_n^\ell)\Delta w_{n+1}^\ell + g^{'}(X_n^\ell)g(X_n^\ell)((\Delta w_{n+1}^\ell)^2 - \Delta t_\ell), \tag{7.12}$$

where $g' \equiv \partial g/\partial x$. The analysis of Lipschitz European payoffs and Asian options with the Milstein scheme is analogous to the EM scheme and it has been proved in Reference 20 that in both these cases $V_\ell = \mathcal{O}(\Delta t_\ell^2)$.

7.3.2.1 Lookback options

For clarity of the exposition, we will express the fine time-step approximation in terms of the coarse timestep, that is $\mathcal{P}^{'}_{\Delta t_\ell} := \{n\Delta t_{\ell-1} : n = 0, \frac{1}{2}, 1, 1 + \frac{1}{2}, 2, \ldots, 2^{\ell-1}\}$. The partition for the coarse approximation is given by $\mathcal{P}_{\Delta t_{\ell-1}} := \{n\Delta t_{\ell-1} : n = 0, 1, 2, \ldots, 2^{\ell-1}\}$. Therefore, $X_n^{\ell-1}$ corresponds to X_n^ℓ for $n = 0, 1, 2, \ldots, 2^{\ell-1}$.

For pricing lookback options with the EM scheme, as an approximation of the minimum of the process we have simply taken $\min_n X_n^\ell$. This approximation could be improved by taking

$$X_{\min}^\ell = \min_n \left(X_n^\ell - \beta^* g(X_n^\ell)\Delta t_\ell^{1/2} \right).$$

Here $\beta^* \approx 0.5826$ is a constant which corrects the $\mathcal{O}(\Delta t_\ell^{1/2})$ leading order error due to the discrete sampling of the path, and thereby restores $\mathcal{O}(\Delta t_\ell)$ weak convergence [23]. However, using this approximation, the difference between the computed minimum values and the fine and coarse paths is $\mathcal{O}(\Delta t_\ell^{1/2})$, and hence the variance V_ℓ is $\mathcal{O}(\Delta t_\ell)$, corresponding to $\beta = 1$. In the previous section, this was acceptable because $\beta = 1$ was the best that could be achieved in general with the Euler path discretization which was used, but we now aim to achieve an improved convergence rate using the Milstein scheme.

In order to improve the convergence, the Brownian Bridge interpolant $\tilde{X}^\ell(t)$ defined in Equation 7.11 is used. We have

$$\begin{aligned}\min_{0\le t<T} \tilde{X}^\ell(t) &= \min_{0\le n\le 2^{\ell-1}-\frac{1}{2}} \left[\min_{n\Delta t_{l-1}\le t<(n+\frac{1}{2})\Delta t_{l-1}} \tilde{X}^\ell(t) \right] \\ &= \min_{0\le n\le 2^{\ell-1}-\frac{1}{2}} X^\ell_{n,\min},\end{aligned}$$

where minimum of the fine approximation over the first half of the coarse timestep is given by [24]

$$X^\ell_{n,\min} = \frac{1}{2}\left(X^\ell_n + X^\ell_{n+\frac{1}{2}} - \sqrt{\left(X^\ell_{n+\frac{1}{2}} - X^\ell_n\right)^2 - 2\,g(X^\ell_n)^2\,\Delta t_l \log U^\ell_n} \right), \tag{7.13}$$

and minimum of the fine approximation over the second half of the coarse timestep is given by

$$X^\ell_{n+\frac{1}{2},\min} = \frac{1}{2}\left(X^\ell_{n+\frac{1}{2}} + X^\ell_{n+1} - \sqrt{\left(X^\ell_{n+1} - X^\ell_{n+\frac{1}{2}}\right)^2 - 2\,g(X^\ell_{n+\frac{1}{2}})^2\,\Delta t_\ell \log U^\ell_{n+\frac{1}{2}}} \right), \tag{7.14}$$

where $U^\ell_n, U^\ell_{n+\frac{1}{2}}$ are uniform random variables on the unit interval. For the coarse path, in order to improve the MLMC variance a slightly different estimator is used (see Equation 7.3). Using the same Brownian increments as we used on the fine path (to guarantee that we stay on the same path), Equation 7.11 is used to define $\tilde{X}^{\ell-1}_{n+\frac{1}{2}} \equiv \tilde{X}^{\ell-1}((n+\frac{1}{2})\Delta t_{\ell-1})$. Given this interpolated value, the minimum value over the interval $[n\Delta t_{\ell-1}, (n+1)\Delta t_{\ell-1}]$ can then be taken to be the smaller of the minima for the two intervals $[n\Delta t_{\ell-1}, (n+\frac{1}{2})\Delta t_{\ell-1})$ and $[(n+\frac{1}{2})\Delta t_{\ell-1}, (n+1)\Delta t_{\ell-1})$,

$$\begin{aligned}X^{\ell-1}_{n,\min} &= \frac{1}{2}\left(X^{\ell-1}_n + \tilde{X}^{\ell-1}_{n+\frac{1}{2}} - \sqrt{\left(\tilde{X}^{\ell-1}_{n+\frac{1}{2}} - X^{\ell-1}_n\right)^2 - 2\,(g(X^{\ell-1}_n))^2\,\frac{\Delta t_{\ell-1}}{2} \log U^\ell_n} \right), \\ X^{\ell-1}_{n+\frac{1}{2},\min} &= \frac{1}{2}\left(\tilde{X}^{\ell-1}_{n+\frac{1}{2}} + X^{\ell-1}_{n+1} - \sqrt{\left(X^{\ell-1}_{n+1} - \tilde{X}^{\ell-1}_{n+\frac{1}{2}}\right)^2 - 2\,(g(X^{\ell-1}_n))^2\,\frac{\Delta t_{\ell-1}}{2} \log U^\ell_{n+\frac{1}{2})}} \right).\end{aligned} \tag{7.15}$$

Note that $g(X_n^{\ell-1})$ is used for both timesteps. It is because we used the Brownian Bridge with diffusion term $g(X_n^{\ell-1})$ to derive both minima. If we changed $g(X_n^{\ell-1})$ to $g(\tilde{X}^{\ell-1}_{n+\frac{1}{2}})$ in $X^{\ell-1}_{n+\frac{1}{2},\min}$, this would mean that different Brownian Bridges were used on the first and second half of the coarse timestep and as a consequence condition 7.3 would be violated. Note also the reuse of the same uniform random numbers U_n^ℓ and $U^\ell_{n+\frac{1}{2}}$ used to compute the fine path minimum. The $\min(X^{\ell-1}_{n,\min}, X^{\ell-1}_{n+\frac{1}{2},\min})$ has exactly the same distribution as $X^{\ell-1}_{n,\min}$, since they are both based on the same Brownian interpolation, and therefore equality 7.3 is satisfied. Giles et al. [20] proved the following theorem.

Theorem 7.2. *The multilevel approximation for a lookback option which is a uniform Lipschitz function of* $x(T)$ *and* $\inf_{[0,T]} x(t)$ *has* $V_l = o(\Delta t_l^{2-\delta})$ *for any* $\delta > 0$.

7.3.3 Conditional Monte Carlo

Giles [7] and Giles et al. [20] have shown that combining conditional Monte Carlo with MLMC results in superior estimators for various financial payoffs.

To obtain an improvement in the convergence of the MLMC variance barrier and digital options, conditional Monte Carlo methods are employed. We briefly describe it here. Our goal is to calculate $\mathbb{E}[P]$. Instead, we can write

$$\mathbb{E}[P] = \mathbb{E}\big[\mathbb{E}[P|Z]\big],$$

where Z is a random vector. Hence $\mathbb{E}[P|Z]$ is an unbiased estimator of $\mathbb{E}[P]$. We also have

$$\mathrm{Var}\,[P] = \mathbb{E}\big[\mathrm{Var}\,[P|Z]\big] + \mathrm{Var}\,\big[\mathbb{E}[P|Z]\big],$$

hence $\mathrm{Var}\,\big[\mathbb{E}[P|Z]\big] \le \mathrm{Var}\,(P)$. In the context of MLMC, we obtain a better variance convergence if we condition on different vectors on the fine and the coarse level. That is on the fine level we take $\mathbb{E}[P^f|Z^f]$, where $Z^f = \{X_n^\ell\}_{0\le n\le 2^\ell}$. On the coarse level instead of taking $\mathbb{E}[P^c|Z^c]$ with $Z^c = \{X_n^{\ell-1}\}_{0\le n\le 2^{\ell-1}}$, we take $\mathbb{E}[P^c|Z^c, \tilde{Z}^c]$, where $\tilde{Z}^c = \{\tilde{X}^{\ell-1}_{n+\frac{1}{2}}\}_{0\le n\le 2^{\ell-1}}$ are obtained from Equation 7.11. Condition 7.3 trivially holds by tower property of conditional expectation

$$\mathbb{E}\left[\mathbb{E}[P^c|Z^c]\right] = \mathbb{E}[P^c] = \mathbb{E}\left[\mathbb{E}[P^c|Z^c, \tilde{Z}^c]\right].$$

7.3.4 Barrier options

The barrier option which is considered a down-and-out option for which the payoff is a Lipschitz function of the value of the underlying at maturity, provided the underlying has never dropped below a value $B \in \mathbb{R}$, is

$$P = f(x(T))\, \mathbf{1}_{\{\tau > T\}}.$$

The crossing time τ is defined as

$$\tau = \inf_t \{x(t) < B\}.$$

This requires the simulation of $(x(T), \mathbf{1}_{\tau > T}))$. The simplest method sets

$$\tau^{\Delta t_\ell} = \inf_n \{X_n^\ell < B\}$$

and as an approximation takes $(X_{2^{\ell-1}}^\ell, \mathbf{1}_{\{\tau^{\Delta t_\ell} > 2^{\ell-1}\}})$. But even if we could simulate the process $\{x(n\Delta t_\ell)\}_{0 \le n \le 2^{\ell-1}}$, it is possible for $\{x(t)\}_{0 \le t \le T}$ to cross the barrier between grid points. Using the Brownian Bridge interpolation, we can approximate $\mathbf{1}_{\{\tau > T\}}$ by

$$\prod_{n=0}^{2^{\ell-1}-\frac{1}{2}} \mathbf{1}_{\{X_{n,\min}^\ell \ge B\}}.$$

This suggests following the lookback approximation in computing the minimum of both the fine and coarse paths. However, the variance would be larger in this case because the payoff is a discontinuous function of the minimum. A better treatment, which is the one used in Reference 25, is to use the conditional Monte Carlo approach to further smooth the payoff. Since the process X_n^ℓ is Markovian, we have

$$\begin{aligned}
&\mathbb{E}\left[f(X_{2^{\ell-1}}^\ell) \prod_{n=0}^{2^{\ell-1}-\frac{1}{2}} \mathbf{1}_{\{X_{n,\min}^\ell \ge B\}} \right] \\
&= \mathbb{E}\left[\mathbb{E}\left[f(X_{2^{\ell-1}}^\ell) \prod_{n=0}^{2^{\ell-1}-\frac{1}{2}} \mathbf{1}_{\{X_{n,\min}^\ell \ge B\}} | X_0^\ell, \ldots, X_{2^{\ell-1}}^\ell \right]\right] \\
&= \mathbb{E}\left[f(X_{2^{\ell-1}}^\ell) \prod_{n=0}^{2^{\ell-1}-\frac{1}{2}} \mathbb{E}\left[\mathbf{1}_{\{X_{n,\min}^\ell \ge B\}} | X_n^\ell, X_{n+1}^\ell \right] \right] \\
&= \mathbb{E}\left[f(X_{2^{\ell-1}}^\ell) \prod_{n=0}^{2^{\ell-1}-\frac{1}{2}} (1 - p_n^\ell) \right],
\end{aligned}$$

where from [24]

$$\begin{aligned}
p_n^\ell &= \mathbb{P}\left(\inf_{n\Delta t_\ell \le t < (n+\frac{1}{2})\Delta t_\ell} \tilde{X}(t) < B | X_n^\ell, X_{n+\frac{1}{2}}^\ell \right) \\
&= \exp\left(\frac{-2\,(X_n^\ell - B)^+ (X_{n+\frac{1}{2}}^\ell - B)^+}{g(X_n^\ell)^2\, \Delta t_\ell} \right),
\end{aligned}$$

and

$$p^{\ell}_{n+\frac{1}{2}} = \mathbb{P}\left(\inf_{(n+\frac{1}{2})\Delta t_\ell \le t < (n+1)\Delta t_\ell} \tilde{X}(t) < B | X^{\ell}_{n+\frac{1}{2}}, X^{\ell}_{n+1} \right)$$
$$= \exp\left(\frac{-2\,(X^{\ell}_{n+\frac{1}{2}} - B)^{+}(X^{\ell}_{n+1} - B)^{+}}{g(X^{\ell}_{n+\frac{1}{2}})^2\, \Delta t_\ell} \right).$$

Hence, for the fine path this gives

$$P^{\mathrm{f}}_{\ell} = f(X^{\ell}_{2^{\ell-1}}) \prod_{n=0}^{2^{\ell-2}-\frac{1}{2}} (1 - p^{\ell}_{n}), \tag{7.16}$$

The payoff for the coarse path is defined similarly. However, in order to reduce the variance, we subsample $\tilde{X}^{\ell-1}_{n+\frac{1}{2}}$, as we did for lookback options, from the Brownian Bridge connecting $X^{\ell-1}_{n}$ and $X^{\ell-1}_{n+1}$

$$\mathbb{E}\left[f(X^{\ell-1}_{2^{\ell-1}}) \prod_{n=0}^{2^{\ell-1}-1} \mathbf{1}_{\{X^{\ell-1}_{n,\min} \ge B\}} \right]$$
$$= \mathbb{E}\left[\mathbb{E}\left[f(X^{\ell-1}_{2^{\ell-1}}) \prod_{n=0}^{2^{\ell-1}-1} \mathbf{1}_{\{X^{\ell-1}_{n,\min} \ge B\}} | X^{\ell-1}_{0}, \tilde{X}^{\ell-1}_{\frac{1}{2}}, \ldots, \tilde{X}^{\ell-1}_{2^{\ell-1}-\frac{1}{2}}, X^{\ell-1}_{2^{\ell-1}} \right] \right]$$
$$= \mathbb{E}\left[f(X^{\ell-1}_{2^{\ell-1}}) \prod_{n=0}^{2^{\ell-1}-1} \mathbb{E}\left[\mathbf{1}_{\{X^{\ell-1}_{n,\min} \ge B\}} | X^{\ell-1}_{n}, \tilde{X}^{\ell-1}_{n+\frac{1}{2}}, X^{\ell-1}_{n+1} \right] \right]$$
$$= \mathbb{E}\left[f(X^{\ell-1}_{2^{\ell-1}}) \prod_{n=0}^{2^{\ell-1}-1} (1 - p^{\ell-1}_{1,n})(1 - p^{\ell-1}_{2,n}) \right],$$

where

$$p^{\ell-1}_{1,n} == \exp\left(\frac{-2\,(X^{\ell-1}_{n} - B)^{+}(\tilde{X}^{\ell-1}_{n+\frac{1}{2}} - B)^{+}}{g(X^{\ell-1}_{n})^2\, \Delta t_\ell} \right)$$

and

$$p^{\ell-1}_{2,n} == \exp\left(\frac{-2\,(\tilde{X}^{\ell-1}_{n+\frac{1}{2}} - B)^{+}(X^{\ell-1}_{n+1} - B)^{+}}{g(X^{\ell-1}_{n})^2\, \Delta t_l} \right).$$

Note that the same $g(X^{\ell-1}_{n})$ is used (rather than using $g(\tilde{X}^{\ell-1}_{n+\frac{1}{2}})$ in $p^{\ell-1}_{2,n}$) to calculate both probabilities for the same reason as we did for lookback options. The final estimator can be written as

$$P^{c}_{\ell-1} = f(X^{\ell-1}_{2^{\ell-1}}) \prod_{n=0}^{2^{\ell-1}-1} (1 - p^{\ell-1}_{1,n})(1 - p^{\ell-1}_{2,n}). \tag{7.17}$$

Giles et al. [20] proved the following theorem.

Theorem 7.3. *Provided* $\inf_{[0,T]} |g(B)| > 0$, *and* $\inf_{[0,T]} x(t)$ *has a bounded density in the neighborhood of* B, *then the multilevel estimator for a down-and-out barrier option has variance* $V_\ell = o(\Delta t_\ell^{3/2-\delta})$ *for any* $\delta > 0$.

The reason the variance is approximately $o(\Delta t_\ell^{3/2-\delta})$ instead of $\mathcal{O}(\Delta t_\ell^2)$ is the following: due to the strong convergence property the probability of the numerical approximation being outside the $\Delta t_\ell^{1-\delta}$-neighborhood of the solution to the SDE 7.4 is arbitrarily small, that is for any $\varepsilon > 0$

$$\mathbb{P}\left(\sup_{0\le n\Delta t_\ell \le T} \|x(n\Delta t_\ell) - X_n^\ell\| \ge \Delta t_\ell^{1-\varepsilon}\right)$$
$$\le \Delta t_\ell^{-p+p\varepsilon}\mathbb{E}\left[\sup_{0\le n\Delta t_\ell \le T} \left\|x(n\Delta t_\ell) - X_n^\ell\right\|^p\right] = \mathcal{O}(\Delta_\ell^{p\varepsilon}). \tag{7.18}$$

If $\inf_{[0,T]} x(t)$ is outside the $\Delta t_\ell^{1/2}$-neighborhood of the barrier B then by Equation 7.18 it is shown that so are numerical approximations. The probabilities of crossing the barrier in that case are asymptotically either 0 or 1 and essentially we are in the Lipschitz payoff case. If the $\inf_{[0,T]} x(t)$ is within the $\Delta t_\ell^{1/2}$-neighborhood of the barrier B, then so are the numerical approximations. In that case it can be shown that $\mathbb{E}[(P_\ell^{\mathrm{f}} - P_{\ell-1}^{\mathrm{c}})^2] = \mathcal{O}(\Delta t^{1-\delta})$ but due to the bounded density assumption, the probability that $\inf_{[0,T]} x(t)$ is within $\Delta t_\ell^{1/2}$-neighborhood of the barrier B is of order $\Delta t_\ell^{1/2-\delta}$. Therefore the overall MLMC variance is $V_\ell = o(\Delta_\ell^{3/2-\delta})$ for any $\delta > 0$.

7.3.5 Digital options

A digital option has a payoff which is a discontinuous function of the value of the underlying asset at maturity, the simplest example being

$$P = \mathbf{1}_{\{x(T)>B\}}.$$

Approximating $\mathbf{1}_{\{x(T)>B\}}$ based only on simulations of $x(T)$ by Milstein scheme will lead to an $\mathcal{O}(\Delta t_\ell)$ fraction of the paths having coarse and fine path approximations to $x(T)$ on either side of the strike, producing $P_\ell - P_{\ell-1} = \pm 1$, resulting in $V_\ell = \mathcal{O}(\Delta t_\ell)$. To improve the variance to $\mathcal{O}(\Delta t_\ell^{3/2-\delta})$ for all $\delta > 0$, the conditional Monte Carlo method is used to smooth the payoff (see Section 7.2.3 in Reference 24). This approach was proved to be successful in Giles et al. [20] and was tested numerically in Reference 25.

If $X_{2^{\ell-1}-\frac{1}{2}}^\ell$ denotes the value of the fine path approximation one timestep before maturity, then the motion thereafter is approximated as Brownian motion with constant drift $f(X_{2^{\ell-1}-\frac{1}{2}}^\ell)$ and volatility $g(X_{2^{\ell-1}-\frac{1}{2}}^\ell)$. The conditional expectation for the payoff is the probability that $X_{2^{\ell-1}}^\ell > B$ after one

TABLE 7.2: Orders of convergence for V_l as observed numerically and proved analytically for Milstein discretizations; δ can be any strictly positive constant

	Milstein	
Option	**Numerical**	**Analysis**
Lipschitz	$\mathcal{O}(\Delta t_\ell^2)$	$\mathcal{O}(\Delta t_\ell^2)$
Asian	$\mathcal{O}(\Delta t_\ell^2)$	$\mathcal{O}(\Delta t_\ell^2)$
Lookback	$\mathcal{O}(\Delta t_\ell^2)$	$\mathcal{O}(\Delta t_\ell^{2-\delta})$
Barrier	$\mathcal{O}(\Delta t_\ell^{3/2})$	$\mathcal{O}(\Delta t_\ell^{3/2-\delta})$
Digital	$\mathcal{O}(\Delta t_\ell^{3/2})$	$\mathcal{O}(\Delta t_\ell^{3/2-\delta})$

further timestep, which is

$$P_\ell^{\mathrm{f}} = \mathbb{E}\left[\mathbf{1}_{\{X^\ell_{2^{\ell-1}}>B\}} | X^\ell_{2^{\ell-1}-\frac{1}{2}}\right] = \Phi\left(\frac{X^\ell_{2^{\ell-1}-\frac{1}{2}} + f(X^\ell_{2^{\ell-1}-\frac{1}{2}})\Delta t_\ell - B}{|g(X^\ell_{2^{\ell-1}-\frac{1}{2}})|\sqrt{\Delta t_\ell}}\right), \tag{7.19}$$

where Φ is the cumulative Normal distribution.

For the coarse path, we note that given the Brownian increment $\Delta w^{\ell-1}_{2^{\ell-1}-\frac{1}{2}}$ for the first half of the last coarse timestep (which comes from the fine path simulation), the probability that $X^\ell_{2^{\ell-1}} > B$ is

$$\begin{aligned} P^{\mathrm{c}}_{\ell-1} &= \mathbb{E}\left[\mathbf{1}_{\{X^{\ell-1}_{2^{\ell-1}}>B\}} | X^{\ell-1}_{2^{\ell-1}-1}, \Delta w^{\ell-1}_{2^{\ell-1}-\frac{1}{2}}\right] \\ &= \Phi\left(\frac{X^{2^{\ell-1}}_{2^{\ell-1}-1} + f(X^{\ell-1}_{2^{\ell-1}-1})\Delta t_{\ell-1} + g(X^{\ell-1}_{2^{\ell-1}-1})\Delta w^{\ell-1}_{2^{\ell-1}-\frac{1}{2}} - B}{|g(X^{2^{\ell-1}}_{2^{\ell-1}-1})|\sqrt{\Delta t_\ell}}\right). \end{aligned} \tag{7.20}$$

The conditional expectation of Equation 7.20 is equal to the conditional expectation of $P^{\mathrm{f}}_{\ell-1}$ defined by Equation 7.19 on level $\ell - \ell$, and so equality 7.3 is satisfied. A bound on the variance of the multilevel estimator is given by the following result.

Theorem 7.4. *Provided $g(B) \neq 0$, and $x(t)$ has a bounded density in the neighborhood of B, then the multilevel estimator for a digital option has variance $V_l = o(\Delta t_l^{3/2-\delta})$ for any $\delta > 0$.*

Results of the above section were tested numerically in Reference 7 and are summarized in Table 7.2.

7.4 Greeks with Multilevel Monte Carlo

Accurate calculation of prices is only one objective of Monte Carlo simulations. Even more important in some ways is the calculation of the sensitivities of the prices to various input parameters. These sensitivities, known collectively as the "Greeks," are important for risk analysis and mitigation through hedging.

Here we follow the results by Burgos et al. [26] to present how MLMC can be applied in this setting. The pathwise sensitivity approach (also known as Infinitesimal Perturbation Analysis) is one of the standard techniques for computing these sensitivities [24]. However, the pathwise approach is not applicable when the financial payoff function is discontinuous. One solution to these problems is to use the Likelihood Ratio Method (LRM) but its weaknesses are that the variance of the resulting estimator is usually $\mathcal{O}(\Delta t_l^{-1})$.

Three techniques are presented that improve MLMC variance: payoff smoothing using conditional expectations [24]; an approximation of the above technique using path splitting for the final timestep [27]; the use of a hybrid combination of pathwise sensitivity and the LRM [28]. We discuss the strengths and weaknesses of these alternatives in different MLMC settings.

7.4.1 Monte Carlo greeks

Consider the approximate solution of the general SDE 7.4 using Euler discretization 7.6. The Brownian increments can be defined to be a linear transformation of a vector of independent unit Normal random variables Z.

The goal is to efficiently estimate the expected value of some financial payoff function $P(x(T))$ and numerous first-order sensitivities of this value with respect to different input parameters such as the volatility or one component of the initial data $x(0)$. In more general cases, P might also depend on the values of process $\{x(t)\}_{0\leq t\leq T}$ at intermediate times.

The pathwise sensitivity approach can be viewed as starting with the expectation expressed as an integral with respect to Z:

$$V_\ell \equiv \mathbb{E}\left[P(X_n^\ell(Z,\theta))\right] = \int P(X_n^\ell(Z,\theta))\, p_Z(Z)\, \mathrm{d}Z. \tag{7.21}$$

Here θ represents a generic input parameter, and the probability density function for Z is

$$p_Z(Z) = (2\pi)^{-d/2} \exp\left(-\|Z\|_2^2/2\right),$$

where d is the dimension of the vector Z.

Let $X_n^\ell = X_n^\ell(Z,\theta)$. If the drift, volatility and payoff functions are all differentiable, Equation 7.21 may be differentiated to give

$$\frac{\partial V_\ell}{\partial \theta} = \int \frac{\partial P(X_n^\ell)}{\partial X_n^\ell}\, \frac{\partial X_n^\ell}{\partial \theta}\, p_Z(Z)\, \Delta Z, \tag{7.22}$$

with $(\partial X_n^\ell)/\partial\theta$ being obtained by differentiating Equation 7.6 to obtain

$$\frac{\partial X_{n+1}^\ell}{\partial\theta} = \frac{\partial X_n^\ell}{\partial\theta} + \left(\frac{\partial f(X_n^\ell,\theta)}{\partial X_n^\ell}\,\frac{\partial X_n^\ell}{\partial\theta} + \frac{\partial f(X_n^\ell,\theta)}{\partial\theta}\right)\Delta t_l$$
$$+ \left(\frac{\partial g(X_n^\ell,\theta)}{\partial X_n^\ell}\,\frac{\partial X_n^\ell}{\partial\theta} + \frac{\partial g(X_n^\ell,\theta)}{\partial\theta}\right)\Delta w_{n+1}^l. \tag{7.23}$$

We assume that $Z \rightarrow \Delta w_{n+1}^l$ mapping does not depend on θ. It can be proved that Equation 7.22 remains valid (i.e., we can interchange integration and differentiation) when the payoff function is continuous and piecewise differentiable, and the numerical estimate obtained by standard Monte Carlo with M independent path simulations

$$M^{-1}\sum_{m=1}^{M}\frac{\partial P(X_n^{\ell,m})}{\partial X_n^\ell}\,\frac{\partial X_n^{\ell,m}}{\partial\theta}$$

is an unbiased estimate for $\partial V/\partial\theta$ with a variance which is $\mathcal{O}(M^{-1})$, if $P(x)$ is Lipschitz and the drift and volatility functions satisfy the standard conditions [12].

Performing a change of variables, the expectation can also be expressed as

$$V_l \equiv \mathbb{E}\left[P(X_n^\ell)\right] = \int P(x)\; p_{X_n^\ell}(x,\theta)\,\mathrm{d}x, \tag{7.24}$$

where $p_{X_n^\ell}(x,\theta)$ is the probability density function for X_n^ℓ which will depend on all of the inputs parameters. Since probability density functions are usually smooth, Equation 7.24 can be differentiated to give

$$\frac{\partial V_\ell}{\partial\theta} = \int P(x)\,\frac{\partial p_{X_n^\ell}}{\partial\theta}\,\mathrm{d}x = \int P(x)\,\frac{\partial(\log p_{X_n^\ell})}{\partial\theta}\,p_{X_n^\ell}\,\mathrm{d}x = \mathbb{E}\left[P(x)\,\frac{\partial(\log p_{X_n^\ell})}{\partial\theta}\right],$$

which can be estimated using the unbiased Monte Carlo estimator

$$M^{-1}\sum_{m=1}^{M} P(X_n^{\ell,m})\,\frac{\partial \log p_{X_n^\ell}(X_n^{\ell,m})}{\partial\theta}.$$

This is the LRM. Its great advantage is that it does not require the differentiation of $P(X_n^\ell)$. This makes it applicable to cases in which the payoff is discontinuous, and it also simplifies the practical implementation because banks often have complicated flexible procedures through which traders specify payoffs. However, it does have a number of limitations, one being a requirement of absolute continuity which is not satisfied in a few important applications such as the LIBOR market model [24].

7.4.2 Multilevel Monte Carlo greeks

The MLMC method for calculating Greeks can be written as

$$\frac{\partial V}{\partial \theta} = \frac{\partial \mathbb{E}(P)}{\partial \theta} \approx \frac{\partial \mathbb{E}(P_L)}{\partial \theta} = \frac{\partial \mathbb{E}(P_0)}{\partial \theta} + \sum_{\ell=1}^{L} \frac{\partial \mathbb{E}(P_\ell^f - P_{\ell-1}^c)}{\partial \theta}. \quad (7.25)$$

Therefore extending Monte Carlo Greeks to MLMC Greeks is straightforward. However, the challenge is to keep the MLMC variance small. This can be achieved by appropriate smoothing of the payoff function. The techniques that were presented in Section 7.3.2 are also very useful here.

7.4.3 European call

As an example, we consider a European call $P = (x(T) - B)^+$ with $x(t)$ being a geometric Brownian motion with Milstein scheme approximation given by

$$X_{n+1}^\ell = X_n^\ell + r\, X_n^\ell \Delta t_\ell + \sigma\, X_n^\ell \Delta w_{n+1}^\ell + \frac{\sigma^2}{2}((\Delta w_{n+1}^\ell)^2 - \Delta t_\ell). \quad (7.26)$$

We illustrate the techniques by computing delta (δ) and vega (ν), the sensitivities to the asset's initial value $x(0)$ and to its volatility σ.

Since the payoff is Lipschitz, we can use pathwise sensitivities. We observe that

$$\frac{\partial}{\partial x}(x - B)^+ = \begin{cases} 0, & \text{for } x < B, \\ 1, & \text{for } x > B. \end{cases}$$

This derivative fails to exist when $x = B$, but since this event has probability 0, we may write

$$\frac{\partial}{\partial x}(x - K)^+ = \mathbf{1}_{\{X > B\}}.$$

Therefore we are essentially dealing with a digital option.

7.4.4 Conditional Monte Carlo for pathwise sensitivity

Using conditional expectation, the payoff can be smooth as we did it in Section 7.3.2. European calls can be treated in exactly the same way as the digital option in Section 7.3.2, that is instead of simulating the whole path, we stop at the penultimate step and then on the last step we consider the full distribution of $(X_{2^l}^\ell | w_0^l, \ldots, w_{2^l-1}^l)$.

For digital options, this approach leads to Equations 7.19 and 7.20. For the call options, we can do analogous calculations. In Reference 26, numerical results for this approach were obtained, with a scalar Milstein scheme used to obtain the penultimate step. The results are presented in Table 7.3. For lookback options, conditional expectations lead to Equations 7.13 and 7.15 and for barriers to Equations 7.16 and 7.17. Burgos et al. [26] applied pathwise sensitivity to these smoothed payoffs, with a scalar Milstein scheme used to obtain the penultimate step, and obtained numerical results that we present in Table 7.4.

TABLE 7.3: Orders of convergence for V_ℓ as observed numerically and corresponding MLMC complexity

	Call		Digital	
Estimator	β	**MLMC complexity**	β	**MLMC complexity**
Value	$\approx$2.0	$\mathcal{O}(\epsilon^{-2})$	$\approx$1.4	$\mathcal{O}(\epsilon^{-2})$
Delta	$\approx$1.5	$\mathcal{O}(\epsilon^{-2})$	$\approx$0.5	$\mathcal{O}(\epsilon^{-2.5})$
Vega	$\approx$2	$\mathcal{O}(\epsilon^{-2})$	$\approx$0.6	$O(\epsilon^{-2.4})$

TABLE 7.4: Orders of convergence for V_ℓ as observed numerically and corresponding MLMC complexity

	Lookback		Barrier	
Estimator	β	**MLMC complexity**	β	**MLMC complexity**
Value	$\approx$1.9	$\mathcal{O}(\epsilon^{-2})$	$\approx$1.6	$\mathcal{O}(\epsilon^{-2})$
Delta	$\approx$1.9	$\mathcal{O}(\epsilon^{-2})$	$\approx$0.6	$\mathcal{O}(\epsilon^{-2.4})$
Vega	$\approx$1.3	$\mathcal{O}(\epsilon^{-2})$	$\approx$0.6	$O(\epsilon^{-2.4})$

7.4.5 Split pathwise sensitivities

There are two difficulties in using conditional expectation to smooth payoffs in practice in financial applications. This first is that conditional expectation will often become a multidimensional integral without an obvious closed-form value, and the second is that it requires a change to the often complex software framework used to specify payoffs. As a remedy for these problems the splitting technique to approximate $\mathbb{E}\left[P(X^\ell_{2^l})|X^\ell_{2^\ell-1}\right]$ and $\mathbb{E}\left[P(X^{\ell-1}_{2^{\ell-1}})|X^{\ell-1}_{2^{\ell-1}-1}, \Delta w^\ell_{2^\ell-2}\right]$ is used. We get numerical estimates of these values by "splitting" every simulated path on the final timestep. At the fine level: for every simulated path, a set of s final increments $\{\Delta w^{\ell,i}_{2^\ell}\}_{i\in[1,s]}$ is simulated, which can be averaged to get

$$\mathbb{E}\left[P(X^\ell_{2^\ell})|X^\ell_{2^\ell-1}\right] \approx \frac{1}{s}\sum_{i=1}^{s} P(X^\ell_{2^\ell-1}, \Delta w^{\ell,i}_{2^\ell}). \tag{7.27}$$

At the coarse level, similar to the case of digital options, the fine increment of the Brownian motion over the first half of the coarse timestep is used,

$$\mathbb{E}\left[P\left(X^{\ell-1}_{2^{\ell-1}}\right)|X^{\ell-1}_{2^{\ell-1}-1}, \Delta w^\ell_{2^\ell-2}\right] \approx \frac{1}{s}\sum_{i=1}^{s} P\left(X^{\ell-1}_{2^{\ell-1}-1}, \Delta w^\ell_{2^\ell-2}, \Delta w^{\ell-1,i}_{2^{\ell-1}}\right). \tag{7.28}$$

This approach was tested in Reference 26, with the scalar Milstein scheme used to obtain the penultimate step, and is presented in Table 7.5. As expected the values of β tend to the rates offered by conditional expectations as s increases and the approximation gets more precise.

TABLE 7.5: Orders of convergence for V_ℓ as observed numerically and the corresponding MLMC complexity

Estimator	s	β	MLMC complexity
Value	10	≈ 2.0	$O(\epsilon^{-2})$
	500	≈ 2.0	$O(\epsilon^{-2})$
Delta	10	≈ 1.0	$O(\epsilon^{-2}(\log \epsilon)^2)$
	500	≈ 1.5	$O(\epsilon^{-2})$
Vega	10	≈ 1.6	$O(\epsilon^{-2})$
	500	≈ 2.0	$O(\epsilon^{-2})$

7.4.6 Optimal number of samples

The use of multiple samples to estimate the value of the conditional expectations is an example of the splitting technique [27]. If w and z are independent random variables, then for any function $P(w, z)$ the estimator

$$Y_{M,S} = M^{-1} \sum_{m=1}^{M} \left(S^{-1} \sum_{i=1}^{S} P(w^m, z^{(m,i)}) \right)$$

with independent samples w^m and $z^{m,i}$ is an unbiased estimator for

$$\mathbb{E}_{w,z}\left[P(w,z)\right] \equiv \mathbb{E}_w\left[\mathbb{E}_z[P(w,z)|w]\right],$$

and its variance is

$$\mathbb{V}[Y_{M,S}] = M^{-1}\, \mathbb{V}_w\left[\mathbb{E}_z[P(w,z)|w]\right] + (MS)^{-1}\, \mathbb{E}_w\left[\mathbb{V}_z[P(w,z)|w]\right].$$

The cost of computing $Y_{M,S}$ with variance $v_1 M^{-1} + v_2 (MS)^{-1}$ is proportional to

$$c_1 M + c_2 MS,$$

with c_1 corresponding to the path calculation and c_2 corresponding to the payoff evaluation. For a fixed computational cost, the variance can be minimized by minimizing the product

$$\left(v_1 + v_2 s^{-1}\right)(c_1 + c_2 s) = v_1 c_2 s + v_1 c_1 + v_2 c_2 + v_2 c_1 s^{-1},$$

which gives the optimum value $s_{\text{opt}} = \sqrt{v_2 c_1 / v_1 c_2}$.

c_1 is $\mathcal{O}(\Delta t_\ell^{-1})$ since the cost is proportional to the number of timesteps, and c_2 is $\mathcal{O}(1)$, independent of Δt_ℓ. If the payoff is Lipschitz, then v_1 and v_2 are both $\mathcal{O}(1)$ and $S_{\text{opt}} = \mathcal{O}(\Delta t_\ell^{-1/2})$.

7.4.7 Vibrato Monte Carlo

The idea of vibrato Monte Carlo is to combine pathwise sensitivity and LRM. Adopting the conditional expectation approach, each path simulation for a particular set of Brownian motion increments $w^\ell \equiv (\Delta w_1^\ell, \Delta w_2^\ell, \ldots, \Delta w_{2^\ell-1}^\ell)$ (excluding the increment for the final timestep) computes a conditional Gaussian probability distribution $p_X(X_{2^\ell}^\ell | w^\ell)$. For a scalar SDE, if μ_{w^ℓ} and σ_{w^ℓ} are the mean and standard deviation for given w^ℓ, then

$$X_{2^l}^\ell(w^\ell, Z) = \mu_{w^\ell} + \sigma_{w^\ell} Z,$$

where Z is a unit Normal random variable. The expected payoff can then be expressed as

$$V_\ell = \mathbb{E}\left[\mathbb{E}\left[P(X_{2^\ell}^\ell)|w^\ell\right]\right] = \int\left\{\int P(x)\, p_{X_{2^\ell}^\ell}(x|w^\ell)\,\mathrm{d}x\right\}\, p_{w^\ell}(y)\,\mathrm{d}y.$$

The outer expectation is an average over the discrete Brownian motion increments, while the inner conditional expectation is averaging over Z.

To compute the sensitivity to the input parameter θ, the first step is to apply the pathwise sensitivity approach for fixed w^l to obtain $\partial\mu_{w^l}/\partial\theta, \partial\sigma_{w^l}/\partial\theta$. We then apply LRM to the inner conditional expectation to get

$$\frac{\partial V_\ell}{\partial\theta} = \mathbb{E}\left[\frac{\partial}{\partial\theta}\mathbb{E}\left[P(X_{2^\ell}^\ell)|w^\ell\right]\right] = \mathbb{E}\left[\mathbb{E}_Z\left[P(X_{2^\ell}^\ell)\,\frac{\partial(\log p_{X_{2^\ell}^\ell})}{\partial\theta}|w^\ell\right]\right],$$

where

$$\frac{\partial(\log p_{X_{2^\ell}^\ell})}{\partial\theta} = \frac{\partial(\log p_{X_{2^\ell}^\ell})}{\partial\mu_{w^\ell}}\,\frac{\partial\mu_{w^\ell}}{\partial\theta} + \frac{\partial(\log p_{X_{2^\ell}^\ell})}{\partial\sigma_{w^\ell}}\,\frac{\partial\sigma_{w^\ell}}{\partial\theta}.$$

This leads to the estimator

$$\begin{aligned}\frac{\partial V_\ell}{\partial\theta} \approx \frac{1}{N_\ell}\sum_{m=1}^{N_\ell}\Bigg(&\frac{\partial\mu_{w^{\ell,m}}}{\partial\theta}\mathbb{E}\left[P\left(X_{2^\ell}^\ell\right)\,\frac{\partial(\log p_{X_{2^\ell}^\ell})}{\partial\mu_{w^\ell}}|w^{\ell,m}\right]\\ &+\frac{\partial\sigma_{\hat{w}^{\ell,m}}}{\partial\theta}\mathbb{E}\left[P\left(X_{2^\ell}^\ell\right)\,\frac{\partial(\log p_{X_{2^\ell}^\ell})}{\partial\sigma_{w^\ell}}|w^{\ell,m}\right]\Bigg).\end{aligned} \tag{7.29}$$

We compute $(\partial\mu_{w^{\ell,m}})/\partial\theta$ and $(\partial\sigma_{w^{\ell,m}})/\partial\theta$ with pathwise sensitivities. With $X_{2^l}^{\ell,m,i} = X_{2^\ell}^\ell(w^{\ell,m}, Z^i)$, we substitute the following estimators into Equation 7.29:

$$\begin{cases}\mathbb{E}\left[P\left(X_{2^\ell}^\ell\right)\,\dfrac{\partial(\log p_{X_{2^\ell}^\ell})}{\partial\mu_{w^\ell}}|w^{\ell,m}\right]\\ \quad\approx \dfrac{1}{s}\sum\limits_{i=1}^{s}\left(P\left(X_{2^\ell}^{\ell,m,i}\right)\,\dfrac{X_{2^\ell}^{2^\ell,m,i} - \mu_{w^{\ell,m}}}{\sigma_{w^{\ell,m}}^2}\right)\\ \mathbb{E}\left[P\left(X_{2^\ell}^\ell\right)\,\dfrac{\partial(\log p_{X_{2^\ell}^\ell})}{\partial\sigma_{w^\ell}}|\hat{w}^{\ell,m}\right]\\ \quad\approx \dfrac{1}{s}\sum\limits_{i=1}^{s}P\left(X_{2^\ell}^{\ell,m,i}\right)\left(-\dfrac{1}{\sigma_{w^{\ell,m}}} + \dfrac{\left(X_{2^\ell}^{\ell,m,i} - \mu_{w^{\ell,m}}\right)^2}{\sigma_{w^{\ell,m}}^3}\right).\end{cases}$$

In a multilevel setting, at the fine level we can use Equation 7.29 directly. At the coarse level, as for digital options in Section 7.3.5, the fine Brownian increments over the first half of the coarse timestep are reused to derive Equation 7.29.

The numerical experiments for the call option with $s = 10$ were obtained [26], with scalar Milstein scheme used to obtain the penultimate step.

Estimator	β	**MLMC complexity**
Value	≈2.0	$O(\epsilon^{-2})$
Delta	≈1.5	$O(\epsilon^{-2})$
Vega	≈2.0	$O(\epsilon^{-2})$

Although the discussion so far has considered an option based on the value of a single underlying value at the terminal time T, it can be shown that the idea extends very naturally to multidimensional cases, producing a conditional multivariate Gaussian distribution, and also to financial payoffs which are dependent on values at intermediate times.

7.5 Multilevel Monte Carlo for Jump-Diffusion Processes

Giles and Xia in Reference 29 investigated the extension of the MLMC method to jump-diffusion SDEs. We consider models with finite rate activity using a jump-adapted discretization in which the jump times are computed and added to the standard uniform discretization times. If the Poisson jump rate is constant, the jump times are the same on both paths and the multilevel extension is relatively straightforward, but the implementation is more complex in the case of state-dependent jump rates for which the jump times naturally differ.

Merton [30] proposed a jump-diffusion process, in which the asset price follows a jump-diffusion SDE:

$$\mathrm{d}x(t) = f(x(t-))\,\Delta t + g(x(t-))\,\Delta w(t) + c(x(t-))\,\Delta J(t), \quad 0 \le t \le T, \tag{7.30}$$

where the jump term $J(t)$ is a compound Poisson process $\sum_{i=1}^{N(t)}(Y_i - 1)$, the jump magnitude Y_i has a prescribed distribution, and $N(t)$ is a Poisson process with intensity λ, independent of the Brownian motion. Due to the existence of jumps, the process is a càdlàg process, that is, having right continuity with left limits. We note that $x(t-)$ denotes the left limit of the process while $x(t) = \lim_{s\to t+} x(t)$. In Reference 30, Merton also assumed that $\log Y_i$ has a normal distribution.

7.5.1 A jump-adapted Milstein discretization

To simulate finite activity jump-diffusion processes, Giles and Xia [29] used the jump-adapted approximation from Platen and Bruti-Liberat [31]. For each path simulation, the set of jump times $\mathbb{J} = \{\tau_1, \tau_2, \ldots, \tau_m\}$ within the time interval $[0, T]$ is added to a uniform partition $\mathcal{P}_{\Delta t_l} := \{n\Delta t_l : n = 0, 1, 2, \cdots, 2^l\}$. A combined set of discretization times is then given by $\mathbb{T} = \{0 = t_0 < t_1 < t_2 < \cdots < t_M = T\}$ and we define the length of the timestep as $\Delta t_l^n = t_{n+1} - t_n$. Clearly, $\Delta t_l^n \leq \Delta t_l$.

Within each timestep, the scalar Milstein discretization is used to approximate the SDE 7.30, and then the jump is simulated when the simulation time is equal to one of the jump times. This gives the following numerical method:

$$
\begin{aligned}
X_{n+1}^{\ell,-} &= X_n^\ell + f(X_n^\ell)\,\Delta t_\ell^n + g(X_n^\ell)\,\Delta w_{n+1}^\ell + \tfrac{1}{2}\,g^{'}(X_n^\ell)\,g(X_n^\ell)\,(\Delta(w_n^\ell)^2 - \Delta t_\ell^n),\\
X_{n+1}^\ell &= \begin{cases} X_{n+1}^{\ell,-} + c(X_{n+1}^{\ell,-})(Y_i - 1), & \text{when } t_{n+1} = \tau_i,\\ X_{n+1}^{\ell,-}, & \text{otherwise,} \end{cases}
\end{aligned}
\tag{7.31}
$$

where $X_n^{\ell,-} = X_{t_n-}^\ell$ is the left limit of the approximated path, Δw_n^ℓ is the Brownian increment, and Y_i is the jump magnitude at τ_i.

7.5.1.1 Multilevel Monte Carlo for constant jump rate

In the case of the jump-adapted discretization, the telescopic sum 7.1 is written down with respect to Δt_ℓ rather than to Δt_ℓ^n. Therefore, we have to define the computational complexity as the expected computational cost since different paths may have different numbers of jumps. However, the expected number of jumps is finite and therefore the cost bound in assumption (iv) in Theorem 7.1 will still remain valid for an appropriate choice of the constant c_3.

The MLMC approach for a constant jump rate is straightforward. The jump times τ_j, which are the same for the coarse and fine paths, are simulated by setting $\tau_j - \tau_{j-1} \sim \exp(\lambda)$.

Pricing European call and Asian options in this setting is straightforward. For lookback, barrier, and digital options, we need to consider Brownian bridge interpolations as we did in Section 7.3.2. However, due to the presence of jumps some small modifications are required. To improve convergence, we will be looking at Brownian bridges between timesteps coming from jump-adapted discretization. In order to obtain an interpolated value $\tilde{X}_{n+\frac{1}{2}}^{2^{\ell-1}}$ for the coarse timestep, a Brownian Bridge interpolation over interval $[k_n, \hat{k}_n]$ is considered, where

$$
\begin{aligned}
k_n &= \max\{n\Delta t_{\ell-1}^n, \max\{\tau \in \mathbb{J} : \tau < (n + \tfrac{1}{2})\Delta t_{\ell-1}^n\}\}\\
\hat{k}_n &= \min\{(n+1)\Delta t_{\ell-1}^n, \min\{\tau \in \mathbb{J} : \tau > (n + \tfrac{1}{2})\Delta t_{\ell-1}^n\}\}.
\end{aligned}
\tag{7.32}
$$

Hence

$$\tilde{X}^{\ell-1}_{n+\frac{1}{2}} = X^{\ell-1}_{k_n} + \lambda_{\ell-1}\,(X^{\ell-1}_{\hat{k}_n} - X^{\ell-1}_{k_n}) \\ + g(X^{\ell-1}_{k_n})\,\Big(w^\ell((n+\tfrac{1}{2})\Delta t_{\ell-1}) - w^\ell(k_n) - \lambda_{\ell-1}\,(w^\ell(\hat{k}_n) - w^\ell(k_n))\Big)$$

where $\lambda_{\ell-1} \equiv ((n+\frac{1}{2})\Delta t^n_{\ell-1} - k_n)/(\hat{k}_n - k_n)$.

In the same way as in Section 7.3.2, the minima over time-adapted discretization can be derived. For the fine timestep, we have

$$X^\ell_{n,\min} = \frac{1}{2}\left(X^\ell_n + X^{\ell,-}_{n+\frac{1}{2}} - \sqrt{\left(X^{\ell,-}_{n+\frac{1}{2}} - X^\ell_n\right)^2 - 2\,g(X^\ell_n)^2\,\Delta t^n_\ell \log U^\ell_n}\right).$$

Notice the use of the left limits $X^{\ell,-}$. Following discussion in the previous sections, the minima for the coarse timestep can be derived using interpolated value $\tilde{X}^{\ell-1}_{n+\frac{1}{2}}$. Deriving the payoffs for lookback and barrier options is now straightforward.

For digital options, due to jump-adapted time grid, in order to find conditional expectations, we need to look at relations between the last jump time and the last timestep before expiry. In fact, there are three cases:

1. The last jump time τ happens before penultimate fixed-time timestep, that is, $\tau < (2^{l-1} - 2)\Delta t_l$.
2. The last jump time is within the last fixed-time timestep, that is, $\tau > (2^{l-1} - 1)\Delta t_l$.
3. The last jump time is within the penultimate fixed-time timestep, that is, $(2^{l-1} - 1)\Delta t_l > \tau > (2^{l-1} - 2)\Delta t_l$.

With this in mind, we can easily write down the payoffs for the coarse and fine approximations as we presented in Section 7.3.5.

7.5.1.2 Multilevel Monte Carlo for path-dependent rates

In the case of a path-dependent jump rate $\lambda(x(t))$, the implementation of the multilevel method becomes more difficult because the coarse and fine path approximations may have jumps at different times. These differences could lead to a large difference between the coarse and fine path payoffs, and hence greatly increase the variance of the multilevel correction. To avoid this, Giles and Xia [29] modified the simulation approach of Glasserman and Merener [32] which uses "thinning" to treat the case in which $\lambda(x(t), t)$ is bounded. Let us recall the thinning property of Poisson processes. Let $(N_t)_{t\geq 0}$ be a Poisson process with intensity l and define a new process Z_t by "thinning" N_t: take all the jump times $(\tau_n, n \geq 1)$ corresponding to N, keep them with probability $0 < p < 1$ or delete them with probability $1 - p$, independently from each other. Now order the jump times that have not been deleted: $(\tau'_n, n \geq 1)$, and define

$$Z_t = \sum_{n \geq 1} \mathbf{1}_{t \geq \tau'_n}.$$

Then the process Z is Poisson process with intensity $p\lambda$.

In our setting, first a Poisson process with a constant rate λ_{sup} (which is an upper bound of the state-dependent rate) is constructed. This gives a set of candidate jump times, and these are then selected as true jump times with probability $\lambda(x(t), t)/\lambda_{\text{sup}}$. The following jump-adapted thinning Milstein scheme is obtained.

1. Generate the jump-adapted time grid for a Poisson process with constant rate λ_{sup}.

2. Simulate each timestep using the Milstein discretization.

3. When the endpoint t_{n+1} is a candidate jump time, generate a uniform random number $U \sim [0, 1]$, and if $U < p_{t_{n+1}} = \dfrac{\lambda(x(t_{n+1}-), t_{n+1})}{\lambda_{\text{sup}}}$, then accept t_{n+1} as a real jump time and simulate the jump.

In the multilevel implementation, the straightforward application of the above algorithm will result in different acceptance probabilities for fine and coarse levels. There may be some samples in which a jump candidate is accepted for the fine path, but not for the coarse path, or vice versa. Because of the first-order strong convergence, the difference in acceptance probabilities will be $\mathcal{O}(\Delta t_\ell)$, and hence there is an $\mathcal{O}(\Delta t_\ell)$ probability of coarse and fine paths differing in accepting candidate jumps. Such differences will give an $\mathcal{O}(1)$ difference in the payoff value, and hence the multilevel variance will be $\mathcal{O}(h)$. A more detailed analysis of this is given in Reference 33.

To improve the variance convergence rate, a change of measure is used so that the acceptance probability is the same for both fine and coarse paths. This is achieved by taking the expectation with respect to a new measure Q:

$$\mathbb{E}_P[P_\ell^{\text{f}} - P_{\ell-1}^{\text{c}}] = \mathbb{E}_Q[P_\ell^{\text{f}} \prod_\tau R_\tau^{\text{f}} - P_{\ell-1}^{\text{c}} \prod_\tau R_\tau^{\text{c}}],$$

where τ are the jump times. The acceptance probability for a candidate jump under the measure Q is defined to be $\frac{1}{2}$ for both coarse and fine paths, instead of $p_\tau = \lambda(X(\tau-), \tau) \,/\, \lambda_{\text{sup}}$. The corresponding Radon–Nikodym derivatives are

$$R_\tau^{\text{f}} = \begin{cases} 2p_\tau^{\text{f}}, & \text{if } U < \dfrac{1}{2}\ ; \\ 2(1 - p_\tau^{\text{f}}), & \text{if } U \geq \dfrac{1}{2}\ , \end{cases} \qquad R_\tau^{\text{c}} = \begin{cases} 2p_\tau^{\text{c}}, & \text{if } U < \dfrac{1}{2}\ ; \\ 2(1 - p_\tau^{\text{c}}), & \text{if } U \geq \dfrac{1}{2}\ . \end{cases}$$

Since $\mathbb{V}[R_\tau^{\text{f}} - R_\tau^{\text{c}}] = \mathcal{O}(\Delta t^2)$ and $\mathbb{V}[\widehat{P}_\ell - \widehat{P}_{\ell-1}] = \mathcal{O}(\Delta t^2)$, this results in the multilevel correction variance $\mathbb{V}_Q[\widehat{P}_\ell \prod_\tau R_\tau^{\text{f}} - \widehat{P}_{\ell-1} \prod_\tau R_\tau^{\text{c}}]$ being $\mathcal{O}(\Delta t^2)$.

If the analytic formulation is expressed using the same thinning and change of measure, the weak error can be decomposed into two terms as follows:

$$\mathbb{E}_Q\left[\widehat{P}_\ell \prod_\tau R^{\mathrm{f}}_\tau - P\prod_\tau R_\tau\right]$$
$$= \mathbb{E}_Q\left[(\widehat{P}_\ell - P)\prod_\tau R^{\mathrm{f}}_\tau\right] + \mathbb{E}_Q\left[P\left(\prod_\tau R^{\mathrm{f}}_\tau - \prod_\tau R_\tau\right)\right].$$

Using Hölder's inequality, the bound $\max(R_\tau, R^{\mathrm{f}}_\tau) \leq 2$ and standard results for a Poisson process, the first term can be bounded using weak convergence results for the constant rate process, and the second term can be bounded using the corresponding strong convergence results [33]. This guarantees that the multilevel procedure does converge to the correct value.

7.5.2 Lévy processes

Dereich and Heidenreich [34] analyzed approximation methods for both finite and infinite activity Lévy-driven SDEs with globally Lipschitz payoffs. They have derived upper bounds for MLMC variance for the class of path-dependent payoffs that are Lipschitz continuous with respect to supremum norm. One of their main findings is that the rate of MLMC variance convergence is closely related to the Blumenthal–Getoor index of the driving Lévy process that measures the frequency of small jumps. In Reference 34, the authors considered SDEs driven by the Lévy process

$$s(t) = \Sigma\, w(t) + L(t) + b\,t,$$

where Σ is the diffusion coefficient, $L(t)$ is a compensated jump process, and b is a drift coefficient. The simplest treatment is to neglect all the jumps with sizes smaller than h. To construct MLMC, they took h^ℓ, that is at level ℓ they neglected jumps smaller than h^ℓ. Then similarly as in the previous section, a uniform time discretization Δt_ℓ augmented with jump times is used. Let us denote by $\Delta L(t) = L(t) - L(t)^-$, the jump-discontinuity at time t. The crucial observation is that for $h' > h > 0$, the jumps of the process $L^{h'}$ can be obtained from those of L^h by

$$\Delta L(t)^{h'} = \Delta L^h_t\, \mathbf{1}_{\{|\Delta L(t)^h| > h'\}},$$

this gives the necessary coupling to obtain a good MLMC variance. We define a decreasing and invertible function $g : (0,\infty) \to (0,\infty)$ such that

$$\int \frac{|x|^2}{h^2} \wedge 1 \nu(\mathrm{d}x) \leq g(h) \quad \text{for all } h > 0,$$

where ν is a Lévy measure, and for $\ell \in \mathbb{N}$ we define

$$\Delta t_\ell = 2^{-\ell} \quad \text{and} \quad h^\ell = g^{-1}(2^\ell).$$

With this choice of Δt_ℓ and h^ℓ, the authors of Reference 34 analyzed the standard EM scheme for Lévy-driven SDEs. This approach gives good results for a Blumenthal–Getoor index smaller than 1. For a Blumenthal–Getoor index bigger than 1, a Gaussian approximation of small jumps gives better results [35].

7.6 Multidimensional Milstein Scheme

In the previous sections, it was shown that by combining a numerical approximation with the strong order of convergence $\mathcal{O}(\Delta t_\ell)$ with MLMC results in a reduction of the computational complexity to estimate expected values of functionals of SDE solutions with a root-mean-square error of ϵ from $\mathcal{O}(\epsilon^{-3})$ to $\mathcal{O}(\epsilon^{-2})$. However, in general, to obtain a rate of strong convergence higher than $O(\Delta t^{1/2})$ requires simulation, or approximation, of Lévy areas. Giles and Szpruch in Reference 18 through the construction of a suitable antithetic multilevel correction estimator showed that we can avoid the simulation of Lévy areas and still achieve an $O(\Delta t^2)$ variance for smooth payoffs, and almost an $O(\Delta t^{3/2})$ variance for piecewise smooth payoffs, even though there is only $O(\Delta t^{1/2})$ strong convergence.

In the previous sections, we have shown that it can be better to use different estimators for the finer and coarser of the two levels being considered, P_ℓ^{f} when level ℓ is the finer level, and P_ℓ^{c} when level ℓ is the coarser level. In this case, we required that

$$\mathbb{E}[P_\ell^{\mathrm{f}}] = \mathbb{E}[P_\ell^{\mathrm{c}}] \quad \text{for } \ell = 1, \ldots, L, \tag{7.33}$$

so that

$$\mathbb{E}[P_L^{\mathrm{f}}] = \mathbb{E}[P_0^{\mathrm{f}}] + \sum_{\ell=1}^{L} \mathbb{E}[P_\ell^{\mathrm{f}} - P_{\ell-1}^{\mathrm{c}}],$$

still holds. For lookback, barrier, and digital options, we showed that we can obtain a better MLMC variance by suitably modifying the estimator on the coarse levels. By further exploiting the flexibility of MLMC, Giles and Szpruch [18] modified the estimator on the fine levels in order to avoid simulation of the Lévy areas.

7.6.1 Antithetic multilevel Monte Carlo estimator

Based on the well-known method of antithetic variates [24], the idea for the antithetic estimator is to exploit the flexibility of the more general MLMC estimator by defining $P_{\ell-1}^{\mathrm{c}}$ to be the usual payoff $P(X^{\mathrm{c}})$ coming from a level $\ell - 1$ coarse simulation X^{c}, and defining P_ℓ^{f} to be the average of the payoffs $P(X^{\mathrm{f}}), P(X^{\mathrm{a}})$ coming from an antithetic pair of level ℓ simulations, X^{f} and X^{a}.

X^{f} will be defined in a way which corresponds naturally to the construction of X^{c}. Its antithetic "twin" X^{a} will be defined so that it has exactly the same distribution as X^{f}, conditional on X^{c}, which ensures that $\mathbb{E}[P(X^{\mathrm{f}})] = \mathbb{E}[P(X^{\mathrm{a}})]$ and hence Equation 7.3 is satisfied, but at the same time

$$\left(X^{\mathrm{f}} - X^{\mathrm{c}}\right) \approx -(X^{\mathrm{a}} - X^{\mathrm{c}})$$

and therefore

$$\left(P(X^{\mathrm{f}}) - P(X^{\mathrm{c}})\right) \approx -(P(X^{\mathrm{a}}) - P(X^{\mathrm{c}})),$$

so that $\frac{1}{2}\left(P(X^{\mathrm{f}}) + P(X^{\mathrm{a}})\right) \approx P(X^{\mathrm{c}})$. This leads to $\frac{1}{2}\left(P(X^{\mathrm{f}}) + P(X^{\mathrm{a}})\right) - P(X^{\mathrm{c}})$ having a much smaller variance than the standard estimator $P(X^{\mathrm{f}}) - P(X^{\mathrm{c}})$.

We now present a lemma which gives an upper bound on the convergence of the variance of $\frac{1}{2}\left(P(X^{\mathrm{f}}) + P(X^{\mathrm{a}})\right) - P(X^{\mathrm{c}})$.

Lemma 7.1. *If $P \in C^2(\mathbb{R}^d, \mathbb{R})$ and there exist constants L_1, L_2 such that for all $x \in \mathbb{R}^d$*

$$\left\|\frac{\partial P}{\partial x}\right\| \leq L_1, \qquad \left\|\frac{\partial^2 P}{\partial x^2}\right\| \leq L_2,$$

then for $p \geq 2$,

$$\mathbb{E}\left[\left(\frac{1}{2}(P(X^{\mathrm{f}}) + P(X^{\mathrm{a}})) - P(X^{\mathrm{c}})\right)^p\right]$$
$$\leq 2^{p-1}\, L_1^p\, \mathbb{E}\left[\left\|\frac{1}{2}(X^{\mathrm{f}} + X^{\mathrm{a}}) - X^{\mathrm{c}}\right\|^p\right] + 2^{-(p+1)}\, L_2^p\, \mathbb{E}\left[\left\|X^{\mathrm{f}} - X^{\mathrm{a}}\right\|^{2p}\right].$$

In the multidimensional SDE applications considered in finance, the Milstein approximation with the Lévy areas set to zero, combined with the antithetic construction, leads to $X^{\mathrm{f}} - X^{\mathrm{a}} = O(\Delta t^{1/2})$ but $\overline{X}^{\mathrm{f}} - X^{\mathrm{c}} = O(\Delta t)$. Hence, the variance $\mathbb{V}[\frac{1}{2}(P_l^{\mathrm{f}} + P_l^{\mathrm{a}}) - P_{l-1}^{\mathrm{c}}]$ is $O(\Delta t^2)$, which is the order obtained for scalar SDEs using the Milstein discretization with its first-order strong convergence.

7.6.2 Clark–Cameron example

The paper of Clark and Cameron [16] addresses the question of how accurately one can approximate the solution of an SDE driven by an underlying multidimensional Brownian motion, using only uniformly spaced discrete Brownian increments. Their model problem is

$$\begin{aligned} \mathrm{d}x_1(t) &= \mathrm{d}w_1(t) \\ \mathrm{d}x_2(t) &= x_1(t)\, \mathrm{d}w_2(t), \end{aligned} \tag{7.34}$$

with $x(0) = y(0) = 0$, and zero correlation between the two Brownian motions $w_1(t)$ and $w_2(t)$. These equations can be integrated exactly over a time interval

$[t_n, t_{n+1}]$, where $t_n = n\,\Delta t$, to give

$$x_1(t_{n+1}) = x_1(t_n) + \Delta w_{1,n}$$
$$x_2(t_{n+1}) = x_2(t_n) + x_1(t_n)\Delta w_{2,n} + \frac{1}{2}\Delta w_{1,n}\Delta w_{2,n} + \frac{1}{2}A_{12,n}, \quad (7.35)$$

where $\Delta w_{i,n} \equiv w_i(t_{n+1}) - w_i(t_n)$, and $A_{12,n}$ is the Lévy area defined as

$$A_{12,n} = \int_{t_n}^{t_{n+1}} \Big(w_1(t) - w_1(t_n)\Big)\,\mathrm{d}w_2(t) - \int_{t_n}^{t_{n+1}} \Big(w_2(t) - w_2(t_n)\Big)\,\mathrm{d}w_1(t).$$

This corresponds exactly to the Milstein discretization presented in Equation 7.7, so for this simple model problem the Milstein discretization is exact.

The point of Clark and Cameron's paper is that for *any* numerical approximation $X(T)$ based solely on the set of discrete Brownian increments Δw,

$$\mathbb{E}[(x_2(T) - X_2(T))^2] \geq \frac{1}{4}T\,\Delta t.$$

Since in this section, we use superscripts f, a, c for fine X^{f}, antithetic X^{a}, and coarse X^{c} approximations, respectively, we drop the superscript ℓ for the clarity of notation.

We define a coarse path approximation X^{c} with timestep Δt by neglecting the Lévy area terms to give

$$X^{\mathrm{c}}_{1,n+1} = X^{\mathrm{c}}_{1,n} + \Delta w^{\ell-1}_{1,n+1}$$
$$X^{\mathrm{c}}_{2,n+1} = X^{\mathrm{c}}_{2,n} + X^{\mathrm{c}}_{1,n}\Delta w^{\ell-1}_{2,n+1} + \frac{1}{2}\,\Delta w^{\ell-1}_{1,n+1}\,\Delta w^{\ell-1}_{2,n+1}. \quad (7.36)$$

This is equivalent to replacing the true Brownian path by a piecewise linear approximation as illustrated in Figure 7.1.

Similarly, we define the corresponding two half-timesteps of the first fine path approximation X^{f} by

$$X^{\mathrm{f}}_{1,n+\frac{1}{2}} = X^{\mathrm{f}}_{1,n} + \Delta w^{\ell}_{1,n+\frac{1}{2}}$$
$$X^{\mathrm{f}}_{2,n+\frac{1}{2}} = X^{\mathrm{f}}_{2,n} + X^{\mathrm{f}}_{1,n}\,\Delta w^{\ell}_{2,n+\frac{1}{2}} + \frac{1}{2}\,\Delta w^{\ell}_{1,n+\frac{1}{2}}\,\Delta w^{\ell}_{2,n+\frac{1}{2}}$$
$$X^{\mathrm{f}}_{1,n+1} = X^{\mathrm{f}}_{1,n+1} + \Delta w^{\ell}_{1,n+1}$$
$$X^{\mathrm{f}}_{2,n+1} = X^{\mathrm{f}}_{2,n+\frac{1}{2}} + X^{\mathrm{f}}_{1,n+\frac{1}{2}}\,\Delta w^{\ell}_{2,n+1} + \frac{1}{2}\,\Delta w^{\ell}_{1,n+1}\,\Delta w^{\ell}_{2,n+1},$$

where $\Delta w^{\ell-1}_{n+1} = \Delta w^{\ell}_{n+\frac{1}{2}} + \Delta w^{\ell}_{n+1}$. Using this relation, the equations for the two fine timesteps can be combined to give an equation for the increment over the coarse timestep,

$$
\begin{aligned}
X^{\mathrm{f}}_{1,n+1} &= X^{\mathrm{f}}_{1,n} + \Delta w^{\ell-1}_{1,n+1} \\
X^{\mathrm{f}}_{2,n+1} &= X^{\mathrm{f}}_{2,n} + X^{\mathrm{f}}_{1,n}\, \Delta w^{\ell-1}_{2,n+1} + \frac{1}{2}\, \Delta w^{\ell-1}_{1,n+1}\, \Delta w^{\ell-1}_{2,n+1} \\
&\quad + \frac{1}{2}\left(\Delta w^{\ell}_{1,n+\frac{1}{2}}\, \Delta w^{\ell}_{2,n+1} - \Delta w^{\ell}_{2,n+\frac{1}{2}}\, \Delta w^{\ell}_{1,n+1} \right). \qquad (7.37)
\end{aligned}
$$

The antithetic approximation X^{a}_n is defined by exactly the same discretization except that the Brownian increments δw_n and $\delta w_{n+\frac{1}{2}}$ are swapped, as illustrated in Figure 7.1. This gives

$$
\begin{aligned}
X^{\mathrm{a}}_{1,n+\frac{1}{2}} &= X^{\mathrm{a}}_{1,n} + \Delta w^{\ell}_{1,n+1}, \\
X^{\mathrm{a}}_{2,n+\frac{1}{2}} &= X^{\mathrm{a}}_{2,n} + X^{\mathrm{a}}_{1,n}\, \Delta w^{\ell}_{2,n+1} + \frac{1}{2}\, \Delta w^{\ell}_{1,n+1}\, \Delta w^{\ell}_{2,n+1}, \\
X^{\mathrm{a}}_{1,n+1} &= X^{\mathrm{a}}_{1,n+\frac{1}{2}} + \Delta w^{\ell}_{1,n+\frac{1}{2}}, \\
X^{\mathrm{a}}_{2,n+1} &= X^{\mathrm{a}}_{2,n+\frac{1}{2}} + X^{\mathrm{a}}_{1,n+\frac{1}{2}}\, \Delta w^{\ell}_{2,n+\frac{1}{2}} + \frac{1}{2}\, \Delta w^{\ell}_{1,n+\frac{1}{2}}\, \Delta w^{\ell}_{2,n+\frac{1}{2}},
\end{aligned}
$$

and hence

$$
\begin{aligned}
X^{\mathrm{a}}_{1,n+1} &= X^{\mathrm{a}}_{1,n} + \Delta w^{\ell-1}_{1,n+1}, \\
X^{\mathrm{a}}_{2,n+1} &= X^{\mathrm{a}}_{2,n} + X^{\mathrm{a}}_{1,n}\, \Delta w^{\ell-1}_{2,n+1} + \frac{1}{2}\, \Delta w^{\ell-1}_{1,n+1}\, \Delta w^{\ell-1}_{2,n+1} \\
&\quad - \frac{1}{2}\left(\Delta w^{\ell}_{1,n+\frac{1}{2}}\, \Delta w^{\ell}_{2,n+1} - \Delta w^{\ell}_{2,n+\frac{1}{2}}\, \Delta w^{\ell}_{1,n+1} \right). \qquad (7.38)
\end{aligned}
$$

Swapping $\Delta w^{\ell}_{n+\frac{1}{2}}$ and Δw^{ℓ}_{n+1} does not change the distribution of the driving Brownian increments, and hence X^{a} has exactly the same distribution as X^{f}.

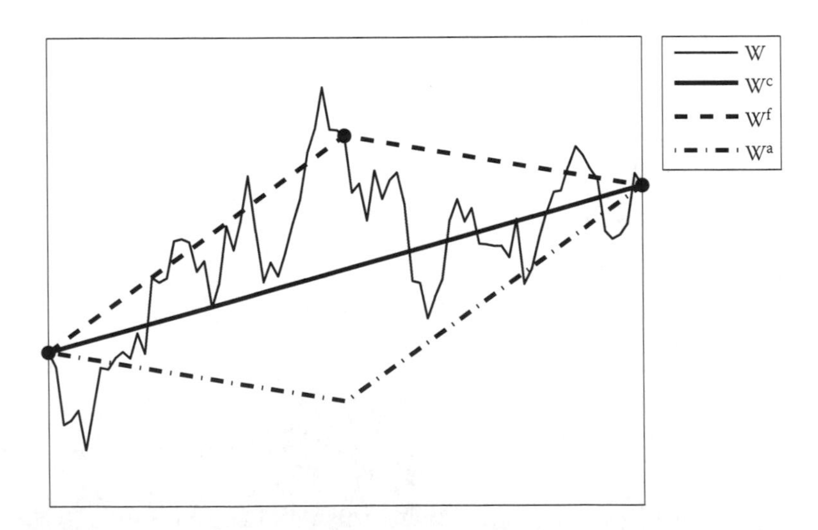

FIGURE 7.1: Brownian path and approximations over one coarse timestep.

Note also the change in sign in the last term in Equation 7.37 compared to the corresponding term in Equation 7.38. This is important because these two terms cancel when the two equations are averaged.

These last terms correspond to the Lévy areas for the fine and antithetic paths, and the sign reversal is a particular instance of a more general result for time-reversed Brownian motion [36]. If $(w_t, 0 \le t \le 1)$ denotes a Brownian motion on the time interval $[0, 1]$, then the time-reversed Brownian motion $(z_t, 0 \le t \le 1)$ defined by

$$z_t = w_1 - w_{1-t} \tag{7.39}$$

has exactly the same distribution, and it can be shown that its Lévy area is equal in magnitude and opposite in sign to that of w_t.

Lemma 7.2. *If X^{f}_n, X^{a}_n, and X^{c}_n are as defined above, then*

$$X^{\mathrm{f}}_{1,n} = X^{\mathrm{a}}_{1,n} = X^{\mathrm{c}}_{1,n}, \qquad \frac{1}{2}\left(X^{\mathrm{f}}_{2,n} + X^{\mathrm{a}}_{2,n}\right) = X^{\mathrm{c}}_{2,n}, \qquad \forall n \le N$$

and

$$\mathbb{E}\left[\left(X^{\mathrm{f}}_{2,N} - X^{\mathrm{a}}_{2,N}\right)^4\right] = \frac{3}{4} T\,(T + \Delta t)\,\Delta t^2.$$

In the following section, we will see how this lemma generalizes to nonlinear multidimensional SDEs 7.4.

7.6.3 Milstein discretization: General theory

Using the coarse timestep Δt, the coarse path approximation X^{c}_n is given by the Milstein approximation without the Lévy area term:

$$\begin{aligned} X^{\mathrm{c}}_{i,n+1} = X^{\mathrm{c}}_{i,n} &+ f_i(X^{\mathrm{c}}_n)\,\Delta t_{\ell-1} + \sum_{j=1}^{m} g_{ij}(X^{\mathrm{c}}_n)\,\Delta w^{\ell-1}_{j,n+1} \\ &+ \sum_{j,k=1}^{m} h_{ijk}(X^{\mathrm{c}}_n)\left(\Delta w_{j,n}\,\Delta w^{\ell-1}_{k,n+1} - \Omega_{jk}\,\Delta t_{\ell-1}\right). \end{aligned}$$

The first fine path approximation X^{f}_n (that corresponds to X^{c}_n) uses the corresponding discretization with timestep $\Delta t/2$,

$$\begin{aligned} X^{\mathrm{f}}_{i,n+\frac{1}{2}} = X^{\mathrm{f}}_{i,n} &+ f_i(X^{\mathrm{f}}_n)\,\Delta t_{\ell-1}/2 + \sum_{j=1}^{m} g_{ij}(X^{\mathrm{f}}_n)\,\Delta w^{\ell}_{j,n+\frac{1}{2}} \\ &+ \sum_{j,k=1}^{m} h_{ijk}(X^{\mathrm{f}}_n)\left(\Delta w^{\ell}_{j,n+\frac{1}{2}}\,\Delta w^{\ell}_{k,n+\frac{1}{2}} - \Omega_{jk}\,\Delta t_{\ell-1}/2\right), \end{aligned} \tag{7.40}$$

$$\begin{aligned} X^{\mathrm{f}}_{i,n+1} = X^{\mathrm{f}}_{i,n+\frac{1}{2}} &+ f_i\left(X^{\mathrm{f}}_{n+\frac{1}{2}}\right)\,\Delta t_{\ell-1}/2 + \sum_{j=1}^{m} g_{ij}\left(X^{\mathrm{f}}_{n+\frac{1}{2}}\right)\,\Delta w^{\ell}_{j,n+1} \\ &+ \sum_{j,k=1}^{m} h_{ijk}\left(X^{\mathrm{f}}_{n+\frac{1}{2}}\right)\left(\Delta w^{\ell}_{j,n+1}\,\Delta w^{\ell}_{k,n+1} - \Omega_{jk}\,\Delta t_{\ell-1}/2\right), \end{aligned} \tag{7.41}$$

where $\Delta w^{\ell-1}_{n+1} = \Delta w^{\ell}_{n+\frac{1}{2}} + \Delta w^{\ell}_{n+1}$.

The antithetic approximation X^{a}_n is defined by exactly the same discretization except that the Brownian increments $Dw^{\ell}_{n+\frac{1}{2}}$ and Δw^{ℓ}_{n+1} are swapped, so that

$$\begin{aligned}
X^{\mathrm{a}}_{i,n+\frac{1}{2}} &= X^{\mathrm{a}}_{i,n} + f_i(X^{\mathrm{a}}_n)\,\Delta t_{\ell-1}/2 + \sum_{j=1}^{m} g_{ij}(X^{\mathrm{a}}_n)\,\delta w_{n+\frac{1}{2}} \\
&\quad + \sum_{j,k=1}^{m} h_{ijk}(X^{\mathrm{a}}_n)\left(\Delta w^{\ell}_{j,n+1}\,\Delta w^{\ell}_{k,n+1} - \Omega_{jk}\,\Delta t_{\ell-1}/2\right), \\
X^{\mathrm{a}}_{i,n+1} &= X^{\mathrm{a}}_{i,n+\frac{1}{2}} + f_i\left(X^{\mathrm{a}}_{n+\frac{1}{2}}\right)\,\Delta t_{\ell-1}/2 + \sum_{j=1}^{m} g_{ij}\left(X^{\mathrm{a}}_{n+\frac{1}{2}}\right)\,\Delta w^{\ell}_{j,n+\frac{1}{2}} \\
&\quad + \sum_{j,k=1}^{m} h_{ijk}\left(X^{\mathrm{a}}_{n+\frac{1}{2}}\right)\left(\Delta w^{\ell}_{j,n+\frac{1}{2}}\,\Delta w^{\ell}_{k,n+\frac{1}{2}} - \Omega_{jk}\,\Delta t_{\ell-1}/2\right).
\end{aligned} \tag{7.42}$$

It can be shown that [18]

Lemma 7.3. *For all integers $p \geq 2$, there exists a constant K_p such that*

$$\mathbb{E}\left[\max_{0\leq n\leq N} \|X^{\mathrm{f}}_n - X^{\mathrm{a}}_n\|^p\right] \leq K_p\,\Delta t^{p/2}.$$

Let us denote the average fine and antithetic path as follows:

$$\overline{X}^{\mathrm{f}}_n \equiv \frac{1}{2}(X^{\mathrm{f}}_n + X^{\mathrm{a}}_n).$$

The main results of [18] is the following theorem.

Theorem 7.5. *For all $p \geq 2$, there exists a constant K_p such that*

$$\mathbb{E}\left[\max_{0\leq n\leq N} \|\overline{X}^{\mathrm{f}}_n - X^{\mathrm{c}}_n\|^p\right] \leq K_p\,\Delta t^{p}.$$

This together with a classical strong convergence result for Milstein discretization allows to estimate the MLMC variance for smooth payoffs. In the case of payoff which is a smooth function of the final state $x(T)$, taking $p=2$ in Lemma 7.1, $p=4$ in Lemma 7.3, and $p=2$ in Theorem 7.5 immediately gives the result that the multilevel variance

$$\mathbb{V}\left[\frac{1}{2}\left(P(X^{\mathrm{f}}_N) + P(X^{\mathrm{a}}_N)\right) - P(X^{\mathrm{c}}_N)\right]$$

has an $O(\Delta t^2)$ upper bound. This matches the convergence rate for the multilevel method for scalar SDEs using the standard first-order Milstein discretization, and is much better than the $O(\Delta t)$ convergence obtained with the EM discretization.

However, very few financial payoff functions are twice differentiable on the entire domain $\mathbb{R}^d$. A more typical 2D example is a call option based on the minimum of two assets,

$$P(x(T)) \equiv \max\left(0, \min(x_1(T), x_2(T)) - K\right),$$

which is piecewise linear, with a discontinuity in the gradient along the three lines (s, K), (K, s), and (s, s) for $s \geq K$.

To handle such payoffs, an assumption which bounds the probability of the solution of the SDE having a value at time T close to such lines with discontinuous gradients is needed.

Assumption 7.1. *The payoff function $P \in C(\mathbb{R}^d, \mathbb{R})$ has a uniform Lipschitz bound, so that there exists a constant L such that*

$$|P(x) - P(y)| \leq L\, |x - y|, \quad \forall\, x, y \in \mathbb{R}^d,$$

and the first and second derivatives exist, are continuous and have uniform bound L at all points $x \notin K$, where K is a set of zero measure, and there exists a constant c such that the probability of the SDE solution $x(T)$ being within a neighborhood of the set K has the bound

$$\mathbb{P}\left(\min_{y \in K} \|x(T) - y\| \leq \varepsilon\right) \leq c\, \varepsilon, \quad \forall\, \varepsilon > 0.$$

In a 1D context, Assumption 7.1 corresponds to an assumption of a locally bounded density for $x(T)$.

Giles and Szpruch in Reference 18 proved the following result.

Theorem 7.6. *If the payoff satisfies Assumption 7.1, then*

$$\mathbb{E}\left[\left(\frac{1}{2}\left(P\left(X_N^{\mathrm{f}}\right) + P\left(X_N^{\mathrm{a}}\right)\right) - P\left(X_N^{\mathrm{c}}\right)\right)^2\right] = o(\Delta t^{3/2-\delta})$$

for any $\delta > 0$.

7.6.4 Piecewise linear interpolation analysis

The piecewise linear interpolant $X^{\mathrm{c}}(t)$ for the coarse path is defined within the coarse timestep interval $[t_k, t_{k+1}]$ as

$$X^{\mathrm{c}}(t) \equiv (1 - \lambda)\, X_k^{\mathrm{c}} + \lambda\, X_{k+1}^{\mathrm{c}}, \quad \lambda \equiv \frac{t - t_k}{t_{k+1} - t_k}.$$

Likewise, the piecewise linear interpolants $X^{\mathrm{f}}(t)$ and $X^{\mathrm{a}}(t)$ are defined on the fine timestep $[t_k, t_{k+\frac{1}{2}}]$ as

$$X^{\mathrm{f}}(t) \equiv (1-\lambda)\, X_k^{\mathrm{f}} + \lambda\, X_{k+\frac{1}{2}}^{\mathrm{f}}, \quad X^{\mathrm{a}}(t) \equiv (1-\lambda)\, X_k^{\mathrm{a}} + \lambda\, X_{k+\frac{1}{2}}^{\mathrm{a}}, \quad \lambda \equiv \frac{t - t_k}{t_{k+\frac{1}{2}} - t_k},$$

and there is a corresponding definition for the fine timestep $[t_{k+\frac{1}{2}}, t_{k+1}]$. It can be shown that [18]

Theorem 7.7. *For all $p \geq 2$, there exists a constant K_p such that*

$$\sup_{0 \leq t \leq T} \mathbb{E}\left[\|X^{\mathrm{f}}(t) - X^{\mathrm{a}}(t)\|^p \right] \leq K_p \, \Delta t^{p/2},$$

$$\sup_{0 \leq t \leq T} \mathbb{E}\left[\left\|\overline{X}^{\mathrm{f}}(t) - X^{\mathrm{c}}(t)\right\|^p \right] \leq K_p \, \Delta t^{p},$$

where $\overline{X}^{\mathrm{f}}(t)$ is the average of the piecewise linear interpolants $X^{\mathrm{f}}(t)$ and $X^{\mathrm{a}}(t)$.

For an Asian option, the payoff depends on the average

$$x_{\mathrm{ave}} \equiv T^{-1} \int_0^T x(t) \, \mathrm{d}t.$$

This can be approximated by integrating the appropriate piecewise linear interpolant which gives

$$X^{\mathrm{c}}_{\mathrm{ave}} \equiv T^{-1} \int_0^T X^{\mathrm{c}}(t) \, \mathrm{d}t = N^{-1} \sum_{n=0}^{N-1} \frac{1}{2} \left(X^{\mathrm{c}}_n + X^{\mathrm{c}}_{n+1} \right),$$

$$X^{\mathrm{f}}_{\mathrm{ave}} \equiv T^{-1} \int_0^T X^{\mathrm{f}}(t) \, \mathrm{d}t = N^{-1} \sum_{n=0}^{N-1} \frac{1}{4} \left(X^{\mathrm{f}}_n + 2X^{\mathrm{f}}_{n+\frac{1}{2}} + X^{\mathrm{f}}_{n+1} \right),$$

$$X^{\mathrm{a}}_{\mathrm{ave}} \equiv T^{-1} \int_0^T X^{\mathrm{a}}(t) \, \mathrm{d}t = N^{-1} \sum_{n=0}^{N-1} \frac{1}{4} \left(X^{\mathrm{a}}_n + 2X^{\mathrm{a}}_{n+\frac{1}{2}} + X^{\mathrm{a}}_{n+1} \right).$$

Due to Hölder's inequality,

$$\begin{aligned} \mathbb{E}\left[\left\|X^{\mathrm{f}}_{\mathrm{ave}} - X^{\mathrm{a}}_{\mathrm{ave}}\right\|^p \right] &\leq T^{-1} \int_0^T \mathbb{E}\left[\left\|X^{\mathrm{f}}(t) - X^{\mathrm{a}}(t)\right\|^p \right] \mathrm{d}t \\ &\leq \sup_{[0,T]} \mathbb{E}\left[\left\|X^{\mathrm{f}}(t) - X^{\mathrm{a}}(t)\right\|^p \right], \end{aligned}$$

and similarly

$$\mathbb{E}\left[\left\| \frac{1}{2}(X^{\mathrm{f}}_{\mathrm{ave}} + X^{\mathrm{a}}_{\mathrm{ave}}) - X^{\mathrm{c}}_{\mathrm{ave}} \right\|^p \right] \leq \sup_{[0,T]} \mathbb{E}\left[\left\|\overline{X}^{\mathrm{f}}(t) - X^{\mathrm{c}}(t)\right\|^p \right].$$

Hence, if the Asian payoff is a smooth function of the average, then we obtain a second-order bound for the multilevel correction variance.

This analysis can be extended to include payoffs which are a smooth function of a number of intermediate variables, each of which is a linear functional of the path $x(t)$ of the form:

$$\int_0^T g^T(t)\, x(t)\, \mu(\mathrm{d}t),$$

for some vector function $g(t)$ and measure $\mu(\mathrm{d}t)$. This includes weighted averages of $x(t)$ at a number of discrete times, as well as continuously weighted averages over the whole time interval.

As with the European options, the analysis can also be extended to payoffs which are Lipschitz functions of the average, and have first and second derivatives which exist, and are continuous and uniformly bounded, except for a set of points K of zero measure.

Assumption 7.2. *The payoff $P \in C(\mathbb{R}^d, \mathbb{R})$ has a uniform Lipschitz bound, so that there exists a constant L such that*

$$|P(x) - P(y)| \le L\, |x - y|\,, \quad \forall\, x, y \in \mathbb{R}^d,$$

and the first and second derivatives exist, are continuous and have uniform bound L at all points $x \notin K$, where K is a set of zero measure, and there exists a constant c such that the probability of x_{ave} being within a neighborhood of the set K has the bound

$$\mathbb{P}\left(\min_{y \in K} \|x_{\text{ave}} - y\| \le \varepsilon\right) \le c\, \varepsilon, \quad \forall\, \varepsilon > 0.$$

Theorem 7.8. *If the payoff satisfies Assumption 7.2, then*

$$\mathbb{E}\left[\left(\frac{1}{2}(P(X^{\text{f}}_{\text{ave}}) + P(X^{\text{a}}_{\text{ave}})) - P(X^{\text{c}}_{\text{ave}})\right)^2\right] = o(\Delta t^{3/2-\delta})$$

for any $\delta > 0$.

We refer the reader to Reference 18 for more details.

7.6.5 Simulations for antithetic Monte Carlo

Here we present numerical simulations for a European option for a process simulated by X^{f}, X^{a}, and X^{c} defined in Section 7.6.2 with initial conditions $x_1(0) = x_2(0) = 1$. The results in Figure 7.2 are for a European call option with terminal time 1 and strike $K = 1$, that is, $P = (x(T) - K)^+$.

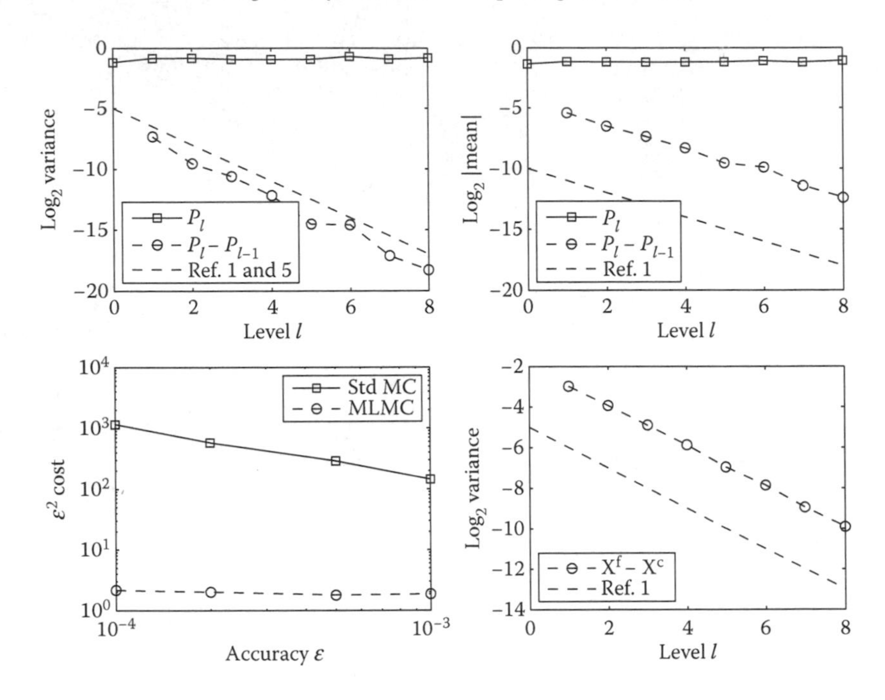

FIGURE 7.2: Call option.

The top left plot shows the behavior of the variance of both P_ℓ and $P_\ell - P_{\ell-1}$. The superimposed reference slope with rate 1.5 indicates that the variance $\mathbb{V}_\ell = \mathbb{V}[P_\ell - P_{\ell-1}] = O(\Delta t_\ell^{1.5})$, corresponding to $O(\epsilon^{-2})$ computational complexity of antithetic MLMC. The top right plot shows that $\mathbb{E}[P_\ell - P_{\ell-1}] = O(\Delta t_\ell)$. The bottom left plot shows the computational complexity C (as defined in Theorem 7.1) with desired accuracy ϵ. The plot is of $\epsilon^2 C$ versus ϵ, because we expect to see that $\epsilon^2 C$ is only weakly dependent on ϵs for MLMC. For standard Monte Carlo, theory predicts that $\epsilon^2 C$ should be proportional to the number of timesteps on the finest level, which in turn is roughly proportional to ϵ^{-1} due to the weak convergence order. For accuracy $\epsilon = 10^{-4}$, the antithetic MLMC is approximately 500 times more efficient than the standard Monte Carlo. The bottom right plot shows that $\mathbb{V}[X_{1.\ell} - X_{1.\ell-1}] = O(\Delta t_\ell)$. This corresponds to the standard strong convergence of order 0.5. We have also tested the algorithm presented in Reference 18 for approximation of Asian options. Our results were almost identical as for the European options. In order to treat the lookback, digital, and barrier options, we found that a suitable antithetic approximation to the Lévy areas is needed. For suitable modification of the antithetic MLMC estimator, we performed numerical experiments where we obtained $O(\epsilon^{-2}\log(\epsilon)^2)$ complexity for estimating barrier, digital, and lookback options. Currently, we are working on theoretical justification of our results.

7.7 Other Uses of Multilevel Method

7.7.1 Stochastic partial differential equations

Multilevel method has been used for a number of parabolic and elliptic SPDE applications [37–39] but the first use for a financial SPDE is in a new paper by Giles and Reisinger [40].

This paper considers an unusual SPDE which results from modelling credit default probabilities,

$$\Delta p = -\mu \frac{\partial p}{\partial x} \Delta t + \frac{1}{2} \frac{\partial^2 p}{\partial x^2} \Delta t - \sqrt{\rho} \frac{\partial p}{\partial x} \Delta M_t, \quad x > 0 \tag{7.43}$$

subject to boundary condition $p(0,t) = 0$. Here $p(x,t)$ represents the probability density function for firms being a distance x from default at time t. The diffusive term is due to idiosyncratic factors affecting individual firms, while the stochastic term due to the scalar Brownian motion M_t corresponds to the systemic movement due to random market effects affecting all firms.

Using a Milstein time discretization with uniform timestep k, and a central space discretization of the spatial derivatives with uniform spacing h gives the numerical approximation

$$\begin{aligned} p_j^{n+1} = p_j^n &- \frac{\mu\, k + \sqrt{\rho\, k}\, Z_n}{2h} \left(p_{j+1}^n - p_{j-1}^n\right) \\ &+ \frac{(1-\rho)\, k + \rho\, k\, Z_n^2}{2h^2} \left(p_{j+1}^n - 2p_j^n + p_{j-1}^n\right), \end{aligned} \tag{7.44}$$

where Z_n are standard Normal random variables so that $\sqrt{h} Z_n$ corresponds to an increment of the driving scalar Brownian motion.

This chapter shows that the requirement for mean-square stability as the grid is refined and $k, h \to 0$ is $k/h^2 \le (1+2\rho^2)^{-1}$, and in addition the accuracy is $O(k, h^2)$. Because of this, the multilevel treatment considers a sequence of grids with $h_\ell = 2\, h_{\ell-1}$, $k_\ell = 4\, k_{\ell-1}$.

The multilevel implementation is very straightforward, with the Brownian increments for the fine path being summed pairwise to give the corresponding Brownian increments for the coarse path. The payoff corresponds to different tranches of a credit derivative that depends on a numerical approximation of the integral

$$\int_0^\infty p(x,t)\ \Delta x.$$

The computational cost increases by factor 8 on each level, and numerical experiments indicate that the variance decreases by a factor of 16. The MLMC Theorem still applies in this case, with $\beta = 4$ and $\gamma = 3$, and so the overall computational complexity to achieve an $O(\epsilon)$ RMS error is again $\mathcal{O}(\varepsilon^{-2})$.

7.7.2 Nested simulation

The pricing of American options is one of the big challenges for Monte Carlo methods in computational finance, and Belomestny and Schoenmakers have recently written a very interesting paper on the use of MLMC for this purpose [41]. Their method is based on Anderson and Broadie's dual simulation method [42] in which a key component at each timestep in the simulation is to estimate a conditional expectation using a number of subpaths.

In their multilevel treatment, Belomestny and Schoenmakers use the same uniform timestep on all levels of the simulation. The quantity which changes between different levels of simulation is the number of subsamples used to estimate the conditional expectation.

To couple the coarse and fine levels, the fine level uses N_ℓ subsamples, and the coarse level uses $N_{\ell-1} = N_\ell/2$ of them. Similar research by N. Chen* found that the multilevel correction variance is reduced if the payoff on the coarse level is replaced by an average of the payoffs obtained using the first $N_\ell/2$ and second $N_\ell/2$ samples. This is similar in some ways to the antithetic approach described in Section 7.6.

In future research, Belomestny and Schoenmakers intend to also change the number of timesteps on each level, to increase the overall computational benefits of the multilevel approach.

7.7.3 Truncated series expansions

Building on earlier work by Broadie and Kaya [43], Glasserman and Kim have recently developed an efficient method [44] of simulating the Heston stochastic volatility model exactly [45].

The key to their algorithm is a method of representing the integrated volatility over a time interval $[0, T]$, conditional on the initial and final values, v_0 and v_T as

$$\left(\int_0^T V_s \, \mathrm{d}s | V_0 = v_0, V_T = v_T \right) \stackrel{d}{=} \sum_{n=1}^{\infty} x_n + \sum_{n=1}^{\infty} y_n + \sum_{n=1}^{\infty} z_n$$

where x_n, y_n, z_n are independent random variables.

In practice, they truncate the series expansions at a level which ensures the desired accuracy, but a more severe truncation would lead to a tradeoff between accuracy and computational cost. This makes the algorithm a candidate for a multilevel treatment in which level ℓ computation performs the truncation at N_ℓ (taken to be the same for all three series, for simplicity).

To give more details, the level ℓ computation would use

$$\sum_{n=1}^{N_\ell} x_n + \sum_{n=1}^{N_\ell} y_n + \sum_{n=1}^{N_\ell} z_n$$

*Unpublished, but presented at the MCQMC12 conference.

while the level $\ell - 1$ computation would use

$$\sum_{n=1}^{N_{\ell-1}} x_n + \sum_{n=1}^{N_{\ell-1}} y_n + \sum_{n=1}^{N_{\ell-1}} z_n$$

with the same random variables x_n, y_n, z_n.

This kind of multilevel treatment has not been tested experimentally, but it seems that it might yield some computational savings even though Glasserman and Kim typically retain only 10 terms in their summations through the use of a carefully constructed estimator for the truncated remainder. In other circumstances requiring more terms to be retained, the savings may be larger.

7.7.4 Mixed precision arithmetic

The final example of the use of multilevel is unusual, because it concerns the computer implementation of Monte Carlo algorithms.

In the latest CPUs from Intel and AMD, each core has a vector unit which can perform 8 single precision or 4 double precision operations with one instruction. Together with the obvious fact that double precision variables are twice as big as single precision variables and so require twice as much time to transfer, in bulk, it leads to single precision computations being twice as fast as double precision computations. On GPUs (graphics processors), the difference in performance can be even larger, up to a factor of 8 in the most extreme cases.

This raises the question of whether single precision arithmetic is sufficient for Monte Carlo simulations. In general, our view is that the errors due to single precision arithmetic are much smaller than the errors due to

- Statistical error due to Monte Carlo sampling
- Bias due to SDE discretization
- Model uncertainty

We have just two concerns with single precision accuracy:

- There can be significant errors when averaging the payoffs unless one uses binary tree summation to perform the summation.
- When computing Greeks using "bumping," the single precision inaccuracy can be greatly amplified if a small bump is used.

Our advice would be to always use double precision for the final accumulation of payoff values and pathwise sensitivity analysis as much as possible for computing Greeks, but if there remains a need for the path simulation to be performed in double precision then one could use a two-level approach in which level 0 corresponds to single precision and level 1 corresponds to double precision.

On both levels one would use the same random numbers. The multilevel analysis would then give the optimal allocation of effort between the single precision and double precision computations. Since it is likely that most of the calculations would be single precision, the computational savings would be a factor of 2 or more compared to standard double precision calculations.

7.8 Multilevel Quasi-Monte Carlo

In Theorem 7.1, if $\beta > \gamma$, so that the rate at which the multilevel variance decays with increasing grid level is greater than the rate at which the computational cost increases, then the dominant computational cost is on the coarsest levels of approximation.

Since coarse levels of approximation correspond to low-dimensional numerical quadrature, it is quite natural to consider the use of quasi-Monte Carlo techniques. This has been investigated by Giles and Waterhouse [46] in the context of scalar SDEs with a Lipschitz payoff. Using the Milstein approximation with a doubling of the number of timesteps on each level gives $\beta = 2$ and $\gamma = 1$. They used a rank-1 lattice rule to generate the quasi-random numbers, randomization with 32 independent offsets to obtain confidence intervals, and a standard Brownian Bridge construction of the increments of the driving Brownian process.

Their empirical observation was that MLMC on its own was better than QMC on its own, but the combination of the two was even better. The QMC treatment greatly reduced the variance per sample for the coarsest levels, resulting in significantly reduced costs overall. In the simplest case of a European call option, shown in Figure 7.3, the top left plot shows the reduction in the variance per sample as the number of QMC points is increased. The benefit is much greater on the coarsest levels than on the finest levels. In the bottom two plots, the number of QMC points on each level is determined automatically to obtain the required accuracy; see Reference 46 for the precise details. Overall, the computational complexity appears to be reduced from $O(\epsilon^{-2})$ to approximately $O(\epsilon^{-1.5})$.

Giles and Waterhouse interpreted the fact that the variance is not reduced on the finest levels as being due to a lack of significant low-dimensional content, that is, the difference in the two payoffs due to neighboring grid levels is due to the difference in resolution of the driving Brownian path, and this is inherently of high dimensionality. This suggests that in other applications with $\beta < \gamma$, which would lead to the dominant cost being on the finest levels, then the use of quasi-Monte Carlo methods is unlikely to yield any benefits.

Further research is needed in this area to investigate the use of other low discrepancy sequences (e.g., Sobol) and other ways of generating the Brownian increments (e.g., PCA). We also refer the reader to Reference 47 for some results for randomized multilevel quasi-Monte Carlo.

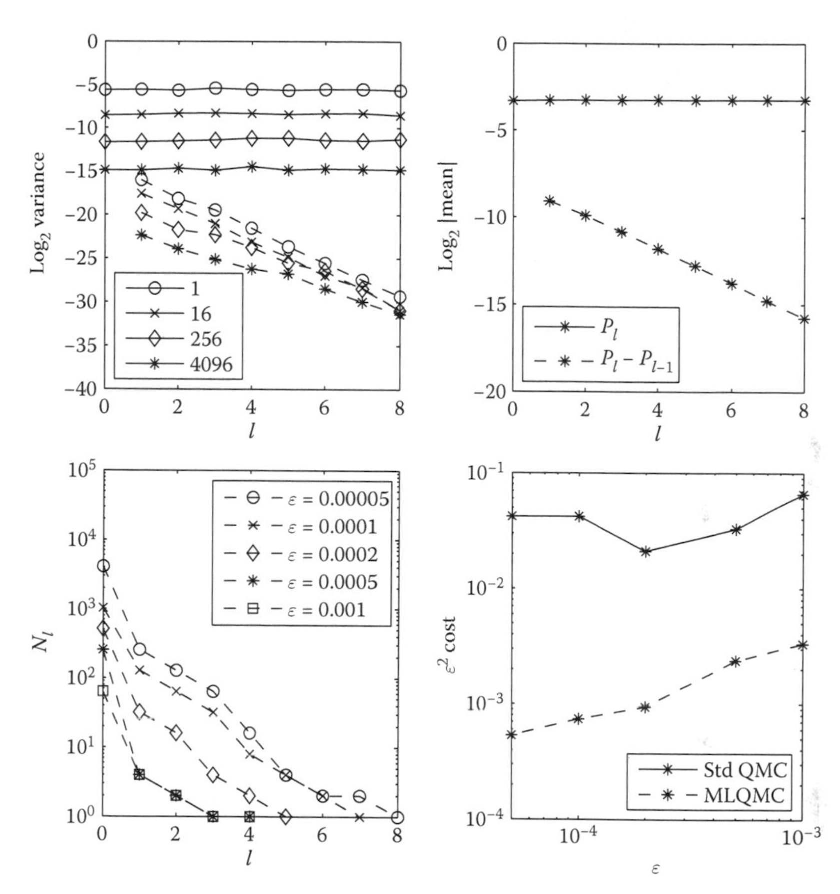

FIGURE 7.3: European call option (From Giles M.B. and Waterhouse B.J. *Advanced Financial Modelling*, Radon Series on Computational and Applied Mathematics, pages 165–181. de Gruyter, 2009.)

7.9 Conclusion

In the past 6 years, considerable progress has been achieved with the MLMC method for financial options based on underlying assets described by Brownian diffusions, jump diffusions, and more general Lévy processes.

The multilevel approach is conceptually very simple. In essence it is a recursive control variate strategy, using a coarse path simulation as a control variate for a fine path simulation, relying on strong convergence properties to ensure a very strong correlation between the two.

In practice, the challenge is to couple the coarse and fine path simulations as tightly as possible, minimizing the difference in the payoffs obtained for each. In doing this, there is considerable freedom to be creative, as shown in the use of Brownian Bridge constructions to improve the variance for lookback and barrier options, and in the antithetic estimators for multidimensional SDEs which would require the simulation of Lévy areas to achieve first-order strong convergence. Another challenge is avoiding large payoff differences due to discontinuous payoffs; here one can often use either conditional expectations to smooth the payoff or a change of measure to ensure that the coarse and fine paths are on the same side of the discontinuity.

Overall, multilevel methods are being used for an increasingly wide range of applications. These biggest savings are in situations in which the coarsest approximation is very much cheaper than the finest. If the finest level of approximation has only 32 timesteps, then there are very limited savings to be achieved, but if the finest level has 256 timesteps, then the potential savings are much larger.

Looking to the future, exciting areas for further research include:

- More research on multilevel techniques for American and Bermudan options
- More investigation of multilevel Quasi Monte Carlo methods
- Use of multilevel ideas for completely new financial applications, such as Gaussian copula and new SPDE models.

Appendix 7A Analysis of Brownian Bridge Interpolation

Let $w^{\alpha}(t) = \alpha t + w(t)$ denote a Brownian motion with drift α. Its running minimum we denote by $m(t)^a = \min_{0\le u\le t} w(u)^{\alpha} = \min_{0\le u\le t}\{\alpha u + w(u)\}$. Using the Girsanov theorem and the Reflection principle of the Brownian motion, we can derive the joint distribution of $(w^{\alpha}(t), m(t)^{\alpha})$ and as as consequence the conditional distribution of $m(t)^{\alpha}$ given $w^{\alpha}(t)$, [48]

$$\mathbb{P}\left(m_t^{\alpha} \le y | w(t)^{\alpha} = z\right) = \exp\left(\frac{2y(z-y)}{t}\right) \tag{7A.1}$$

Brownian Bridge 7.11 for $t \in [s, u]$

$$\begin{aligned} x(t) = x(s) + (t-s)/(u-s)(x(u) - x(s)) + g(s)(w(t) - w(s) \\ - (t-s)/(u-s)(w(u) - w(s))) \end{aligned} \tag{7A.2}$$

can be obtained by considering arithmetic Brownian motion on the time interval $[s, t]$

$$\begin{aligned} x(t) &= x(s) + f(s)(t-s) + g(s)(w(t) - w(s)) \\ &= x(s) + g(s)\left(\frac{f(s)}{g(s)}(t-s) + (w(t) - w(s))\right) \\ &= x(s) + g(s)w_{t-s}^{\alpha(s)}, \end{aligned}$$

with $\alpha(s) = f(s)/g(s)$. Similarly, the minimum $y_{s,t}$ of the process $(x(t))$ on the time interval $[s,t]$ is given by

$$y_{s,t} = x(s) + g(s)m_{t-s}^{\alpha(s)}.$$

Hence, by Equation 7A.1

$$\begin{aligned}\mathbb{P}\left[y_{s,t} \le y | x(s), x(t)\right] &= \mathbb{P}\left[x(s) + g(s)m_{t-s}^{\alpha(s)} \le y | x(s) + g(s)w_{t-s}^{\alpha(s)} = x(t)\right] \\ &= \mathbb{P}\left[m_{t-s}^{\alpha(s)} \le \frac{y - x(s)}{g(s)} | w_{t-s}^{\alpha(s)} = \frac{x(t) - x(s)}{g(s)}\right] \\ &= \exp\left(-\frac{2(x(s) - y)(x(t) - y)}{(g(s))^2(t-s)}\right) \qquad (7A.3)\end{aligned}$$

Now imagine we want to derive these probabilities over time interval $[s,u]$, where $t \le u$ conditioned on $x(s)$ and $x(u)$. The first strategy would be to take Equation 7A.2 connecting $x(s)$ and $x(u)$ and calculate the conditional distribution as we did in Equation 7A.4. The second strategy is as follows: (a) first we sample a point $x(t)$ from the BB connecting $x(s)$ and $x(u)$; (b) we calculate the conditional distribution of the minimum of BB (Equation 7A.2) conditioned first on $x(s)$, $x(t)$, and then on $x(t)$, $x(u)$. However in order to make sure both strategies give us results that are equivalent in distribution we are only allow to use the same Brownian bridge as we have used the first strategy. This has a consequence in calculating conditional distribution of the minimum given $x(t)$ and $x(u)$:

$$\begin{aligned}\mathbb{P}\left[y_{s,t} \le y | x(t), x(u)\right] &= \mathbb{P}\left[x(t) + g(s)m_{u-t}^{\alpha(s)} \le y | x(t) + g(s)w_{u-t}^{\alpha(s)} = x(u)\right] \\ &= \exp\left(-\frac{2(x(t) - y)(x(u) - y)}{(g(s))^2(u-t)}\right). \qquad (7A.4)\end{aligned}$$

Notice that we have not changed $g(s)$ to $g(t)$ and hence we have used the same Brownian bridge for both strategies.

Another implication of conditional distribution 7A.1 is that we can find the minimum $y_{s,t}$ explicitly. If we know the probability function $F(z)$ of a continuous random variable Z, we can generate random variable Z using uniformly distributed random variable U. Let $U \sim U([0,1])$, then $F^{-1}(U) = Z$, where F^{-1} is an inverse function. It is straightforward to see that from Equation 7A.1 we have

$$m_t^{\alpha} = \frac{1}{2}\left(z - \sqrt{z^2 - 2t \log U}\right) \quad \text{in distribution.}$$

Now

$$
\begin{aligned}
y_{s,t} &= x(s) + g(s)m_{t-s}^{\alpha(s)} \\
&= x(s) + \frac{1}{2}\gamma(s)\left(\frac{x(t)-x(s)}{g(s)} - \sqrt{\left(\frac{x(t)-x(s)}{g(s)}\right)^2 - 2(g(s))^2(t-s)\log U}\right) \\
&= \frac{1}{2}\left(x(t) + x(s) - \sqrt{(x(t)-x(s))^2 - 2(g(s))^2(t-s)\log U}\right)
\end{aligned}
$$

In order to find minima in the case where additionally subsample from Brownian bridge we just need to invert appropriate probabilities.

References

1. Heinrich, S. *Multilevel Monte Carlo Methods*, volume 2179 of *Lecture Notes in Computer Science*, pages 58–67. Springer-Verlag, 2001.

2. Giles, M.B. Multilevel Monte Carlo path simulation. *Operations Research*, 56(3):607–617, 2008.

3. Kebaier, A. Statistical Romberg extrapolation: A new variance reduction method and applications to options pricing. *Annals of Applied Probability*, 14(4):2681–2705, 2005.

4. Speight, A.L. A multilevel approach to control variates. *Journal of Computational Finance*, 12:1–25, 2009.

5. Speight, A.L. Multigrid techniques in economics. *Operations Research*, 58(4):1057–1078, 2010.

6. Pagès, G. Multi-step Richardson–Romberg extrapolation: Remarks on variance control and complexity. *Monte Carlo Methods and Applications*, 13(1):37–70, 2007.

7. Giles, M.B. Improved multilevel Monte Carlo convergence using the Milstein scheme. In Keller, A., Heinrich, S., and Niederreiter, H., editors, *Monte Carlo and Quasi-Monte Carlo Methods 2006*, pages 343–358. Springer-Verlag, 2008.

8. Kloeden, P. and Neuenkirch, A. Convergence of numerical methods for stochastic differential equations in mathematical finance. In *Recent Developments in Computational Finance: Foundations, Algorithms and Applications*, pp. 49–80, 2013.

9. Mao, X. and Szpruch, L. Strong convergence rates for backward Euler–Maruyama method for nonlinear dissipative-type stochastic differential equations with super-linear diffusion coefficients. *Stochastics An International Journal of Probability and Stochastic Processes*, 85(1):144–171, 2013.

10. Szpruch, L., Mao, X., Higham, D.J., and Pan, J. Numerical simulation of a strongly nonlinear Ait-Sahalia-type interest rate model. *BIT Numerical Mathematics*, 51(2):405–425, 2011.

11. Kloeden, P.E., Neuenkirch, A., and Pavani, R. Multilevel Monte Carlo for stochastic differential equations with additive fractional noise. *Annals of Operations Research*, 189(1):255–276, 2011.

12. Kloeden, P.E. and Platen, E. *Numerical Solution of Stochastic Differential Equations*. Springer, Berlin, 1992.

13. Gaines, J.G. and Lyons, T.J.. Random generation of stochastic integrals. *SIAM Journal of Applied Mathematics*, 54(4):1132–1146, 1994.

14. Rydén, T. and Wiktorsson, M. On the simulation of iterated Itô integrals. *Stochastic Processes and their Applications*, 91(1):151–168, 2001.

15. Wiktorsson, M. Joint characteristic function and simultaneous simulation of iterated Itô integrals for multiple independent Brownian motions. *Annals of Applied Probability*, 11(2):470–487, 2001.

16. Clark, J.M.C. and Cameron, R.J. The maximum rate of convergence of discrete approximations for stochastic differential equations. In Grigelionis, B., editor, *Stochastic Differential Systems Filtering and Control*, pp. 162–171. Springer, Berlin, Heidelberg, 1980.

17. Müller-Gronbach, T. *Strong Approximation of Systems of Stochastic Differential Equations*. Habilitation thesis, TU Darmstadt, 2002.

18. Giles, M.B. and Szpruch, L. Antithetic Multilevel Monte Carlo estimation for multi-dimensional SDEs without Lévy area simulation. *Arxiv preprint arXiv:1202.6283*, 2012.

19. Giles, M.B., Higham, D.J., and Mao, X. Analysing multilevel Monte Carlo for options with non-globally Lipschitz payoff. *Finance and Stochastics*, 13(3):403–413, 2009.

20. Debrabant, K., Giles, M.B., and Rossler, A. Numerical analysis of multilevel Monte Carlo path simulation using Milstein discretization: Scalar case. *Technical Report*, 2011.

21. Müller-Gronbach, T. The optimal uniform approximation of systems of stochastic differential equations. *The Annals of Applied Probability*, 12(2):664–690, 2002.

22. Avikainen, R. On irregular functionals of SDEs and the Euler scheme. *Finance and Stochastics*, 13(3):381–401, 2009.

23. Broadie, M., Glasserman, P., and Kou, S. A continuity correction for discrete barrier options. *Mathematical Finance*, 7(4):325–348, 1997.

24. Glasserman, P. *Monte Carlo Methods in Financial Engineering*. Springer, New York, 2004.

25. Giles, M.B. Monte Carlo evaluation of sensitivities in computational finance. *Technical Report* NA07/12, 2007.

26. Burgos, S. and Giles, M.B. Computing Greeks using multilevel path simulation. In Plaskota, L. and Woźniakowski, H., editors, *Monte Carlo and Quasi-Monte Carlo Methods 2010*. pp. 281–296. Springer, Berlin, Heidelberg, 2012.

27. Asmussen, A. and Glynn, P. *Stochastic Simulation*. Springer, New York, 2007.

28. Giles, M.B. Multilevel Monte Carlo for basket options. In Winter Simulation Conference, pp. 1283–1290. Winter Simulation Conference, 2009.

29. Xia, Y. and Giles, M.B. Multilevel path simulation for jump-diffusion SDEs. In Plaskota, L. and Woźniakowski, H., editors, *Monte Carlo and Quasi-Monte Carlo Methods 2010*. pp. 695–708. Springer, Berlin, Heidelberg, 2012.

30. Merton, R.C. Option pricing when underlying stock returns are discontinuous. *Journal of Finance*, 3:125–144, 1976.

31. Platen, E. and Bruti-Liberati, N. *Numerical Solution of Stochastic Differential Equations with Jumps in Finance*. Springer, 2010.

32. Glasserman, P. and Merener, N. Convergence of a discretization scheme for jump-diffusion processes with state-dependent intensities. *Proceedings of the Royal Society of London A*, 460:111–127, 2004.

33. Xia, Y. Multilevel Monte Carlo method for jump-diffusion SDEs. *Arxiv preprint arXiv:1106.4730*, 2011.

34. Dereich, S. and Heidenreich, F. A multilevel Monte Carlo algorithm for Lévy-driven stochastic differential equations. *Stochastic Processes and their Applications*, 121(7):1565–1587, 2011.

35. Dereich, S. Multilevel Monte Carlo algorithms for Lévy-driven SDEs with Gaussian correction. *Annals of Applied Probability*, 21(1):283–311, 2011.

36. Karatzas, I. and Shreve, S.E. *Brownian Motion and Stochastic Calculus*. Graduate Texts in Mathematics, Vol. 113. Springer, New York, 1991.

37. Barth, A., Schwab, C., and Zollinger, N.. Multi-level Monte Carlo finite element method for elliptic PDEs with stochastic coefficients. *Numerische Mathematik*, 119(1):123–161, 2011.

38. Cliffe, K.A., Giles, M.B., Scheichl, R., and Teckentrup, A. Multilevel Monte Carlo methods and applications to elliptic PDEs with random coefficients. *Computing and Visualization in Science*, 14(1):3–15, 2011.

39. Graubner, S. Multi-level Monte Carlo Method für stochastiche partial Differentialgleichungen. Diplomarbeit, TU Darmstadt, 2008.

40. Giles, M.B. and Reisinger, C. Stochastic finite differences and multilevel Monte Carlo for a class of SPDEs in finance. *SIAM Journal of Financial Mathematics*, 3:572–592, 2012.

41. Belomestny, D. and Schoenmakers, J. Multilevel dual approach for pricing American style derivatives. Preprint 1647, WIAS, 2011.

42. Andersen, L. and Broadie, M. A primal–dual simulation algorithm for pricing multi-dimensional American options. *Management Science*, 50(9):1222–1234, 2004.

43. Broadie, M. and Kaya, O. Exact simulation of stochastic volatility and other affine jump diffusion processes. *Operations Research*, 54(2):217–231, 2006.

44. Glasserman, P. and Kim K.-K. Gamma expansion of the Heston stochastic volatility model. *Finance and Stochastics*, 15(2):267–296, 2011.

45. Heston, S.I. A closed-form solution for options with stochastic volatility with applications to bond and currency options. *Review of Financial Studies*, 6:327–343, 1993.

46. Giles, M.B. and Waterhouse, B.J. Multilevel quasi-Monte Carlo path simulation. In *Advanced Financial Modelling*, Radon Series on Computational and Applied Mathematics, pages 165–181, 2009.

47. Gerstner, T. and Noll, M. Randomized multilevel quasi-Monte Carlo path simulation. In *Recent Developments in Computational Finance: Foundations, Algorithms and Applications*, pp. 349–369, 2013.

48. Shreve, S.E. *Stochastic Calculus for Finance: Continuous-Time Models*, Vol. 2. Springer, Berlin, Heidelberg, 2004.

Chapter 8

Fourier and Wavelet Option Pricing Methods

Stefanus C. Maree, Luis Ortiz-Gracia, and Cornelis W. Oosterlee

CONTENTS

8.1 Introduction

In this overview chapter, we will discuss the use of exponentially converging option pricing techniques for option valuation. We will focus on the pricing of European options, and they are the basic instruments within a calibration procedure when fitting the parameters in asset dynamics. The numerical solution is governed by the solution of the discounted expectation of the pay-off function. For the computation of the expectation, we require knowledge about the corresponding probability density function, which is typically not available for relevant stochastic asset price processes. Many publications regarding highly efficient pricing of these contracts are available, where computation often takes place in the Fourier space. Methods based on quadrature and the Fast Fourier Transform (FFT) [1–3] and methods based on Fourier cosine expansions [4,5] have therefore been developed because for relevant *log-asset price processes*, the characteristic function appears to be available. The characteristic function is defined as the Fourier transform of the density function. Here, we wish to extend the overview by discussing the recently presented highly promising class of wavelet option pricing techniques, based on either B-splines or Shannon wavelets.

Cosines in corresponding Fourier cosine expansions form a global basis, and that comes with disadvantages, as, especially for long maturity options, round-off errors may accumulate, and, for short maturity options, many cosine terms are needed for an accurate representation of a highly peaked density function. Local wavelet bases have been considered by Ortiz-Gracia and Oosterlee [6,7] which rely on Haar and B-spline wavelets. These local bases give flexibility and enhance robustness when treating long maturity options and heavy tailed asset processes, but at the cost of a more involved computation and certain loss of accuracy for the standard cases, where the COS method [4] exhibits exponential convergence. Employing these local wavelet bases, Kirkby [8] computes the density coefficients by means of Parseval's identity instead of relying on Cauchy's integral theorem in Ortiz-Gracia and Oosterlee [6] and used an FFT algorithm to speed up the method. Shannon wavelets, based on the sinus cardinal (sinc) function, are very interesting alternatives, as we can benefit from the local features of the approximation of the density function, but the convergence of the method is exponential due to the regularity of the employed wavelet basis. Long- and short-term maturity options are priced robustly and accurately, as well as fat-tailed asset price distributions. The resulting option pricing method is called SWIFT (Shannon Wavelet Inverse Fourier Technique) and also relies heavily on the use of the FFT.

We will confirm by numerical experiments that the COS and SWIFT methods exhibit exponential convergence. It is our opinion that the fastest converging methods should be implemented on high-performance computing (HPC) platforms. It is well known that mainly with inferior methods a tremendous speedup can be obtained on these platforms. It is much harder to speed up highly efficient computational methods governed by low operation counts. The required HPC speedup can also be achieved by the methods advocated here, mainly in the context of the calibration exercise. During the calibration, European options need to be valued for many different strike prices, and these multiple strike computations can be performed simultaneously and independently. So, rather than reducing the computation time of an individual option valuation, highly efficient parallelization should take place in the "strike direction," see also [9], which can easily be done under COS and SWIFT pricing.

8.1.1 European option pricing problem

The pricing of European options in computational finance is governed by the numerical solution of partial (integro-)differential equations. The corresponding solution, being the option value at time t, can also be found by means of the Feynman–Kac formula as a discounted expectation of the option value at final time $t = T$, the so-called payoff function. Here we consider the *risk-neutral option valuation formula*,

$$v(x,t) = e^{-r(T-t)}\mathbb{E}^{\mathbb{Q}}[\,v(y,T)|x] = e^{-r(T-t)}\int_{\mathbb{R}} v(y,T)f(y|x)\,\mathrm{d}y, \tag{8.1}$$

where v denotes the option value, T the maturity, t the initial date, $\mathbb{E}^{\mathbb{Q}}$ the expectation operator under the risk-neutral measure $\mathbb{Q}$, x and y the state variables at time t and T, respectively, $f(y|x)$ the probability density function of y given x, and r the deterministic risk-neutral interest rate.

Whereas f is typically not known, the characteristic function is often available. We represent the option values as function of the scaled log-asset prices, and denote these prices by,

$$x = \ln(S_t/K) \quad \text{and} \quad y = \ln(S_T/K),$$

where S_t is the underlying price at time t and K the strike price.

The payoff $v(y,T)$ for European options in log-asset space is then given by,

$$v(y,T) = [\alpha \cdot K(e^y - 1)]^+, \quad \text{with } \alpha = \begin{cases} 1, & \text{for a call,} \\ -1, & \text{for a put.} \end{cases} \tag{8.2}$$

The strategy to follow to determine the price of the option consists of an approximation of the density function f in terms of a series expansion, where the series coefficients can be efficiently recovered using the characteristic function.

8.2 COS Method

The COS method for European options, introduced by Fang and Oosterlee [4], is based on the insight that the Fourier cosine series coefficients of $f(y|x)$ are closely related to its characteristic function. Since the density function $f(y|x)$ decays rapidly as $y \to \pm\infty$, we can truncate the infinite integration range in the risk-neutral valuation formula without loosing significant accuracy. Suppose that we have, with $[a,b] \subset \mathbb{R}$,

$$\int_{\mathbb{R}\setminus[a,b]} f(y|x)\,\mathrm{d}y < \text{TOL},$$

for some given tolerance TOL. Then we can approximate $v(x,t)$ in Equation 8.1 by,

$$v(x,t) \approx v_1(x,t) = e^{-r(T-t)} \int_a^b v(y,T) f(y|x) \, dy. \tag{8.3}$$

(The intermediate terms v_i are used to distinguish approximation errors.) As a second step, we replace the (unknown) density function $f(y|x)$ by its Fourier cosine expansion over $[a,b]$,

$$f(y|x) = {\sum_{k=0}^{\infty}}' D_k(x) \cos\left(k\pi \frac{y-a}{b-a}\right),$$

$$\text{where } D_k(x) = \frac{2}{b-a} \int_a^b f(y|x) \cos\left(k\pi \frac{y-a}{b-a}\right) dy, \tag{8.4}$$

where the apostrophe (') after the summation sign denotes that the first term of the summation is divided by 2. We will refer to $D_k(x)$ as the (Fourier cosine) *density coefficients.*

Inserting the Fourier cosine expansion of $f(y|x)$ into Equation 8.3, using Fubini's Theorem, gives,

$$v_1(x,t) = e^{-r(T-t)} {\sum_{k=0}^{\infty}}' D_k(x) \left[\int_a^b v(y,T) \cos\left(k\pi \frac{y-a}{b-a}\right) dy \right], \tag{8.5}$$

where we note that the integral at the right-hand side is equal to the Fourier coefficients of $v(y,T)$ in y (except for a constant). We therefore define the *payoff coefficients* V_k as the Fourier cosine series coefficients of $v(y,T)$ as

$$V_k := \frac{2}{b-a} \int_a^b v(y,T) \cos\left(k\pi \frac{y-a}{b-a}\right) dy, \tag{8.6}$$

and obtain,

$$v_1(x,t) = \frac{b-a}{2} e^{-r(T-t)} {\sum_{k=0}^{\infty}}' D_k(x) V_k.$$

Due to the rapid decay of the payoff and density coefficients, we can further truncate the series summation to obtain,

$$v(x,t) \approx v_2(x,t) = \frac{b-a}{2} e^{-r(T-t)} {\sum_{k=0}^{N-1}}' D_k(x) V_k.$$

8.2.1 Density coefficients

The strength of the COS method is the insight that the Fourier cosine coefficients $D_k(x)$ are closely related to the conditional characteristic function. Let the Fourier transform of a function f, whenever it exists, be given by,

$$\hat{f}(\omega) := \int_{\mathbb{R}} f(y)e^{-i\omega y}\,\mathrm{d}y,$$

then we can define the conditional characteristic function $\check{f}(\omega; x)$ related to the density function $f(y|x)$ as $\check{f}(\omega; x) := \hat{f}(-\omega; x)$.

Since we assume that the density function $f(y|x)$ is an $L^2(\mathbb{R})$-function, the characteristic function $\check{f}(\omega; x) := \hat{f}(-\omega; x)$ is also in $L^2(\mathbb{R})$, thus it can approximated well by,

$$\check{f}(\omega; x) = \int_{\mathbb{R}} f(y|x)e^{i\omega y}\,\mathrm{d}y \approx \int_a^b f(y|x)e^{i\omega y}\,\mathrm{d}y.$$

Using the approximation of the characteristic function, we can derive an approximation for the density coefficients,

$$\begin{aligned}
D_k(x) &= \frac{2}{b-a}\int_a^b f(y|x)\cos\left(k\pi\frac{y-a}{b-a}\right)\mathrm{d}y \\
&= \frac{2}{b-a}\mathrm{Re}\left\{e^{-ik\pi\frac{a}{b-a}}\int_a^b f(y|x)\exp\left(i\frac{k\pi}{b-a}y\right)\mathrm{d}y\right\} \\
&\approx \frac{2}{b-a}\mathrm{Re}\left\{e^{-ik\pi\frac{a}{b-a}}\int_{\mathbb{R}} f(y|x)\exp\left(i\frac{k\pi}{b-a}y\right)\mathrm{d}y\right\} \\
&= \frac{2}{b-a}\mathrm{Re}\left\{\check{f}\left(\frac{k\pi}{b-a}; x\right)e^{-ik\pi\frac{a}{b-a}}\right\} =: D_k^*(x).
\end{aligned} \tag{8.7}$$

In a final step, we replace $D_k(x)$ by its approximation $D_k^*(x)$ in $v_2(x,t)$ to obtain the general COS pricing formula,

$$v(x,t) \approx v_3(x,t) = e^{-r(T-t)}\sum_{k=0}^{N-1}{}' \,\mathrm{Re}\left\{\check{f}\left(\frac{k\pi}{b-a}; x\right)e^{-ik\pi\frac{a}{b-a}}\right\}V_k, \tag{8.8}$$

where the payoff coefficients V_k depend on the option type.

8.2.2 Plain vanilla payoff coefficients

The payoff coefficients V_k, as defined in Equation 8.6, for a European call (or put) with payoff function as in Equation 8.2 are given by,

$$V_k = \frac{2}{b-a} \int_a^b [\alpha \cdot K(e^y - 1)]^+ \cos\left(k\pi \frac{y-a}{b-a}\right) \mathrm{d}y.$$

Let us consider a European call option, $\alpha = 1$. For a put, the steps are similar. We distinguish two different cases. If $a < b < 0$, the integral equals zero, and $V_k = 0$ for all k. In the other case, set $\bar{a} = \max(0, a)$. We can then rewrite V_k as,

$$V_k = K \left[\frac{2}{b-a} \int_{\bar{a}}^b e^y \cos\left(k\pi \frac{y-a}{b-a}\right) \mathrm{d}y - \frac{2}{b-a} \int_{\bar{a}}^b \cos\left(k\pi \frac{y-a}{b-a}\right) \mathrm{d}y \right],$$

where the first term within the brackets represents the Fourier cosine coefficient of the function e^y and the second term the Fourier cosine coefficient of the constant function 1. Both of them can be solved analytically using basic calculus, and for a proof the reader is referred to [4].

8.2.3 Domain truncation

A next step in the derivation of the COS pricing formula is to determine the truncation interval $[a, b] \subset \mathbb{R}$. It is important that the interval contains almost all "mass" of the distribution function f, that is, $\int_a^b f(y|x)\,\mathrm{d}y \approx 1$. This interval might be hard to determine for distribution functions with fat tails or when little is known about the underlying distribution.

A heuristic solution proposed by Fang and Oosterlee [4] is to make use of the cumulants of the underlying distribution. Let c_n denote the nth cumulant of $y = \ln(S_T/K)$ and let L be a scaling parameter, then,

$$[a, b] := \left[c_1 - L\sqrt{c_2 + \sqrt{c_4}}, c_1 + L\sqrt{c_2 + \sqrt{c_4}} \right], \tag{8.9}$$

where L is suggested to be chosen in the range $[7.5, 10]$. Cumulants for a number of common underlying models are given by Fang and Oosterlee [4].

8.2.4 Pricing multiple strikes

It is worth mentioning that Equation 8.8 is greatly simplified for the Lévy and Heston models, so that options for many strike prices can be computed simultaneously. We denote vectors with bold-faced characters. For a vector of strikes $\mathbf{K}$, the V_k formulas for European options can be factored as $\mathbf{V}_k = \mathbf{K}U_k$, where U_k is, independent of the strike, given by,

$$U_k = \left[\frac{2}{b-a}\int_{\bar{a}}^{b} e^y \cos\left(k\pi\frac{y-a}{b-a}\right) dy - \frac{2}{b-a}\int_{\bar{a}}^{b} \cos\left(k\pi\frac{y-a}{b-a}\right) dy\right].$$

For Lévy processes, whose characteristic functions can be represented by,

$$\check{f}(\omega;\mathbf{x}) \stackrel{d}{=} \check{f}_{\text{levy}}(\omega)e^{i\omega\mathbf{x}}, \quad \text{with } \check{f}_{\text{levy}}(\omega) := \check{f}(\omega;0),$$

the pricing formula is simplified to,

$$v(\mathbf{x},t) \approx e^{-r(T-t)}\mathbf{K}\sum_{k=0}^{N-1}{}' \operatorname{Re}\left\{\check{f}_{\text{levy}}\left(\frac{k\pi}{b-a}\right)e^{ik\pi\frac{\mathbf{x}-a}{b-a}}U_k\right\}, \tag{8.10}$$

where the summation can be written as a matrix–vector product if $\mathbf{K}$ (and therefore $\mathbf{x}$) is a vector. We see that the evaluation of the characteristic function is independent of the strike. In general, the evaluation of the characteristic function is more expensive than the other computations.

In the section with numerical results, we show that with very small N we can achieve highly accurate results.

8.3 Wavelet Series

As for the COS method, the wavelet methods discussed in the consecutive sections use a series expansion, this time in terms of wavelet bases, to approximate the density function. In this section, we briefly introduce Multi Resolution Analysis (MRA), a general framework for wavelets. Extensive theory on MRA and wavelets in general can be found in the work by Daubechies [10]. In the next two sections, we describe the WA$^{[a,b]}$ method [6], which is a wavelet-based method using B-splines. After, we discuss the SWIFT method [11], a similar approach, but based on Shannon wavelets.

The starting point of MRA is a family of closed nested subspaces,

$$\cdots \subset V_{-2} \subset V_{-1} \subset V_0 \subset V_1 \subset V_2 \subset \cdots,$$

in $L^2(\mathbb{R})$, where,

$$\bigcap_{j\in\mathbb{Z}} V_j = \{0\}, \quad \overline{\bigcup_{j\in\mathbb{Z}} V_j} = L^2(\mathbb{R}),$$

and,

$$f(x) \in V_j \Leftrightarrow f(2x) \in V_{j+1}.$$

If these conditions are met, then a function $\phi \in V_0$ exists, such that $\{\phi_{j,k}\}_{k\in\mathbb{Z}}$ forms an orthonormal basis of V_j, where,

$$\phi_{j,k}(x) = 2^{j/2}\phi(2^j x - k).$$

In other words, the function ϕ, called the *scaling* function or *father* wavelet, generates an orthonormal basis for each V_j subspace.

Let us define W_j in such a way that $V_{j+1} = V_j \oplus W_j$. That is, W_j is the space of functions in V_{j+1} but not in V_j, and so, $L^2(\mathbb{R}) = \sum_j \oplus W_j$. Then a function $\psi \in W_0$ exists, the mother wavelet, such that by defining,

$$\psi_{j,k}(x) = 2^{j/2}\psi(2^j x - k),$$

the wavelet family $\{\psi_{j,k}\}_{k\in\mathbb{Z}}$ gives rise to an orthonormal basis of W_j and $\{\psi_{j,k}\}_{j,k\in\mathbb{Z}}$ is a wavelet basis of $L^2(\mathbb{R})$.

For any $f \in L^2(\mathbb{R})$, a projection map of $\mathcal{P}_m : L^2(\mathbb{R}) \rightarrow V_m$ is defined by means of,

$$\begin{aligned}
\mathcal{P}_m f(x) &= \sum_{j=-\infty}^{m-1} \sum_{k\in\mathbb{Z}} d_{j,k}\psi_{j,k}(x) = \sum_{k\in\mathbb{Z}} c_{m,k}\phi_{m,k}(x), \\
\text{where } d_{j,k} &= \int_{\mathbb{R}} f(x)\psi_{j,k}(x)\mathrm{d}x, \\
c_{m,k} &= \int_{\mathbb{R}} f(x)\phi_{m,k}(x)\mathrm{d}x.
\end{aligned} \tag{8.11}$$

Note that the first part in Equation 8.11 is a truncated wavelet series. If j were allowed to go to $+\infty$, we would have the full wavelet series. The second part in Equation 8.11 gives us an equivalent sum in terms of the scaling functions $\phi_{m,k}$. When m tends to infinity, by the theory of MRA, the truncated wavelet series converges to f.

As opposed to the Fourier series in Equation 8.4, wavelets can be translated (by means of k) and stretched or compressed (by means of j) to accurately represent local properties of a function.

8.4 WA$^{[a,b]}$ Method

The WA$^{[a,b]}$ method is based on a wavelet series expansion using the jth order cardinal B-splines as the scaling function for $j = 0$ and $j = 1$. The B-splines of order zero are defined as,

$$\phi^0(x) := \begin{cases} 1, & \text{if } x \in [0,1), \\ 0, & \text{otherwise,} \end{cases}$$

and they are called Haar wavelets.

Higher order B-splines are defined recursively by a convolution,

$$\phi^j(x) = \int_0^1 \phi^{j-1}(x-t)\mathrm{d}t, \quad j \geq 1,$$

but the resulting wavelet family $\{\phi^j_{m,k}\}_{m,k\in\mathbb{Z}}$ does not form an orthonormal basis of $L^2(\mathbb{R})$. However, they do form a Riesz basis, a relaxation or orthonormality, which still allows us to apply MRA. For details about Riesz basis, see [10].

Following the MRA framework, we define the wavelet family $\{\phi^j_{m,k}\}_{m,k\in\mathbb{Z}}$ with wavelets,

$$\phi^j_{m,k}(x) = 2^{m/2}\phi^j(2^m x - k),$$

for a fixed wavelet scale m. We discuss the choice of $m \in \mathbb{N}$ in the numerical section at the end of this chapter.

Cardinal B-spline functions are compactly supported, with support $[0, j+1]$, and their Fourier transform is,

$$\hat{\phi}^j(\omega) = \left(\frac{1-e^{-i\omega}}{i\omega}\right)^{j+1}.$$

Since splines are only piecewise polynomial functions (see Figure 8.1), they are very easy to implement.

In Ortiz-Gracia and Oosterlee [6], two methods for approximating the density function are described. We focus on the method that applies a Wavelet Approximation on a bounded interval $[a,b]$, the WA$^{[a,b]}$ method.

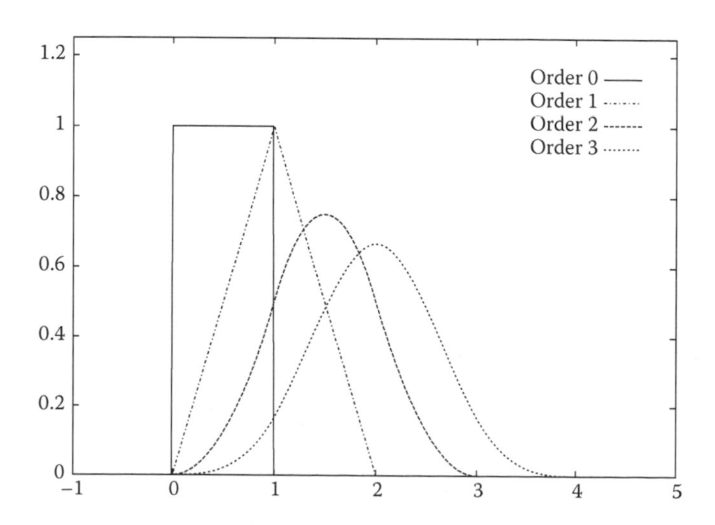

FIGURE 8.1: Cardinal B-splines of orders $j = 0, 1, 2, 3$. For $j = 0$, we have the scaling function of the Haar wavelet system.

We assume that the density function $f(y|x)$ is an $L^2(\mathbb{R})$-function, and thus the mass in the tails tends to zero when $y \to \pm\infty$, so that it can be well approximated in a finite interval $[a, b]$ by,

$$f^c(y|x) = \begin{cases} f(y|x), & \text{if } x \in [a, b], \\ 0, & \text{otherwise.} \end{cases}$$

Following the theory of MRA on a bounded interval[1], we can approximate $f^c \approx f^c_{m,j}$ for all $y \in [a, b]$, where,

$$f^c_{m,j}(y|x) = \sum_{k=0}^{(j+1)\cdot(2^m-1)} D^j_{m,k}(x)\phi^j_{m,k}\left((j+1)\cdot\frac{y-a}{b-a}\right), \quad j \geq 0, \tag{8.12}$$

with convergence in the $L^2(\mathbb{R})$-norm and $D^j_{m,k}(x)$ are the wavelet density coefficients.

We use the cumulants method of Section 8.2.3 to determine the interval $[a, b]$ and obtain the truncated risk-neutral option valuation formula for $v_1(x, t)$ as in Equation 8.3. Then, by substituting $\hat{f}$ by $\hat{f}^c_{m,j}$, we obtain by interchange of integration and summation,

$$\begin{aligned} v_2(x, t) &= e^{-r(T-t)} \int_a^b v(y, T) f^c_{m,j}(y|x)\,\mathrm{d}y \\ &= e^{-r(T-t)} \sum_{k=0}^{(j+1)\cdot(2^m-1)} D^j_{m,k}(x) V^j_{m,k}, \end{aligned} \tag{8.13}$$

where we defined the payoff coefficients as,

$$V^j_{m,k} = \int_a^b v(y, T)\phi^j_{m,k}\left((j+1)\cdot\frac{y-a}{b-a}\right)\mathrm{d}y. \tag{8.14}$$

8.4.1 Density coefficients

As in Ortiz-Gracia and Oosterlee [6], we use Cauchy's integral formula to find an expression for the density coefficients $D^j_{m,k}(x)$. An alternative approach, based on Parseval's identity, is described by Kirkby [8].

The main idea behind the Wavelet Approximation method is to approximate $\hat{f}$ by $\hat{f}^c_{m,j}$ and then to compute the coefficients $D^j_{m,k}$ by inverting the

[1]Scaling function in a bounded interval is discussed in detail by Chui [12].

Fourier Transform. Proceeding this way, we have,

$$\hat{f}(\omega;x) = \int_{\mathbb{R}} f(y|x)e^{-i\omega y}\,\mathrm{d}y \approx \int_{\mathbb{R}} f^c_{m,j}(y|x)e^{-i\omega y}\,\mathrm{d}y$$
$$= \sum_{k=0}^{(j+1)\cdot(2^m-1)} D^j_{m,k}(x)\left[\int_{\mathbb{R}} \phi^j_{m,k}\left((j+1)\cdot\frac{y-a}{b-a}\right)e^{-i\omega y}\,\mathrm{d}y\right].$$

Introducing a change of variables, $u = (j+1)\cdot\frac{u-a}{b-a}$, gives us,

$$\hat{f}(\omega;x) \approx \frac{b-a}{j+1}\cdot e^{-ia\omega}\sum_{k=0}^{(j+1)\cdot(2^m-1)} D^j_{m,k}(x)\left[\int_{\mathbb{R}} \phi^j_{m,k}(u)e^{-i\omega\frac{b-a}{j+1}u}\mathrm{d}u\right]$$
$$= \frac{b-a}{j+1}\cdot e^{-ia\omega}\sum_{k=0}^{(j+1)\cdot(2^m-1)} D^j_{m,k}(x)\hat{\phi}^j_{m,k}\left(\frac{b-a}{j+1}\cdot\omega\right).$$

Taking into account that $\hat{\phi}^j_{m,k}(\omega) = 2^{-\frac{m}{2}}\hat{\phi}^j(\frac{\omega}{2^m})e^{-i\frac{k}{2^m}\omega}$ and performing a change of variables, $z = e^{-i\frac{b-a}{2^m(j+1)}\omega}$, we find by rearranging terms,

$$P^j_m(z;x) \approx Q^j_m(z;x), \tag{8.15}$$

where,

$$P^j_m(z;x) := \sum_{k=0}^{(j+1)\cdot(2^m-1)} D^j_{m,k}(x)z^k$$
$$Q^j_m(z;x) := \frac{2^{\frac{m}{2}}(j+1)z^{-\frac{2^m(j+1)a}{b-a}}\hat{f}\left(\frac{2^m(j+1)}{b-a}i\cdot\log(z)\right)}{(b-a)\hat{\phi}^j\left(i\cdot\log(z)\right)}.$$

Since $P^j_m(z;x)$ is a polynomial (in z), it is (in particular) analytic inside a disc of the complex plane $\{z \in \mathbb{C} : |z| < \rho\}$ for $\rho > 0$. We can obtain expressions for the coefficients $D^j_{m,k}(x)$ by means of Cauchy's integral formula. This is,

$$D^j_{m,k}(x) = \frac{1}{2\pi i}\int_{\gamma}\frac{P^j_m(z;x)}{z^{k+1}}\mathrm{d}z, \quad k = 0,\ldots,(j+1)\cdot(2^m-1),$$

where γ denotes a circle of radius ρ, $\rho > 0$, about the origin. We set $\rho = 0.9995$ [13]. Considering now the change of variables $z = \rho e^{iu}$, and the approximation $P^j_m(z;x) \approx Q^j_m(z;x)$ gives us,

$$D^j_{m,k}(x) \approx \frac{1}{2\pi\rho^k}\int_0^{2\pi} Q^j_m(\rho e^{iu};x)e^{-iku}\,\mathrm{d}u. \tag{8.16}$$

We approximate the above integral with the Trapezoidal Rule over the grid points $u_n = n\frac{2\pi}{N}$ for $N = 2^m(j+1)$ and $n = 0,1,2,\ldots,N-1$. Thus the final approximation for the density coefficients is,

$$D^j_{m,k}(x) \approx D^{j,*}_{m,k}(x) := \frac{1}{N\rho^k} \mathrm{Re} \left\{ \sum_{n=0}^{N-1} Q^j_m(\rho e^{iu_n}; x) e^{-i\frac{2\pi}{N}kn} \right\}. \tag{8.17}$$

Note that we can directly apply the FFT algorithm to compute the whole vector of coefficients $\{D^j_{m,k}\}_{k=0}^{N-1}$ with a computational complexity of just $\mathcal{O}(N \cdot \log_2 N)$.

The resulting B-splines wavelet pricing formula for general European options is,

$$v(x,t) \approx e^{-r(T-t)} \sum_{k=0}^{N} D^{j,*}_{m,k}(x) V^j_{m,k}, \tag{8.18}$$

where the density coefficients $D^{j,*}_{m,k}(x)$ are given by Equation 8.17 . The payoff coefficients depend on the type of contract, which we discuss in the following section for plain vanilla options.

8.4.2 Plain vanilla payoff coefficients

We derive the payoff coefficients for a European call (or put) with payoff as in Equation 8.2. The payoff coefficients, as defined in Equation 8.14, are given by,

$$V^j_{m,k} = \int_a^b [\alpha \cdot K(e^y - 1)]^+ \, \phi^j_{m,k} \left((j+1) \cdot \frac{y-a}{b-a} \right) \mathrm{d}y. \tag{8.19}$$

Let us consider a European call option, $\alpha = 1$. For a put, the steps are similar.

We distinguish two different cases. If $a < b < 0$, the integral equals zero, and $V^j_{m,k} = 0$ for all k. In the other case, set $\overline{a} = \max(0, a)$. We can then rewrite $V^j_{m,k}$ as,

$$V^j_{m,k} = K \left[\int_{\overline{a}}^b e^y \phi^j_{m,k} \left((j+1) \cdot \frac{y-a}{b-a} \right) \mathrm{d}y - \int_{\overline{a}}^b \phi^j_{m,k} \left((j+1) \cdot \frac{y-a}{b-a} \right) \mathrm{d}y \right]. \tag{8.20}$$

Both of the integrals can be solved analytically using basic calculus, and for a proof the reader is referred to [6].

8.5 SWIFT method

The SWIFT method, introduced by Ortiz-Gracia and Oosterlee [11] and extended by Colldeforns-Papiol et al. [14] and Maree et al. [15] is similar to the $\mathrm{WA}^{[a,b]}$ method in methodology; the main novelty is the type of wavelet; the so-called Shannon wavelet is used, resulting in its name, the "Shannon Wavelets Inverse Fourier Technique" (SWIFT) method. Shannon wavelet approximations are appealing due to their exponential convergence for the smooth density functions that occur in finance.

The Shannon scaling function is given by,

$$\phi(x) = \mathrm{sinc}(x) := \begin{cases} \frac{\sin(\pi x)}{\pi x}, & \text{if } x \neq 0, \\ 0, & x = 0, \end{cases}$$

and the mother wavelet is,

$$\psi(x) = \frac{\sin\left(\pi(x - \frac{1}{2})\right) - \sin\left(2\pi(x - \frac{1}{2})\right)}{\pi(x - \frac{1}{2})},$$

and following the theory of MRA, we define the wavelet family $\{\phi_{m,k}\}_{m,k\in\mathbb{Z}}$ with wavelets,

$$\phi_{m,k}(x) = 2^{m/2}\phi(2^m x - k).$$

The Shannon scaling function and mother wavelet are shown in Figure 8.2.

In terms of time frequency analysis, the Shannon scaling function is the opposite of the Haar wavelet ϕ^0 we saw in the previous section. The Shannon wavelet, a regular function in the time domain, is a compact-supported

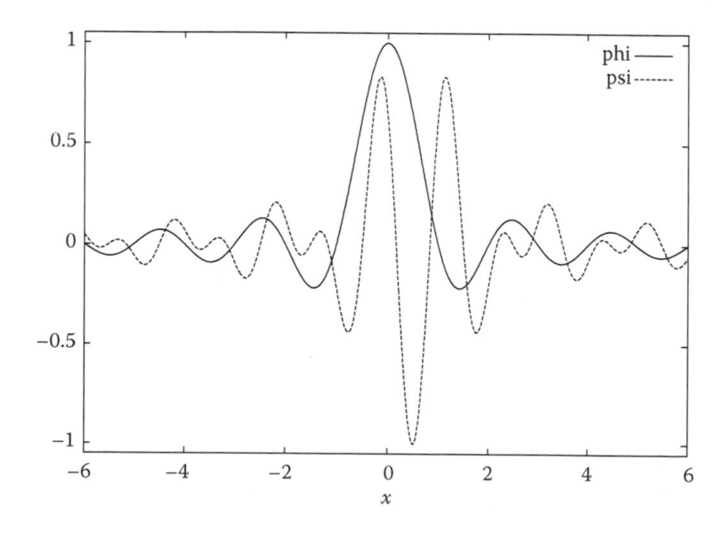

FIGURE 8.2: Shannon scaling function $\phi(x)$ (phi, thick line) and wavelet $\psi(x)$ (psi, dashed line).

rectangle in the frequency domain, given by,

$$\hat{\phi}_{m,k}(\omega) = 2^{-m/2} e^{-i\frac{k}{2^m}\omega} \text{rect}\left(\frac{\omega}{2^{m+1}\pi}\right),$$

where the rectangle function is defined as,

$$\text{rect}(x) = \begin{cases} 1, & \text{if } |x| < 1/2, \\ 1/2, & \text{if } |x| = 1/2, \\ 0, & \text{if } |x| > 1/2. \end{cases}$$

Following the MRA framework, the truncated Shannon wavelet expansion of the density function $f(y|x)$ is given by,

$$f(y|x) \approx \mathcal{P}_m f(y|x) = \sum_{k \in \mathbb{Z}} D_{m,k}(x)\phi_{m,k}(y),$$
$$\text{with } D_{m,k}(x) := \int_{\mathbb{R}} f(y|x)\phi_{m,k}(y)\text{d}y, \tag{8.21}$$

where the scaling functions are defined from $\phi(x) = \text{sinc}(x)$ for a fixed wavelet scale $m \in \mathbb{N}$.

Since Shannon wavelets have infinite support, we take a different approach in truncating the wavelet series. We note that for $h \in \mathbb{Z}$,

$$f\left(\frac{h}{2^m}\bigg| x\right) \approx \mathcal{P}_m f\left(\frac{h}{2^m}\bigg| x\right) = 2^{\frac{m}{2}} \sum_{k \in \mathbb{Z}} D_{m,k}(x)\delta_{k,h} = 2^{\frac{m}{2}} D_{m,h}(x).$$

Now, since $f \in L^2(\mathbb{R})$ and it is nonnegative, and if we assume that $\lim_{x \to \pm\infty} f(x) = 0$ then we conclude that $D_{m,k}$ vanishes as well as $k \to \pm\infty$. We therefore approximate the infinite series in Equation 8.21 by a finite summation without loss of considerable density mass,

$$f(y|x) \approx f_m(y|x) := \sum_{k=k_1}^{k_2} D_{m,k}(x)\phi_{m,k}(y), \tag{8.22}$$

for conveniently chosen integers $k_1 < k_2$. When setting $\mathcal{I}_m = [\frac{k_1}{2^m}, \frac{k_2}{2^m}]$, the option pricing formula becomes,

$$\begin{aligned} v(x,t) &= e^{-r(T-t)} \int_{\mathbb{R}} v(y,T) f(y|x)\text{d}y \\ &\approx e^{-r(T-t)} \sum_{k=k_1}^{k_2} D_{m,k}(x) V_{m,k}, \end{aligned} \tag{8.23}$$

where the payoff coefficients are defined as,

$$V_{m,k} := \int_{\mathcal{I}_m} v(y,T)\phi_{m,k}(y)\text{d}y. \tag{8.24}$$

We define the truncation parameters k_1 and k_2 as the smallest integers (in absolute sense), such that when a and b are the truncation parameters of

Section 8.2.3, determined by the cumulants of the density function, we have,

$$\frac{k_1}{2^m} \leq a < b \leq \frac{k_2}{2^m},$$

which implies that $[a, b] \subset \mathcal{I}_m$.

8.5.1 Density coefficients

In Ortiz-Gracia and Oosterlee [11] present two methods are presented to compute the density coefficients $D_{m,k}$ in Equation 8.21. We discuss the approach based on Vieta's formula, as this approach allows us to control the numerical error.

Theorem 8.1. *For $J \in \mathbb{N}$, we can approximate the* sinc *function by,*

$$\operatorname{sinc}(t) \approx \operatorname{sinc}^*(t) := \frac{1}{2^{J-1}} \sum_{j=1}^{2^{J-1}} \cos\left(\frac{2j-1}{2^J}\pi t\right), \tag{8.25}$$

where the absolute error is bounded by,

$$|\operatorname{sinc}(t) - \operatorname{sinc}^*(t)| \leq \frac{(\pi c)^2}{2^{2(J+1)} - (\pi c)^2}, \tag{8.26}$$

for $t \in [-c, c]$, where $c \in \mathbb{R}$, $c > 0$, and $J \geq \log_2(\pi c)$.

Proof. We show how to find the expression for $\operatorname{sinc}^*(t)$. The proof of the error bound is Lemma 2 in Ortiz-Gracia and Oosterlee [11]. As shown by Vieta, and described by Gearhart and Shultz [16], the sinc function can be written as an infinite product,

$$\operatorname{sinc}(t) = \prod_{j=1}^{\infty} \cos\left(\frac{\pi t}{2^j}\right), \tag{8.27}$$

and by truncating the infinite product to a finite product with J factors, we can apply the cosine product-to-sum identity described by Quine and Abrarov [17]. This gives the desired result,

$$\operatorname{sinc}(t) \approx \prod_{j=1}^{J} \cos\left(\frac{\pi t}{2^j}\right) = \frac{1}{2^{J-1}} \sum_{j=1}^{2^{J-1}} \cos\left(\frac{2j-1}{2^J}\pi t\right) =: \operatorname{sinc}^*(t).$$

□

If we write out the definition of the coefficients $D_{m,k}$ in Equation 8.21, we get,

$$D_{m,k}(x) = 2^{\frac{m}{2}} \int_{\mathbb{R}} f(y|x)\phi(2^m y - k)\mathrm{d}y.$$

Using Vieta's approximation $\mathrm{sinc}(x)$ by $\mathrm{sinc}^*(x)$ from Theorem 8.1 gives us,

$$D_{m,k}(x) \approx D^*_{m,k}(x) := \frac{2^{m/2}}{2^{J-1}} \sum_{j=1}^{2^{J-1}} \int_{\mathbb{R}} f(y|x) \cos\left(\frac{2j-1}{2^J}\pi(2^m y - k)\right) \mathrm{d}y.$$

We now note the resemblance between the integral in the right-hand side of the equation above and the integral in the COS method in Equation 8.7. In a similar way, we replace the integral over the unknown density function by its Fourier transform,

$$\begin{aligned}
&\int_{\mathbb{R}} f(x) \cos\left(\frac{2j-1}{2^J}\pi(2^m x - k)\right) \mathrm{d}x \\
&\qquad = \mathrm{Re}\left\{\int_{\mathbb{R}} f(x) \exp\left(-i\frac{2j-1}{2^J}\pi(2^m x - k)\right) \mathrm{d}x\right\} \\
&\qquad = \mathrm{Re}\left\{e^{i\frac{2j-1}{2^J}\pi k} \int_{\mathbb{R}} f(x) \exp\left(-i\frac{(2j-1)\pi 2^m}{2^J} x\right) \mathrm{d}x\right\} \\
&\qquad = \mathrm{Re}\left\{e^{i\frac{2j-1}{2^J}\pi k} \hat{f}\left(\frac{(2j-1)\pi 2^m}{2^J}\right)\right\}. \qquad (8.28)
\end{aligned}$$

Inserting this into the density coefficients gives us an expression for the density coefficients,

$$D^*_{m,k}(x) = \frac{2^{m/2}}{2^{J-1}} \sum_{j=1}^{2^{J-1}} \mathrm{Re}\left\{\hat{f}\left(\frac{(2j-1)\pi 2^m}{2^J}; x\right) e^{\frac{ik\pi(2j-1)}{2^J}}\right\}. \qquad (8.29)$$

A strategy for choosing J follows from Theorem 8.1, which implies that when we set $M_{m,k} = \max(|2^m a - k|, |2^m b + k|)$ and $M_m := \max_{k_1<k<k_2} M_{m,k}$ then, we set $J = \jmath := \lceil \log_2(\pi M_m) \rceil$, where $\lceil x \rceil$ denotes the smallest integer greater than or equal to x. For a proof, the reader is referred to [11].

Although for every k another J could be chosen, we decide to fix a $\jmath$ for all k, such that we can benefit from the efficiency of the FFT algorithm to compute the vector of density coefficients $\{D_{m,k}(x)\}_{k=k_1}^{k_2}$ at once, as described by Ortiz-Gracia and Oosterlee [11].

8.5.2 Payoff coefficients

We show how to compute the payoff coefficients for a European call (or put), based on Ortiz-Gracia and Oosterlee [11]. In contrast to the COS and

$\text{WA}^{[a,b]}$ methods, we do not have an analytic expression for the payoff coefficients, but we can once more benefit from the FFT algorithm for an efficient approximation.

We look for an expression for the payoff for a European call. The steps for deriving a formula for the European put are similar. Recall that the payoff coefficients for a European call are defined by,

$$V_{m,k} = K\left[\int_{\mathcal{I}_m} e^y v \phi_{m,k}(y)\mathrm{d}y - \int_{\mathcal{I}_m} \phi_{m,k}(y)\mathrm{d}y\right], \tag{8.30}$$

as in Equation 8.24. Let us define,

$$I^1_{j,k}(a,b) := \int_a^b e^y \cos(\omega_j(2^m y - k))\mathrm{d}y$$

and,

$$I^2_{j,k}(a,b) := \int_a^b \cos(\omega_j(2^m y - k))\mathrm{d}y,$$

where $\omega_j := \frac{2j-1}{2^J}\pi$. Note that these integrals are just a change of variables of integrals we had to solve for the COS payoff coefficients. For a proof the reader is referred to [11].

When $k_2 \leq 0$, the payoff coefficients vanish, that is, $V_{m,k} = 0$ for every k. In case $0 < k_2$, we can write Equation 8.30 as,

$$V_{m,k} \approx V^*_{m,k} := K\frac{2^{m/2}}{2^{J-1}}\sum_{j=1}^{2^{J-1}}\left[I^1_{j,k}\left(\frac{\bar{k}_1}{2^m}, \frac{k_2}{2^m}\right) - I^2_{j,k}\left(\frac{\bar{k}_1}{2^m}, \frac{k_2}{2^m}\right)\right]. \tag{8.31}$$

Remark 8.1. *As in the case of the density coefficients, we consider a constant J for all k, which we call $\bar{\jmath}$, defined by $\bar{\jmath} := \lceil \log_2(\pi N)\rceil$, where $N := \max_{k_1<k<k_2} N_k$ and $N_k := \max(|\bar{k}_1 - k|, |k_2 - k|)$. This allows us to compute the whole vector of payoff coefficients $\{V^*_{m,k}\}_{k=k_1}^{k_2}$ with the help of the FFT algorithm [11].*

The resulting SWIFT pricing formula for general European options is,

$$v(x,t) \approx e^{-r(T-t)}\sum_{k=k_1}^{k_2} D^*_{m,k}(x)V^*_{m,k}, \tag{8.32}$$

where the density coefficients are given by Equation 8.29, and the payoff coefficients $V^*_{m,k}$ for a European call are given by Equation 8.31.

8.5.3 Pricing multiple strikes

As for the COS method, both the WA$^{[a,b]}$ and SWIFT methods can be applied to efficiently price multiple strikes simultaneously. We describe the approach for the SWIFT method when the underlying process is a Lévy process or the Heston model. For these processes, the density function can be written as,

$$f(y|x) = \frac{1}{2\pi} \int_{\mathbb{R}} \hat{f}(\omega; x) e^{i\omega y} \mathrm{d}\omega = \frac{1}{2\pi} \int_{\mathbb{R}} \hat{f}(\omega; 0) e^{-i\omega x} e^{i\omega y} \mathrm{d}\omega = f(y - x|0).$$

Applying the Shannon wavelet expansion to the density function $f(y|0)$, instead of $f(y|x)$, gives,

$$f(y|x) = f(y - x|0) = \sum_{k=k_1}^{k_2} D^*_{m,k}(0) \phi_{m,k}(y - x),$$

where $D^*_{m,k}(x)$ are the density coefficients as in Equation 8.29, evaluated at $x = 0$. The SWIFT option pricing formula then becomes,

$$\begin{aligned} v(x,t) =& e^{-r(T-t)} \sum_{k=k_1}^{k_2} D^*_{m,k}(0) V^*_{m,k}(x), \\ &\text{where } V^*_{m,k}(x) := \int_{\mathcal{I}_m} v(y,T) \phi_{m,k}(y - x) \mathrm{d}y. \end{aligned}$$

Compared to the original SWIFT pricing formula in Equation 8.32, the dependence on x has been moved from the density coefficients to the payoff coefficients. The density coefficients have to be computed only once. The payoff coefficients now depend on x, and they are generally cheaper to compute, especially for the WA$^{[a,b]}$ method.

8.6 Numerical Results

In this section, we give examples of pricing options by the COS, WA$^{[a,b]}$, and SWIFT methods. All examples are implemented in Matlab and run on an Intel Core i5-4250U CPU @ 1.30 GHz × 4 with 8 GB of memory.

In BENCHOP [18], option pricing methods based on Monte Carlo, Fourier, finite difference, and radial basis functions are compared for different option styles and underlying processes. When the characteristic function of the underlying method is known, the Fourier methods in general, but the COS method especially, are shown to be extremely competitive methods in terms of both accuracy and computational time.

TABLE 8.1: CPU times (in milliseconds) for a European put on CGMY dynamics at different scales for a corresponding absolute price error

COS			SWIFT			Haar		
N	**Error**	**Time**	**Scale**	**Error**	**Time**	**Scale**	**Error**	**Time**
32	1.36e−02	0.41	1	1.58e−01	0.36	5	8.59e−02	0.42
64	3.32e−05	0.34	2	2.27e−04	0.42	7	1.41e−04	0.49
128	3.42e−09	0.39	3	5.61e−07	0.48	9	5.13e−07	1.10
256	4.44e−14	0.53	4	5.80e−11	0.70	11	1.73e−08	2.16

Note: Reference price by the COS method.

8.6.1 Computational time

The COS, SWIFT, and $\mathrm{WA}^{[a,b]}$ methods all follow the same approach. First, the density coefficients are recovered from the characteristic function. For the COS method, this is a single evaluation of the characteristic function (see Equation 8.7). The SWIFT and $\mathrm{WA}^{[a,b]}$ methods require the application of the FFT for an efficient evaluation of the density coefficients with a computational complexity of $\mathcal{O}(N \log_2 N)$. Thus the computation of the coefficients of the SWIFT and $\mathrm{WA}^{[a,b]}$ methods is generally more involved, compared to the COS method. In Table 8.1, we price a European call with strike $K = 100$ and $T = 1$ on Carr-Geman-Madan-Yor (CGMY) dynamics [19] with parameters $(C, G, M, Y) = (1, 5, 5, 1.5)$ and $S_0 = 100, r = 0.1$. We compare the three methods, COS, SWIFT, and $\mathrm{WA}^{[a,b]}$ ($j = 0$, Haar basis), and observe that all of the methods are capable of reaching engineering accuracy in milliseconds.

An advantage of the $\mathrm{WA}^{[a,b]}$ method over the COS method is its robustness, as described in the following section.

8.6.2 Robustness of $\mathbf{WA}^{[a,b]}$

The $\mathrm{WA}^{[a,b]}$ method cannot compete against the COS method in terms of computational time, but it is attractive due to its robustness. Since wavelets series represent functions locally, we can easily *adjust* coefficients to match local *difficulties*.

We demonstrate this robustness by pricing a very long maturity $T = 100$ call option, which arises in economy and real options. Due to the unbounded payoff of a call option, extreme payoffs occur on the right-hand side of the computation domain.

Since the COS method uses a global basis, all coefficients are affected by this extreme payoff, and as is shown in Figure 8.3a, payoff coefficients of magnitude 10^{12} are multiplied by density coefficients of magnitude 10^{-15}, causing such round-off errors that the final option value has an absolute error of 10^{+4}.

We price the same long maturity call using $\mathrm{WA}^{[a,b]}$ with the Haar basis and scale $m = 5$ (32 coefficients), and the coefficients are shown in Figure 8.3b.

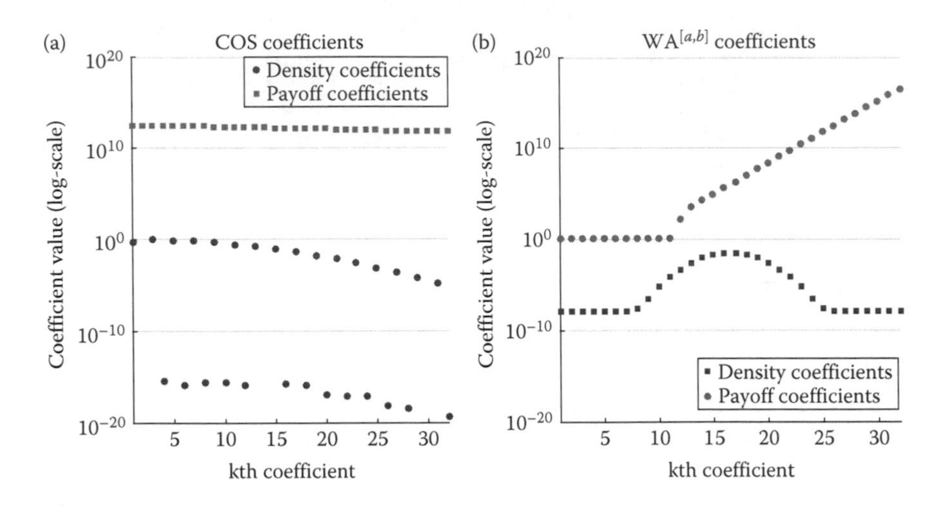

FIGURE 8.3: Coefficients for COS ($N = 32$) and WA$^{[a,b]}$ (Haar, $m = 5$) methods arising from the pricing of a long maturity call on GBM with $S_0 = 100, K = 100, r = 0.1, q = 0, \sigma = 0.25, T = 100$, and $L = 10$. Reference price by the Black–Scholes formula.

Wavelets form a local basis and, as can be seen from the figure, each coefficient $D^{j,*}_{m,k}$ only affects the points of the density locally, in the interval $[\frac{k}{2^m}, \frac{k+1}{2^m}]$.

We can avoid big round-off errors by removing the payoff coefficients that cause very big round-off errors at the right-hand side of the domain. We therefore consider the truncated series,

$$v_{\kappa_m}(x, t) := e^{-r(T-t)} \sum_{k=0}^{\kappa_m} D^{j,*}_{m,k}(x) V^j_{m,k}, \tag{8.33}$$

with $D^{j,*}_{m,k}$ as in Equation 8.17, and by choosing κ_m such that $v_{\kappa_m}(x, t) < S_0$, using that S_0 is an upper bound for the value of a call, we find an error of about 10^{-1}.

8.6.3 Rate of convergence

The COS method and the SWIFT method are extremely powerful when the underlying asset is driven by a geometric Brownian motion. In that case, the COS and SWIFT methods converge exponentially, while the WA$^{[a,b]}$ method for $j = 0$ (Haar) and $j = 1$ (linear B-splines) converges algebraically. The rate of convergence for a European call option under geometric Brownian motion is shown in Figure 8.4.

The probability density function corresponding to an asset driven by geometric Brownian motion is approximated very well by a Fourier cosine expansion and a Shannon wavelet expansion. The rate of convergence deteriorates

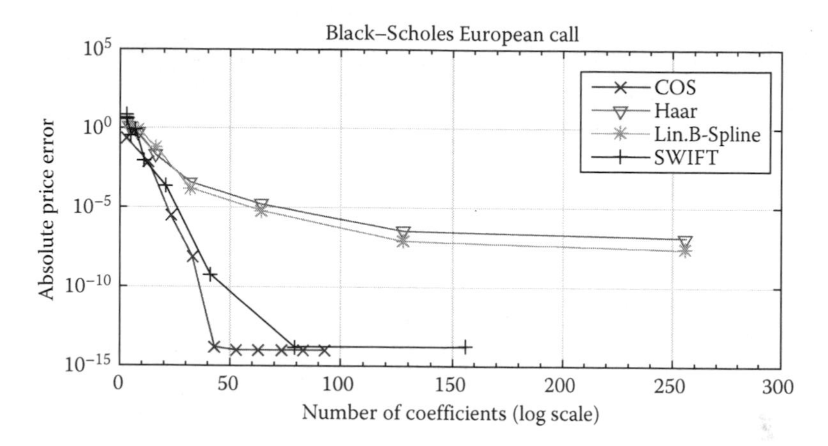

FIGURE 8.4: A European call with strike $K = 110$ on an asset driven by geometric Brownian motion with parameters $S_0 = 100, \sigma = 0.15, r = 0.1$, and maturity $T = 1$. Reference price by the Black–Scholes formula.

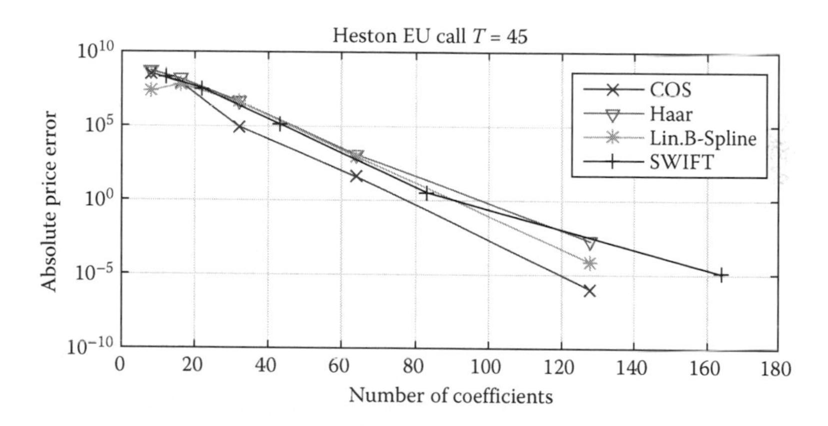

FIGURE 8.5: A long maturity $T = 45$ call option under Heston dynamics. Parameters as in Ortiz-Gracia and Oosterlee [6]. All methods show a similar rate of convergence.

to algebraic convergence for example when we price a long maturity option under the Heston model [20], as shown in Figure 8.5. Long maturity optionalities occur in economy and real option pricing. In that case, the $\mathrm{WA}^{[a,b]}$ method is competitive to the SWIFT and COS methods. Note that we can further improve the accuracy of the $\mathrm{WA}^{[a,b]}$ method by using the truncated series from Equation 8.33.

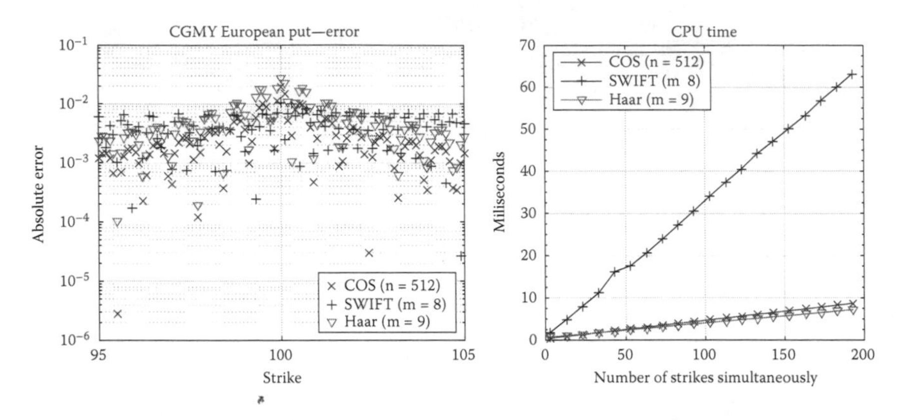

FIGURE 8.6: A short to maturity $T = 0.01$ European put options on CGMY dynamics for a range of strikes $K = 95, \ldots, 105$.

8.6.4 Multiple strike pricing

Generally, asset models are calibrated such that the resulting option prices match quoted option prices. Therefore, a whole range of options with different strike prices has to be priced for each parameter set.

When the underlying asset is driven by an exponential Lévy process, all of the discussed methods can be reformulated such that the density coefficients only have to be computed once and the payoff coefficients once for every option.

As mentioned before, the COS method has a computational complexity of $\mathcal{O}(N)$, while $\mathrm{WA}^{[a,b]}$ method has a complexity of $\mathcal{O}(N \log_2 N)$. However, when pricing M options simultaneously, the complexity of the COS method is $\mathcal{O}(MN)$, while the $\mathrm{WA}^{[a,b]}$ method has a complexity of $\mathcal{O}(N \log_2 N + MN)$, and when M is large compared to N, the two methods have essentially the same computational complexity. The $\mathrm{WA}^{[a,b]}$ method benefits then from the very simple representation of the payoff coefficients.

In the following example, we price a range of put options on an asset driven by the CGMY model [19] with parameters $(C, G, M, Y) = (1, 5, 5, 0.5)$ with a very short maturity $T = 0.01$, initial asset price $S_0 = 100$, and strikes $K = 95, 80.1, \ldots, 119.9, 105$. Thus, in total, 100 options are priced simultaneously. In Figure 8.6, on the left, we see the price errors corresponding to each strike price.

Remark 8.2. *The SWIFT method has a slight disadvantage in multiple strike pricing, as both the payoff and density coefficients are computed using the FFT, and thus the resulting computational complexity is* $\mathcal{O}((M+1)N \log_2 N)$.

Concluding this chapter on Fourier pricing methods, we can state that Fourier as well as wavelets based option pricing methods are extremely fast and thus very efficient for European option pricing under Levy models and generally for log-asset models from the affine jump-diffusion class, like the Heston model. By wavelets, and in particular by the SWIFT method, we can enhance the robustness of Fourier methods, when dealing with fat-tailed distributions or very long (and very short) maturity options. Parallelization techniques may be employed in the context of the calibration framework, where option pricing computations need to be performed for multiple strike prices. It is the strike direction which lends itself well for parallelization, leading to a truly high-performance calibration.

References

1. Carr, P.P. and Madan, D.B. Option valuation using the fast Fourier transform. *Journal of Computational Finance*, 2:61–73, 1999.

2. Lee, R.W. Option pricing by transform methods: Extensions, unification, and error control. *Journal of Computational Finance*, 7:51–86, 2004.

3. Lindström, E., Ströjby, J., Brodén, M., Wiktorsson, M., and Holst, J. Sequential calibration of options. *Computational Statistics & Data Analysis*, 52:2877–2891, 2008.

4. Fang, F. and Oosterlee, C.W. A novel option pricing method based on Fourier-cosine series expansions. *SIAM Journal on Scientific Computing*, 31(2):826–848, 2008.

5. Ruijter, M. and Oosterlee, C.W. Two-dimensional Fourier cosines series expansion method for pricing financial options. *SIAM Journal on Scientific Computing*, 34:642–671, 2012.

6. Ortiz-Gracia, L. and Oosterlee, C.W. Robust pricing of European options with wavelets and the characteristic function. *SIAM Journal on Scientific Computing*, 35(5):B1055–B1084, 2013.

7. Ortiz-Gracia, L. and Oosterlee, C.W. Efficient VaR and expected shortfall computations for nonlinear portfolios within the delta-gamma approach. *Applied Mathematics and Computation*, 244:16–31, 2014.

8. Kirkby, J.L. Efficient option pricing by frame duality with the fast Fourier transform. *SIAM Journal on Financial Mathematics*, 6(1):713–747, 2016.

9. Zhang, B. and Oosterlee, C.W. Acceleration of option pricing technique on graphics processing units. *Concurrency and Computation: Practice and Experience*, 29(9):1626–1639, 2014.

10. Daubechies, I. *Ten Lectures on Wavelets.* Society for Industrial and Applied Mathematics, Philadelphia, PA, USA, 1992.

11. Ortiz-Gracia, L. and Oosterlee, C.W. A highly efficient Shannon wavelet inverse Fourier technique for pricing European options. *SIAM Journal on Scientific Computing*, 38(1):B118–B143, 2016.

12. Chui, C.K. *An Introduction to Wavelets*. Academic Press, Cambridge, Massachusetts, USA, 1992.

13. Ortiz-Gracia, L. and Masdemont, J.J. Peaks and jumps reconstruction with B-splines scaling functions. *Journal of Computational and Applied Mathematics*, 272:258–272, 2014.

14. Colldeforns-Papiol, G., Ortiz-Gracia, L., and Oosterlee, C.W. Two-dimensional Shannon wavelet inverse Fourier technique for pricing European options. *Applied Numerical Mathematics*, 117:115–138, 2017.

15. Maree, S.C., Ortiz-Gracia, L., and Oosterlee, C.W. Pricing early-exercise and discrete barrier options by Shannon wavelet expansions. *Numerische Mathematik*, 136(4):1035–1070, 2017.

16. Gearhart, W.B. and Shultz, H.S. The function $\sin(x)/x$. *The College Mathematics Journal*, 21(2):90–99, 1990.

17. Quine, B.M. and Abrarov, S.M. Application of the spectrally integrated Voigt function to line-by-line radiative transfer modelling. *Journal of Quantitative Spectroscopy & Radiative Transfer*, 244:37–48, 2013.

18. von Sydow, L. et al. BENCHOP—The BENCH marking project in option pricing. *International Journal of Computer Mathematics*, 92:12, 2015.

19. Carr, P.P., Geman, H., Madan, D.B., and Yor, M. The fine structure of asset returns: An empirical investigation. *Journal of Business*, 75:305–332, 2002.

20. Heston, S. A closed-form solution for options with stochastic volatility with applications to bond and currency options. *The Review of Financial Studies*, 6:327–343, 1993.

Chapter 9

A Practical Robust Long-Term Yield Curve Model

M. A. H. Dempster, Elena A. Medova, Igor Osmolovskiy, and Philipp Ustinov

CONTENTS

9.1 Introduction

Since the 2007–2008 financial crisis, low interest rates have prevailed in all the world's major developed economies and were presaged by more than a decade in Japan. This has posed a problem for the widespread use of diffusion-based yield curve models for derivative and other structured financial product pricing and for forward rate simulation for systematic investment, asset liability management (ALM), and economic forecasting. Indeed, while Gaussian models remained sufficiently accurate for pricing and discounting in relatively high-rate environments, their tendency to produce an unacceptable proportion of negative forward rates at short maturities with Monte Carlo scenario simulation from initial conditions in low-rate environments has called their current use into question. The implications for this question of negative nominal rates in deflationary regimes remain to be seen, as does the necessity for currently fashionable multicurve models. Be that as it may, beginning with work in the Bank of Japan in the early 2000s, there has recently been a flurry of research in universities, central banks, and financial services firms to develop yield curve models whose simulation produces nonnegative rate scenarios.

All this work is based on a posthumously published suggestion of Fisher Black (1995) to apply a call option payoff with zero strike to the model instantaneous short rate which leads to a piecewise linear nonlinearity in standard Gaussian affine yield curve model formulae for zero coupon (discount) bond prices and the corresponding yields, and precludes their explicit closed-form solution. As a result, most of the published solutions to Black-corrected yield curve models to date are approximations and even these require high-performance computing (HPC) techniques for numerical solution. We shall study here an obvious approximation which works extremely well and is amenable to cloud computing for speed up.

In practice, there are a variety of approaches to yield curve modeling which are driven by the intended use of the models. The literature is devoted predominately to the needs of investment banks in pricing and hedging fixed income derivative and other structured products. Model calibration is short term, to current forward market data at pricing and hedging time, and is updated for rehedging. A specific model is evaluated by its realized hedging profit and loss.

A second approach is that of central bank forecasting for monetary policy making. Here calibration uses long-term historical data for medium-term forecasts and is updated for the next forecast. Model accuracy appears from the open literature to be mainly evaluated by in-sample fit to the historical data employed, with little out-of-sample forecasting evaluation reported.

The approach of interest here supports consultants' and fund managers' advice to institutional and individual clients regarding product pricing, investment, and asset liability management over long horizons. It involves long-term calibration of historical market data, often using filtering techniques, which is updated for decision points such as restructurings or portfolio rebalances.

Models are evaluated by the consistency of their forecasts with out-of-sample market data, for example, prices and returns.

This paper describes the preliminary development of a robust long-term yield curve model which is a (mildly) nonlinear version of our workhorse Gaussian three-factor affine yield curve model, the Economic Factor Model (EFM),which we have used for Monte Carlo scenario generation over many years in practical structured derivative pricing, investment, and asset liability management. We have employed the EFM model in the past using time steps from daily to semiannual in the five major currency areas: EUR, CHF, GBP, USD, and JPY. Its 14 parameters are calibrated to market data using the expectation–maximization (EM) algorithm which (Dempster et al. 1977) alternates the linear Kalman filter (KF) with maximum likelihood parameter estimation (MLE) to convergence. Here we implement a Black (1995) corrected version (Black EFM) of the EFM model using the nonlinear unscented Kalman filter (UKF) (Julier et al. 1995; Julier and Uhlmann 1997) in place of the ordinary KF. This represents an approximation to the mathematical Black model in the presence of Black's piecewise linear 0-strike call option nonlinearity which suppresses negative rates. Indeed, while we use the EFM affine closed-form expressions for yields of all maturities, it should be noted that the zero lower bound (ZLB) will be inactive for all but those of short, but not necessarily the shortest, maturities.

Our approach to calibrating the Black-corrected EFM model is promising both in calibration and forecasting accuracy relative to market data, even at short maturities, but further work is necessary in both empirical testing and, particularly, in the theoretical development of the UKF.

The remainder of this chapter is structured as follows. After enunciating our guiding design requirements, Section 9.2 briefly summarizes recent yield curve (term structure) models developed for both pricing and forecasting. It illustrates the nature of the difficulties encountered with each model in terms of our requirements and due to the current low-rate environment. In Section 9.3, the basic EFM model is set out and the Black correction is defined. Section 9.4 surveys recent contributions to approximating the Black-corrected affine yield curve models whose calibrations are all (too) heavily computationally intensive. These models are often (in our view mis-) designated "shadow rate models," a term introduced by Black for the underlying Gaussian model which is to be corrected for nonnegative rates. Our approximate, but accurate, Black EFM model is described in Section 9.5, which outlines the UKF, including our HPC implementation. Section 9.6 presents the empirical evaluation of the model in terms of both in-sample calibration and out-of-sample scenario generation, using data for the five major currency areas. Although there remains room for deeper understanding and improvement, the Black EFM model is found to be sufficiently accurate for both purposes. Conclusions are presented in Section 9.7, in which it is noted that on average the runtime of its computationally intensive calibration is only double that of the original EFM model—a very significant improvement on the current alternatives in the literature. An appendix gives a step of the EM algorithm for the EFM model in pseudo code.

9.2 Multifactor Yield Curve Models and Their Drawbacks

The range of yield curve models discussed in the literature is vast. The task of choosing a suitable model for a variety of purposes, including trading, systematic investment, asset liability management, and structured product valuation, is nontrivial. However, there are relatively few papers in the literature that measure (as opposed to just discussing) the comparative advantages of more than a handful of different models, as the implementation of the more complex ones is a time-consuming process. Therefore in developing a suitable new model, it is important to start with a clear formulation of model requirements, so that the set of possible suitable choices is limited.

9.2.1 Requirements for model development

The principal applications of the model we envisage are the following:

- *Scenario simulation* for a diverse set of (predominantly long-term) ALM problems for multiple currencies
- *Valuation* of complex structured derivatives and loans (which often have embedded derivatives) in multiple currencies
- *Risk assessment* of portfolios and structured products

The problem of model selection has been discussed in Dempster et al. (2010, 2014). We borrow some of the requirements for the new yield curve model from these works and extend them here:

- A continuous-time framework to allow the flexibility of using different time steps, including uneven time steps
- Mean-reverting behavior
- Allow both pricing and dynamic evolution under the market (real world) measure, that is, the model should reflect the market risk premium[1]
- Reproduce a wide range of yield curve shapes and dynamics (to allow for realistic risk assessment, for example), including steepening, flattening,[2] inverted and humped yield curves

[1] As argued in Nawalkha and Rebonato (2011), this is especially relevant for the buy-side practitioner. For sell-side banks, it usually suffices to do pricing and initial hedging calculations under the risk-neutral measure from the forward market data on the day. Having an exact fit to the observed yields is thus more important for the sell side.

[2] Products based on these properties of the yield curve are traded on the NYSE, for example, US Treasury Flattener ETN (ticker: FLAT) and iPath US treasury Steepener ETN (ticker: STPP), although admittedly these are not very popular.

- Incorporate realistic modeling of the ZLB empirically observed for zero-coupon bond yields
- Feasible and efficient bond price calculation in closed form or numerically
- Parameter estimation by efficient model calibration to market data
- Allow estimation and use of the model for multiple correlated yield curves and currency exchange rates
- Parsimony in parameters
- Time homogeneity

Clearly, the requirements related to ease and speed of calculations contradict the requirements for model realism, so a compromise is obviously necessary. The choice of model made when prevailing short-term yields are near-zero could be different from that in a high-rate environment, but ideally the chosen model should cover all rate environments. However, the present global low-rate environment is the principal motivation for our work to improve on our workhorse affine yield curve model (Dempster et al. 2010). The most important enhancement required for our existing EFM model (see Section 9.3) is a better way of dealing with the ZLB for initial low short rates.[3] If we assume that both normal and low-rate environments are probable in the medium- to long-term future, and that once the situation reverts to long-term nominal levels away from zero it will be reasonably similar to the nominal rate environment before the crisis, a prudent decision would be to try to construct a model that is suitable for both environments.

9.2.2 Available multifactor yield curve models

In order to briefly survey available model choices, we can divide yield curve models into three broad overlapping classes:

- Short rate models
- Models in the Heath–Jarrow–Morton (HJM) framework
- Market models

Some authors also categorize stochastic volatility models (such as SABR) separately. Short rate models are based on modeling a process for the instantaneous interest rate which is then used to derive zero-coupon bond prices, that is, discount factors, or their yields at different maturities. This class of models is the oldest and probably the most extensively researched.

The HJM models start from modeling forward rates directly. A feature of this framework is using the no-arbitrage property to derive constraints on

[3]We have previously used penalties added to the model likelihood function.

the structure of forward rates. This framework is very general and convenient for studying arbitrage-free properties in theory. However, some of the models in this framework may be non-Markovian and most practical models coming from the HJM framework are either well-known short rate models or market models.

The class of market models is focused on describing the dynamics of the observable quantities (e.g., LIBOR and SWAP market models). They are especially useful for derivative pricing. However, under the current actually occurring low interest rate conditions, the popular LIBOR Market Model (see, e.g., de Jong et al. 2001) may require parameter estimates that are unrealistic (e.g., simulated cash returns more volatile than actual equity returns, with significant probability assigned to interest rates of more than 10,000%).

Model and computational complexity considerations, as well as the applications envisaged, suggest that short rate models are the most suitable class for our needs.

There are other factors influencing our considerations. First, we have had long successful experience with utilization of the EFM model (see Section 9.3) in situations in which the rate ZLB is not binding. We have confidence in the performance of the EFM yield curve model in these situations, so we would prefer our new model not to deviate too far from it.

Secondly, most of the current research on ZLBs is done in the framework of short rate models. Having a way of estimating the level of "shadow" rates may be useful, not least because some policy makers appear to monitor them. There have been attempts in the research literature to use the "shadow" short rate and its distance to zero as a forecast of the estimated time until the low-rate regime is lifted, see Ueno et al. (2006) and Wu and Xia (2014). The Federal Reserve Bank of Atlanta publishes the Wu-Xia Shadow Federal Funds Rate based on the Wu and Xia paper. It should be noted, however, that the level of the shadow rate is not a very reliable indicator, as it is strongly model dependent (see Bauer and Rudebusch 2013; Christensen and Rudebusch 2013).

A review of the literature on short rate models shows that most popular sub-class of short rate models in empirical research and applications are the Affine Term Structure Models (ATSMs), due to their analytical tractability, flexibility, and empirical efficiency. This class of models includes Vasicek (1977), Dothan (1978), Cox-Ingersoll-Ross (1985), Ho-Lee (1986), Hull-White (1990), and many other one- and multifactor models.

To illuminate the analysis that we undertake below for the more complex multifactor models, we first discuss the characteristics of the simpler one-factor models. The stochastic differential equations (SDEs) governing the evolution of the short rate under the risk-neutral or pricing measure Q for the respective models are:[4]

1. Vasicek (1977)

$$\mathrm{d}\boldsymbol{X}_t = \lambda(\theta - \mathrm{X}_t)\,\mathrm{d}t + \sigma\,\mathrm{d}\boldsymbol{W}_t \tag{9.1}$$

[4]We use boldface type in the sequel to denote stochastic entities, here conditionally.

2. Dothan (1978)

$$\mathrm{d}\boldsymbol{X}_t = -\lambda \mathrm{X}_t \,\mathrm{d}t + \sigma X_t \,\mathrm{d}\,\boldsymbol{W}_t \tag{9.2}$$

3. Cox-Ingersoll-Ross (1985)

$$\mathrm{d}\boldsymbol{X}_t = \lambda(\theta - \mathrm{X}_t)\,\mathrm{d}t + \sigma\sqrt{X_t}\,\mathrm{d}\,\boldsymbol{W}_t \tag{9.3}$$

4. Ho-Lee (1986)

$$\mathrm{d}\boldsymbol{X}_t = \theta_t \,\mathrm{d}t + \sigma\,\mathrm{d}\,\boldsymbol{W}_t \tag{9.4}$$

5. Hull-White (1990)

$$\mathrm{d}\boldsymbol{X}_t = \lambda(\theta_t - \mathrm{X}_t)\,\mathrm{d}t + \sigma_t\,\mathrm{d}\,\boldsymbol{W}_t. \tag{9.5}$$

It is easy to see that Hull–White (also called extended Vasicek) is the most general of these models. It can fit any term structure exactly because of the time-dependent equilibrium drift coefficients θ_t. However, having time-dependent parameters, as in the Ho–Lee and Hull–White models, contradicts our requirements of parsimony and time homogeneity. The Ho–Lee model also lacks the desired mean-reversion property and the Vasicek, Ho–Lee, and Hull–White diffusion models can all produce negative yields. The Dothan and Cox–Ingersoll–Ross models produce positive yields, but the short rate in these models never hits the ZLB. In other words, the ZLB in these models is repelling instead of absorbing. This is not consistent with the historical data recently observed in developed countries.[5] The Vasicek, Dothan, and Cox–Ingersoll–Ross models do not satisfy our requirement on yield curve shapes in that the shapes attainable with these models are constrained. On the other hand, choosing a time-homogeneous structure for our model, by not using nonstationary parameters in the corresponding SDE, means that exact matching of the yield curve is not possible with a small number of factors.

The number of factors necessary for adequate modeling of the whole term structure has been analyzed in Litterman and Scheinkman (1991). Their principal component analysis of U.S. Treasury data showed that 99% of the variance can be captured by three factors.

It is well known (see, e.g., Nawalkha and Rebonato 2011) that single-factor and two-factor time-homogenous models deviate significantly from the initially observed bond prices. However, three to five factors produce a close fit. The Nelson-Siegel (1987) model widely employed in central banks uses three factors to estimate the entire yield curve but has time inhomogeneous parameters, except in the Diebold–Rudebusch arbitrage-free version of the model (see Diebold and Rudebusch 2013; Rebonato 2015). Rebonato and Cooper (1995) argued that a two-factor affine or quadratic model cannot reproduce a realistic correlation structure of interest rate changes, but that three to five factors are sufficient for this purpose. So it seems that a reasonable choice

[5]Note that the *multifactor* square root (CIR) model and *quadratic Gaussian models* (QGMs) are also unable to reproduce the absorbing ZLB.

(taking into account additional computational complexities connected with introducing the ZLB property) would be an affine short rate model with three factors.

9.2.3 Classification of three-factor affine short rate models

Duffie and Kan (1996) derive necessary and sufficient conditions on the SDEs to have an affine representation and Dai and Singleton (2000) analyze the different subfamilies of ATSMs. The analysis of Dai and Singleton for the three-factor cases shows that some affine subfamilies explain historical interest rate behavior better than others.

The SDEs for their factors are of the form

$$\mathrm{d}\boldsymbol{X}_t = \Lambda\,(\Theta - X_t)\;\mathrm{d}t + \Sigma\sqrt{S_t}\,\mathrm{d}\boldsymbol{W}_t, \tag{9.6}$$

with $\boldsymbol{X}$ the K-dimensional state vector; $\boldsymbol{W}$ K-dimensional Brownian motion; Θ a fixed point in K-dimensional space; Λ, S_t and Σ $K \times K$ matrices, and S_t a diagonal matrix with diagonal elements satisfying

$$[S_t]_{ii} = \alpha_i + \beta_i^{'} X_t, \quad i = 1, \ldots, K, \tag{9.7}$$

where prime denotes transpose.

Zero coupon *bond prices* in terms of expectations of the *instantaneous short rate* $\boldsymbol{r}_t$ under the risk-neutral Q measure are given by

$$P_t(\tau) = E_t^Q\left[\exp\left(-\int_t^{t+\tau} \boldsymbol{r}_s\,\mathrm{d}s\right)\right]. \tag{9.8}$$

For an *admissable parametrization*, the bond prices can be calculated as

$$P_t(\tau) = e^{A(\tau)+B(\tau)'X_t}, \tag{9.9}$$

where A and B are solutions of certain ODEs (see, e.g., James and Webber 2000).

The instantaneous short rate is also an *affine* function of the state

$$r_t = \phi_0^Q + \phi_X^{Q^{'}} X_t. \tag{9.10}$$

Zero coupon bond yields to maturity, termed *rates*, are linked with the bond prices by

$$y_t(\tau) = -\log P_t(\tau)/\tau. \tag{9.11}$$

There are models that lack affine structure (and thus forfeit simple formulae for bond prices) but a vector of K rates R_t of specified maturities may sometimes still be recovered as the numerical solution of the *Ricatti* equation

$$\frac{\partial R_t(\tau)}{\partial \tau} = \Lambda R_t(\tau) - \frac{1}{2} R_t(\tau)\Sigma S \Sigma' R_t(\tau)' + r_t 1, \tag{9.12}$$

where 1 is the K vector of ones (Dempster et al. 2014).

Dai and Singleton (2000) denote different affine subfamilies by $A_m(n)$ with n the number of factors and $m \leq n$ the number of bounded factors. They perform empirical tests on the different subclasses for n equal to 3. Dempster et al. (2014) also analyzed various three-factor affine models with requirements similar to ours to uncover a variety of shortcomings with the models evaluated. In particular, they studied the three-factor *extended Vasicek* model specified under the market measure P in Equation 9.6 by

$$
\begin{aligned}
\Lambda &:= \begin{pmatrix} \lambda_{11} & 0 & 0 \\ \lambda_{21} & \lambda_{22} & 0 \\ \lambda_{31} & \lambda_{32} & \lambda_{33} \end{pmatrix} \\
\Theta &:= \begin{pmatrix} \theta_1 \\ \theta_2 \\ \theta_3 \end{pmatrix} \\
\Sigma &:= \begin{pmatrix} \sigma_1 & 0 & 0 \\ 0 & \sigma_2 & 0 \\ 0 & 0 & \sigma_3 \end{pmatrix} \\
S &:= \begin{pmatrix} 1 & 0 & 0 \\ 0 & 1 & 0 \\ 0 & 0 & 1 \end{pmatrix} \\
r(t) &:= \delta_0 + \delta_1 y_1(t) + \delta_2 y_2(t) + \delta_3 y_3(t).
\end{aligned} \tag{9.13}
$$

This Dai and Singleton $A_0(3)$ model with 16 parameters (also known as the Hull–White model) is not econometrically identified under P (i.e., different values of Θ can give the same paths of the factor process $\boldsymbol{X}$) unless $\Theta := 0$, which is only appropriate to the pricing measure Q, and has other difficulties as well.

Dempster et al. (2014) were led to introduce a Black-corrected affine model which always produces nonnegative rates.[6] This was based on the recent (2011) Joslin–Singleton–Zhu (JSZ) three-factor affine Gaussian yield curve model whose continuous evolution of the three factors Y is given by

$$
\mathrm{d}\,\boldsymbol{Y}(t) = \Lambda(\theta - Y(t) + \Pi(t))\,\mathrm{d}t + \Sigma\sqrt{S(t)}\,\mathrm{d}\,\boldsymbol{W}(t), \tag{9.14}
$$

where

$$
\boldsymbol{\Pi}(t) = k_0 + K_1 \boldsymbol{Y}(\boldsymbol{t}) \tag{9.15}
$$

is the affine state-dependent market price of risk (excess factor return) vector. JSZ estimate the parameters of the discrete time version of their model with three observed yield curve points (rates) fit exactly and a few extra rates fit approximately by least squares by means of two standard econometric vector autoregression (VAR) models given, respectively, under the market (real-world) and pricing (risk-neutral) measures P and Q. Dempster et al. (2014) use only a single extra yield curve point fit by least squares.

[6] We shall describe the Black correction in the following section.

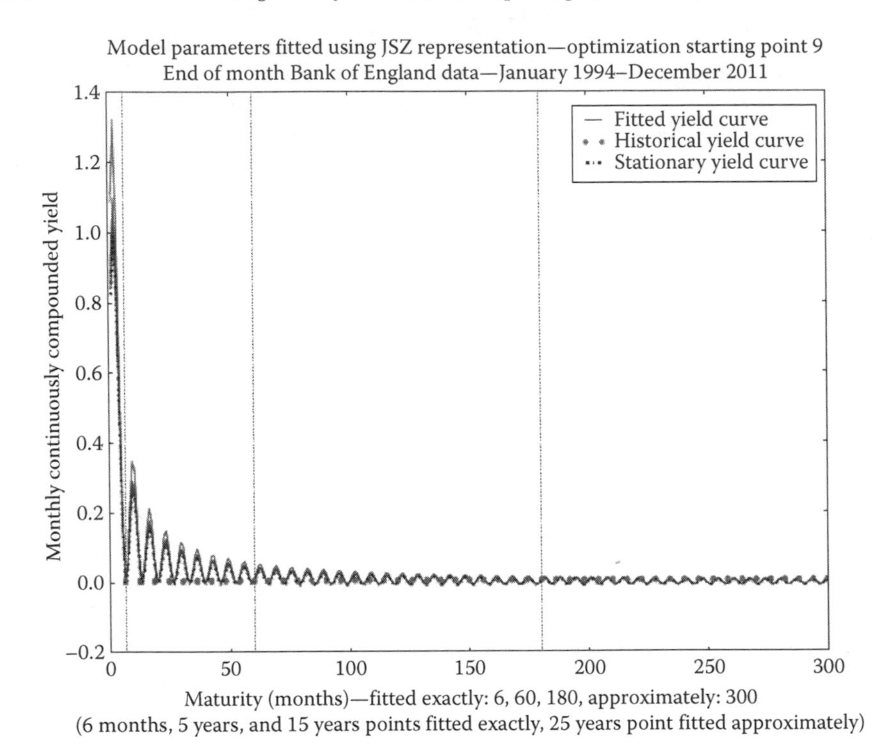

FIGURE 9.1: Joslin–Singleton–Zhu affine model fit numerical instability.

9.2.4 Difficulties with Gaussian affine models

Besides the problem of negative simulated rates, Dempster et al. (2014) encountered a number of unexpected difficulties with existing models not discussed in the literature. For example, the JSZ affine model calibration implementation was initially subject to numerical instability of in-sample fit as illustrated in Figure 9.1, which shows extreme oscillating monthly fitting errors dwarfing a typical UK yield curve up to 25-year maturity.

While this difficulty was relatively easily overcome by parameter bounding in the calibration, Figure 9.2 shows a much more fundamental difficulty with the JSZ model. Namely, while the properly calibrated model fits well a selection of historical yield curves in-sample, yield curve extrapolation out-of-sample projects, that all such curves are monotonically declining to eventually have negative nominal rates within about a 100-year maturity and on average at about 80 years.

Figure 9.3 illustrates the worrying difficulty with three-factor models of the Gaussian $A_0(3)$ affine class in low-rate environments with the EFM model described in the following section. The figure shows the quantiles of 10,000 30-year forward simulations of 10-year euro bond rates for the EFM yield curve model which has been calibrated to daily data from January 2, 2001, to

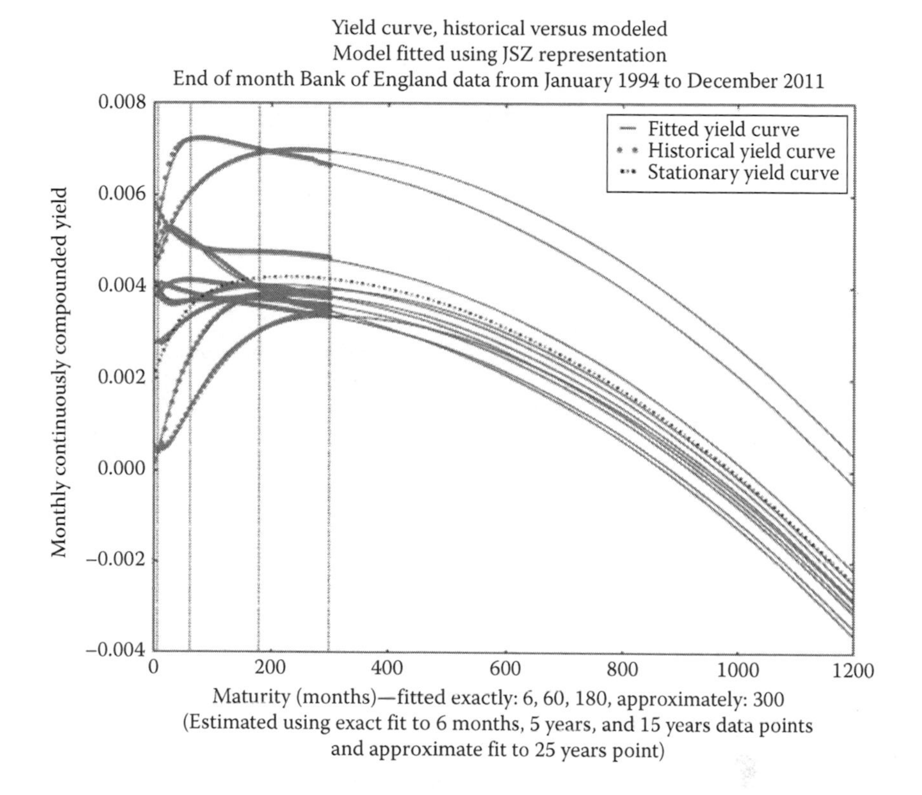

FIGURE 9.2: Joslin–Singleton–Zhu 25-year UK yield curve out-of-sample projections.

January 2, 2012. This model gives a 25% probability of future 10-year negative rates within 3.5 years starting from an initial value of about 3.2% which it predicts will remain the median value over the 30-year forecast horizon.

In summary, we have stated a number of desirable requirements for a practical long-term yield curve model and have briefly surveyed the range of models available in the literature. We determined that the short rate class is the most suitable for our needs and within this class it seems that the most reasonable decision *a priori* is to evaluate a model with three factors, in particular, within the $A_0(3)$ affine class. We have illustrated some of the potential drawbacks of such models which do not have a correction for simulated nonnegative rates.

9.3 Nonlinear Black Correction for the EFM Model

Around the turn of the last century, a famous Austrian economist, Eugen von Bohm-Bawerk (1851–1914), declared that the cultural level of a nation is

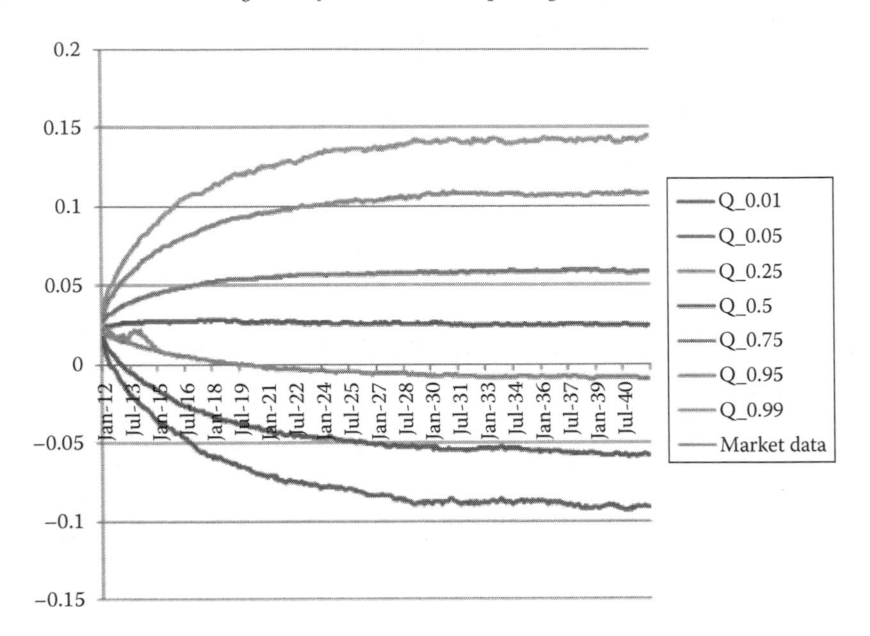

FIGURE 9.3: EFM model Euro 10-year rate forecast for 30 years.

mirrored by its rate of interest: the higher a people's intelligence and moral strength, the lower the rate of interest (Homer and Sylla 2005).

As a low-rate environment has prevailed in most major developed countries since 2008, and in Japan since the early 1990s, it is crucial to realistically model rates behavior in these circumstances. We will present a Black-corrected version of the EFM model discussed in Medova et al. (2005), Yong (2007), and Dempster et al. (2010).

9.3.1 Three-factor basic EFM model

We first describe the original *EFM* model of the yield curve, which we have used previously in a variety of applications in the five principal currency areas with various time steps from daily to quarterly.[7]

The evolution under the risk-neutral measure Q of the three unobservable (i.e., *latent*) factors of the model is governed by the SDEs

$$\begin{aligned} \mathrm{d}\boldsymbol{X}_t &= \lambda_X\,(\theta_X - X_t)\,\mathrm{d}t + \sigma_X\,\mathrm{d}\boldsymbol{W}_t^X \\ \mathrm{d}\boldsymbol{Y}_t &= \lambda_Y\,(\theta_Y - Y_t)\,\mathrm{d}t + \sigma_Y\,\mathrm{d}\boldsymbol{W}_t^Y \\ \mathrm{d}\boldsymbol{R}_t &= k\,(X_t + Y_t - R_t)\,\mathrm{d}t + \sigma_R\,\mathrm{d}\boldsymbol{W}_t^R, \end{aligned} \tag{9.16}$$

[7]It is interesting to note that this model originated at Long-Term Capital Management and was first brought to our attention by Lehman Brothers under the auspices of Pioneer Investments of UniCredit Bank.

with fixed pairwise correlations of the standard Brownian motion innovations given by

$$(\rho_{XY}\,\mathrm{d}t, \rho_{XR}\,\mathrm{d}t, \rho_{YR}\,\mathrm{d}t). \tag{9.17}$$

The stochastic evolution of the three factors under the market (i.e., real-world) measure P involving the *market prices of risk* of the factors is governed by

$$\begin{aligned}
\mathrm{d}\boldsymbol{X}_t &= \lambda_X \left(\theta_X + \frac{\gamma_X \sigma_X}{\lambda_X} - X_t\right) dt + \sigma_X \,\mathrm{d}\boldsymbol{W}_t^X \\
\mathrm{d}\boldsymbol{Y}_t &= \lambda_Y \left(\theta_Y + \frac{\gamma_Y \sigma_Y}{\lambda_Y} - Y_t\right) \mathrm{d}t + \sigma_Y \,\mathrm{d}\boldsymbol{W}_t^Y \\
\mathrm{d}\boldsymbol{R}_t &= k \left(X_t + Y_t + \frac{\gamma_R \sigma_R}{k} - R_t\right) \mathrm{d}t + \sigma_R \,\mathrm{d}\boldsymbol{W}_t^R.
\end{aligned} \tag{9.18}$$

The first factor $\boldsymbol{X}_t$ represents the long rate and the third factor $\boldsymbol{R}_t$ the short rate. The second factor $\boldsymbol{Y}_t$ represents minus the slope of the yield curve between the long rate and the unobservable instantaneous short rate. Thus the sum of the first two factors $\boldsymbol{X}_t + \boldsymbol{Y}_t$ represents the unobservable stochastic instantaneous short rate about which the observable short rate $\boldsymbol{R}_t$ mean reverts.

Note that as the $\boldsymbol{X}$ and $\boldsymbol{Y}$ equations have the same form, the factor dynamics under Q given in Equation 9.18 are not econometrically identified, that is, the parameters are not uniquely determined in that different sets will generate the same factor dynamics. However, the factor dynamics under P given in Equation 9.18 are identified by virtue of expressing the fixed factor market prices of risk in volatility units. Also note that the rates of mean reversion of the three factors are identical under P and Q. As a result, the parameters of the dynamics must be estimated from market data under the P measure and the resulting market price of risk estimates set to 0 to generate the dynamics of the Q measure for pricing.

The SDEs (9.16 and 9.18) have explicit solutions. Substituting the explicit solutions of the SDEs (9.16) into the sum of the first two factors and using Equations 9.10 and 9.11 produces closed-form formulae for bond prices and yields in terms of affine functions of the SDE parameters (see, e.g., Medova et al. 2005; Yong 2007 for the factor covariance matrix). In particular, denoting the three latent factors at time t in vector form by $x_t := (X_t, Y_t, R_t)'$, the yields of the K different maturity zero coupon bonds are given in affine form by

$$y_t = Bx_t + d, \tag{9.19}$$

where B and d are closed-form deterministic affine functions (matrix or vector-valued, respectively) of the SDE parameters.

9.3.2 EFM model calibration

Calibration of the EFM model is a nontrivial task even without the Black correction. The parameters of the model are estimated using a version of the

EM algorithm (Dempster et al. 1977) which iterates to parameter convergence the *KF* to generate sample paths and *MLE* of parameters for each path. Given a fixed set of parameters, the KF produces estimates for the unobserved states of the factors and prediction for the yields from Equation 9.19. These are then used as the observed sample for the next numerical parameter optimization step of MLE.

Note that for this version of the EFM model in *state–space form*, MLE is trying to fit all of the observed rates *approximately*, in contrast to some other approaches discussed previously which fit a small number of rates (equal to the number of factors) *exactly*.

KF transition equation

Taking the discretization time step $\Delta t := 1$, the Euler approximation of the SDEs for the three-factor *state variables* becomes the state variable *transition equation*

$$\boldsymbol{x}_{t+1} = Ax_t + c + \boldsymbol{\eta}_t, \tag{9.20}$$

where $\boldsymbol{\eta}_t \sim N(0, G)$ is the Gaussian innovation, with A, c, and G deterministic matrix or vector-valued functions of the SDE coefficients.

KF measurement equation

The corresponding *measurement equation* is

$$\boldsymbol{y}_t^{\text{obs}} = Bx_t + d + \boldsymbol{\varepsilon}_t, \tag{9.21}$$

where $\boldsymbol{y}_t^{\text{obs}}$ corresponds to the yields observed in the market and B and d are defined above. The centered *measurement error* process $\boldsymbol{\varepsilon}_t$ is a K-vector of serially independent Gaussian noise with covariance matrix H.[8]

Given a data series for the observed yields $\boldsymbol{y}_t^{\text{obs}}$, the KF generates an estimated expected path of the Gaussian state variables and their conditional covariance matrix $\Sigma_{t|t-1}$.

Initialization

The filter is initialized using unconditional moments which following Harvey (1993) gives initial values

$$\begin{aligned} \hat{x}_0 &:= (I - A)^{-1} c \\ vec(\Sigma_0) &:= (I - A \otimes A)^{-1} vec(G), \end{aligned} \tag{9.22}$$

where $\otimes$ is the Kronecker product, $vec(.)$ is the operation of writing out a matrix as a vector, and G is the covariance matrix of the factor dynamics

[8]But see Dempster and Tang (2011) regarding handling measurement error serial correlation, which we intend to implement in future research.

innovations $\boldsymbol{\eta}$. The matrix A and the vector c are the entities in the transition equation 9.20 and the elements of Σ_0 can be computed analytically.

KF prediction

$$\begin{aligned} \hat{x}_{t|t-1} &= A\hat{x}_{t-1} + c \\ \hat{y}_{t|t-1} &= B\hat{x}_{t|t-1} + d \\ \Sigma_{t|t-1} &= A\Sigma_{t-1}A^{\mathrm{T}} + G. \end{aligned} \tag{9.23}$$

KF update

$$\begin{aligned} v_t := y_t^{\mathrm{obs}} - \hat{y}_{t|t-1} &= y_t^{\mathrm{obs}} - B\hat{x}_{t|t-1} - d \\ F_t &= B\Sigma_{t|t-1}B' + H \\ \hat{x}_t &= \hat{x}_{t|t-1} + \Sigma_{t|t-1}B^{\mathrm{T}}F_t^{-1}v_t \\ \Sigma_t &= \Sigma_{t|t-1} - \Sigma_{t|t-1}B'F_t^{-1}B\Sigma_{t|t-1}. \end{aligned} \tag{9.24}$$

Quasi-maximum likelihood parameter estimation

Let Θ denote the 14 SDE model parameters of the transition equation and define $\psi := \{\Theta, H\}$. Then the *log-likelihood* is given by

$$\log \mathrm{L}(\Theta, H) = -\frac{TK}{2}\log 2\pi - \frac{1}{2}\sum_{t=1}^{T}\log\det F_t - \frac{1}{2}\sum_{t=1}^{T} v_t' F_t^{-1} v_t, \tag{9.25}$$

where K is the total number of maturities used, T is the number of time steps, and v and F are computed using Equation 9.24.

The maximization of the log-likelihood is performed in two steps, alternatively optimizing Θ and H to convergence. There are two phases of the numerical optimization: a *global* phase using the DIRECT global optimization algorithm (Jones et al. 1993) to locate the *region* of the maximum, followed by a *local* phase using an approximate conjugate gradient algorithm (Powell 1964) to locate the *maximum* itself.

9.3.3 Black correction for negative rates

The distribution of the instantaneous short rate $\boldsymbol{r}_t$ is Gaussian in most yield curve models. Therefore it is easy to see that it can become negative when initialized at a low level. Black (1995) suggested a way of solving this problem. He argued that nominal rates cannot become negative, because there always exists the option of investing in the 0-yielding currency. Black started from a process $\boldsymbol{s}_t$ which can take negative values, which he called the *shadow short rate*, and the *nominal short rate* is then defined as

$$\boldsymbol{r}_t := \max(0, \boldsymbol{s}_t) = 0 \vee \boldsymbol{s}_t, \tag{9.26}$$

where $\vee$ denotes meet in the natural lattice order of the real line. This modification makes all the yields calculated through the bond price nonnegative. Unfortunately, the shadow short rates implied by affine models lose

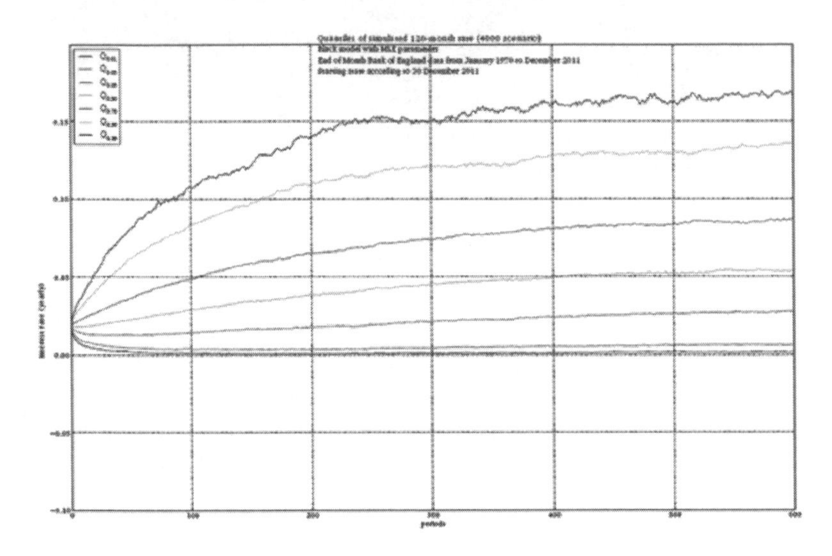

FIGURE 9.4: Black JSZ model 10-year gilt rate 50-year prediction.

their linearity when modified using this idea. This makes the resulting models difficult to calibrate. We discuss different approaches to calibration in the following section.

9.3.4 Stylized properties of Black models

Most recent papers considering Black-corrected models have been based on the $A_0(3)$ class. The EFM three-factor model which has proven itself in a variety of different applications also belongs to this class. Dempster et al. (2014) summarized in a table the stylized features satisfied by the alternative models they evaluated, which is reproduced here as Table 9.1.

Figure 9.4, also reproduced from their paper, shows for the Black-corrected JSZ model the quantiles of the 50-year predicted distribution for the 10-year UK gilt rate from December 30, 2011, based on 10,000 scenarios with no nonnegative rates.

Recent literature has also investigated the implications of Black-corrected affine yield curve models for the long-term reversion from near-zero rates to more normal levels (see Swanson and Williams 2013; Krippner 2014; Christensen 2015) as well as nonlinear models (Feldhutter et al. 2015).

9.4 HPC Approaches to Calibrating Black Models

Although Black's idea was proposed in the 90s, the first implementation only followed 9 years later in the work of Gorovoi and Linetsky (2004). Active

TABLE 9.1: Properties of evaluted yield curve models with regard to stylized facts

	Yield curve model					
Stylized fact properties	**CIR** $A_3(3)$	**BDFS** $A_3(3)$	**Vasicek** $A_3(3)$	**HW/JSZ** $A_1(3)$	**HW/JSZ/BRW** $A_0(3)$	**Black** $A_0(3)$
Mean reverting rates	Yes	Yes	Yes	Yes	No	Yes
Nonnegative rates	Yes	No	No	No	No	Yes
Stochastic rate volatility	Yes	Yes	No	No	No	Yes[a]
Closed-form bond prices	Yes	Yes	Yes	Yes	Yes	No
Replicates all observed curves	No	Yes	Yes	Yes	Yes	Yes
Good for long-term simulations	No	No	No	No	No	Yes
State-dependent risk premia	No	No	No	Yes	Yes	Yes
+ve Rate/volatility correlation	No	No	No	No	No	Yes
Effective in low-rate regimes	No	No	No	No	No	Yes

[a]Rate volatilities are piecewise constant punctuated by random jumps to 0 at rate 0 boundary hitting points.

work on extensions to multifactor models started only after the crisis of 2008. There are two main reasons for such a timeline. First, the ZLB was not observed in the United States from the Great Depression until 2008. Only in Japan from the mid-1990s did rates come near zero in a major economy. Perhaps more importantly, the implementation of the Black correction is considerably more difficult (both theoretically and computationally) than implementation of the usual ATSMs. The main problem is the lack of a closed-form formula for the bond price given by

$$P_t(\tau) = E_t^Q \left[\exp\left(- \int_t^{t+\tau} (0 \vee s_u)\, \mathrm{d}u \right) \right]. \tag{9.27}$$

9.4.1 Three-factor Black model calibration

Calibration of nonlinear Black models with any underlying three-factor Gaussian shadow rate is much more computationally intensive than for the underlying affine shadow rate model and requires *HPC* facilities to be undertaken successfully. This is because even linear Kalman filtering is not computationally insignificant, likelihood functions are multiextremal and discount bond prices or yields must be calculated for all maturities at each discrete time point, for example, daily, over the model simulation horizon.

The various current approaches to calibrating Black models based on affine three-factor shadow rate models may be categorized in terms of how they handle the three steps crucial to the process. These are:

- Method of *inferring states* of the latent three factors from observed market rates
  - Inverse mapping or least squares
  - Extended or iterated extended Kalman filter (EKF or IEKF) using piecewise linearization
  - Unscented Kalman filter (UKF) using averaged multiple displaced KF paths
- Method of *parameter estimation* for a given factor path
  - Method of moments
  - Maximum likelihood estimation (MLE) or quasi-maximum likelihood estimation (QMLE)
- Method of *bond price calculation*
  - Monte Carlo simulation
  - PDE (partial differential equation) solution
  - Approximation

Filtering

Christensen and Rudebusch (2013), Bauer and Rudebusch (2014), and Lemke and Vladu (2014) use the EKF for parameter estimates. However, Krippner (2013a) uses IEKF to fit his shadow rate approximation for the case of two and three factors, because he found that using the EKF was not robust. Priebsch (2013) uses the UKF. Christoffersen et al. (2014) perform a series of comparisons of EKF with UKF and the particle filter. They conclude that the UKF significantly outperforms the EKF and performs well compared to the significantly more computationally expensive particle filter.

Likelihood maximization

Most of the papers on shadow rate models omit discussion of the optimization methods used for this purpose. Richard (2013) mentions that he maximizes the likelihood function by using Nelder-Mead (1965) global search combined with Powell (1964) local search.

9.4.2 Monte Carlo bond pricing

Dempster et al. (2014) use cloud facilities for bond pricing in a three-factor Black-corrected JSZ 4 yield curve point model using a combination of analytical closed-form calculations of yields in the affine form given by Equations 9.9 and 9.11 for short maturities (Joslin et al. 2011) and Monte Carlo simulation (10,000 paths) for longer maturities. In more detail, the averages of Monte Carlo forward simulated paths are used for long rates, which automatically take account of the convexity adjustments otherwise required for this model. They suggest using the full unconditional likelihood function, cf. (9.25), and multiple starting points for parameter optimization, as there will be numerous local optima for the likelihood. With this approach filtering a multicurrency EFM for over the counter (OTC) structured derivative valuation becomes very computationally intensive.

Bauer and Rudebusch (2014) use Monte Carlo simulations (circa 500 paths of the shadow short rate) to calculate the bond prices. They employ the EKF to infer the states from the observed yields. However, they report using the same workaround as Bomfim (2003), that is, estimating the parameters of the model on the subset of data where the ZLB is not important, to compare the shadow rate affine and Black models practically. Lemke and Vladu (2014) have applied the Monte Carlo bond price calculation method and the EKF to construct yield curves in the Eurozone. Krippner (2013b) suggests using his framework results as control variates in Monte Carlo simulations for calculating true Black model bond prices. Christensen and Rudebusch (2013, 2015) apply the Krippner framework to estimate a three-factor shadow rate model. They argue that the divergence of the Krippner approach from the fully arbitrage-free Black approach is not very significant and is for example well compensated for by much greater tractability. Wu and Xia (2014) apply an approach equivalent to the Krippner framework in discrete time.

9.4.3 PDE bond pricing

A possible key to calibration of, for example, both the JSZ and EFM models is the efficient solution for discount bond prices $P_t(\tau)$ of all maturities τ at each time t of a three-dimensional (3D) parabolic quasi-linear PDE of the form

$$\partial P_t(\tau)/\partial\tau = \sum_{i,j=1}^{3} a_{ij}\partial^2 P_t(\tau)/\partial y_i \partial y_j + \sum_{i=1}^{3} b_i \partial P_i(\tau)/\partial y_i + cP_t(\tau). \quad (9.28)$$

Kim and Singleton's 2D alternating direction implicit (ADI) solution method will not cope well with the 3D case (Kim and Singleton 2011; Lipton et al. 2014). In future, we intend to investigate applying a fast robust 3D PDE solver based on an interpolating wavelet-specified irregular mesh implicit method that we have developed for complex derivative valuation (Jameson 1998; Carton de Wiart and Dempster 2011) and which is expected to form a part of the Numerical Algorithms Group library in future.

9.4.4 Black model calibration progress

The investigations of Black models in the literature naturally began with one-factor models and proceeded to the more generally accepted three-factor models.

Gorovoi and Linetsky (2004) showed that for a shadow short rate following an 1D diffusion process, the zero-coupon bond price can be calculated as the Laplace transform (at the unit value of the transform parameter) of the area functional of the shadow rate process. They applied the method of eigen function expansions (see Linetsky 2002; Davydov and Linetsky 2003; Linetsky 2004) to derive the quasi-analytic formulae (relying on Weber-Hermite parabolic cylinder functions) for the bond prices in the Vasicek and shifted CIR process cases. Unfortunately their method only works in the scalar case. Gorovoi and Linetsky applied their method to estimating yield curve models for only a single time point. However, their method was used for the Japanese market by Ueno et al. (2006) who applied the method to a dynamic model with a market price of risk.

The shadow rate $\boldsymbol{s}_t$ in these models is given by a diffusion, therefore in state–space form the single discretized transition equation takes the form

$$\boldsymbol{x}_{t+1} = a x_t + c + \boldsymbol{\eta}_t. \quad (9.29)$$

The mapping that links observed yields and the shadow rate is no longer linear, so it takes the piecewise linear form

$$y_t^{\text{obs}} = h(x_t). \quad (9.30)$$

Ueno et al. (2006) applied the KF with conditional linearization of Equation 9.30 to calibrate the model. However, it was clear from their results that further work in developing the shadow rate models would be needed. For example, the shadow rates in their analysis reach the implausibly low levels of −15%, which suggests model misspecification (see Ichiue and Ueno, 2006, 2007).

Both Bomfim (2003) and Kim and Singleton (2011) relied on a numerical (finite-difference) method for solving a 2D parabolic quasi-linear bond price PDE given by

$$\frac{\partial P_t}{\partial \tau} - \frac{1}{2} tr\left(\frac{\partial^2 P_t}{\partial x \partial x'} \Sigma\Sigma'\right) - \frac{\partial P_t}{\partial x} K(\theta - x) + \max\left[0, s(x)\right] P_t = 0 \qquad (9.31)$$

with boundary condition $P(\tau = 0, x) = 1$. Here the short rate $s(x)$ is an affine function of the 2D state x. Bomfim (2003) estimated the parameters of his model on the subset of data where rates were safely above zero, using an analytical approximation, that is, the usual affine model. Kim and Singleton (2011) used the EKF with quasi-maximum likelihood to estimate the parameters. Ueno et al. (2006) performed a sensitivity analysis of the two-factor Black-corrected model without estimating the parameters.

Kim and Singleton and Ueno et al. report superior performance of the full shadow rate models compared to their standard affine term structure equivalents (used for the shadow rates). Two-factor models also produce more plausible levels of the shadow rate. However, the analysis of Section 9.2 suggests that three factors would be preferred for realistic modeling. Unfortunately, the ADI finite difference scheme used by Kim and Singleton cannot easily be extended to the corresponding PDE in three dimensions as noted above.

Krippner (2013b) applies a different method, which can be seen as an *approximation* to the Black model. The advantage of his method is that the forward rates have closed-form formulae. In the Black model, the price of a bond can be expressed as

$$P_t(\tau) = P_t^{\mathrm{S}}(\tau) - C_t^{\mathrm{A}}(\tau, \tau; 1), \qquad (9.32)$$

where $P_t^{\mathrm{S}}(\tau)$ is the shadow bond price (i.e., the price of a bond in a market where currency is not available) and $C_t^{\mathrm{A}}(\tau, \tau; 1)$ is the value of an American call option at time t with maturity in τ years and strike 1, written on the shadow bond maturing in τ years. There is no analytic formula for $C_t^{\mathrm{A}}(\tau, \tau; 1)$, but Krippner argues that the American option can be approximated by an analytically tractable European one and introduces an auxiliary bond price equation

$$P_t^{\mathrm{aux}}(\tau, \tau + \delta) = P_t^{\mathrm{S}}(\tau + \delta) - C_t^{\mathrm{E}}(\tau, \tau + \delta; 1), \qquad (9.33)$$

where $C_t^{\mathrm{E}}(\tau, \tau + \delta; 1)$ is the value of a European call option at time t with maturity at time $t + \tau$ and strike 1 written on a shadow bond maturing at $t + \tau + \delta$. Krippner then takes the limit with $\delta \to 0$ to obtain the nonnegative (due to future currency availability immediately after maturity) instantaneous forward rate as

$$\underline{f_t}(\tau) = \lim_{\delta \to 0} \left[-\frac{\partial}{\partial \delta} \ln P_t^{\text{aux}}(\tau, \tau + \delta) \right]. \tag{9.34}$$

The nonnegative yield with maturity τ in Krippner's framework is calculated as

$$\underline{y_t}(\tau) = \frac{1}{\tau} \int_t^{t+\tau} \underline{f_t}(s)\, \mathrm{d}s = y_t^{\mathrm{S}}(\tau) + \frac{1}{\tau} \int_t^{t+\tau} \lim_{\delta \to 0} \left[\frac{\partial}{\partial \delta} \frac{C_t^{\mathrm{E}}(\tau, \tau + \delta; 1)}{P_t(s + \delta)} \right] \mathrm{d}s. \tag{9.35}$$

Here $y_t^{\mathrm{S}}(\tau)$ are the shadow bond yields. Unfortunately, closed-form analytic expressions for the bond prices and yields are still not available, but they can be evaluated through calculating integrals that are numerically tractable. More importantly, Krippner's approach is not fully arbitrage-free. The short rates are identical under the market measure P in the Black and Krippner frameworks, but different under the risk-neutral measure Q. Krippner's approach is extendible to three factors. Priebsch (2013) proposes to view the quantity

$$\log P_t(\tau) = \log E_t^Q \left[\exp \left(- \int_t^{t+\tau} (0 \vee \boldsymbol{s}_u)\, \mathrm{d}u \right) \right]. \tag{9.36}$$

as the value at -1 of the conditional cumulant-generating function of the random variable $S_t(\tau) = \int_t^{t+\tau} \max(0, s_u)\, du$ under Q. It can be expanded as

$$\log E_t^Q \left[\exp\left(-\boldsymbol{S}_t(\tau) \right) \right] = \sum_{j=1}^{\infty} (-1)^j \frac{\kappa_j^Q}{j!}, \tag{9.37}$$

where κ_j^Q is the jth cumulant of $\boldsymbol{S}_t(\tau)$ under Q and an approximation can be computed by taking a finite number of terms in this series. The method of Ichiue and Ueno (2013) is equivalent to using the first-term approximation in Equation 9.28. Priebsch (2013) evaluates both one- and two-term approximations by analytically deriving the expression for the first two moments of $\boldsymbol{S}_t(\tau)$ (see also Kim and Priebsch 2013). He shows that this technique is sufficiently fast and accurate to fit the term structure within a half basis point for a single time step. Priebsch notes that the Krippner (2012) approximation tends to underestimate the arbitrage-free yields of the Black model, while the first-order cumulant approximation tends to overestimate these yields, suggesting a systematic error. The errors of second-order cumulant approximation do not appear to have a discernible bias in any direction. Andreasen and Meldrum (2014) compare cumulant approximation to shadow rate models with quadratic term structure to find that shadow rate models are better at out-of-sample forecasts.

Using HPC techniques, Richard (2013) estimates a full Black-corrected shadow rate model by solving the 3D PDE (9.28) for bond prices using an implicit numerical scheme and notes that calibration "requires a long time, literally a month, on a large and fast computer to estimate (model parameters)." In the latest version of his paper, the search time has been reduced to 3 days on the Cornell supercomputer, but the methods used to achieve this are not specified.

In summary, implementing the Black correction leads to nonlinearity of the measurement equation, that is, of the mapping of factors/states to yields, so that the classical KF is no longer applicable.

9.5 UKF EM Algorithm HPC Implementation

Taking account of the information in the literature reviewed above, we will use the *UKF* (Julier et al. 1995; Julier and Uhlmann 1997) for our shadow rate model. To calculate the bond yields, we will use the measurement equation *approximation*

$$y_t^{\text{obs}} = 0 \vee (Bx_t + d) + \varepsilon_t, \tag{9.38}$$

where $\vee$ now denotes *coordinate-wise* maximum at each step of the UKF dynamics. It will be demonstrated in the sequel that the computational times for the HPC implementation of our approach are very acceptable relative to those of the basic linear Kalman filtering algorithm and the nonlinear KF alternatives.

9.5.1 UKF for the Black EFM model

We initialize the filter at the unconditional mean and variance of the state variables under the P measure in the EFM model. This can be justified by the fact that most of our datasets start before the onset of low-rate regimes. Since only the measurement equation is nonlinear, the state prediction step is the same as that of the linear Kalman filter in Equation 9.23.

For the factor path update step of the UKF, the state is first augmented with the expected measurement error (here 0) of the linear KF to give

$$x_{t|t-1}^a = \left[\hat{x}_{t|t-1}', E\left[\varepsilon_t'\right]\right]', \tag{9.39}$$

and the state innovation conditional covariance matrix is augmented with the measurement error covariance matrix to give

$$\Sigma_{t|t-1}^a = \begin{bmatrix} \Sigma_{t|t-1} & 0 \\ 0 & H \end{bmatrix}. \tag{9.40}$$

Next, a set of perturbed *sigma points* is constructed as

$$\begin{aligned} \chi_{t|t-1}^0 &= \hat{x}_{t|t-1} \\ \chi_{t|t-1}^j &= \hat{x}_{t|t-1} + \left(\sqrt{(L+\lambda)\,\Sigma_{t|t-1}^a}\right)_j, \qquad j = 1, \ldots, L \\ \chi_{t|t-1}^j &= \hat{x}_{t|t-1} - \left(\sqrt{(L+\lambda)\,\Sigma_{t|t-1}^a}\right)_{j-L}, \quad j = L+1, \ldots, 2L, \end{aligned} \tag{9.41}$$

where $\sqrt{}$ denotes the matrix square root of the symmetric positive definite augmented matrix (9.40), whose jth column augments the conditional state vector to give the *augmented conditional* state vector. Here L is the *dimension* of the augmented state and the scalar parameter λ is defined as

$$\lambda := \alpha^2(L + \kappa) - L, \tag{9.42}$$

where α and κ control the *spread* of the sigma points in an elliptical configuration around the conditional augmented state vector. The choice of these parameters is very important for the results of the calibration. We used an (NAG) code which sets α equal to 1 and κ equal to 0, but we shall see that this is probably not the best choice.

Next, the (here piecewise linear) measurement equation is evaluated at the $2L$ $(= 34)$ sigma points to obtain $2L$ estimates of the augmented observations as

$$\gamma^j_{t|t-1} = h(\chi^j_{t|t-1}) = 0 \vee (B\chi^j_{t|t-1} + d), \quad j = 1, \ldots, 2L. \tag{9.43}$$

These $2L$ sigma point results are then combined to obtain the predicted (here yield) measurements, measurements covariance matrix, and predicted state-measurement cross-covariance matrix

$$\begin{aligned} \hat{y}_{t|t-1} &= \sum_{j=0}^{2L} W^j_s \gamma^j_t \\ \Sigma_{y_t y_t} &= \sum_{j=0}^{2L} W^j_c \left[\gamma^j_t - \hat{y}_{t|t-1}\right]\left[\gamma^j_t - \hat{y}_{t|t-1}\right]' \\ \Sigma_{x_t y_t} &= \sum_{j=0}^{2L} W^j_c \left[\chi^j_{t|t-1} - \hat{x}_{t|t-1}\right]\left[\gamma^j_t - \hat{y}_{t|t-1}\right]', \end{aligned} \tag{9.44}$$

where the weights W^j for combining sigma point estimates (predictions) are potentially different for the state vector and the covariance matrices. They are given by

$$\begin{aligned} &W^0_s := \tfrac{\lambda}{L+\lambda} \quad W^0_c := \tfrac{\lambda}{L+\lambda} + (1 - \alpha^2 + \beta) \\ &W^j_s := W^j_c := \tfrac{1}{2(L+\lambda)} \quad j = 1, \ldots, 2L. \end{aligned} \tag{9.45}$$

Here β is related to the higher moments of the state vector distribution and is usually set to 2, which is optimal for Gaussian innovations.

These results are used to compute UKF Kalman *gain*

$$K_t := \Sigma_{x_t y_t} \Sigma^{-1}_{y_t y_t}, \tag{9.46}$$

which, defining $v_t := y^{\text{obs}}_t - \hat{y}_{t|t-1} = y^{\text{obs}}_t - B\hat{x}_{t|t-1} - d$, gives the updated state estimate in observation prediction error feedback form as

$$\hat{x}_t := \hat{x}_{t|t-1} + K_t v_t \tag{9.47}$$

with updated state covariance matrix

$$\Sigma_t = \Sigma_{t|t-1} - K_t \Sigma_{y_t y_t} K_t. \tag{9.48}$$

As noted above, the choice of parameters (α, κ, β) for the UKF is very important and is not usually detailed in the literature (but see Turner and Rasmussen 2012). Some nonlinear models are known to exhibit UKF algorithm divergence with certain parameter values. The other problem is inefficient estimation because of excessive spread of the sigma points. The first issue is not a problem here, but we begin to address the last issue in the following section.

9.5.2 Quasi-maximum likelihood estimation

Parameter estimates in the approximate Black-corrected EM algorithm are calibrated from the current UHF data path prediction as before by maximizing the log-likelihood function (9.21)

$$\log \mathrm{L}(\Theta, H) = -\frac{TK}{2}\log 2\pi - \frac{1}{2}\sum_{t=1}^{T}\log\det F_t - \frac{1}{2}\sum_{t=1}^{T} v_t' F_t^{-1} v_t,$$

by alternating between the parameters Θ and H, except that now the measurement prediction errors in the last term of the log-likelihood are those of the UKF and the calculation of the F_t terms from Equations 9.24 and 9.25 uses the UKF state covariance matrices Σ_t of Equation 9.48.

9.5.3 Technical implementation

The Numerical Algorithms Group Ltd (NAG) routine g13 ejc with key parameter setting flexibility was used for the UKF implementation. The EM calibration code was implemented in C++ with quasi-MLE optimization using global search with DIRECT (Jones et al. 1993) followed by a local (approximate) conjugate gradient algorithm (Powell 1964) which does not require derivatives.

The Black EFM model calibration run times for currencies with 12 years of daily data are around 4.5 hours (scaling linearly with data length) on the system described next. This is approximately twice the time the basic KF takes for EFM model calibration to the same data on this system.

9.5.4 HPC implementation

The development was coded in C and C++ under Linux with the use of MPI functionality. Calculations are performed on 5 compute nodes with 32 cores in total and the following hardware:

- Node 1:
  Memory: 16 GB

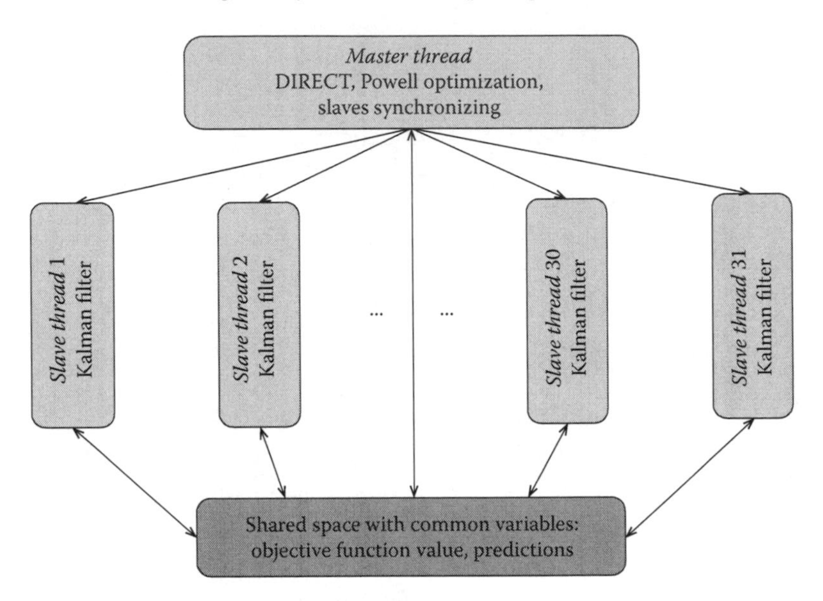

FIGURE 9.5: Parallelization schema.

4 x CPU Xeon (X5550) 2.66 GHz quad core
OS is Centos 5.7

- Nodes 2 to 5:
  4 x CPU Xeon (TM) 3 GHz
  Memory: 16 GB
  OS Centos 5.7

The DIRECT global optimization algorithm to cope with the nonunimodal likelihood function was implemented in parallel in a master–slave configuration with synchronicity. The Powell local optimization algorithm is not parallelizable.

The functionality actually parallelized was the full linear KF algorithm path estimation which is the basis of iteration step of the EM algorithm for the EKF algorithm. The *master* thread controls the optimization calculation synchronizing 31 *slave* threads (number of cores = number of threads = 32), that is, providing them at each optimization step with the current best step values of the log-likelihood *objective* function and calculated filter predictions (Figure 9.5).

When the Black EFM model UKF implementation is fully developed, we will migrate it to the cloud. We have already experimented with using the Amazon cloud for this model and have consulted previously on a compute intensive earlier Black model yield curve commercial development which uses this cloud (Dempster et al. 2014).

9.6 Empirical Evaluation of the Model In- and Out-of-Sample

This section contains the preliminary empirical evaluation of our approach to developing a robust long-term nonnegative yield curve model from the EFM Gaussian model using the Black correction. We will evaluate the Black-corrected EFM yield curve model against the original EFM model for the market data used to calibrate both models, both in-sample, for goodness-of-fit, and out-of-sample, for prediction accuracy.[9]

9.6.1 Data

We use a combination of LIBOR data and fixed interest rate swap rates (the ISDA fix) for each of 4 currency areas (EUR, GBP, USD, JPY) to bootstrap the yield curve *daily* for 14 maturities:

3 months, 6 months, 1 year, 2 years, 3 years, 4 years, 5 years, 6 years, 7 years, 8 years, 9 years, 10 years, 20 years, 30 years

In the case of the Swiss franc (CHF), only 12 maturities are available:

3 months, 6 months, 1 year, 2 years, 3 years, 4 years, 5 years, 6 years, 7 years, 8 years, 9 years, 10 years

The calibration periods used for these 5 currencies are the following:

EUR: 02.01.2001 to 02.01.2012
CHF: 02.01.2001 to 31.05.2013
GBP: 07.10.2008 to 31.05.2013
USD: 02.01.2001 to 31.05.2013
JPY: 30.03.2009 to 31.05.2013

After the 2012 Libor scandals, ICAP (formerly InterCapital Brokers) lost to ICE Benchmark Administration Limited its role as administrator for the ISDA fix rates, data collection, and calculation. Major reforms in the calculation methodology are being implemented (changing sources from polls of contributing banks to actual transaction quotes). This transfer process is not without difficulties for data providers.

The data was obtained from Bloomberg (indices US000**, BP000**, EE000**, JY000**, SF000** for LIBOR rates and USISDA**, BPISDB**, JYISDA**, SFISDA** for ISDAFIX rates).

[9] Longer term out-of-sample yield curve prediction has recently been independently found to be superior to the arbitrage-free Nelson–Siegel model of Christensen et al. (2011) widely used by central banks.

TABLE 9.2: Comparative model in-sample goodness-of-fit

Currency	Observation	Calibration	log-Likelihood	Sample Fit MSE (vol)(bp)
EUR	2817	EFM	232,652	15
		EFM UKF	252,500	17
		Black EFM $\sigma := 0.0025$	259,436	15
CHF	3100	EFM	232,100	8
		EFM UKF	250,391	10
		Black EFM $\sigma := 1.0$	253,095	8
GBR	1171	EFM	98,021	16
		EFM UKF	103,529	15
		Black EFM $\sigma := 0.0001$	105,368	14
USD	3093	EFM	279,114	15
		EFM UKF	280,745	25
		Black EFM $\sigma := 0.001$	292,954	22
JPY	950	EFM	91,014	6
		EFM UKF	84,564	28
		Black EFM $\sigma := 0.006$	102,544	6

9.6.2 Yield curve bootstrapping

Given the vector of parameters θ the Gaussian EFM model has rates (zero coupon bond yields) for maturity $\tau := T - t$ of the form

$$y(t,T) = \tau^{-1}[A(\tau,\theta)R_t + B(\tau,\theta)X_t + C(\tau,\theta)Y_t + D(\tau,\theta)]. \tag{9.49}$$

For each currency jurisdiction, we interpolate the appropriate market swap curve linearly to obtain swap rates at all maturities and then use 1-, 3-, and 6-month LIBOR rates and the swap curve to recursively back out a zero coupon bond yield curve for each day from the basic swap pricing equation (Ron 2000). This gives the input data for model calibration to give the parameter estimates $\hat{\theta}$.

9.6.3 In-sample goodness-of-fit

First, let us consider statistics for overall goodness-of-fit across the entire sample period for the five currency areas EUR, CHF, GPB, USD, and JPY, ordered by average rates in the data period from highest to lowest average rate. Table 9.2 shows the comparative goodness-of-fit, in terms of optimal log-likelihood and standard deviation (vol) of the sample measurement errors across all yields at the data maturities and all observations, of three models: the affine EFM estimated with both the KF and UKF and the Black EFM estimated with the UKF.

The table allows an overall comparison of the fitting errors of the original and Black-corrected models and also of the size of the fitting error relative

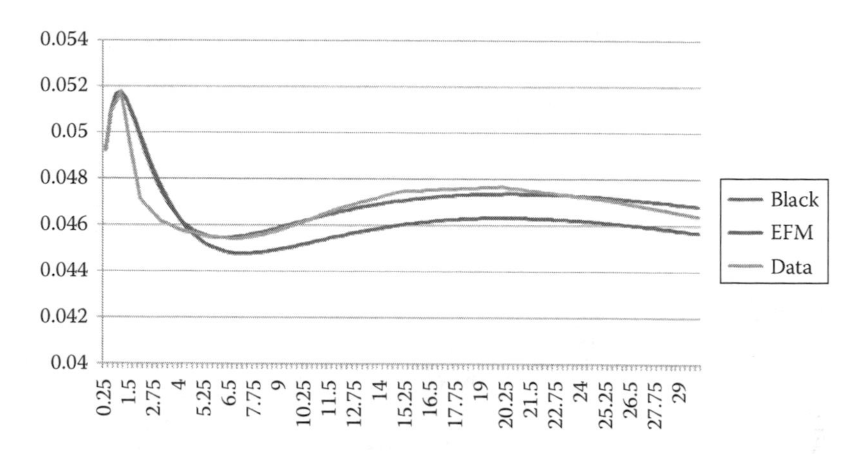

FIGURE 9.6: In-sample EUR yield curves on August 12, 2008.

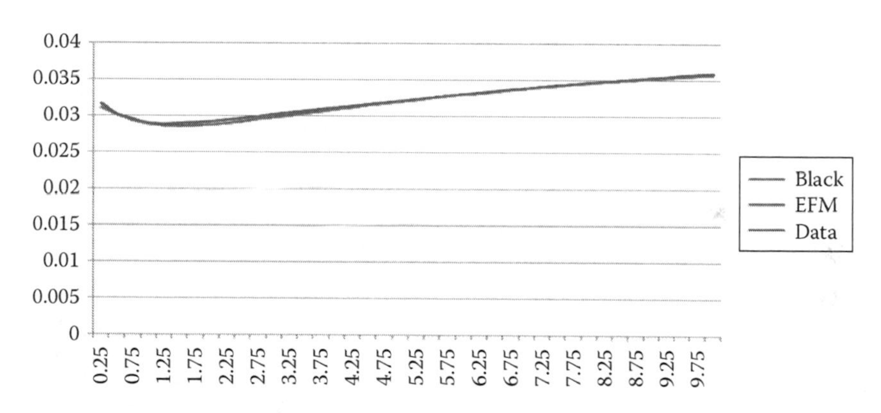

FIGURE 9.7: In-sample CHF yield curves on August 20, 2001.

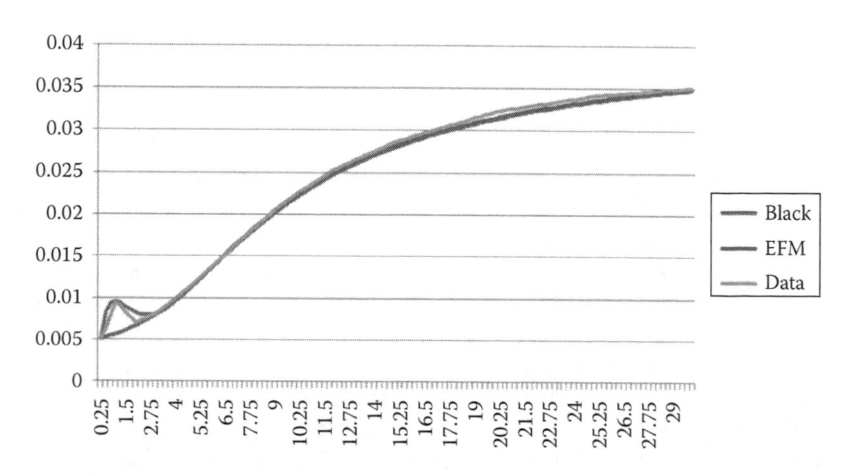

FIGURE 9.8: In-sample GBP yield curves on February 18, 2013.

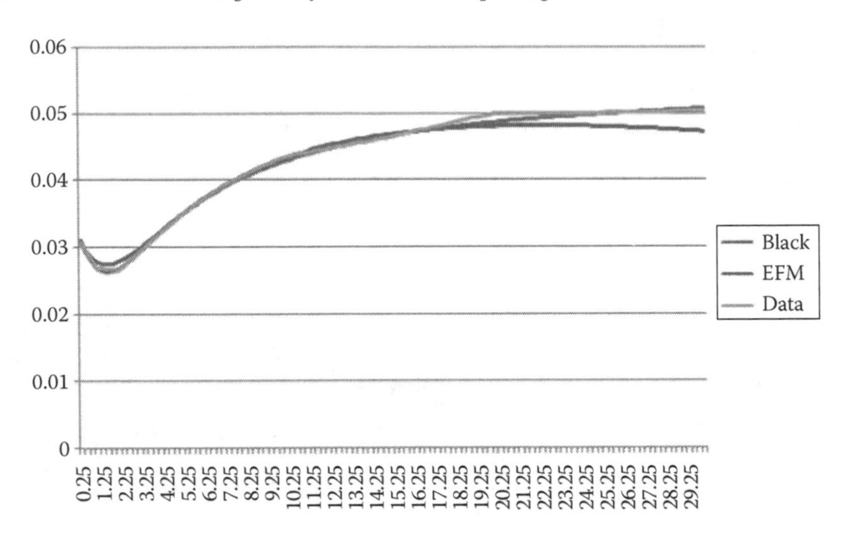

FIGURE 9.9: In-sample USD yield curves on October 14, 2008.

to the average level of rates in the currency area. From Table 9.2 and Figures 9.6 through 9.10, we can see that the total measurement error volatility of the best model fit is very respectably small for all currency areas, EUR, GBP, and USD being the highest and CHF and JPY the lowest. Moreover, for all but USD the Black-corrected EFM model goodness-of-fit equals or exceeds that of the EFM shadow rate model.

Although the three models in Table 9.2 all have the same parameter set, their likelihoods are not generally comparable as the models are not nested in the statistical sense. However, the likelihoods of the affine EFM model estimated with the KF and the UKF *are* comparable and in all cases, except for Japan, the UKF likelihood exceeds the KF likelihood, a reflection of the general power of the UKF widely attested to in the literature.

We may nevertheless compare the likelihoods achieved with the UKF for the affine EFM and nonlinear Black EFM models. Here the Black EFM likelihood exceeds that of the EFM in all cases. In terms of total measurement error standard deviation, the two models are also close, but the Black EFM again gives the lowest values.

We found that the NAG UKF code we used required careful tuning of the α parameter of Equation 9.42 which adjusts the displacement of sigma points from the central path for each data set. The calibration column of Table 9.2 shows that for all currencies except CHF small α parameter values closer to the generally recommended 10^{-3} are appropriate for the simple piecewise linear option "hockey stick" nonlinearity being handled here with the UHF. We are currently at a loss to explain the anomalous case with $\alpha = 1$, particularly since JPY has an even better overall fit than CHF. This suggests that in future it may be fruitful to consider the reparametrization of sigma point displacement and more generally to study the properties of the algorithm theoretically—which appears to be a lacunae in the literature to date. To this end, further

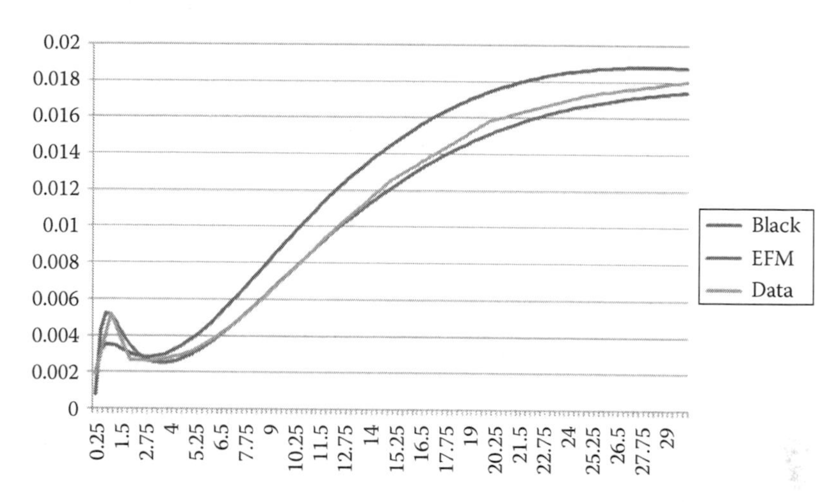

FIGURE 9.10: In-sample JPY yield curves on November 12, 2012.

experimentation with the current algorithm is called for. In these experiments, we should probably also consider using a nonzero β parameter in Equation 9.42 to reflect the positive skew in the Black EFM yield distributions, cf. Figures 9.7 through 9.11.

Turning to yield curve fits on specific days, Figures 9.6 through 9.10 show yield curve fits of the Black EFM model on representative days throughout the data period for all five currency areas relative to both the data and the original linear EFM alternative. (We have in fact developed software that can show these yield curve fits stepping through every (daily) observation in the data.) Each figure shows a typical good fit of the Black EFM model for a single currency. Root mean square error (in basis points) based on quarterly evaluation of the yield curve rates over 30 years (10 for CHF) is calculated using the model expression for the yield at each quarterly maturity in terms of the estimated parameters. These individual day figures are much smaller than the overall figures for the Black EFM model in Table 9.2 due to a relatively few very bad fits (for all three models) on certain days. Note however the generally nontextbook shapes of the yield curves in Figures 9.6 through 9.10, where the Black EFM outperforms the affine EFM by following yield curve distortions more closely.

Overall, the Black EFM model is broadly an improvement on the original EFM model in terms of in-sample yield curve fit in all five jurisdictions. However, USD is in general fit worse than EUR and GBP by both models and much worse than CHF and JPY. However, for GBP, USD, and JPY, the Black model fits the short-end kinks (see Figures 9.8 through 9.10) significantly better than the original model (although both models are based on three factors). It should be noted that such nontextbook yield curve shapes in the data period may reflect a behavioral market excess demand for short-term bonds or have resulted from market manipulation, or both.

EUR 10-year rate EFM with market data up to 15 July 2015

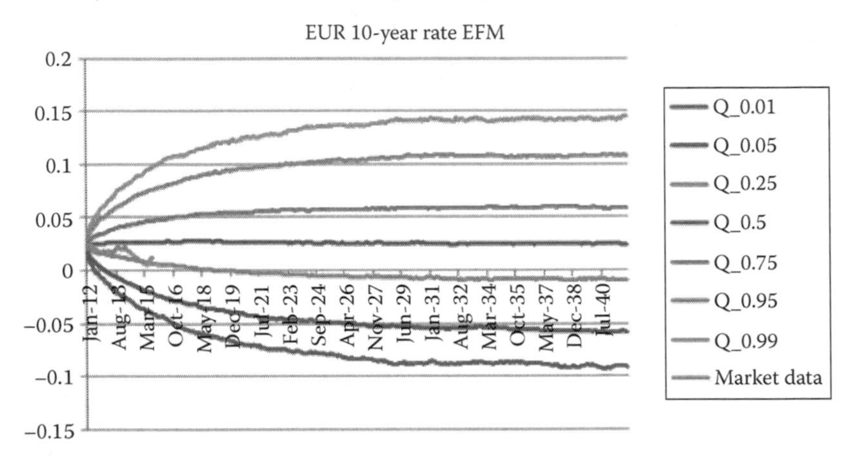

EUR 10-year rate Black EFM with market data up to 15 July 2015

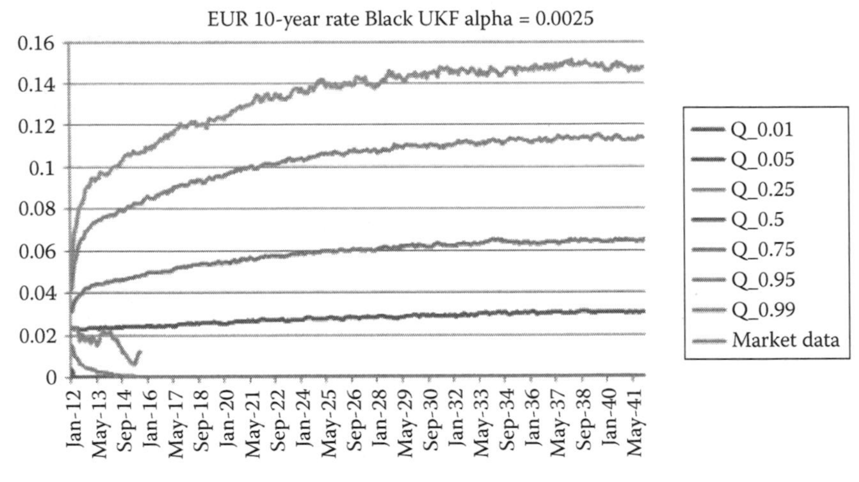

EUR 10-year rate forecast RMSE over 44 months

Black median	0.89%
EFM median	1.13%

FIGURE 9.11: Out-of-sample EUR 10-year rate 30 year projections on May 31, 2013.

9.6.4 Out-of-sample Monte Carlo projection

Figures 9.11 through 9.14 show the results of monthly out-of-sample Monte Carlo scenario projection over a 30-year horizon using the EFM and Black EFM models calibrated to the last day of the data period, May 31, 2013. These are for 10-year rates in all 5 currency areas. The figures show the evolution

CHF 10-year rate EFM with market data up to 11 December 2014

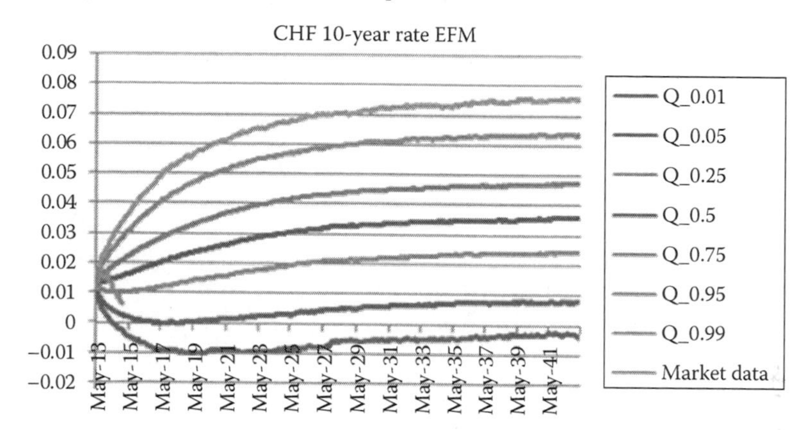

CHF 10-year rate Black EFM with market data up to 11 December 2014

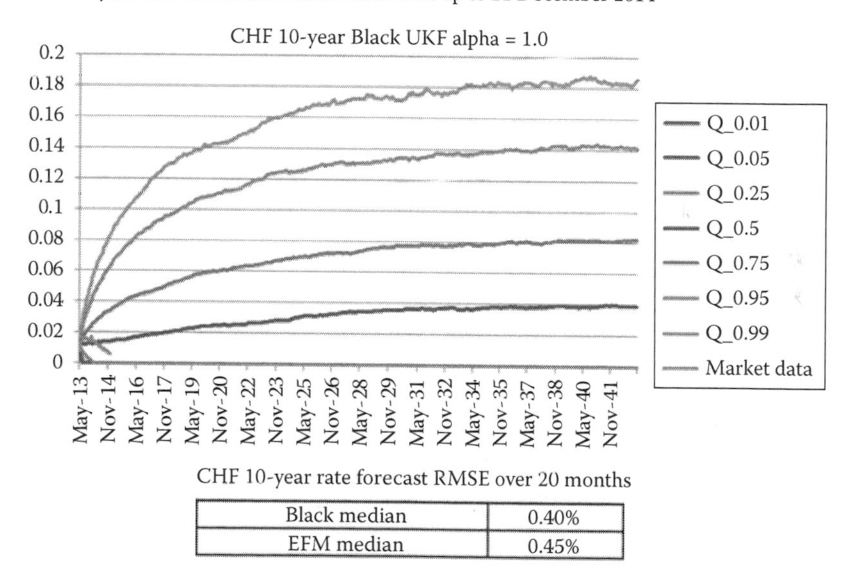

CHF 10-year rate forecast RMSE over 20 months

Black median	0.40%
EFM median	0.45%

FIGURE 9.12: Out-of-sample CHF 10-year rate 30 year projections on May 31, 2013.

of the paths of the quartiles and 1% and 5% tails of the 10,000 scenario distribution. The actual market data evolution is also plotted on these figures up to a more recent date for each currency area: EUR and CHF, December 11, 2014; GPB and USD, January 15, 2015; and JPY, January 24, 2014.[10] The out-of-sample 10-year rate median forecast root mean square error relative to the market realization is also shown in the figures for EUR, CHF,

[10]After which date, Bloomberg dropped JPY CMS swap data.

GBP 10-year rate EFM with market data up to 15 July 2015

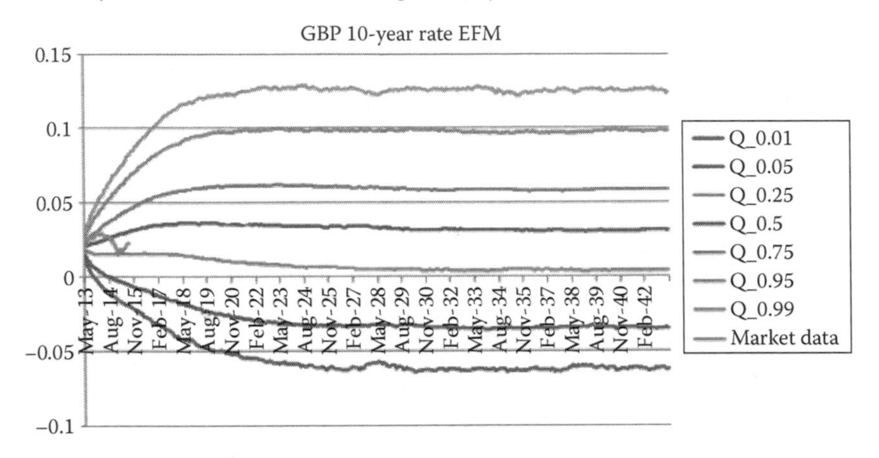

GBP 10-year rate Black EFM with market data up to 15 July 2015

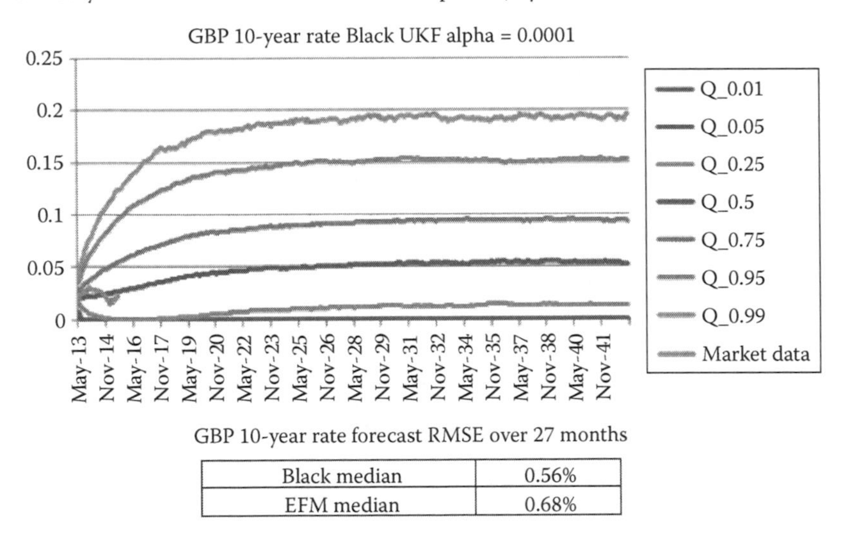

GBP 10-year rate forecast RMSE over 27 months

Black median	0.56%
EFM median	0.68%

FIGURE 9.13: Out-of-sample GBP 10-year rate 30 year projections on May 31, 2013.

GBP, and USD, which are naturally higher than the comparable in-sample figures. Surprisingly these are best for the poorest in-sample fitting USD, for which, as for CHF, the Black EFM model prediction is superior to that of EFM.[11]

These figures demonstrate the basic negative scenario generation problem with the Gaussian EFM model (cf. Dempster et al. 2014) and the primary

[11] The data series for JPY was deemed too short at 8 months to be significant.

USD 10-year rate EFM with market data up to 15 July 2015

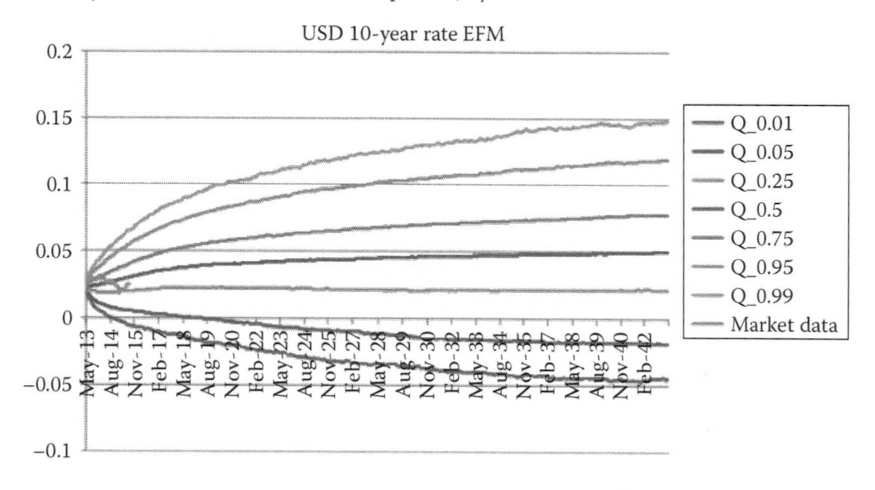

USD 10-year rate Black EFM with market data up to 15 July 2015

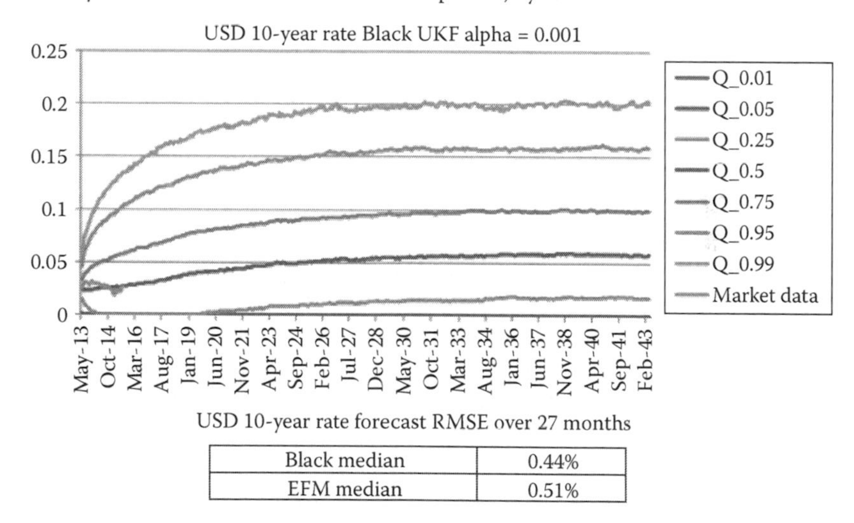

USD 10-year rate forecast RMSE over 27 months

Black median	0.44%
EFM median	0.51%

FIGURE 9.14: Out-of-sample USD 10-year rate 30 year projections on May 31, 2013.

effectiveness of the nonnegative Black correction for the long-standing low-rate Japanese economy (Figure 9.15). By way of comparison of the two- model stochastic predictions, it should be noted that the spread of the 10-year rate scenario distributions as time evolves for the Black EFM model produces a wider, perhaps more realistic, spread than the diffusion-based affine EFM for the 10-year rate over a 30-year horizon for all economies except the Japanese. Intuitively, this is likely due to the pushing up to nonnegativity of the negative scenarios of the shadow EFM model in the Black EFM model, but a deeper theoretical understanding is left to future research.

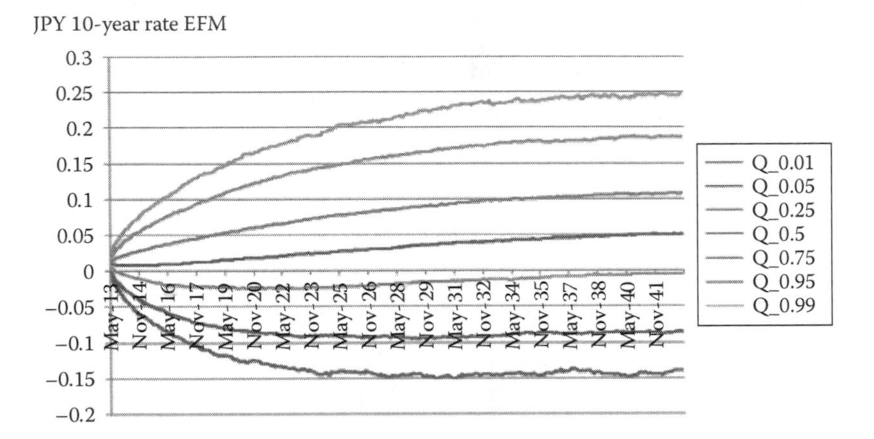

JPY 10-year rate Black EFM

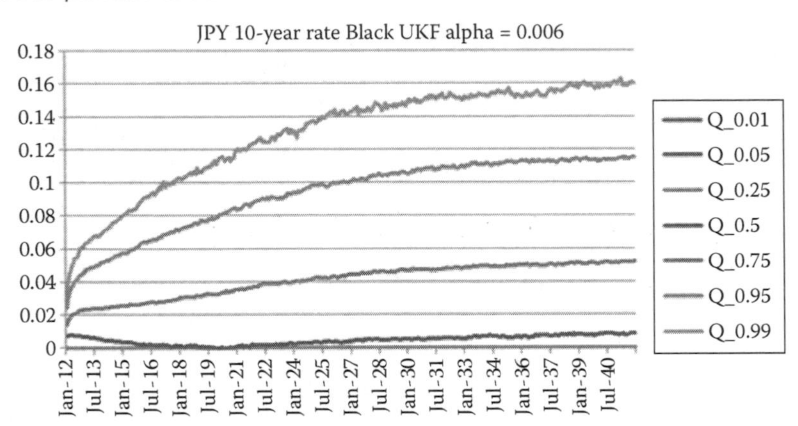

FIGURE 9.15: Out-of-sample JPY 10-year rate 30 year projections on May 31, 2013.

9.7 Conclusion

This chapter explains the initial development and evaluation of a new approximation of the Black (1995) correction to ensure nonnegative nominal rates at all maturities for a practically effective Gaussian three-factor affine yield curve model—the EFM model. Perhaps the most important feature of this novel approach is the demonstrated fact that the HPC calibration of the Black EFM model can be effected in only about double the runtime of that of the underlying shadow rate EFM model. Although some issues with using the UKF code have been identified here for further work, the results presented in this paper are promising, both in- and out-of-sample. We are confident that addressing the identified issues in future research can result in a deeper understanding of both the Black correction and the UKF.

Acknowledgments

The research leading to these results has received funding from the European Union Seventh Framework Programme (FP7/2007–13) under grant agreement no. 289032 (HPC Finance). The authors wish to acknowledge support and helpful comments from John Holden and Martyn Byng of the Numerical Algorithms Group, Grigorios Papamanousakis of Aberdeen Asset Management and, particularly, Giles Thompson, senior associate of Cambridge Systems Associates.

Appendix 9A

Given an initial set of parameters (Θ_0, H_0), the EM algorithm for estimating the parameters of the EFM model from market data using the Kalman filter alternates between generating paths with the filter for the log-likelihood function and optimizing this function in the model parameters.

Defining $O(\Theta, \mathrm{H}) := \log \mathrm{L}(\Theta, \mathrm{H})$, a single step of the EM algorithm for quasi-MLE can be presented in pseudo code as follows:

Calculation of the log-likelihood function

1. **Input** (Θ_0, H_0)
2. **for** $t = 1$ to T **do**
3. KF predictions (9.19)
4. KF update (9.20)
5. Calculate a term of the log-likelihood function (9.21)
6. **end for**
7. Compute $O(\Theta, \mathrm{H})$ (9.21)
8. **Output** $O(\Theta, \mathrm{H})$

Optimization of the log-likelihood function

The two-phase optimization algorithm is as follows:

1. Initialize parameters (Θ, H) from previous EM algorithm step
2. **while** $|\Delta O(\Theta, \mathrm{H})| \geq tolerance_1$ **do**
3. DIRECT optimization step of $O(\Theta, \mathrm{H})$ with H fixed
4. DIRECT optimization step of $O(\Theta, \mathrm{H})$ with Θ fixed
5. **end while**

6. **while** $|\Delta O(\Theta, \mathrm{H})| \geq tolerance_2$ **do**
7. Powell optimization of $O(\Theta, \mathrm{H})$ with H fixed
8. Powell optimization of $O(\Theta, \mathrm{H})$ with Θ fixed
9. **end while**

References

Andreasen, M. A. and Meldrum, A. 2014. Dynamic term structure models: The best way to enforce the zero lower bound. CREATES Research Paper 2014–47.

Bauer, M. D. and Rudebusch, G. D. 2013. The shadow rate, Taylor rules and monetary policy lift-off. Working Paper, Federal Reserve Bank of San Francisco, February 2013. https://www.economicdynamics.org/meetpapers/2013/paper_691.pdf

Bauer, M. D. and Rudebusch, G. D. 2014. Monetary Policy Expectations at the Zero Lower Bound: Federal Reserve Bank of San Francisco Working Paper 2013–18.

Black, F. 1995. Interest rates as options. *Journal of Finance* 50, 1371–1376.

Bomfim, A. N. 2003. Interest rates as options: Assessing the markets view of the liquidity trap. Working Paper 2003–45, Finance and Economics Discussion Series, Federal Reserve Board, Washington, DC.

Carton de Wiart, B. and Dempster, M. A. H. 2011. Wavelet optimized valuation of financial derivatives. *International Journal of Theoretical and Applied Finance* 14, 1113–1137.

Christensen, J. H. E. 2015. A regime-switching model of the yield curve at the zero bound. Federal Reserve Bank of San Francisco Working Paper 2013–34.

Christensen, J., Diebold, F., and Rudebusch, G. D. 2011. The affine arbitrage-free class of Nelson–Siegel term structure models. Working Paper, Federal Reserve Bank of San Francisco. *Journal of Econometrics* 164, 4–20.

Christensen, J. H. E. and Rudebusch, G. D. 2013. Modeling Yields at the Zero Lower Bound: Are Shadow Rates the Solution. Federal Reserve Bank of San Francisco, working paper, December 2013, http://www.frbsf.org/economic-research/files/wp2013-39.pdf

Christensen, J. H. E. and Rudebusch, G. D. 2015. Estimating shadow-rate term structure models with near-zero yields. *Journal of Financial Econometrics* 13, 226–259.

Christoffersen, P., Dorion, C., Jacobs, K., and Karoui, L. 2014. Nonlinear Kalman filtering in affine term structure models: CREATES Research Paper 14–04, Aarhus University, January 2014.

Cox, J., Ingersoll, J., and Ross, S. 1985. A theory of the term structure of interest rates. *Econometrica* 53, 363–384.

Dai, Q. and Singleton, K. J. 2000. Specification analysis of affine term structure models. *Journal of Finance* 50, 1943–1978.

Davydov, D. and Linetsky, V. 2003. Pricing options on scalar diffusions: An eigenfunction expansion approach. *Operations Research* 51185209.

Dempster, A. P., Laird, N. M., and Rubin, D. B. 1977. Maximum likelihood for incomplete data via the EM algorithm. *Journal of the Royal Statistical Society* 39, 1–38.

Dempster, M. A. H., Evans, J., and Medova, E. A. 2014. Developing a practical yield curve model: An odyssey. In *New Developments in Macro-Finance Yield Curves.* Chadha, J., Durre, A., Joyce, M., and Sarnio, L., eds., Cambridge University Press, Cambridge, pp. 251–290.

Dempster, M. A. H., Medova, E. A., and Villaverde, M. 2010. Long-term interest rates and consol bond valuation. *Journal of Asset Management* 11, 113–135.

Dempster, M. A. H. and Tang, K. 2011. Estimating exponential affine models with correlated measurement errors: Applications to fixed income and commodities. *Journal of Banking and Finance* 35(3), 639–652.

Diebold, F. X. and Rudebusch, G. D. 2013. *Yield Curve Modelling and Forecasting—The Dynamic Nelson–Siegel Approach.* Princeton University Press, Princeton.

Dothan, M. 1978. On the term structure of interest rates. *Journal of Financial Economivs* 7, 229–264.

Duffie, J. D. and Kan, R. 1996. A yield-factor model of interest rates. *Mathematical Finance* 6, 379–406.

Feldhutter, P., Heyerdahl-Larsen, C., and Illeditsch, P. 2015. Risk premia and volatilities in a non-linear term structure model. Working Paper, The Wharton School, University of Pennsylvania. http://www.ssrn.com/abstracts=2242280

Gorovoi, V. and Linetsky, V. 2004. Black's model of interest rates as options, eigenfunction expansions and Japanese interest rates. *Mathematical Finance* 14(1), 49–78.

Harvey, A. C. 1993. *Time Series Models,* Second edition. Harvester-Wheatsheaf, Hemel Hempstead.

Ho, T. and Lee, S. 1986. Term structure movements and pricing interest rate contingent claims. *Journal of Finance* 41, 1011–1029.

Homer, S. and Sylla, R. 2005. *A History of Interest Rates.* Wiley, Hoboken, NJ, p.1.

Hull, J. and White, A. 1990. Pricing interest rate derivative securities. *Review of Financial Studies* 3, 573–592.

Ichiue, H. and Ueno, Y. 2006. Monetary policy and the yield curve at zero interest: The macro-finance model of interest rates as options. Working Paper 06-E-16, Bank of Japan. https://www.boj.or.jp/en/research/wps_rev/wps_2006/data/wp06e16.pdf

Ichiue, H. and Ueno, Y. 2007. Equilibrium interest rates and the yield curve in a low interest rate environment. Working Paper 07-E-18, Bank of Japan.

Ichiue, H. and Ueno, Y. 2013. Estimating term premia at zero bound: An analysis of Japanese, US and UK yields. Working Paper 13-E-8, Bank of Japan.

Jameson, L. 1998. A wavelet-optimized, very high order adaptive grid and order numerical method. *SIAM Journal on Scientific Computing* 19, 1980–2013.

James, J. and Webber, N. 2000. *Interest Rate Modelling.* Wiley, Chichester.

Jones, D. R., Perttunen, C. D., and Stuckmann, B. E. 1993. Lipschitzian optimization without the Lipschitz constant. *Journal of Optimization Theory and Applications* 79, 157–181.

Jong, F. de, Driessen, J., and Pelsser, A. 2001. Libor market models versus swap market models for pricing interest rate derivatives: An empirical analysis. *European Finance Review* 5, 201–237.

Joslin, S., Singleton, K. J., and Zhu, H. 2011. A new perspective on Gaussian dynamic term structure models. *Review of Financial Studies* 24, 926–970.

Julier, S. J., Uhlmann, J. K., and Durrant-Whyte, H. 1995. A new approach for filtering nonlinear systems. In *Proceedings of the American Control Conference*, 1628–1632.

Julier, S. J. and Uhlmann, J. K. 1997. A new extension of the Kalman filter to nonlinear systems. *International Symposium on Aerospace/Defense Sensing, Simulation and Control, Signal Processing, Sensor Fusion, and Target Recognition VI* 3, 182–193.

Kim, D. H. and Priebsch, M. A. 2013. Estimation of multi-factor shadow-rate term structure models. Working Paper, Federal Reserve Board, October 9, 2013.

Kim, D. H. and Singleton, K. J. 2011. Term structure models and the zero bound: An empirical investigation of Japanese yields. *Journal of Econometrics* 170, 32–49.

Krippner, L. 2012. Modifying Gaussian term structure models when interest rates are near the zero lower bound. Discussion Paper 2012/02, Reserve Bank of New Zealand.

Krippner, L. 2013a. A tractable framework for zero lower bound Gaussian term structure models. CAMA Working Paper No. 49/2013, Australian National University.

Krippner, L. 2013b. Faster solutions for Black zero lower bound term structure models. CAMA Working Paper No. 66/2013, Australian National University.

Krippner, L. 2014. Measuring the stance of monetary policy in conventional and unconventional environments. CAMA Working Paper No. 6/2014, Australian National University.

Lemke, W. and Vladu, A. L. 2014. A shadow-rate term structure for the Euro Area. www.ecb.europa.eu/events/pdf/conferences/140908/lemke_vladu.pdf.

Linetsky, V. 2002. Exotic spectra. *RISK*, April 2002, 85–89.

Linetsky, V. 2004. The spectral decomposition of the option value. *International Journal of Theoretical and Applied Finance* 7, 337–384.

Lipton, A., Gal, A., and Lasis, A. 2014. Pricing of vanilla nad first-generation exotic options in the local stochastic volatility framework: Survey and results. *Quantitative Finance* 14, 1899–1922.

Litterman, R. and Scheinkman, J. A. 1991. Common factors affecting bond returns. *Journal of Fixed Income* 1, 54–61.

Medova, E. A., Rietbergen, M. I., Villaverde, M., and Yong, Y. S. 2005. Modelling the long term dynamics of yield curves. Working Paper 09/2005, Centre for Financial Research, Judge Business School, University of Cambridge.

Nawalkha, S. K. and Rebonato, R. 2011. What interest rate models to use? Buy side versus sell side. *SSRN Electronic Journal*, 01/2011.

Nelder, J. A. and Mead, R. 1965. A simplex method for function minimization. *Computer Journal* 7, 308–313.

Nelson, C. R. and Siegel, A. F. 1987. Parsimonious modeling of yield curves. *Journal of Business* 60, 473–489.

Powell, J. D. 1964. An efficient method of finding the minimum of a function of several variables without calculating derivatives. *Computer Journal* 11, 302–304.

Priebsch, M. A. 2013. Computing arbitrage-free yields in multi-factor Gaussian shadow-rate term structure models. Working Paper 2013–63, Finance and Economics Discussion Series, Federal Reserve Board.

Rebonato, R. 2015. Review of Yield Curve Modelling and Forecasting—The Dynamic Nelson–Siegel Approach by Diebold, F. X. and Rudebusch, G. D. *Quantitative Finance* 15, 1609–1612.

Rebonato, R. and Cooper, I. 1995. The limitations of simple two-factor interest rate models. *Journal of Financial Engineering* 5, 1–16.

Richard, S. F. 2013. A non-linear macroeconomic term structure model. Working Paper, University of Pennsylvania.

Ron, U. 2000. A practical guide to swap curve construction. Working Paper 2000–17, Bank of Canada, August 2000.

Swanson, E. T. and Williams, J. C. 2013. Measuring the effect of the zero lower bound on medium- and longer-term interest rates. Working Paper, Federal Reserve Bank of San Francisco, January 2013. http://www.frbsf.org/economic-research/files/wp12-02bk.pdf

Turner, R. and Rasmussen, C. E. 2012. Model based learning of sigma points in unscented Kalman filtering. *Neurocomputing* 80, 47–53.

Ueno, Y., Baba, N., and Sakurai, Y. 2006. The use of the Black model of interest rates as options for monitoring the JGB market expectations. Working Paper 06E15, Bank of Japan. https://www.boj.or.jp/en/research/wps_rev/wps_2006/data/wp06e15.pdf

Vasicek, O. 1977. An equilibrium characterization of the term structure. *Journal of Financial Economics* 5, 177–188.

Wu, J. C. and Xia, F. D. 2014. Measuring the macroeconomic impact of monetary policy at the zero lower bound. Working Paper, Booth School of Business, University of Chicago, July 2014. http://faculty.chicagobooth.edu/jing.wu/research/pdf/wx.pdf

Yong, Y. S. 2007. Scenario Generation for Dynamic Fund Management. PhD Thesis, Centre for Financial Research, Judge Business School, University of Cambridge.

Chapter 10

Algorithmic Differentiation

Uwe Naumann, Jonathan Hüser, Jens Deussen, and Jacques du Toit

CONTENTS

10.1 Motivation

Inspired by Giles and Glasserman [1], Algorithmic Differentiation (AD) [2,3] has been gaining popularity in computational finance over recent years. Adjoint AD (AAD) in particular facilitates a paradigm shift in financial modelling through provision of first-order sensitivities at a relative computational cost which is independent of the number of sensitivities asked for.

For illustration, we consider a simple European call option written on an underlying driven by a local volatility process. Let $S = (S_t)_{t\geq 0}$ be the solution to the stochastic differential equation (SDE)

$$\mathrm{d}S_t = rS_t\,\mathrm{d}t + \sigma\left(\log(S_t), t\right) S_t\,\mathrm{d}W_t, \tag{10.1}$$

where $W = (W_t)_{t \geq 0}$ is a standard Brownian motion, $r > 0$ is the risk-free interest rate, and σ is the local volatility function. The price of the call option is then given by

$$V = e^{-rT}\mathbb{E}(S_T - K)^+ \tag{10.2}$$

for maturity $T > 0$ and strike $K > 0$. The local volatility $\sigma = \sigma(x, t)$ is computed from the market observed implied volatility surface using bicubic spline interpolation.

To compute the call price V from Equation 10.2, we apply a simple Euler–Maruyama scheme to the log process $X_t = \log(S_t)$ which satisfies the SDE

$$\mathrm{d}X_t = \left(r - \tfrac{1}{2}\sigma^2(X_t, t)\right)\mathrm{d}t + \sigma(X_t, t)\,\mathrm{d}W_t. \tag{10.3}$$

Setting $\Delta = T/M$ for some integer M and defining a sequence of Monte Carlo time steps $t_i = i\Delta$ for $i = 1, 2, \ldots, M$, we set

$$X_{t_{i+1}} = X_{t_i} + \left(r - \tfrac{1}{2}\sigma^2(X_{t_i}, t_i)\right)\Delta + \sigma(X_{t_i}, t_i)\sqrt{\Delta}Z_i, \tag{10.4}$$

where each Z_i is a standard normal random number and $X_{t_0} = \log(S_0)$. N sample paths of the log process are generated and used in a Monte Carlo integral of Equation 10.2 to estimate the price V and to obtain a confidence interval.

In Figure 10.1, we compare run times of central finite difference (CFD) approximation and AAD for the computation of sensitivities of V with respect to the input parameters K, T, r, S_0, and a set of $n-4$ market observed implied volatilities for $N = 10^5$, $M = 360$, and growing values of n. The cost of CFD relative to the cost of a single price calculation scales with n. The AAD solution

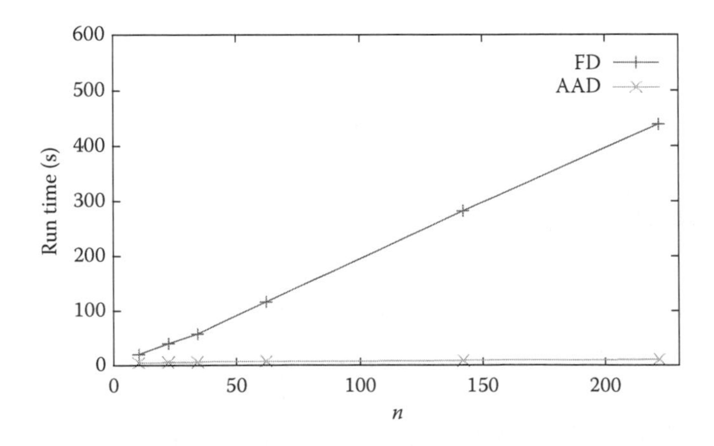

FIGURE 10.1: Motivation.

exhibits an essentially constant relative cost of less than 10 price calculations independent of the size of n. The AD tool dco/c++ [4] is used throughout this chapter.

10.2 Introduction

AD is a semantic transformation[1] of a given computer code called the *primal code* or *primal function*. In addition to computing the primal function value, the transformed code also computes the derivatives of the primal function with respect to a specified set of parameters.

Consider a computer implementation of a function F mapping $\mathbb{R}^n \times \mathbb{R}^{\tilde{n}}$ into $\mathbb{R}^m \times \mathbb{R}^{\tilde{m}}$. We are interested in derivatives of the first m outputs of F (the *active outputs*) with respect to the first n inputs (the *active inputs*). The second $\tilde{m}$ outputs and second $\tilde{n}$ inputs of F are termed the *passive outputs* and *passive inputs*, respectively. For example, an active output may be the Monte Carlo price of an option while a passive output may be the corresponding confidence interval. An active input may be the initial asset price S_0, while a passive input may be the set of random numbers used in the Monte Carlo simulation. Without loss of generality and to keep the notation simple, we restrict the discussion to scalar active outputs, that is, $m = 1$.[2] We therefore consider multivariate functions of the type

$$F : \mathbb{R}^n \times \mathbb{R}^{\tilde{n}} \to \mathbb{R} \times \mathbb{R}^{\tilde{m}}, \quad (y, \tilde{\mathbf{y}}) = F(\mathbf{x}, \tilde{\mathbf{x}}), \tag{10.5}$$

where we assume that F and its computer implementation are differentiable up to the required order.[3] We are interested in (semi-)automatic ways to generate the vector of all partial derivatives of the active output y with respect to the active inputs $\mathbf{x}$, that is, the gradient

$$\nabla F = \nabla F(\mathbf{x}, \tilde{\mathbf{x}}) \equiv \left(\frac{\partial y}{\partial x_i}\right)^T_{i=0,\ldots,n-1} \in \mathbb{R}^{1\times n}, \tag{10.6}$$

along with the values of all active and passive outputs as functions of the active and passive inputs. Similarly, we look for the Hessian of all second partial derivatives of y with respect to $\mathbf{x}$, that is,

$$\nabla^2 F = \nabla^2 F(\mathbf{x}, \tilde{\mathbf{x}}) \equiv \left(\frac{\partial^2 y}{\partial x_j \partial x_i}\right)_{j,i=0,\ldots,n-1} \in \mathbb{R}^{n\times n}. \tag{10.7}$$

[1] That is, changes the meaning.

[2] The modifications for $m > 1$ are straightforward.

[3] Differentiability is a crucial prerequisite for (algorithmic) differentiation. Nondifferentiability at selected points can be handled by smoothing techniques as well as by combinations of AD with local finite difference approximations (see Section 10.3.3).

10.2.1 Tangent mode AD

In the following, we use the notation from Reference 3. *Tangent (also: forward) mode AD* yields the tangent function

$$F^{(1)} : \mathbb{R}^n \times \mathbb{R}^n \times \mathbb{R}^{\tilde{n}} \to \mathbb{R} \times \mathbb{R} \times \mathbb{R}^{\tilde{m}}.$$

A corresponding tangent code implements $(y, y^{(1)}, \tilde{\mathbf{y}}) := F^{(1)}(\mathbf{x}, \mathbf{x}^{(1)}, \tilde{\mathbf{x}})$, where

$$y^{(1)} := \nabla F(\mathbf{x}, \tilde{\mathbf{x}}) \cdot \mathbf{x}^{(1)} \tag{10.8}$$

is the directional derivative of F in direction $\mathbf{x}^{(1)}$. We use superscripts $^{(1)}$ to denote first-order tangents. The operator $:=$ represents the assignment of imperative programming languages, not to be confused with equality $=$ in the mathematical sense. The entire gradient can be calculated entry by entry with n runs of the tangent code through a process known as *seeding* and *harvesting*. The vector $\mathbf{x}^{(1)}$ in Equation 10.8 is successively set equal to each of the Cartesian basis vectors in $\mathbb{R}^n$ (it is *seeded*), the tangent code is run, and the corresponding gradient entry is *harvested* from $y^{(1)}$. The gradient is computed with machine accuracy while the computational cost is $O(n) \cdot \text{Cost}(F)$, the same as that of a finite difference approximation.

10.2.2 Adjoint mode AD

Adjoint (also: reverse) mode AD yields the adjoint function

$$F_{(1)} : \mathbb{R}^n \times \mathbb{R}^n \times \mathbb{R}^{\tilde{n}} \times \mathbb{R} \to \mathbb{R} \times \mathbb{R}^{\tilde{m}} \times \mathbb{R}^n.$$

A corresponding adjoint code implements $(y, \tilde{\mathbf{y}}, \mathbf{x}_{(1)}) := F_{(1)}(\mathbf{x}, \mathbf{x}_{(1)}, \tilde{\mathbf{x}}, y_{(1)})$, where

$$\mathbf{x}_{(1)} := \mathbf{x}_{(1)} + \nabla F(\mathbf{x}, \tilde{\mathbf{x}})^T \cdot y_{(1)}. \tag{10.9}$$

The adjoint code therefore increments given adjoints $\mathbf{x}_{(1)}$ of the active inputs with the product of the gradient and a given adjoint $y_{(1)}$ of the active output (see Reference 3 for details). We use subscripts $_{(1)}$ to denote first-order adjoints. Initializing (seeding) $\mathbf{x}_{(1)} = \mathbf{0}$ and $y_{(1)} = 1$ yields the gradient in $\mathbf{x}_{(1)}$ from a single run of the adjoint code. Again, the gradient is computed with machine accuracy. The computational cost no longer depends on the size of the gradient n.

10.2.3 Second derivatives

The second derivative is the first derivative of the first derivative. Second derivative code can therefore be obtained by any of the four combinations of tangent and adjoint modes. In *tangent-over-tangent mode AD* tangent mode is applied to Equation 10.8 yielding the second-order tangent function

$$F^{(1,2)} : I\!R^n \times I\!R^n \times I\!R^n \times I\!R^n \times I\!R^{\tilde{n}} \to I\!R \times I\!R \times I\!R \times I\!R \times I\!R^{\tilde{m}}.$$

A corresponding second-order tangent code implements

$$(y, y^{(2)}, y^{(1)}, y^{(1,2)}, \tilde{\mathbf{y}}) := F^{(1,2)}(\mathbf{x}, \mathbf{x}^{(2)}, \mathbf{x}^{(1)}, \mathbf{x}^{(1,2)}, \tilde{\mathbf{x}}),$$

where

$$\begin{aligned} y^{(2)} &:= \nabla F(\mathbf{x}, \tilde{\mathbf{x}}) \cdot \mathbf{x}^{(2)} \\ y^{(1,2)} &:= \mathbf{x}^{(1)^T} \cdot \nabla^2 F(\mathbf{x}, \tilde{\mathbf{x}}) \cdot \mathbf{x}^{(2)} + \nabla F(\mathbf{x}, \tilde{\mathbf{x}}) \cdot \mathbf{x}^{(1,2)}. \end{aligned} \tag{10.10}$$

The computational cost of Hessian accumulation is of the same order as that of a second-order finite difference approximation. Obviously, the accuracy of the latter is typically far from satisfactory, particularly for calculations performed in single precision floating-point arithmetic.

Symmetry of the Hessian implies mathematical equivalence of the three remaining second-order adjoint modes (see Reference 3 for details). We therefore focus on one of them. In *tangent-over-adjoint mode AD* tangent mode is applied to Equation 10.9 yielding the second-order adjoint function

$$F_{(1)}^{(2)} : I\!R^n \times I\!R^n \times I\!R^n \times I\!R^n \times I\!R^{\tilde{n}} \times I\!R \times I\!R \to I\!R \times I\!R \times I\!R^{\tilde{m}} \times I\!R^n \times I\!R^n.$$

A corresponding second-order adjoint code implements

$$\left(y, y^{(2)}, \tilde{\mathbf{y}}, \mathbf{x}_{(1)}, \mathbf{x}_{(1)}^{(2)}\right) := F_{(1)}^{(2)}\left(\mathbf{x}, \mathbf{x}^{(2)}, \mathbf{x}_{(1)}, \mathbf{x}_{(1)}^{(2)}, \tilde{\mathbf{x}}, y_{(1)}, y_{(1)}^{(2)}\right),$$

where

$$\begin{aligned} y^{(2)} &:= \nabla F(\mathbf{x}, \tilde{\mathbf{x}}) \cdot \mathbf{x}^{(2)} \\ \mathbf{x}_{(1)}^{(2)} &:= \mathbf{x}_{(1)}^{(2)} + y_{(1)} \cdot \nabla^2 F(\mathbf{x}, \tilde{\mathbf{x}}) \cdot \mathbf{x}^{(2)}. \end{aligned} \tag{10.11}$$

The full Hessian can be obtained by n runs of the second-order adjoint code. The reduction in computational complexity due to the initial application of adjoint mode to the primal code is therefore carried over to the second-order adjoint. Sparsity of the Hessian can and should be exploited [5].

10.2.4 Review of AD in finance

Starting with the paper of Giles and Glasserman [1] in 2006, AAD has increasingly been adopted in computational finance applications. Since then there have been several contributions to the literature, utilizing AAD for both the calculation of Greeks and for calibration.

Leclerc et al. extended the pathwise approach of Giles and Glasserman to Greeks for Bermudan-style derivatives [6]. Denson and Joshi [7] applied the pathwise AAD method for Greeks to a LIBOR market model and Joshi has several other contributions in the field with different collaborators [8–11]. Capriotti et al. applied AAD for fast Greeks in various different applications including PDEs, credit risk, Bermudan-style options, and XVA [12–16]. Antonov also used AAD for XVA and callable exotics [17,18]. Contributions in the area of calibration include those of Turinici [19], Kaebe et al. [20], Schlenkrich [21], and Henrard [22]. For Greeks in the context of discontinuous payoffs, Giles introduced the Vibrato Monte Carlo method [23]. The problem of discontinuity is also treated in works related to second-order Greeks [24,25].

The above is not meant as a complete review of AAD applications in finance. Further related work can be found in the references within the cited publications.

10.3 Implementation

AD constitutes a set of rules for deriving first- and higher-order tangent and adjoint versions of a primal numerical simulation code. Two fundamental modes of implementing AD are distinguished between source transformation and operator overloading.

Source transformation rewrites the given primal code yielding a corresponding derivative code usually in the same programming language. For example, consider an implementation of $f : \mathbb{R}^2 \rightarrow \mathbb{R}$, $y = f(\mathbf{x}) = e^{\sin(x_0 \cdot x_1)}$ as

```
void f(double x, double& y) { x*=y; y=exp(sin(x)) }
```

where x $\hat{=}\ x_0$ and y $\hat{=}\ x_1$ on input and with output y $\hat{=}\ y$. The first-order tangent version returns the directional derivative

$$y^{(1)} := \frac{\partial y}{\partial \mathbf{x}} \cdot \mathbf{x}^{(1)}$$

in addition to the function value, for example,

```
void tf(double x, double& tx, double& y, double &ty) {
  tx=tx*y+x*ty; x*=y;
  y=\exp(\sin(x)); ty=y*\cos(x)*tx.
}
```

Each arithmetic statement is augmented with its directional derivative (also tangent).

A first-order adjoint version returns the adjoint directional derivative

$$\mathbf{x}_{(1)} := \mathbf{x}_{(1)} + \frac{\partial y}{\partial \mathbf{x}}^{\mathrm{T}} \cdot y_{(1)}$$

in addition to the function value as

```
void af(double x, double& ax, double& y, double &ay) {
  // augmented forward section
  double s=x; x*=y; double v=exp(sin(x));
  // reverse section
  double av=ay*v*cos(x); ax+=av*y; ay=av*s; y=v.
}
```

Adjoints of all arithmetic statements are executed in reverse order (see reverse section of the adjoint code). They require intermediate values computed in the augmented forward section of the adjoint code. Some of these values may need to be recorded prior to getting lost due to overwriting, for example, s. An in-depth discussion of adjoint code generation rules is beyond the scope of this article. Refer to Reference 3 for a more detailed description. Manual source transformation turns out to be tedious, error-prone, and hard to maintain from a software evolution perspective. Preprocessing tools have been developed for many years providing reasonable coverage of Fortran and C in addition to various simpler special-purpose scripting languages [26].

Currently, there is no mature source transformation tool for C++. The method of choice for implementing AD for C++ programs is based on operator and function overloading typically combined with advanced meta-programming techniques using the dynamic typing mechanism provided by C++ templates. Instead of parsing the primal source code followed by unparsing a differentiated version, the semantics of operators and intrinsic functions are redefined. Overloading for a custom active data type yields augmented operations. For example, in basic tangent mode the active data type consists of a value and a directional derivative component. All operations are overloaded for such pairs yielding, for example, $(y, y^{(1)}) := (\sin(x), \cos(x) \cdot x^{(1)})$. In *basic adjoint mode*, the operations are overloaded to record a tape of the primal computation. Conceptually, the tape can be regarded as a directed acyclic graph with vertices representing the inputs to the program as well as all operations performed to compute its outputs. Edges represent data dependencies. They can be labeled with the local partial derivative of the value represented by its target vertex with respect to the value represented by its source. An example is shown in Figure 10.2a. Adjoints are computed by interpretation of the tape. The tape interpreter eliminates all intermediate vertices in reverse topological order by introducing new edges connecting all predecessor vertices with all successors. New edges are labeled with the product of the local partial derivatives on the corresponding incoming and outgoing edges. Parallel edges are merged by adding their labels. The vertex to be eliminated is removed together with its incident edges; see Figure 10.2b and c for elimination of vertices 3 and 2 from the tape in Figure 10.2a. Tape interpretation amounts to a sequence of fused multiply–add (fma; the elemental operation of the chain rule) operations whose length is of the order of the number of edges in the tape. The resulting bipartite graph contains only edges representing nonzero Jacobian/gradient entries.

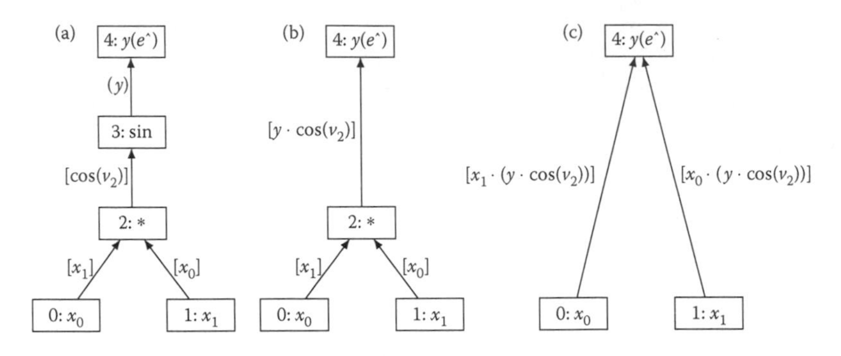

FIGURE 10.2: AAD by overloading and tape interpretation.

Basic adjoint mode does not exploit special structural nor mathematical properties of the given primal code. It automatically records the sequence of operations including data dependencies and required values on a tape followed by purely sequential interpretation of the tape. Real-world adjoints feature sophisticated extensions to be discussed in further detail in the following.

The efficiency (or relative cost) of AD codes is usually measured as the ratio

$$\mathcal{R} = \frac{\text{Run time of given AD code}}{\text{Run time of primal code}}. \tag{10.12}$$

For basic adjoint mode, the ratio $\mathcal{R}$ typically ranges between 5 and ∞ (out of memory) depending on the level of optimization applied to the primal code, its computational complexity, the amount of tape memory available, and the speed of tape accesses. Compared to tangent mode AD or finite differences, $\mathcal{R}$ gives an indication of how large the gradient must be before adjoint methods become attractive.

Basic adjoint mode is likely to yield infeasible tape memory requirement for real-world applications. Checkpointing techniques have been proposed to overcome this problem. They can be regarded as special cases of a more general approach to handling *gaps* in the tape. Let therefore the primal function

$$(y, \tilde{\mathbf{y}}) = F(\mathbf{x}, \tilde{\mathbf{x}}) = (F^q \circ \cdots \circ F^1)(\mathbf{x}, \tilde{\mathbf{x}})$$

be decomposable into a sequence of *elemental* function[4] evaluations $\mathbf{x}^i = F^i(\mathbf{x}^{i-1})$ for $i = 1, \ldots, q$ and where $\mathbf{x}^0 = \mathbf{x}$ and $y = \mathbf{x}^q$. Adjoint AD computes

$$\mathbf{x}_{(1)} := \mathbf{x}_{(1)} + \nabla F^{1^{\mathrm{T}}} \cdot (\ldots \cdot \overbrace{(\nabla F^{k^{\mathrm{T}}} \cdot \underbrace{(\nabla F^{k+1^{\mathrm{T}}} \cdot \ldots \cdot (\nabla F^{q^{\mathrm{T}}} \cdot \mathbf{x}^q_{(1)}) \ldots))}_{\mathbf{x}^k_{(1)}}}^{\mathbf{x}^{k-1}_{(1)}} \ldots)$$

[4]In basic adjoint mode, elemental functions are the arithmetic operations built into the given programming language. In general, elemental functions can be arbitrarily complex subfunctions of F, such as single Monte Carlo path evaluations, individual time steps in an integration scheme, numerical methods for solving systems of linear or nonlinear equations, or calls to black-box routines.

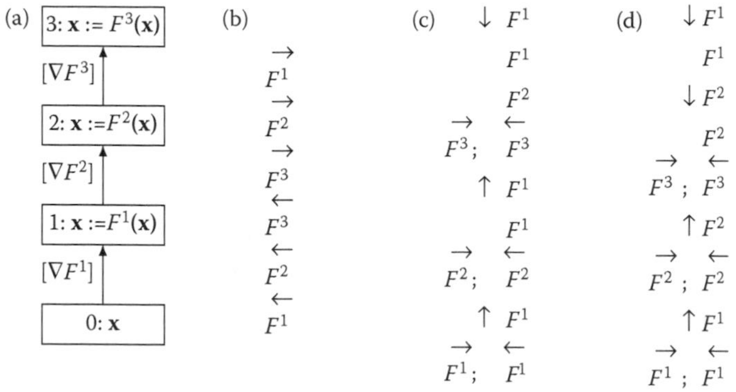

(a) 3: $\mathbf{x} := F^3(\mathbf{x})$ | $[\nabla F^3]$ | 2: $\mathbf{x} := F^2(\mathbf{x})$ | $[\nabla F^2]$ | 1: $\mathbf{x} := F^1(\mathbf{x})$ | $[\nabla F^1]$ | 0: $\mathbf{x}$

(b) $\overrightarrow{F^1}$ $\overrightarrow{F^2}$ $\overrightarrow{F^3}$ $\overleftarrow{F^3}$ $\overleftarrow{F^2}$ $\overleftarrow{F^1}$

(c) $\downarrow F^1$ F^1 F^2 $\overrightarrow{F^3};\ \overleftarrow{F^3}$ $\uparrow F^1$ F^1 $\overrightarrow{F^2};\ \overleftarrow{F^2}$ $\uparrow F^1$ $\overrightarrow{F^1};\ \overleftarrow{F^1}$

(d) $\downarrow F^1$ F^1 $\downarrow F^2$ F^2 $\overrightarrow{F^3};\ \overleftarrow{F^3}$ $\uparrow F^2$ $\overrightarrow{F^2};\ \overleftarrow{F^2}$ $\uparrow F^1$ $\overrightarrow{F^1};\ \overleftarrow{F^1}$

FIGURE 10.3: Adjoint of evolution: tape (a); store-all (b); recompute-all (c); and checkpointing (d).

assuming availability of adjoint elementals $\mathbf{x}_{(1)}^{i-1} = \nabla F^{i^{\mathrm{T}}} \cdot \mathbf{x}_{(1)}^{i}$ for $i = q, \ldots, 1$. Let $\mathbf{x}_{(1)}^{k-1} = \nabla F^{k^{\mathrm{T}}} \cdot \mathbf{x}_{(1)}^{k}$ for some $k \in \{1, \ldots, q\}$ not be treated in basic adjoint mode. A gap is induced in the tape to be filled by some alternative approach to computing $\mathbf{x}_{(1)}^{k-1}$ given $\mathbf{x}_{(1)}^{k}$. Potential scenarios include checkpointing, adjoints of implicit functions, smoothing of locally nondifferentiable functions, and coupling with hand-written adjoint code potentially running on accelerators such as GPUs. Brief discussions of these topics including references for further reading are presented in the following sections.

10.3.1 Checkpointing

Consider Figure 10.3 for motivation. Basic adjoint mode applied to $x := F^i(x)$ for $i = 1, \ldots, q$ ($q = 3$ in Figure 10.3) uses a store-all strategy. It generates a tape of size q assuming unit tape size for the individual F^i. The total primal operations count[5] adds up to q assuming unit primal operations count per F^i. A tape similar to Figure 10.3a is generated for all F^i by running $\overrightarrow{F^i}$ for $i = 1, \ldots, q$ followed by its interpretation by $\overleftarrow{F^j}$ for $j = q, \ldots, 1$ (see Figure 10.3b). The tape memory requirement reaches its minimum $1 + \epsilon$ in a recompute-all strategy by checkpointing the original inputs of size $\epsilon \ll 1$ ($\downarrow F^1$) followed by the evaluation of F^i for $i = 1, \ldots, q-1$, the generation of the tape for F^q ($\overrightarrow{F^q}$) and its interpretation ($\overleftarrow{F^q}$). Repeated accesses to the checkpoint ($\uparrow F^1$) enable the recursive application of this data flow reversal scheme for $i = q-1, \ldots, 1$ (see Figure 10.3c). The primal operations count grows quadratically with q yielding 6 for $q = 3$. Figure 10.3d illustrates a data

[5]Number of evaluations of the primal function. The adjoint operations count is invariant with respect to different data flow reversal schemes as the adjoint of each primal operation is evaluated exactly once.

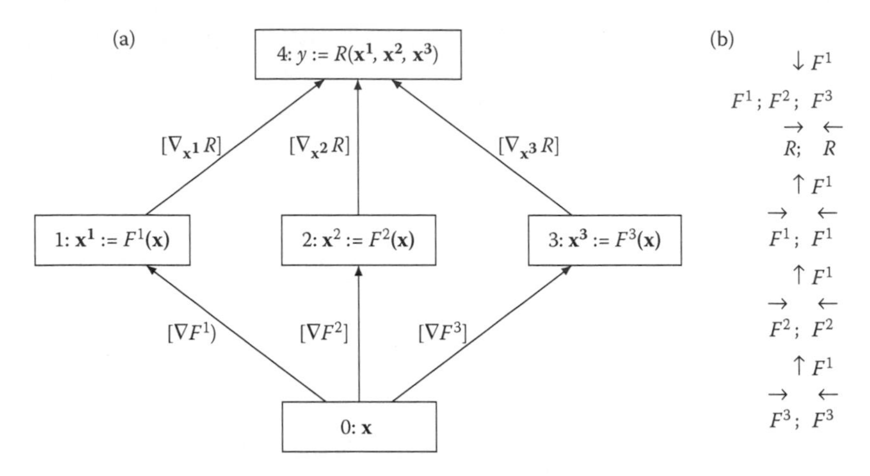

FIGURE 10.4: Adjoint of ensemble: tape (a) and pathwise adjoints (b).

flow reversal scheme built on two checkpoints. It reduces the primal operations count to 5 at the expense of additional memory required to store the second checkpoint yielding a total memory requirement of $1 + 2\epsilon$.

Single- and multilevel checkpointing schemes have been proposed in the literature [27]. The fundamental combinatorial optimization problem of minimizing the primal operations count given an upper bound on the available memory for storing tape and checkpoints is known to be NP-complete [28]. Efficient algorithms for its solution exist for relevant special cases such as evolutions [29] similar to the previous example. They form the core of many iterative algorithms including Crank–Nicolson schemes used in the context of finite difference methods for solving parabolic partial differential equations (see also Section 10.4).

A second special case with particular relevance to finance is Monte Carlo sampling for solving SDEs as, for example, in Section 10.1 (see also Section 10.4). Refer to Figure 10.4 for illustration. Adjoints of such ensembles can be computed very efficiently through exploiting the missing data dependencies among the individual paths (F^i) drawing from a common set of active inputs $\mathbf{x}$. Their results are typically computed in parallel followed by a reduction to a (often scalar) value y by some function R. A gap in the tape is induced by checkpointing $\mathbf{x}$ (with size in memory equal to ϵ) as an input to the F^i (e.g., $\downarrow F^1$) followed by a passive evaluation of the primal ensemble and the generation and interpretation of the tape for R. Adjoints can be computed individually for each path after recovering $\mathbf{x}$ (e.g., $\uparrow F^1$). The maximum memory requirement is limited to $1+\epsilon$ under the assumption that a single path has unit memory requirement exceeding the memory occupied by the tape of R. The primal operations count is roughly doubled, that is, approximately equal to $2q$, where again $q = 3$ in Figure 10.4. Parallelization of pathwise adjoint computation turns out to be straightforward. Potential conflicts need to be handled when writing to $\mathbf{x}_{(1)}$.

10.3.2 Implicit functions

In this section, we consider implementations of $F : \mathbb{R}^n \rightarrow \mathbb{R}$ as $y = F(\mathbf{x}) = F^3(F^2(F^1(\mathbf{x})))$, where F^2 is a numerical method for evaluating an implicit function defined, for example, by systems of linear or nonlinear equations or as an unconstrained convex nonlinear optimization problem.

10.3.2.1 Linear systems

Let $F^1 : \mathbb{R}^n \rightarrow \mathbb{R}^{n\times n} \times \mathbb{R}^n : (A, \mathbf{b}) = F^1(\mathbf{x})$, $F^2 : \mathbb{R}^{n\times n} \times \mathbb{R}^n \rightarrow \mathbb{R}^n : \mathbf{s} = F^2(A, \mathbf{b})$ such that $\mathbf{s}$ is the solution of the system of linear equations $A \cdot \mathbf{s} = \mathbf{b}$, and $F^3 : \mathbb{R}^n \rightarrow \mathbb{R} : y = F^3(\mathbf{s})$. Symbolic adjoint differentiation of the solution $\mathbf{s}$ with respect to A and $\mathbf{b}$ in direction $\mathbf{s}_{(1)}$ yields the linear system

$$A^{\mathrm{T}} \cdot \mathbf{b}_{(1)} = \mathbf{s}_{(1)} \tag{10.13}$$

see, for example, Reference 30. The adjoint system matrix turns out to have unit rank,

$$A_{(1)} = -\mathbf{b}_{(1)} \cdot \mathbf{s}^{\mathrm{T}}. \tag{10.14}$$

Equation 10.14 uses the primal solution $\mathbf{s}$. Hence, its evaluation needs to be preceded by a single run of the primal solver to obtain (a sufficiently precise approximation of) $\mathbf{s}$ and a factorization of the system matrix A if a direct solver is used. This factorization $(LU, QR, \ldots)$ can be reused for the solution of Equation 10.13 [31]. Consequently, both memory requirement and operations count can be reduced from $O(n^3)$ when using basic adjoint AD to $O(n^2)$ when differentiating the linear system symbolically.

10.3.2.2 Nonlinear systems

Consider $F^1 : \mathbb{R}^n \rightarrow \mathbb{R}^n \times \mathbb{R}^k : (\mathbf{s}^0, \boldsymbol{\lambda}) = F^1(\mathbf{x})$, $F^2 : \mathbb{R}^n \times \mathbb{R}^k \rightarrow \mathbb{R}^n : \mathbf{s} = F^2(\mathbf{s}^0, \boldsymbol{\lambda})$ with $\mathbf{s}$ being the solution of the system of parameterized nonlinear equations $N(\mathbf{s}, \boldsymbol{\lambda}) = 0$, and $F^3 : \mathbb{R}^n \rightarrow \mathbb{R} : y = F^3(\mathbf{s})$. Let F^2 be implemented by Newton's method with initial estimate of the solution $\mathbf{s}^0$, parameter vector $\boldsymbol{\lambda} \in \mathbb{R}^k$, and ν denoting the number of Newton iterations performed. Basic adjoint AD applied to F^2 yields both memory requirement and operations count of $O(\nu \cdot n^3)$; a direct solver is assumed to be used for the solution of the Newton system, which can be differentiated symbolically as shown in Section 10.3.2.1. Alternatively, according to Reference 32 the adjoint Newton solver needs to compute a solution to the linear system

$$\frac{\partial N}{\partial \mathbf{x}}(\mathbf{x}, \boldsymbol{\lambda})^{\mathrm{T}} \cdot \mathbf{z} = -\mathbf{x}_{(1)}$$

followed by a single call of the adjoint of N seeded with the solution $\mathbf{z}$ which gives

$$\boldsymbol{\lambda}_{(1)} = \boldsymbol{\lambda}_{(1)} + \frac{\partial N}{\partial \boldsymbol{\lambda}}(\mathbf{x}, \boldsymbol{\lambda})^{\mathrm{T}} \cdot \mathbf{z}.$$

Again, the (sufficiently accurate) primal solution is required. Both memory requirement $(O(n^2))$ and operations count $(O(n^3))$ of computing $\boldsymbol{\lambda}_{(1)}$ are reduced significantly. The derivative of $\mathbf{s}$ with respect to $\mathbf{s}^0$ vanishes identically at the solution.

10.3.2.3 Convex optimization

Consider $F^1 : \mathbb{R}^n \to \mathbb{R}^n \times \mathbb{R}^k : (\mathbf{s}^0, \boldsymbol{\lambda}) = F^1(\mathbf{x})$, $F^2 : \mathbb{R}^n \times \mathbb{R}^k \to \mathbb{R}^n : \mathbf{s} = F^2(\mathbf{s}^0, \boldsymbol{\lambda})$ with $\mathbf{s}$ being the solution of the unconstrained convex nonlinear optimization problem $\operatorname{argmin}_{\mathbf{s}\in\mathbb{R}^n} G(\mathbf{s}, \boldsymbol{\lambda})$, and $F^3 : \mathbb{R}^n \to \mathbb{R} : y = F^3(\mathbf{s})$. F^2 can be regarded as root finding for the first-order optimality condition $\frac{\partial}{\partial \mathbf{s}} G(\mathbf{s}, \boldsymbol{\lambda}) = 0$. Consequently, the computation of $\boldsymbol{\lambda}_{(1)}$ amounts to solving the linear system

$$\frac{\partial^2 G}{\partial \mathbf{s}^2}(\mathbf{s}, \boldsymbol{\lambda})^{\mathrm{T}} \cdot \mathbf{z} = -\mathbf{s}_{(1)}$$

followed by a single call of the second-order adjoint version of F at the solution $\mathbf{z}$ to obtain

$$\boldsymbol{\lambda}_{(1)} := \boldsymbol{\lambda}_{(1)} + \frac{\partial^2 G}{\partial \boldsymbol{\lambda} \partial \mathbf{s}}(\mathbf{s}, \boldsymbol{\lambda})^{\mathrm{T}} \cdot \mathbf{z}.$$

Savings in computational complexity are obtained as in Section 10.3.2.2. Similar comments apply. See Reference 33 for a discussion of this method in the context of calibration.

10.3.3 Smoothing

AD is based on the assumption that the given implementation of the target function $F : \mathbb{R}^n \to \mathbb{R}^m$ is continuously differentiable at all points of interest. This prerequisite is likely to be violated in many practical applications. Generalized derivatives have been proposed to overcome this problem; see, for example, References 34 and 35 for recent work in this area.

In pricing of financial derivatives, nonsmoothness is often induced by branches in the flow of control depending, for example, on a strike price. An option may be exercised or not. Any data flow dependence of the predicted price or payoff p on strike K is lost suggesting independence of the sensitivity of p from K, which is obviously false. Local finite differencing as well as various smoothing techniques can be used to potentially overcome this problem. For example, in sigmoidal smoothing [36] the nonsmooth function

$$f(S) = \begin{cases} f_1(S), & S < K, \\ f_2(S), & \text{otherwise} \end{cases}$$

is replaced by $f(S) = (1 - \sigma_s(S))f_1(S) + \sigma_s(S)f_2(S)$, where

$$\sigma_s(S) = \frac{1}{1 + e^{-\frac{S-K}{\alpha}}}$$

and the width of the transition α controls the quality of the approximation. A case study is discussed in Section 10.4.2.

10.3.4 Preaccumulation

Preaccumulation is a technique for speeding up the computation of adjoints while at the same time reducing the size of the tape. It comes in various flavors. As an example, we consider adjoint versions of implementations of a simulation $F : \mathbb{R}^n \rightarrow \mathbb{R}^m$ as $\mathbf{y} = F(\mathbf{x}) = F^3(F^2(F^1(\mathbf{x})))$, where $F^2 : \mathbb{R}^k \rightarrow \mathbb{R}^l$ is assumed to yield a tape with q edges (local partial derivatives). Without loss of generality, the Jacobian $\nabla F^2 \in \mathbb{R}^{l \times k}$ is assumed to be dense.[6] Hence, the number of scalar fma operations required for its evaluation in tangent mode AD is $k \cdot q$. Accumulation of the overall Jacobian $\nabla F \in \mathbb{R}^{m \times n}$ in adjoint mode induces a local computational cost of $m \cdot q$ due to m interpretations of the tape of F^2. Preaccumulation of ∇F^2 in tangent mode yields a tape with $k \cdot l$ edges. The contribution of F^2 to the total cost of accumulating ∇F without preaccumulation $(m \cdot q)$ potentially exceeds the cumulative cost of preaccumulation of ∇F^2 $(k \cdot q)$ followed by interpretation of the compressed tape (adding $m \cdot k \cdot l$). Moreover, no tape is required for preaccumulating ∇F^2 yielding a reduction of the overall tape size by $q - k \cdot l$ assuming unit memory size per tape entry.

Alternative scenarios for preaccumulation include the repeated use of a local Jacobian as part of an iteration in the enclosing adjoint computation. In this case, the local Jacobian should be preaccumulated (cached) potentially yielding significant savings in terms of tape size and run time. An exponential number (in terms of the size of the tape) of different scenarios for preaccumulation result from the associativity of the chain rule. Determining the optimal method turns out to be computationally intractable [37]. Further theory is discussed in Reference 38.

10.3.5 Further issues

Variety and complexity of available hardware platforms and of corresponding software yield a large number of further issues to be taken into account for the design of adjoint solutions. Relevant topics include the integration of (hand-written) adjoint source code into tape-based solutions, AD of parallel code, AD on accelerators (e.g., GPUs), adjoint numerical libraries, and the handling of black-box code in the context of AD. In any case, the augmentation of a large simulation software package with AD capabilities remains a challenging conceptual as well as software engineering task. AD tools can simplify this process tremendously depending on their levels of robustness and efficiency and the flexibility of their application programming interface. Once started, AD needs to become an element of the overall software development strategy.

---

[6] Jacobian compression methods based on coloring techniques enter the scene in case of sparsity [5].

10.4 Case Studies

The given selection of case studies presented in this section is by no means complete. Our intention is to give the reader some impression of potential use cases for AAD in computational finance.

10.4.1 European option

The material in this section is based on a more extensive discussion in Reference 4. We consider the computation of first- and second-order Greeks for the simple option pricing problem described in Section 10.1. In addition to the previously outlined Monte Carlo approach, we investigate a Crank–Nicolson method for the solution of the corresponding initial and boundary value problem for a partial differential equation (PDE). To recast the pricing problem in a PDE setting, we extend the value V into a value function $V : \mathbb{R} \times [0,T] \to \mathbb{R}_+$ given by

$$V(x,t) = e^{-r(T-t)}\mathbb{E}_{x,t}\left(e^{X_T} - K\right)^+, \tag{10.15}$$

where $\mathbb{E}_{x,t}$ denotes expectation with respect to the measure under which the Markov process X starts at time $t \in [0,T]$ at the value $x \in \mathbb{R}$. Standard results from the theory of Markov processes then show that V satisfies the parabolic PDE

$$\begin{aligned} 0 = {} & \frac{\partial}{\partial t}V(x,t) + \left(r - \tfrac{1}{2}\sigma^2(x,t)\right)\frac{\partial}{\partial x}V(x,t) \\ & + \tfrac{1}{2}\sigma^2(x,t)\frac{\partial^2}{\partial x^2}V(x,t) - rV(x,t) \quad \text{for } (x,t) \in \mathbb{R} \times [0,T), \end{aligned} \tag{10.16}$$

$$(e^x - K)^+ = V(x,T) \quad \text{for all } x \in \mathbb{R}. \tag{10.17}$$

To these we add the asymptotic boundary conditions

$$\lim_{x \to -\infty} V(x,t) = 0 \quad \text{for all } t \in [0,T], \tag{10.18}$$

$$\lim_{x \to \infty} V(x,t) = e^{-r(T-t)}(e^x - K) \quad \text{for all } t \in [0,T]. \tag{10.19}$$

The system is solved by a Crank–Nicolson scheme as described in Reference 39. We use AAD to compute the same sensitivities as before.

Table 10.1 compares peak memory requirements and elapsed run times for the Monte Carlo solutions without and with (ensemble) checkpointing for $N = 10^5$ sample paths at $M = 360$ time steps each with CFDs on our reference computer with 3 GB of main memory available and swapping to disk disabled. Basic adjoint mode (`mc/a1s`) yields infeasible memory requirements for gradient sizes $n \geq 22$. Exploiting the ensemble property in `mc/a1s_ensemble` yields adjoints with machine precision at the expense $\mathcal{R} < 10$ primal function evaluations for all gradient sizes.

TABLE 10.1: Run times and peak memory requirements as a function of gradient size n for `dco/c++` first-order adjoint code vs. central finite differences for the monte carlo code from section 10.1

n	mc/primal	mc/cfd	mc/a1s	mc/a1s_ensemble	$\mathcal{R}$
10	0.3 s	6.1 s	1.8 s (2 GB)	1.3 s (1.9 MB)	4.3
22	0.4 s	15.7 s	(>3 GB)	2.3 s (2.2 MB)	5.7
34	0.5 s	29.0 s	(>3 GB)	3.0 s (2.5 MB)	6.0
62	0.7 s	80.9 s	(>3 GB)	5.1 s (3.2 MB)	7.3
142	1.5 s	423.5 s	(>3 GB)	12.4 s (5.1 MB)	8.3
222	2.3 s	1010.7 s	(>3 GB)	24.4 s (7.1 MB)	10.6

Note: Basic adjoint mode fails for $n \geq 22$ due to prohibitive memory requirement. The relative computational cost $\mathcal{R}$ is given for `mc/a1s_ensemble`. Although theoretically constant, $\mathcal{R}$ is sensitive to specifics such as compiler flags, memory hierarchy and cache sizes, and level of optimization of the primal code.

TABLE 10.2: Accuracy of selected (ith) forward and central finite difference gradient entries vs. AD for the monte carlo code with scenario $n = 10$

i	mc/ffd	mc/cfd
0	**0.982097**033091	**0.9820970**84181
7	−**0.07167**05322265	−**0.071666**955947
4	**0.3461**74240112	**0.3461**31324768
i	mc/t1s	mc/a1s_ensemble
0	**0.98209708315948**5	**0.98209708315948**4
7	−**0.071666056824648**2	−**0.071666056824648**4
4	**0.3461268202393**18	**0.3461268202393**24

Note: Top row shows best case (rel. err. ≈ 1e−8), bottom row worst case (rel. err. ≈ 6e−5) while middle row is a representative value (rel. err. ≈ 2e−5).

Table 10.2 shows the accuracy of gradient entries computed via finite differences (forward and central) compared with AD. Figures are for the smallest problem (gradient size $n = 10$).

Table 10.3 compares peak memory requirements and elapsed run times for the PDE solutions without and with (evolution) checkpointing for $N = 10^4$ spatial grid points and $M = 360$ time steps on our reference computer. Basic adjoint mode (`pde/a1s`) yields infeasible memory requirements for all gradients of size $n \geq 10$.

10.4.2 American option pricing

The following case study is presented in further detail in Reference 40.

An American-put option on a single asset priced by the Longstaff–Schwartz algorithm [41] is considered. For the generation of the stock price paths under the risk neutral measure and without a dividend yield, Equation 10.4 for a

TABLE 10.3: Run time and peak memory requirements as a function of the gradient size n of the basic and checkpointed adjoint codes vs. central finite differences

n	pde/primal	pde/cfd	pde/a1s	pde/a1s_checkpointing	$\mathcal{R}$
10	0.3 s	6.5 s	(>3 GB)	5.2 s (205 MB)	17.3
22	0.5 s	19.6 s	(>3 GB)	8.3 s (370 MB)	16.6
34	0.6 s	37.7 s	(>3 GB)	11.6 s (535 MB)	19.3
62	1.0 s	119.5 s	(>3 GB)	18.7 s (919 MB)	18.7
142	2.6 s	741.2 s	(>3 GB)	39 s (2 GB)	15.0
222	4.1 s	1857.3 s	(>3 GB)	60 s (3 GB)	14.6

Note: The checkpointing used is equidistant (every 10th time step). The basic adjoint ran out of memory even for the smallest problem size. The relative computational cost $\mathcal{R}$ is given for pde/a1s_checkpointing. Again, this theoretically constant value is typically rather sensitive to specifics of the target computer architecture and software stack.

constant volatility is used.

$$X_t = X_0 + \left(r - \frac{1}{2}\sigma^2\right) t \cdot \Delta + \sigma \sum_{i=0}^{t} Z_i. \tag{10.20}$$

First-order sensitivities are obtained by AAD for five active inputs, the stock price, volatility, time to maturity, risk-free interest rate, and strike price, respectively. Second-order sensitivities are computed for Δ with respect to the five active inputs.

A computation of the test case with the number of paths N_P equal to 10^6 and 10^2 exercise opportunities N_T yields enormous memory requirements in basic adjoint mode. Moreover, it can be seen that some of the second-order sensitivities including Γ are computed to be zero by using the AD technique, due to its inability to capture control flow dependencies. After each regression, a decision occurs whether to exercise the current option or to hold it (see line 11 of Algorithm 10.1). This decision leads to zero adjoints of the local cash flow with respect to the exercise boundary.

For the memory reduction evolution, checkpointing introduced in Section 10.3.1 is applied to each iteration in line 2 of Algorithm 10.1. A checkpoint consists of the values of the inputs, the local cash flow, and the time of the respective loop cycle.

To capture the control flow dependency of the exercise decision sigmoidal smoothing is used as introduced in Section 10.3.3. Therefore the exercise boundary is chosen to be the center of the transition. Then, the "if" statement in line 11 and the payoff function in line 12 of the algorithm are replaced by the assignment (10.21) in which σ_s is the sigmoid function and v_p denotes the payoff of the current path.

$$v_p := (1 - \sigma_s) \cdot (K - S_p) + \sigma_s \cdot v_p. \tag{10.21}$$

Checkpointing yields a reduction in memory requirement by approximately 85% compared to basic adjoint mode. Larger test cases can be computed at a

Algorithm 10.1 Longstaff–Schwartz algorithm for put options

In:

→ initial stock price $S_0 \in \mathbb{R}$, strike price $K \in \mathbb{R}$, time to maturity $T \in \mathbb{R}$, volatility $\sigma \in \mathbb{R}$, risk-free interest rate $r \in \mathbb{R}$, number of paths $N_P \in \mathbb{N}$, number of time steps $N_T \in \mathbb{N}$, accumulated random numbers $\mathbf{Z} \in \mathbb{R}^{N_P \times N_T}$

→ implementation of the stock price generation for a given time and path: $h : \mathbb{R}^4 \times \mathbb{N}^2 \times \mathbb{R} \to \mathbb{R}, \; S_{p,t} = h(S_0, T, \sigma, r, t, N_T, Z_{p,t})$

→ implementation of the regression for the set of paths in the money $\mathbf{I}$, the vector of stock prices for a given time $\mathbf{S_t}$ and the vector of discounted cash flows $\mathbf{v}$:
$R : \mathbb{N}^i \times \mathbb{R}^i \times \mathbb{R}^i \to \mathbb{R}, \; b = R(\mathbf{I}, \mathbf{S_t}, \mathbf{v})$

Out:

← option price: $V \in \mathbb{R}$

Algorithm:

1: Initialization
2: **for** $t = N_T - 2$ **to** 1 **do**
3: $\quad \mathbf{I} \leftarrow \{\}$
4: $\quad$ **for** $p = 1$ **to** N_P **do**
5: $\quad\quad v_p \leftarrow v_p \cdot \exp\left(-r \cdot T/N_T\right)$
6: $\quad\quad S_{p,t} = h(S_0, T, \sigma, r, t, N_T, Z_{p,t})$
7: $\quad\quad$ **if** $S_{p,t} < K$ **then**
8: $\quad\quad\quad \mathbf{I} \leftarrow \mathbf{I} \cup \{p\}$;
9: $\quad b = R(\mathbf{I}, \mathbf{S_t}, \mathbf{v})$
10: $\quad$ **for all** $p \in \mathbf{I}$ **do**
11: $\quad\quad$ **if** $K - S_{p,t} > b$ **then**
12: $\quad\quad\quad v_p \leftarrow K - S_{p,t}$
13: $V \leftarrow \sum_p v_p \cdot \exp\left(-r \cdot T/N_T\right)/N_P$

slightly higher computational cost. Run times and memory requirements are shown in Table 10.4.

Those second-order sensitivities which could not be calculated satisfactorily due to the missing control flow dependencies are approximated by the smoothing approach for the exercise decision. All other sensitivities are similar to the values that are computed with the AAD method without smoothing. The option price as well as the sensitivities are given in Table 10.5. The quality of the smoothing depends on the transition parameter α often determined through experiments in practice.

By assuming the missing control flow dependencies to be negligible, the time and the path loop can be switched and the algorithm for the sensitivity

TABLE 10.4: Run times and required tape memory for a single pricing calculation and the basic and the checkpointed adjoint methods

	Run time (s)			Memory requirement (GB)		
N_T	Pricer	Basic adjoint	Checkpointed adjoint	Pricer	Basic adjoint	Checkpointed adjoint
100	22	192 (8.7)	228 (10.4)	0.80	84.78	10.93
500	113	–	1011 (8.9)	3.78	>100	49.90
1000	226	–	2245 (9.9)	7.51	>100	98.47

Note: Relative run times are given in brackets.

TABLE 10.5: Value and sensitivities of the test cases for the algorithmic differentiation methods applied to the basic and smoothed version (subscript s) of the Longstaff–Schwartz algorithm for $S_0 = 1$, $K = 1$, $T = 1$, $r = 4\%$, $\sigma = 20\%$, and $\alpha = 0.005$

N_T	V	V_s	Δ	Δ_s	ν	ν_s	Γ	Γ_s
100	0.06361	0.06353	−0.41962	−0.41761	0.37653	0.37921	0	0.82227
500	0.06378	0.06345	−0.42414	−0.41749	0.37688	0.38275	0	0.86813
1000	0.06369	0.06318	−0.42465	−0.41869	0.37611	0.38515	0	0.73878

Note: In Reference 44, analytical reference values for V and Δ are given as $V_{\text{ref}} = 0.064$ and $\Delta_{\text{ref}} = -0.416$.

computation can be simplified. This pathwise adjoint approach reduces the memory requirement further and it enables parallelization [6].

10.4.3 Nearest correlation matrix

Monte Carlo simulation of multiple correlated underlyings requires the sampling of jointly distributed random variables. The n-dimensional Gaussian copula model can be used to sample jointly normal random variables $\mathbf{z} = (z_1, \ldots, z_n)^{\mathrm{T}}$ by starting from a sample $\tilde{\mathbf{z}} = (\tilde{z}_1, \ldots, \tilde{z}_n)^{\mathrm{T}}$ of independent standard normal variates. The Cholesky decomposition $G = AA^{\mathrm{T}}$ of the correlation matrix $G \in \mathbb{R}^{n \times n}$ is then used to correlate the components of $\tilde{\mathbf{z}}$ by performing the matrix vector product $\mathbf{z} = A \cdot \tilde{\mathbf{z}}$.

Real-world correlation data are often not consistent in the sense that the matrix G of pairwise correlations is not positive definite as required by the Cholesky decomposition. To arrive at a positive definite matrix $C \in \mathbb{R}^{n \times n}$ that is a correlation matrix with a given lower bound on eigenvalues and that is closest to G in the Frobenius norm, we can solve a variation of the nearest correlation matrix (NCM) problem as a preprocessing step [42].

Due to Qi and Sun [43], a generalized Newton method can be used to find the root of the first-order optimality condition for the unconstrained convex optimization problem formulation of the NCM

$$h(\mathbf{y}, G) = \mathrm{Diag}((G + \mathrm{diag}(\mathbf{y}))_+) - \mathbf{e} = 0. \tag{10.22}$$

Notation is as follows: $\text{Diag}(\cdot)$ gives the diagonal of a matrix, $\text{diag}(\cdot)$ gives a matrix with the given vector as diagonal, $(\cdot)_+$ is the projection onto the set of positive semidefinite matrices, and $\mathbf{e} = (1 \ldots 1)^{\mathrm{T}} \in \mathbb{R}^n$. To get the NCM, we solve Equation 10.22 for $\mathbf{y} = \mathbf{y}^* \in \mathbb{R}^n$ and set

$$C := (G + \text{diag}(\mathbf{y}^*))_+. \tag{10.23}$$

AAD has been introduced as an efficient way of quantifying correlation risk in a copula-based Monte Carlo setting by Capriotti and Giles [13]. If we want to compute correlation risk through the NCM step to get sensitivities with regard to the actual model inputs, an adjoint of the NCM algorithm is required.

As described in Section 10.3.2.2, basic AAD applied to a Newton algorithm yields a tape memory requirement of $O(\nu \cdot n^3)$, where ν is the number of Newton iterations. We examine how known derivatives and the implicit function theorem can be used to arrive at a symbolic adjoint for the NCM algorithm with much better performance. Since the involved $(\cdot)_+$ operator is not differentiable where the argument has an eigenvalue that is zero, a smoothed version can be used to get sensitivities comparable to those obtained by finite differences.

A matrix is projected onto the set of positive semidefinite matrices by setting all negative eigenvalues to zero. We can compute both tangents and adjoints of the $(\cdot)_+$ operator implicitly by using the symbolic matrix derivative results of the eigen decomposition provided in Reference 30 and a derivative of the max function. Given $C_{(1)}$ the adjoints $G_{(1)}, {\mathbf{y}^*}_{(1)}$ for the assignment (10.23) follow directly from the adjoint of the $(\cdot)_+$ operator.

Since Equation 10.22 already gives the first-order optimality condition, the adjoint $G_{(1)}$ of the function $\mathbf{y}^* = \mathbf{y}^*(G)$ implicitly defined by $h(\mathbf{y}^*, G) = 0$ can be computed by first solving the linear system

$$\frac{\partial h}{\partial \mathbf{y}}(\mathbf{y}^*, G)^{\mathrm{T}} \cdot \mathbf{z} = -{\mathbf{y}^*}_{(1)} \tag{10.24}$$

and then setting

$$G_{(1)} := G_{(1)} + \frac{\partial h}{\partial G}(\mathbf{y}^*, G)^{\mathrm{T}} \cdot \mathbf{z}, \tag{10.25}$$

where the adjoint of $h(\mathbf{y}, G)$ again follows from the adjoint of the $(\cdot)_+$ operator.

With the above method, the adjoint $G_{(1)}$ as a function of $C_{(1)}$ can be computed with a memory complexity of only $O(n^2)$ and a run time complexity of $O(n^3)$ but its accuracy now depends on the Newton residual, that is, the quality of the NCM solution itself.

In Table 10.6, we give the run time results of the primal NCM computation and adjoints acquired by basic AAD, finite differences and the symbolic adjoint approach described earlier. We also give the memory requirements for the basic adjoint tape. Test matrices are correlation matrices with a few perturbed elements resulting in small negative eigenvalues.

TABLE 10.6: Run time results for computing the NCM and three different adjoint routines and the required tape memory for the basic adjoint

	Run time (s)				Tape memory (GB)
n	NCM primal	Basic adjoint	Symbolic adjoint	Finite differences	Basic adjoint
20	0.0014	0.12	0.0015	0.85	0.038
50	0.0078	1.6	0.0094	38	0.53
100	0.045	12	0.056	890	4.4

The basic adjoint run times clearly show the advantages of AAD for computing Greeks as opposed to using a finite differences based approach. The symbolic adjoint of the NCM presents a significant improvement as it is multiple orders of magnitude faster than the basic adjoint and is not restricted by memory even for large or hard problem instances.

10.5 Summary and Conclusion

This chapter presented AD and its adjoint mode in particular as one of the fundamental elements of the high-performance computational finance toolbox. Improvements in computational cost by an order of complexity over classical finite difference approximation of first-order Greeks motivate the widespread use of adjoint AD. An introduction to its basic principles was followed by a discussion of implementation and related conceptual challenges. Illustration was provided by three case studies.

Getting started with basic adjoint AD on relatively simple problems is typically rather straightforward. A growing set of AD tools is available to support this process; see, for example, `www.autodiff.org`. Much more challenging is the application of AD to large financial libraries including the need for checkpointing, symbolic differentiation of implicit functions, smoothing, preaccumulation, integration of adjoint source code, support for MPI and/or OpenMP, acceleration using GPUs, and so on. Second- and higher-order adjoints might be needed. We expect large-scale high-performance adjoints to require a substantial amount of user interaction for the foreseeable future. AD tools need to expose the corresponding flexibility through an intuitive and well-documented user interface.

References

1. Giles, M. and Glasserman, P. Smoking adjoints: Fast Monte Carlo greeks. *Risk*, 19:88–92, 2006.

2. Griewank, A. and Walter, A. *Evaluating Derivatives. Principles and Techniques of Algorithmic Differentiation* (2nd Edition). SIAM, Philadelphia, 2008.

3. Naumann, U. *The Art of Differentiating Computer Programs. An Introduction to Algorithmic Differentiation.* Number 24 in Software, Environments, and Tools. SIAM, Philadelphia, 2012.

4. Naumann, U. and Du Toit, J. Adjoint algorithmic differentiation tool support for typical numerical patterns in computational finance. *NAG Technical Report No. TR3/14*, 2014. *Journal of Computational Finance* (to appear).

5. Gebremedhin, A., Manne, F., and Pothen, A. What color is your Jacobian? Graph coloring for computing derivatives. *SIAM Review*, 47:629–705, 2005.

6. Leclerc, M., Liang, Q., and Schneider, I. Fast Monte Carlo Bermudan greeks. *Risk*, 22(7):84–88, 2009.

7. Denson, N. and Joshi, M. Flaming logs. *Wilmott Journal*, 1:259–262, 2009.

8. Joshi, M. and Pitt, D. Fast sensitivity computations for Monte Carlo valuation of pension funds. *Astin Bulletin*, 40:655–667, 2010.

9. Joshi, M. and Yang, C. Fast and accurate pricing and hedging of long-dated CMS spread options. *International Journal of Theoretical and Applied Finance*, 13:839–865, 2010.

10. Joshi, M. and Yang, C. Algorithmic hessians and the fast computation of cross-gamma risk. *IIE Transactions*, 43:878–892, 2011.

11. Joshi, M. and Yang, C. Fast delta computations in the swap-rate market model. *Journal of Economic Dynamics and Control*, 35:764–775, 2011.

12. Capriotti, L. Fast greeks by algorithmic differentiation. *Journal of Computational Finance*, 14:3–35, 2011.

13. Capriotti, L. and Giles, M. Fast correlation greeks by adjoint algorithmic differentiation. *Risk*, 23:79–83, 2010.

14. Capriotti, L. and Giles, M. Adjoint greeks made easy. *Risk*, 25:92, 2012.

15. Capriotti, L., Jiang, Y., and Macrina, A. Real-time risk management: An AAD–PDE approach. *International Journal of Financial Engineering*, 2:1550039, 2015.

16. Capriotti, L., Jiang, Y., and Macrina, A. AAD and Least Squares Monte Carlo: Fast Bermudan-style options and XVA greeks. *Algorithmic Finance*, 6(1–2):35–49, 2017.

17. Antonov, A. Algorithmic differentiation for callable exotics. 2017. Available at SSRN: https://ssrn.com/abstract=2839362 or http://dx.doi.org/10.2139/ssrn.2839362

18. Antonov, A., Issakov, S., Konikov, M., McClelland, A., and Mechkov, S. PV and XVA greeks for callable exotics by algorithmic differentiation. 2017. Available at

SSRN: https://ssrn.com/abstract=2881992 or http://dx.doi.org/10.2139/ssrn.2881992

19. Turinici, G. Calibration of local volatility using the local and implied instantaneous variance. *Journal of Computational Finance*, 13(2):1, 2009.

20. Käbe, C., Maruhn, J. H., and Sachs, E. W. Adjoint-based Monte Carlo calibration of financial market models. *Finance and Stochastics*, 13(3):351–379, 2009.

21. Schlenkrich, S. Efficient calibration of the Hull–White model. *Optimal Control Applications and Methods*, 33:352–362, 2012.

22. Henrard, M. Calibration in finance: Very fast greeks through algorithmic differentiation and implicit function. *Procedia Computer Science*, 18:1145–1154, 2013.

23. Giles, M. Vibrato Monte Carlo sensitivities. In L'Ecuyer, P. and Owen, A. editors. *Monte Carlo and Quasi Monte Carlo Methods*, 369–382. Springer, 2009.

24. Capriotti, L. Likelihood ratio method and algorithmic differentiation: Fast second order greeks. *Algorithmic Finance*, 4:81–87, 2015.

25. Pironneau, O. et al. Vibrato and automatic differentiation for high order derivatives and sensitivities of financial options. *arXiv preprint arXiv:1606.06143*, 2016.

26. Utke, J., Naumann, U., Fagan, M., Tallent, N., Strout, M., Heimbach, P., Hill, C., and Wunsch, C. OpenAD/F: A modular open-source tool for automatic differentiation of Fortran codes. *ACM Transactions on Mathematical Software*, 34:18:1–18:36, July 2008.

27. Stumm, P. and Walther, A. Multi-stage approaches for optimal offline checkpointing. *SIAM Journal of Scientific Computing*, 31:1946–1967, 2009.

28. Naumann, U. DAG reversal is NP-complete. *Journal of Discrete Algorithms*, 7:402–410, 2009.

29. Griewank, A. and Walther, A. Algorithm 799: Revolve: An implementation of checkpoint for the reverse or adjoint mode of computational differentiation. *ACM Transactions on Mathematical Software*, 26(1):19–45, 2000. Also appeared as Technical Report IOKOMO-04-1997, Technical University of Dresden.

30. Giles, M. Collected matrix derivative results for forward and reverse mode algorithmic differentiation. In Bischof, C., Bücker, M., Hovland, P., Naumann, U., and Utke, J., editors. *Advances in Automatic Differentiation, Volume 64 of Lecture Notes in Computational Science and Engineering*, pages 35–44. Springer, 2008.

31. Naumann, U. and Lotz, J. Algorithmic differentiation of numerical methods: Tangent-linear and adjoint direct solvers for systems of linear equations. Technical Report AIB-2012-10, LuFG Inf. 12, RWTH Aachen, June 2012.

32. Naumann, U., Lotz, J., Leppkes, K., and Towara, M. Algorithmic differentiation of numerical methods: Tangent and adjoint solvers for parameterized systems of nonlinear equations. *ACM Transactions on Mathematical Software*, 41:26:1–26:21, 2015.

33. Henrard, M. Adjoint algorithmic differentiation: Calibration and implicit function theorem. *Journal of Computational Finance*, 17(4):37–47, 2014.

34. Griewank, A. On stable piecewise linearization and generalized algorithmic differentiation. *Optimization Methods and Software*, 28:1139–1178, 2013.

35. Khan, K. and Barton, P. A vector forward mode of automatic differentiation for generalized derivative evaluation. *Optimization Methods and Software*, 30(6):1–28, 2015.

36. Schneider, J. and Kirkpatrick, S. *Stochastic Optimization*. Springer, 2006.

37. Naumann, U. Optimal Jacobian accumulation is NP-complete. *Mathematical Programming*, 112:427–441, 2008.

38. Naumann, U. Optimal accumulation of Jacobian matrices by elimination methods on the dual computational graph. *Mathematical Programming*, 99:399–421, 2004.

39. Andersen, L. B. G. and Brotherton-Ratcliffe, R. The equity option volatility smile: An implicit finite difference approach. *Journal of Computational Finance*, 2000.v

40. Deussen, J., Mosenkis, V., and Naumann, U. Fast Estimates of Greeks from American Options: A Case Study in Adjoint Algorithmic Differentiation. Technical Report AIB-2018-02, RWTH Aachen University, January 2018.

41. Longstaff, F. A. and Schwartz, E. S. Valuing American options by simulation: A simple Least-Squares approach. *Review of Financial Studies*, 14:113–147, 2001.

42. Higham, N. J. Computing the nearest correlation matrix—A problem from finance. *IMA Journal of Numerical Analysis*, 22(3):329–343, 2002.

43. Qi, H. and Sun, D. A quadratically convergent Newton method for computing the nearest correlation matrix. *SIAM Journal Matrix Analysis Applications*, 28:360–385, 2006.

44. Geske, R. and Johnson, H. E. The American put option valued analytically. *Journal of Finance*, 39(5):1511–1524, 1984.

Chapter 11

Case Studies of Real-Time Risk Management via Adjoint Algorithmic Differentiation (AAD)

Luca Capriotti and Jacky Lee

CONTENTS

11.1 Introduction

The renewed emphasis of the financial industry on quantitatively sound risk management practices comes with formidable computational challenges. In fact, standard approaches for the calculation of risk require repeating the calculation of the P&L of the portfolio under hundreds of market scenarios. As a result, in many cases these calculations cannot be completed in a practical amount of time, even employing a vast amount of computer power, especially for risk management problems requiring computationally intensive Monte Carlo (MC) simulations. Since the total cost of the through-the-life risk management can determine whether it is profitable to execute a new trade, solving this technology problem is critical to allow a securities firm to remain competitive.

Following the introduction of adjoint methods in Finance [1], a computational technique dubbed *adjoint algorithmic differentiation* (AAD) [2–4] has recently emerged as tremendously effective for speeding up the calculation of sensitivities in MC in the context of the so-called pathwise derivative method [5].

Algorithmic differentiation (AD) [6] is a set of programming techniques for the efficient calculation of the derivatives of functions implemented as computer programs. The main idea underlying AD is that any such function—no matter how complicated—can be interpreted as a composition of basic arithmetic and intrinsic operations that are easy to differentiate. What makes AD particularly attractive, when compared to standard (finite-difference) methods for the calculation of derivatives, is its computational efficiency. In fact, AD exploits the information on the structure of the computer code in order to optimize the calculation. In particular, when one requires the derivatives of a small number of outputs with respect to a large number of inputs, the calculation can be highly optimized by applying the chain rule through the instructions of the program in opposite order with respect to their original evaluation [6]. This gives rise to the adjoint (mode of) algorithmic differentiation (AAD).

Surprisingly, even if AD has been an active branch of computer science for several decades, its impact in other research fields has been fairly limited until recently. Interestingly, in a twist with the usual situation in which well-established ideas in Applied Maths or Physics have been often "borrowed" by quants, AAD has been introduced in MC applications in Natural Science [7] only after its "rediscovery" in Quantitative Finance.

In this chapter, we discuss three particularly significant applications of AAD to risk management, interest rate products, counterparty credit risk management (CCRM), and volume credit products, that illustrate the power and generality of this groundbreaking numerical technique.

11.2 Adjoint Algorithmic Differentiation: A Primer

Reference 6 contains a detailed discussion of the computational cost of AAD. Here we will only recall the main results in order to clarify how this technique can be beneficial for the applications discussed in this chapter. The interested reader can find in References 2–4 several examples illustrating the intuition behind these results.

To this end, consider a function

$$Y = \texttt{FUNCTION}(X) \tag{11.1}$$

mapping a vector X in $\mathbb{R}^n$ in a vector Y in $\mathbb{R}^m$ through a sequence of steps

$$X \to \cdots \to U \to V \to \cdots \to Y. \tag{11.2}$$

Here, each step can be a distinct high-level function or even an individual instruction.

The adjoint mode of AD results from propagating the derivatives of the final result with respect to all the intermediate variables—the so-called *adjoints*—until the derivatives with respect to the independent variables are formed. Using the standard AD notation, the adjoint of any intermediate variable V_k is defined as

$$\bar{V}_k = \sum_{j=1}^{m} \bar{Y}_j \frac{\partial Y_j}{\partial V_k}, \tag{11.3}$$

where $\bar{Y}$ is vector in $\mathbb{R}^m$. In particular, for each of the intermediate variables U_i, using the chain rule we get,

$$\bar{U}_i = \sum_{j=1}^{m} \bar{Y}_j \frac{\partial Y_j}{\partial U_i} = \sum_{j=1}^{m} \bar{Y}_j \sum_k \frac{\partial Y_j}{\partial V_k} \frac{\partial V_k}{\partial U_i},$$

which corresponds to the adjoint mode equation for the intermediate function $V = V(U)$

$$\bar{U}_i = \sum_k \bar{V}_k \frac{\partial V_k}{\partial U_i}, \tag{11.4}$$

namely a function of the form $\bar{U} = \bar{V}(U, \bar{V})$. Starting from the adjoint of the outputs, $\bar{Y}$, we can apply this to each step in the calculation, working from right to left,

$$\bar{X} \leftarrow \cdots \leftarrow \bar{U} \leftarrow \bar{V} \leftarrow \cdots \leftarrow \bar{Y} \tag{11.5}$$

until we obtain $\bar{X}$, that is, the following linear combination of the rows of the Jacobian of the function $X \to Y$:

$$\bar{X}_i = \sum_{j=1}^{m} \bar{Y}_j \frac{\partial Y_j}{\partial X_i}, \tag{11.6}$$

with $i = 1, \ldots, n$.

In the adjoint mode, the cost does not increase with the number of inputs, but it is linear in the number of (linear combinations of the) rows of the Jacobian that need to be evaluated independently. In particular, if the full Jacobian is required, one needs to repeat the adjoint calculation m times, setting the vector $\bar{Y}$ equal to each of the elements of the canonical basis in $\mathbb{R}^m$. Furthermore, since the partial (branch) derivatives depend on the values of the intermediate variables, one generally first has to compute the original calculation storing the values of all of the intermediate variables such as U and V, before performing the adjoint mode sensitivity calculation.

One particularly important theoretical result [6] is that given a computer program performing some high-level function 11.1, the execution time of its adjoint counterpart

$$\bar{X} = \texttt{FUNCTION\_b}(X, \bar{Y}) \tag{11.7}$$

(with suffix `_b` for "backward" or "bar") calculating the linear combination 11.6 is bounded by approximately four times the cost of execution of the original one, namely

$$\frac{\mathrm{Cost}[\texttt{FUNCTION\_b}]}{\mathrm{Cost}[\texttt{FUNCTION}]} \leq \omega_A, \tag{11.8}$$

with $\omega_A \in [3,4]$. Thus one can obtain the sensitivity of a single output, or of a linear combination of outputs, to an unlimited number of inputs for a little more work than the original calculation.

As also discussed at length in References 2–4, AAD can be straightforwardly implemented by starting from the output of an algorithm and proceeding backwards applying systematically the adjoint composition rule 11.4 to each intermediate step, until the adjoints of the inputs 11.6 are computed. As already noted, the execution of such *backward sweep* requires information that needs to be computed and stored by executing beforehand the steps of the original algorithm—the so-called *forward sweep.*

11.2.1 Adjoint design paradigm

The propagation of the adjoints according to the steps 11.5, being mechanical in nature, can be automated. Several AD tools are available[1] that given a procedure of the form 11.1 generate the adjoint function 11.7. Unfortunately, the application of such automatic AD tools on large inhomogeneous computer codes, like the one used in financial practices, is challenging. Indeed, pricing applications are rarely available as self-contained packages, for example, that can be easily parsed by an automatic AD tool. In practice, even simple option pricing software is almost never implemented as a self-contained component. In fact, to ensure consistency and leverage reusability, it generally relies on auxiliary objects representing the relevant market information, for example,

[1] An excellent source of information can be found at www.autodiff.org.

a volatility surface or an interest rate curve, that are shared across different pricing applications.

Fortunately, the principles of AD can be used as a programming paradigm for any algorithm. An easy way to illustrate the adjoint design paradigm is to consider again the arbitrary computer function in Equation 11.1 and to imagine that this represents a certain high-level algorithm that we want to differentiate. By appropriately defining the intermediate variables, any such algorithm can be abstracted in general as a composition of functions as in Equation 11.2. In the following section, we give a very simple example illustrating this idea. The interested reader can find in Reference 3 a practical step-by-step guide.

11.2.1.1 A simple example

As a simple example of AAD implementation, we consider an algorithm mapping a set of inputs $(\theta_1, \ldots, \theta_n)$ into a single output P, according to the following steps:

Step 1 Set $X_i = \exp(-\theta_i^2/2 + \theta_i Z)$, for $i = 1, \ldots n$, where Z is a constant.

Step 2 Set $P = (\sum_{i=1}^{n} X_i - K)^+$, where K is a constant.

The corresponding adjoint algorithm consists of Steps 1–2 (forward sweep), plus a backward sweep consisting of the adjoints of Steps $\bar{2}$ and $\bar{1}$, respectively:

Step $\bar{2}$ Set $\bar{X}_i = \bar{P}\ \mathbb{I}(\sum_{i=1}^{n} X_i - K)$, for $i = 1, \ldots, n$. Here $\mathbb{I}(x)$ is the Heaviside function.

Step $\bar{1}$ Set $\bar{\theta}_i = X_i(-\theta_i + Z)$, for $i = 1, \ldots, n$.

It is immediate to verify that the output of the adjoint algorithm above gives for $\bar{P} = 1$ the full set of sensitivities with respect to the inputs, $\bar{\theta}_i = \partial P/\partial \theta_i$. Note that, as described in the main text, the backward sweep requires information that is computed during the execution of the forward sweep, Steps 1 and 2, for example, to compute the indicator $\mathbb{I}(\sum_{i=1}^{n} X_i - K)$ and the value of X_i. Finally, simple inspection shows that both the forward and backward sweeps have a computation complexity $O(n)$, that is, all the components of the gradient of P can be obtained at a cost that is of the same order of the cost computing P, in agreement with the general result 11.8. It is easy to recognize in this example a stylized representation of the calculation of the pathwise estimator for vega (volatility sensitivity) of a call option on a sum of lognormal assets.

11.3 Real-Time Risk Management of Interest Rate Products

An important application of AAD is the efficient implementation of the so-called pathwise derivative method [5] in MC applications. We begin by briefly recalling the main ideas underlying this method. Then we discuss its application to the simulation of the Libor Market Model [8] in the context of risk management of interest rate products. This example is of particular significance being the first financial application of adjoint methods, due to the seminal paper by Giles and Glasserman [1], inspiring the subsequent development of AAD [2,3] in a financial context.

11.3.1 Pathwise derivative method

Option pricing problems can be typically formulated in terms of the calculation of expectation values of the form

$$V = \mathbb{E}_{\mathbb{Q}}[P(X(T_1), \ldots, X(T_M))]. \tag{11.9}$$

Here $X(t)$ is a N-dimensional vector and represents the value of a set of underlying market factors (e.g., stock prices, interest rates, foreign exchange pairs, and so on) at time t. $P(X(T_1), \ldots, X(T_M))$ is the discounted payout function of the priced security and depends in general on M observations of those factors. In the following, we will indicate the collection of such observations with a $d = N \times M$-dimensional state vector $X = (X(T_1), \ldots, X(T_M))^t$.

The expectation value in Equation 11.9 can be estimated by means of MC by sampling a number N_{MC} of random replicas of the underlying state vector $X[1], \ldots, X[N_{\mathrm{MC}}]$, sampled according to the distribution $\mathbb{Q}(X)$, and evaluating the payout $P(X)$ for each of them. This leads to the estimate of the option value V as

$$V \simeq \frac{1}{N_{\mathrm{MC}}} \sum_{i_{\mathrm{MC}}=1}^{N_{\mathrm{MC}}} P\left(X[i_{\mathrm{MC}}]\right). \tag{11.10}$$

The pathwise derivative method allows the calculation of the sensitivities of the option price V (Equation 11.9) with respect to a set of N_θ parameters $\theta = (\theta_1, \ldots, \theta_{N_\theta})$, with a single simulation. This can be achieved by first expressing the expectation value in Equation 11.9 as being over $\mathbb{P}(Z)$, the distribution of the independent random numbers Z used in the MC simulation to generate the random samples of $\mathbb{Q}(X)$, so that

$$V = \mathbb{E}_{\mathbb{Q}}[P(X)] = \mathbb{E}_{\mathbb{P}}[P(X(Z))]. \tag{11.11}$$

The point of this subtle change is that $\mathbb{P}(Z)$ does not depend on the parameters θ whereas $\mathbb{Q}(X)$ does. Indeed, whenever the payout function is regular enough, for example, Lipschitz-continuous, and under additional conditions that are often satisfied in financial pricing (see, e.g., [9]), one can write the sensitivity $\langle\bar{\theta}_k\rangle \equiv \partial V/\partial\theta_k$ as

$$\langle\bar{\theta}_k\rangle = \mathbb{E}_{\mathbb{Q}}\left[\frac{\partial P_\theta(X)}{\partial\theta_k}\right]. \tag{11.12}$$

In general, the calculation of Equation 11.12 can be performed by applying the chain rule and averaging on each MC path the so-called pathwise derivative estimator

$$\bar{\theta}_k \equiv \frac{\partial P_\theta(X)}{\partial\theta_k} = \sum_{j=1}^{d} \frac{\partial P_\theta(X)}{\partial X_j} \times \frac{\partial X_j}{\partial\theta_k} + \frac{\partial P_\theta(X)}{\partial\theta_k}. \tag{11.13}$$

For nonpath-dependent options in the context of Libor Market Models, Giles and Glasserman [1] have shown that the pathwise derivative method can be efficiently implemented by expressing the calculation of the estimator 11.13 in terms of linear algebra operations, and utilize adjoint methods to reduce the computational complexity by rearranging appropriately the order of the calculations.

Unfortunately, algebraic adjoint approach requires in general a fair amount of analytical work and is difficult to generalize to both path-dependent payouts and multiassets simulations. Instead, as we discuss in the following section, AAD overcomes these difficulties.

11.3.1.1 Libor market model simulation

In order to make the connection with previous algebraic implementations of adjoint methods [1,10,11], we discuss the implementation of AAD in the Libor Market Model. Here we indicate with T_i, $i = 1, \ldots, N+1$, a set of $N+1$ bond maturities, with spacings $\delta = T_{i+1} - T_i$, assumed constant for simplicity. The dynamics of the Libor rate as seen at time t for the interval $[T_i, T_{i+1})$, $L_i(t)$ takes the form:

$$\frac{\mathrm{d}L_i(t)}{L_i(t)} = \mu_i(L(t))\,\mathrm{d}t + \sigma_i(t)^{\mathrm{T}}\,\mathrm{d}W_t, \tag{11.14}$$

$0 \leq t \leq T_i$, and $i = 1, \ldots, N$, where W_t is a d_W-dimensional standard Brownian motion, $L(t)$ is the N-dimensional vector of Libor rates, and $\sigma_i(t)$ the d_W-dimensional vector of volatilities, at time t. Here the drift term in the spot measure, as imposed by the no arbitrage conditions [8], reads

$$\mu_i(L(t)) = \sum_{j=\eta(t)}^{i} \frac{\sigma_i^{\mathrm{T}}\sigma_j\delta L_j(t)}{1+\delta L_j(t)}, \tag{11.15}$$

where $\eta(t)$ denotes the index of the bond maturity immediately following time t, with $T_{\eta(t)-1} \leq t < T_\eta(t)$. As is common in the literature, to keep this example as simple as possible, we take each vector σ_i to be a function of time to maturity

$$\sigma_i(t) = \sigma_{i-\eta(t)+1}(0) = \lambda(i - \eta(t) + 1). \tag{11.16}$$

Equation 11.14 can be simulated by applying an Euler discretization to the logarithms of the forward rates, for example, by dividing each interval $[T_i, T_{i+1})$ into N_s steps of equal width, $h = \delta/N_s$. This gives

$$\frac{L_i(t_{n+1})}{L_i(t_n)} = \exp\left[\left(\mu_i(L(t_n)) - ||\sigma_i(t_n)||^2/2\right) h + \sigma_i^{\mathrm{T}}(n) Z(t_n)\sqrt{h}\right], \tag{11.17}$$

for $i = \eta(nh), \ldots, N$, and $L_i(t_{n+1}) = L_i(t_n)$ if $i < \eta(nh)$. Here Z is a d_W-dimensional vector of independent standard normal variables.

In a recent paper, Denson and Joshi [10] extended the original Adjoint implementation to the propagation of the Libor under the predictor–corrector drift approximation, consisting in replacing the drift in Equation 11.15 with

$$\mu_i^{pc}(L(t_n)) = \frac{1}{2} \sum_{j=\eta(nh)}^{i} \left(\frac{\sigma_i^{\mathrm{T}} \sigma_j \delta L_j(t_n)}{1 + \delta L_j(t_n)} + \frac{\sigma_i^{\mathrm{T}} \sigma_j \delta \hat{L}_j(t_{n+1})}{1 + \delta \hat{L}_j(t_{n+1})} \right), \tag{11.18}$$

where $\hat{L}_j(t_{n+1})$ is calculated from $L_j(t_n)$ using the evolution 11.17, that is, with the simple Euler drift 11.15.

The pseudocode for the propagation of the Libor rates for $d_W = 1$, corresponding to a function `PROP` implementing the Euler step 11.17, is shown in Figure 11.1. Here, as discussed in Reference 1, the computational cost of implementing Equation 11.17 is minimized by first evaluating

$$v_i(t_n) = \sum_{j=\eta(nh)}^{i} \frac{\sigma_j \delta L_j(t_n)}{1 + \delta L_j(t_n)}, \tag{11.19}$$

as a running sum for $i = \eta(nh), \ldots, N$, so that $\mu_i = \sigma_i^{\mathrm{T}} v_i$.

The algebraic formulation discussed in Reference 10 comes with a significant analytical effort. Instead, as illustrated in Figure 11.2, the AAD implementation is quite straightforward. According to the general design of AAD, this simply consists of the adjoints of the instructions in the forward sweep executed in reverse order. In this example, the information computed by `PROP` that is required by `PROP_b` is stored in the vectors `scra` and `hat_scra`. By inspecting the structure of the pseudocode, it also appears clear that the computational cost of `PROP_b` is of the same order as evaluating the original function `PROP`.

As a standard test case in the literature, here we have considered contracts with expiry T_n to enter in a swap with payments dates $T_{n+1}, \ldots, T_{N+1}$, with the holder of the option paying a fixed rate K

$$V(T_n) = \sum_{i=n+1}^{N+1} B(T_n, T_i)\delta(S_n(T_n) - K)^+, \tag{11.20}$$

```
PROP(n, L[,], Z, lambda[], L0[])

  if(n=0)
    for(i= 1 .. N)
      L[i,n] = L0[i]; Lhat[i,n] = L0[i];

  for (i = 1 .. eta[n]-1)
    L[i,n+1] = L[i,n];  // settled rates

  sqez = sqrt(h)*Z;
  v = 0.; v_pc = 0.;
  for (i = eta[n] .. N)
    lam = lambda[i-eta[n]+1];
    c1 = del*lam; c2 = h*lam;
    v += (c1*L[i,n])/(1.+del*L[i,n]);
    vrat = exp(c2*(-lam/2.+v)+lam*sqez);
    // standard propagation with the Euler drifts
    Lhat[i,n+1] = L[i,n]*vrat;
    // (n + 1) drift term
    v_pc += (c1*Lhat[i,n+1])/(1.+del*Lhat[i,n+1]);
    vrat_pc = exp(c2*(-lam/2.+(v_pc+v)/2.)+lam*sqez);
    // actual propagation using the average drift
    L[i,n+1]  = L[i,n]*vrat_pc;
    // store what is needed for the reverse sweep
    hat_scra[i,n+1] = vrat*((v-lam)*h+sqez);
    scra[i,n+1] = vrat_pc*(((v_pc+v)/2.-lam)*h+sqez);
```

FIGURE 11.1: Pseudocode implementing the propagation method $\texttt{PROP}_n$ for the Libor Market Model of Equation 11.14 for $d_W = 1$, under the predictor corrector Euler approximation 11.18, and the volatility parameterization 11.16.

where $B(T_n, T_i)$ is the price at time T_n of a bond maturing at time T_i

$$B(T_n, T_i) = \prod_{l=n}^{i-1} \frac{1}{1 + \delta L_l(T_n)}, \tag{11.21}$$

and the swap rate reads

$$S_n(T_n) = \frac{1 - B(T_n, T_{N+1})}{\delta \sum_{l=n+1}^{N+1} B(T_n, T_l)}. \tag{11.22}$$

```
PROP_b(n, L[,], Z, lambda[], L0[], lambda_b[], L0_b[])

  v_b = 0.; v_pc_b = 0.;
  for (i=N .. eta[n])
    lam = lambda[i-eta[n]+1];
    c1 = del*lam; c2 = lam*h;
    //L[i,n+1] = L[i,n]*vrat_pc
    vrat_pc = L[i,n+1]/L[i,n];
    vrat_pc_b = L[i,n]*L_b[i,n+1];
    L_b[i,n] = vrat_pc*L_b[i,n+1];
    // vrat_pc = exp(c2*(-lam/2.+(v_pc+v)/2.)+lam*sqez)
    lambda_b[i-eta[n]+1] += scra[i,n+1]*vrat_pc_b;
    v_pc_b += vrat_pc*lam*h*vrat_pc_b/2.;
    v_b += vrat_pc*lam*h*vrat_pc_b/2.;
    // v_pc += (c1*Lhat[i,n+1])/(1.+del*Lhat[i,n+1])
    rpip = 1./(del*Lhat[i,n+1]+1.);
    Lhat_b[i,n+1] += (c1-c1*Lhat[i,n+1]*del*rpip)*rpip*v_pc_b;
    c1_b = Lhat[i,n+1]*rpip*v_pc_b;
    // Lhat[i,n+1] = L[i,n]*vrat
    vrat_b = L[i,n]*Lhat_b[i,n+1];
    vrat = Lhat[i,n+1]/L[i,n];
    L_b[i,n] += vrat*Lhat_b[i,n+1];
    // vrat = exp(lam*h*(-lam/2.+v)+lam*sqez)
    lambda_b[i-eta[n]+1] += hat_scra[i,n+1]*vrat_b;
    v_b += vrat*lam*h*vrat_b;
    // v += (c1*L[i,n])/(1.+del*L[i,n])
    rpip = 1./(del*L[i,n]+1.);
    L_b[i,n] += (c1-c1*L[i,n]*del*rpip)*rpip*v_b;
    c1_b += L[i+n]*rpip*v_b;
    // lam = lambda[i-eta[n]+1]; c1 = del*lam
    lambda_b[i-eta[n]+1] += del*c1_b;

  for (i=eta[n]-1 .. 1)
    // L[i,n+1] = L[i,n]
    L_b[i,n] += L_b[i,n];

  if(n=0)
    for(i=1 .. N)
      // L[i,n] = L0[i]
      L0_b[i] = L0[i,0];
```

FIGURE 11.2: Adjoint of the propagation method $\texttt{PROP\_b}_n$ for the Libor Market Model of Equation 11.14 for $d_W = 1$, under the predictor corrector Euler approximation 11.18, and the volatility parameterization 11.16. The corresponding forward method is shown in Figure 11.1. The instructions commented are the forward counterpart to the adjoint instructions immediately after.

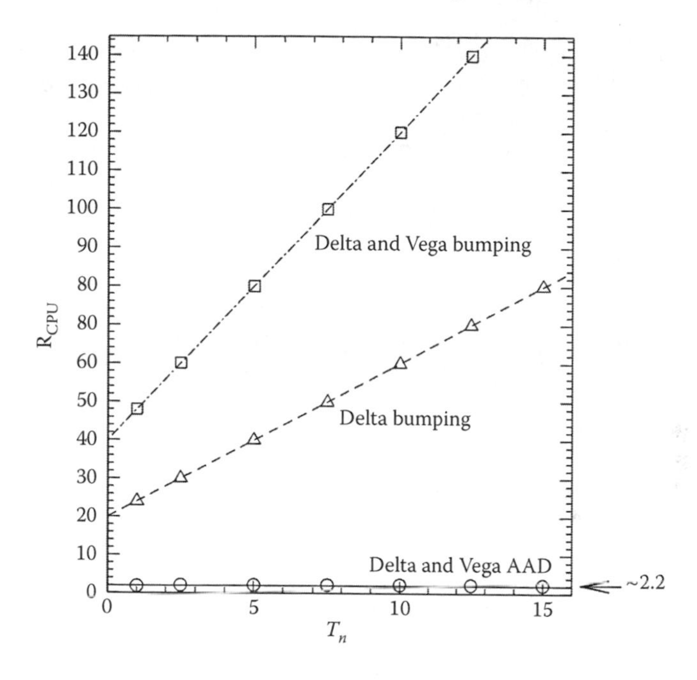

FIGURE 11.3: Ratio of the CPU time required for the AAD calculation of the Delta and Vega and the time to calculate the option value for the swaption in Equation 11.20 as a function of the option expiry T_n. The time to calculate Delta and Vega using bumping is also shown. Lines are guides for the eye.

Here we consider European style payouts. The extension to Bermuda options of Leclerc and collaborators [11] can be obtained with a simple modification of the original algorithm.

The remarkable computational efficiency of the implementation discussed earlier is illustrated in Figure 11.3. Here we plot the execution time for the calculation of all the Delta, $\partial V/\partial L_i(0)$, and Vega, $\partial V/\partial \sigma_i(n)$, relative to the calculation of the swaption value as obtained with the implementation above and by finite differences. As the maturity of the swaption increases, the number of risk to compute also increases. For typical applications, AAD results in orders of magnitude speedups with respect to bumping.

11.4 Real-Time Counterparty Credit Risk Management

One of the most active areas of risk management today is CCRM. Managing counterparty risk is particularly challenging because it requires the simultaneous evaluation of all the trades facing a given counterparty. For multiasset portfolios, this typically comes with extraordinary computational challenges.

Indeed, with the exclusion of the simplest portfolios of vanilla instruments, computationally intensive MC simulations are often the only practical tool available for this task. Standard approaches for the calculation of risk require repeating the calculation of the P&L of the portfolio under hundreds of market scenarios. As a result, in many cases these calculations cannot be completed in a practical amount of time, even employing a vast amount of computer power. Since the total cost of the through-the-life risk management can determine whether it is profitable to execute a new trade, solving this technology problem is critical to allow a securities firm to remain competitive.

In this section, we demonstrate how this powerful technique can be used for a highly efficient computation of price sensitivities in the context of CCRM [12].

11.4.1 Counterparty credit risk management

As a typical task in the day-to-day operation of a CCRM desk, here we consider the calculation of the credit valuation adjustment (CVA) as the main measure of a dealer's counterparty credit risk. For a given portfolio of trades facing the same investor or institution, the CVA aims to capture the expected loss associated with the counterparty defaulting in a situation in which the position, netted for any collateral agreement, has a positive mark-to-market for the dealer. This can be evaluated at time $T_0 = 0$ as

$$V_{\mathrm{CVA}} = \mathbb{E}\Big[\mathbb{I}(\tau_c \leq T) D(0,\tau_c) L_{\mathrm{GD}}(\tau_c)\Big(NPV(\tau_c) - C(R(\tau_c^-))\Big)^+\Big], \tag{11.23}$$

where τ_c is the default time of the counterparty, $NPV(t)$ is the net present value of the portfolio at time t from the dealer's point of view, $C(R(t))$ is the collateral outstanding, typically dependent on the rating R of the counterparty, $L_{\mathrm{GD}}(t)$ is the loss given default, $D(0,t)$ is the discount factor for the interval $[0,t]$, and $\mathbb{I}(\tau_c \leq T)$ is the indicator that the counterparty's default happens before the longest deal maturity in the portfolio, T. Here for simplicity of notation we consider the unilateral CVA, the generalization to bilateral CVA [13] is straightforward. The quantity in Equation 11.23 is typically computed on a discrete time grid of "horizon dates" $T_0 < T_1 < \cdots < T_{N_O}$ as, for instance,

$$V_{\mathrm{CVA}} \simeq \sum_{i=1}^{N_O} \mathbb{E}\Big[\mathbb{I}(T_{i-1} < \tau_c \leq T_i) D(0,T_i) L_{\mathrm{GD}}(T_i)\Big(NPV(T_i) - C\left(R(T_i^-)\right)\Big)^+\Big]. \tag{11.24}$$

In general, the quantity above depends on several correlated random market factors, including interest rate, counterparty's default time and rating, recovery amount, and all the market factors the net present value of the portfolio depends on. As such, its calculation requires an MC simulation.

To simplify the notation and generalize the discussion beyond the small details that might enter in a dealer's definition of a specific credit charge, here we consider expectation values of the form

$$V = \mathbb{E}_{\mathbb{Q}}[P(R, X)], \tag{11.25}$$

with "payout" given by

$$P = \sum_{i=1}^{N_O} P\left(T_i, R(T_i), X(T_i)\right), \tag{11.26}$$

where

$$P\left(T_i, R(T_i), X(T_i)\right) = \sum_{r=0}^{N_R} \tilde{P}_i(X(T_i); r)\, \delta_{r,R(T_i)}. \tag{11.27}$$

Here the rating of the counterparty entity including default, $R(t)$, is represented by an integer $r = 0, \ldots, N_R$ for simplicity; $X(t)$ is the realized value of the M market factors at time t. $\mathbb{Q} = \mathbb{Q}(R, X)$ represents a probability distribution according to which $R = (R(T_1), \ldots, R(T_{N_0}))^t$ and $X = (X(T_1), \ldots, X(T_{N_0}))^t$ are distributed; $\tilde{P}_i(\cdot; r)$ is a rating-dependent payout at time T_i.[2]

The expectation value in Equation 11.25 can be estimated by means of MC by sampling a number N_{MC} of random replicas of the underlying rating and market state vector, $R[1], \ldots, R[N_{\mathrm{MC}}]$ and $X[1], \ldots, X[N_{\mathrm{MC}}]$, according to the distribution $\mathbb{Q}(R, X)$, and evaluating the payout $P(R, X)$ for each of them.

In the following, we will make minimal assumptions on the particular model employed to describe the dynamics of the market factors. In particular, we will only assume that for a given MC sample the value at time T_i of the market factors can be obtained from their value at time T_{i-1} by means of a mapping of the form $X(T_i) = F_i(X(T_{i-1}), Z^X)$ where Z^X is an N^X-dimensional vector of correlated standard normal random variates, $X(T_0)$ is today's value of the market state vector, and F_i is a mapping regular enough for the pathwise derivative method to be applicable [9], as it is generally the case for practical applications.

As an example of a counterparty rating model generally used in practice, here we consider the rating transition Markov chain model of Jarrow et al. [14] in which the rating at time T_i can be simulated as

$$R(T_i) = \sum_{r=1}^{N_R} \mathbb{I}\left(\tilde{Z}_i^R > Q(T_i, r)\right), \tag{11.28}$$

where $\tilde{Z}_i^R$ is a standard normal variate and $Q(T_i, r)$ is the quantile-threshold corresponding to the transition probability from today's rating to a rating r at time T_i. Note that the discussion below is not limited to this particular

[2]The discussion below applies also to the case in which the payout at time T_i depends on the history of the market factors X up to time T_i.

model, and it could be applied with minor modifications to other commonly used models describing the default time of the counterparty, and its rating [15]. Here we consider the rating transition model 11.28 for its practical utility, as well as for the challenges it poses in the application of the pathwise derivative method, because of the discreteness of its state space.

In this setting, MC samples of the payout estimator in Equation 11.10 can be generated according to the following standard algorithm. For $i = 1, \ldots, N_O$:

Step 1 Generate a sample of $N^X + 1$ jointly normal random variables $(Z_i^R, Z_i^X) \equiv (Z_i^R, Z_{i,1}^X, \ldots, Z_{i,N_X}^X)^t$ distributed according to $\phi(Z_i^R, Z^X; \rho_i)$, an $(N_X + 1)$-dimensional standard normal probability density function with correlation matrix ρ_i, for example, with the first row and column corresponding to the rating factor.

Step 2 Iterate the recursion: $X(T_i) = F_i(X(T_{i-1}), Z^X)$.

Step 3 Set $\tilde{Z}_i^R = \sum_{j=1}^{i} Z_j^R / \sqrt{i}$ and compute $R(T_i)$ according to Equation 11.28.[3]

Step 4 Compute the time T_i payout estimator $P(T_i, R(T_i), X(T_i))$ in Equation 11.27, and add this contribution to the total estimator in Equation 11.26.

The calculation of risk can be obtained in an highly efficient way by implementing the pathwise derivative method [5] according to the principles of AAD [2–4]. In particular, the pathwise derivative estimator reads in this case

$$\bar{\theta}_k \equiv \frac{\partial P_\theta(R, X)}{\partial \theta_k} = \sum_{i=1}^{N_O} \sum_{l=1}^{M} \frac{\partial P_\theta(R, X)}{\partial X_l(T_i)} \times \frac{\partial X_l(T_i)}{\partial \theta_k} + \frac{\partial P_\theta(R, X)}{\partial \theta_k}, \qquad (11.29)$$

where we have allowed for an explicit dependence of the payout on the model parameters. Due to the discreteness of the state space of the rating factor, the pathwise estimator for its related sensitivities is not well defined. However, as we will show below, one can express things in such a way that the rating sensitivities are incorporated in the explicit term $\partial P_\theta(R, X)/\partial \theta_k$.

In the following, we will show how the calculation of the pathwise derivative estimator 11.29 can be implemented efficiently by means of AAD.

11.4.2 Adjoint algorithmic differentiation and the counterparty credit risk management

When applied to the pathwise derivative method, AAD allows the simultaneous calculation of the pathwise derivative estimators for an arbitrarily

[3] Here we have used the fact that the payout 11.27 depends on the outturn value of the rating at time T_i and not on its history.

large number of sensitivities at a *small fixed cost*. Here we describe in detail the AAD implementation of the pathwise derivative estimator 11.29 for the CCRM problem 11.23.

As noted above, the sensitivities with respect to parameters affecting the rating dynamics need special care due to discrete nature of the state space. However, setting these sensitivities aside for the moment, the AAD implementation of the pathwise derivative estimator consists of Steps 1–4 described earlier plus the following steps of the backward sweep. For $i = N_O, \ldots, 1$:

Step $\bar{4}$ Evaluate the adjoint of the payout,

$$(\bar{X}(T_i), \bar{\theta}) = \bar{P}(T_i, R(T_i), X(T_i), \theta, \bar{P}),$$

with $\bar{P} = 1$.

Step $\bar{3}$ Nothing to do: the parameters θ do not affect this nondifferentiable step.

Step $\bar{2}$ Evaluate the adjoint of the propagation rule in Step 2.

$$(\bar{X}(T_{i-1}), \bar{\theta}) += \bar{F}_i(X(T_{i-1}), \theta, Z^X, \bar{X}(T_i), \bar{\theta}),$$

where $+=$ is the standard addition assignment operator.

Step $\bar{1}$ Nothing to do: the parameters θ do not affect this step.

A few comments are in order. In Step $\bar{4}$, the adjoint of the payout function is defined while keeping the discrete rating variable constant. This provides the derivatives $\bar{X}_l(T_i) = \partial P_\theta / \partial X_l(T_i)$ and $\bar{\theta}_k = \partial P_\theta / \partial \theta_k$. In defining the adjoint in Step $\bar{2}$, we have taken into account that the propagation rule in Step 2 is explicitly dependent on both $X(T_i)$ and the model parameters θ. As a result, its adjoint counterpart produces contributions to both $\bar{\theta}$ and $\bar{X}(T_i)$. Both the adjoint of the payout and of the propagation mapping can be implemented following the principles of AAD as discussed in References 2 and 3. In many situations, AD tools can be also used as an aid or to automate the implementation, especially for simpler, self-contained functions. In the backward sweep above, Steps $\bar{1}$ and $\bar{3}$ have been skipped because we have assumed for simplicity of exposition that the parameters θ do not affect the correlation matrices ρ_i, and the rating dynamics. If correlation risk is instead required, Step $\bar{2}$ produces also the adjoint of the random variables Z^X, and Step $\bar{1}$ contains the adjoint of the Cholesky decomposition, possibly with the support of the binning technique, as described in Reference 4.

11.4.2.1 Rating transition risk

The risk associated with the rating dynamics can be treated by noting that Equation 11.27 can be expressed more conveniently as

$$P\left(T_i, \tilde{Z}_i^R, X(T_i)\right) = \tilde{P}_i(X(T_i); 0) + \sum_{r=1}^{N_R} \left(\tilde{P}_i(X(T_i); r) - \tilde{P}_i(X(T_i); r{-}1)\right) \times \mathbb{I}\left(\tilde{Z}_i^R > Q(T_i, r; \theta)\right), \quad (11.30)$$

so that the singular contribution to the pathwise derivative estimator reads

$$\partial_{\theta_k} P(T_i, \tilde{Z}_i, X(T_i)) = -\sum_{r=1}^{N_R} \left(\tilde{P}_i(X(T_i); r) - \tilde{P}_i(X(T_i); r - 1)\right) \times \delta\left(\tilde{Z}_i^R = Q(T_i, r; \theta)\right) \times \partial_{\theta_k} Q(T_i, r; \theta). \quad (11.31)$$

This estimator cannot be sampled in this form with MC. Nevertheless, it can be integrated out using the properties of Dirac's delta along the lines of [16], giving after straightforward computations,

$$\bar{\theta}_k = -\sum_{r=1}^{N_R} \frac{\phi(Z^\star, Z_i^X, \rho_i)}{\sqrt{i}\,\phi(Z_i^X, \rho_i^X)} \partial_{\theta_k} Q(T_i, r; \theta)\left(\tilde{P}_i(X(T_i); r) - \tilde{P}_i(X(T_i); r{-}1)\right), \quad (11.32)$$

where $Z^\star$ is such that $(Z^\star + \sum_{j=1}^{i-1} Z_j^R)/\sqrt{i} = Q(T_i, r; \theta)$ and $\phi(Z_i^X, \rho_i^X)$ is a N_X-dimensional standard normal probability density function with correlation matrix ρ_i^X obtained by removing the first row and column of ρ_i; here $\partial_{\theta_k} Q(T_i, r; \theta)$ is not stochastic and can be evaluated (e.g., using AAD) once per simulation. The final result is rather intuitive as it is given by the probability weighted sum of the discontinuities in the payout.

11.4.3 Results

As a numerical test, we present here results for the calculation of risk on the CVA of a portfolio of swaps on commodity Futures. For the purpose of this illustration, we consider a simple one-factor lognormal model for the Futures curve of the form

$$\frac{\mathrm{d}F_T(t)}{F_T(t)} = \sigma_T \exp(-\beta(T - t))\,\mathrm{d}W_t, \quad (11.33)$$

where W_t is a standard Brownian motion; $F_T(t)$ is the price at time t of a Futures contract expiring at T; σ_T and β define a simple instantaneous volatility function that increases approaching the contract expiry, as empirically observed for many commodities. The value of the Futures' price $F_T(t)$

can be simulated exactly for any time t so that the propagation rule in Step 2 reads for $T_i \leq T$

$$F_T(T_i) = F_T(T_{i-1}) \exp\left(\sigma_i \sqrt{\Delta T_i} Z - \frac{1}{2}\sigma_i^2 \Delta T_i\right), \tag{11.34}$$

where $\Delta T_i = T_i - T_{i-1}$, and

$$\sigma_i^2 = \frac{\sigma_T^2}{2\beta \Delta T_i} e^{-2\beta T}\left(e^{2\beta T_i} - e^{2\beta T_{i-1}}\right)$$

is the outturn variance. In this example, we will consider deterministic interest rates. As underlying portfolio for the CVA calculation, we consider a set of commodity swaps, paying on a strip of Futures (e.g., monthly) expiries t_j, $j = 1, \ldots, N_e$ the amount $F_{t_j}(t_j) - K$. The time t net present value for this portfolio reads

$$NPV(t) = \sum_{j=1}^{N_e} D(t, t_j)\left(F_{t_j}(t) - K\right). \tag{11.35}$$

Note that although we consider here for simplicity of exposition a linear portfolio, the method proposed applies to an arbitrarily complex portfolio of derivatives, for which in general the NPV will be a nonlinear function of the market factors $F_{t_j}(t)$ and model parameters θ.

For this example, the adjoint propagation rule in Step $\bar{2}$ simply reads

$$\bar{F}_T(T_i - 1) += \bar{F}_T(T_i) \exp\left(\sigma_i \sqrt{\Delta T_i} Z - \frac{1}{2}\sigma_i^2 \Delta T_i\right),$$
$$\bar{\sigma}_i = \bar{F}_T(T_i) F(T_i)\left(\sqrt{\Delta T_i}\, Z - \sigma_i \Delta T_i\right)$$

with $\bar{\sigma}_i$ related to this step's contribution to the adjoint of the Futures' volatility $\bar{\sigma}_T$ by

$$\bar{\sigma}_T += \frac{\bar{\sigma}_i}{\sqrt{2\beta \Delta T_i}} \sqrt{e^{-2\beta T}\left(e^{2\beta T_i} - e^{2\beta T_{i-1}}\right)}.$$

At the end of the backward path, $\bar{F}_T(0)$ and $\bar{\sigma}_T$ contain the pathwise derivative estimator 11.29 corresponding, respectively, to the sensitivity with respect to today's price and volatility of the Futures contract with expiry T.

The remarkable computational efficiency of the AAD implementation is clearly illustrated in Figure 11.4. Here we plot the speedup produced by AAD with respect to the standard finite-difference method. On a fairly typical trade horizon of 5 years, for a portfolio of 5 swaps referencing distinct commodities Futures with monthly expiries, the CVA bears nontrivial risk to over 600 parameters: 300 Futures prices ($F_T(0)$), and at the money volatilities (σ_T), (say) 10 points on the zero rate curve, and 10 points on the CDS curve of the counterparty used to calibrate the transition probabilities of the rating transition model 11.28. As illustrated in Figure 11.4, the CPU time required for the calculation of the CVA, and its sensitivities, is less than 4 times the

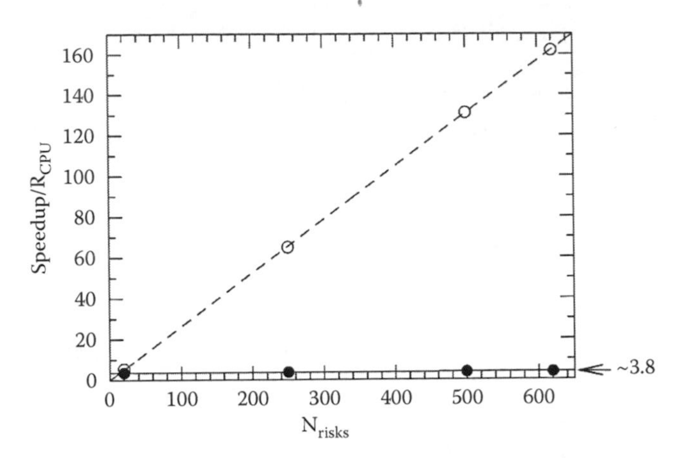

FIGURE 11.4: Speedup in the calculation of risk for the CVA of a portfolio of 5 commodity swaps over a 5-year horizon, as a function of the number of risks computed (empty dots). The full dots are the ratio of the CPU time required for the calculation of the CVA, and its sensitivities, and the CPU time spent for the computation of the CVA alone. Lines are guides for the eye.

TABLE 11.1: Variance reduction (VR) on the sensitivities with respect to the thresholds $Q(1, r)$ ($N_R = 3$) for a call option with a rating-dependent strike

δ	$VR[Q(1,1)]$	$VR[Q(1,2)]$	$VR[Q(1,3)]$
0.1	24	16	12
0.01	245	165	125
0.001	2490	1640	1350

Note: δ Indicates the perturbation used in the finite-difference estimators of the sensitivities.[4]

CPU time spent for the computation of the CVA alone, as predicted by Equation 11.8. As a result, even for this very simple application, AAD produces risk over 150 times faster than finite differences, that is, for a CVA evaluation taking 10 seconds, AAD produces the full set of sensitivities in less than 40 seconds, while finite differences require approximately 1 hour and 40 minutes.

Moreover, as a result of the analytic integration of the singularities introduced by the rating process, the risk produced by AAD is typically less noisy than the one produced by finite differences. This is clearly illustrated in Table 11.1 showing the variance reduction on the sensitivities with respect to the thresholds $Q(T_i, r)$ for a simple test case. Here we have considered the calculation of a call option of the form $(F_T(T_i) - C(R(T_i)))^+$ with a strike $C(R(T_i))$ linearly dependent on the rating, and $T_i = 1$. The variance reduction displayed in the table can be thought of as a further speedup factor because

[4]The Specification of the Parameters used for This Example is Available upon Request.

it corresponds to the reduction in the computation time for a given statistical uncertainty on the sensitivities. This diverges as the perturbation in the finite-difference estimators δ tends to zero and may be very significant even for a fairly large value of δ.

In conclusion, these numerical results illustrate how AAD allows an extremely efficient calculation of counterparty credit risk valuations in MC. In fact, AAD allows one to perform in minutes risk runs that would take otherwise several hours or could not even be performed overnight without large parallel computers.

11.5 Real-Time Risk Management of Flow Credit Products

The aftermath of the recent financial crisis has seen a dramatic shift in the credit derivatives markets, with a conspicuous reduction of demand for complex, capital intensive products, like bespoke collateralized debt obligations (CDO), and a renewed focus on simpler and more liquid derivatives, such as credit default indices and swaptions [17].

In this background, dealers are quickly adapting to a business model geared toward high-volume, lower margin products for which managing efficiently the trading inventory is of paramount importance. As a result, the ability to produce risk in real time is rapidly becoming one of the keys to running a successful trading operation.

In this section, we demonstrate how AAD can be extremely effective also for simpler credit products, typically valued by means of faster semianalytical techniques. We will show how AAD provides orders of magnitude savings in computational time and makes the computation of risk in real time—with no additional infrastructure investment—a concrete possibility [18].

11.5.1 Pricing of credit derivatives

The key concept for the valuation of credit derivatives, in the context of the models generally used in practice, is the *hazard rate*, λ_u, representing the probability of default per unit time of the reference entity between times u and $u + du$, conditional on survival up to time u. By modeling the default event of a reference entity i as the first arrival time of a Poisson process with deterministic intensity λ_u^i, the survival probability, $Q_i(t, T)$, is given by

$$Q(t, T; \lambda^i) = \exp\left[- \int_t^T du \ \lambda_u^i \right] . \tag{11.36}$$

In the hazard rate framework, the price of a credit derivative can be expressed mathematically as

$$V(\theta) = V(\boldsymbol{\lambda}(\theta), \theta), \tag{11.37}$$

where $\boldsymbol{\lambda} = (\lambda^1, \ldots, \lambda^N)$ are the hazard rate functions for N credit entities referenced in a given contract. Here we have indicated generically with $\theta = (\theta_1, \ldots, \theta_{N_\theta})$ the vector of model parameters, for example, credit spreads, recovery rates, volatilities, correlation and the market prices of the interest rate instruments used for the calibration of the discount curve.

In general, the valuation of a credit derivative can be separated in a

Calibration Step:

$$\theta \to \boldsymbol{\lambda}(\theta)$$

for the construction of the hazard rate curve given liquidly traded CDS prices, a term structure of recoveries and a given discount curve, and a

Pricing Step:

$$\theta \to V(\boldsymbol{\lambda}(\theta), \theta)$$

mapping the hazard rate curves and the other parameters to which the pricing model is explicitly dependent on, to the price of the credit derivative. The pricing step is obviously specific to the particular credit derivative under valuation. Instead, the calibration step is the same for any derivative priced within the hazard rate framework. For the purpose of the discussion below, it is useful to recall the main steps involved in the calibration of a hazard rate curve.

11.5.1.1 Calibration step

The hazard rate function λ_u in Equation 11.36 is commonly parameterized as piecewise constant with M knot points at time $(t_1, \ldots, t_M)$, $\lambda = (\lambda_1, \ldots, \lambda_M)$, such that

$$\lambda_u = \lambda_{n-1} = \frac{1}{t_n - t_{n-1}} \ln\left(\frac{Q(t, t_{n-1}; \lambda)}{Q(t, t_n; \lambda)}\right)$$

for $t_{n-1} \leq u < t_n$ and t_0 equal to the valuation date. In the calibration step, the hazard rate knot points are calibrated from the price, or equivalently the credit (par) spreads $(s_1, \ldots, s_M)$, of a set of liquidly traded CDS with maturities $T_1, \ldots, T_M$, for example, using the standard bootstrap algorithm [17].

Such calibration can be expressed mathematically as solving a system of M equations

$$G_j(\lambda, \theta) = 0, \tag{11.38}$$

$j = 1, \ldots M$, with

$$G_j(\lambda, \theta) = s_j - \frac{\mathcal{L}(t, T_j; \lambda, \theta)}{\mathcal{A}(t, T_j; \lambda, \theta)}, \tag{11.39}$$

where $\mathcal{L}(t, T_j; \lambda, \theta)$ and $\mathcal{A}(t, T_j; \lambda, \theta)$ are, respectively, the expected loss and *credit risky* annuity for a T_j maturity CDS contract starting at time t.[5] These are defined as

$$\mathcal{L}(t, T; \lambda, \theta) = \int_t^T \mathrm{d}u \; Z(t, u; \theta) \, (1 - R_u) \left(- \frac{\mathrm{d}Q(t, u; \lambda)}{\mathrm{d}u} \right), \tag{11.40}$$

and (e.g., for continuously paid coupons)

$$\mathcal{A}(t, T; \lambda, \theta) = \int_t^T \mathrm{d}u \; Z(t, u; \theta) Q(t, u; \lambda). \tag{11.41}$$

Here, $Z(t, u; \theta)$ is the discount factor from time t to time u, $Q(t, u; \lambda)$ (resp. $-\mathrm{d}Q(t, u; \lambda)$) is the probability that the reference entity survives up to (resp. defaults in an infinitesimal interval around) time u, and R_u is the expected percentage recovery upon default at time u. The latter is generally expressed as a piecewise constant function with the same discretization of the hazard rate function, say $R = (R_1, \ldots, R_M)$.

The calibration Equations 11.38 and 11.39 are based on the definition of par spread s_i as break-even coupons c making the value of CDS

$$V_{\mathrm{CDS}}(t, T; \theta) = \mathcal{L}(t, T; \lambda, \theta) - c \, \mathcal{A}(t, T; \lambda, \theta), \tag{11.42}$$

worth zero.[6] Since both the expected loss and the risky annuity at time T_i depend on hazard rate points λ_j with $j \leq i$, the calibration equations can be solved iteratively starting from $i = 1$, by keeping fixed the hazard rate knot points λ_j with $j < i$, and solving for λ_i.

Through the calibration process, the system of M Equation 11.38 defines *implicitly* the function $\lambda = \lambda(\theta)$, linking the hazard rate to the credit spreads, the term structure of expected recovery and the discount factors. These are in turn a function of the market instruments that are used for the calibration of the discount curve.

11.5.2 Challenges in the calculation of credit risk

The computation of the sensitivities of the price of the credit derivative 11.37 with respect to the model parameters θ can be performed by means of

[5] Note that although the credit spreads s_j are contained in the model parameter vector θ, the risky annuity and the expected loss do not depend explicitly on them.

[6] Note that since the standardization of CDS contracts in 2008, liquidly traded CDS are characterized by a standard coupon and are generally quoted in terms of upfronts or quote spreads. Both mark types can be mapped to a dollar value of a CDS contract by means of a market standard parameterization [19], and hazard rates can be equivalently bootstrapped from these marks using Equation 11.42. Credit (par) spreads remain nonetheless commonly used in the market practice as risk factors for credit derivatives. The analysis of this paper can be easily formulated in terms of quote spreads or upfronts.

the chain rule

$$\frac{\mathrm{d}V}{\mathrm{d}\theta_k} = \frac{\partial V}{\partial \theta_k} + \sum_{j=1}^{M} \frac{\partial V}{\partial \lambda_j} \frac{\partial \lambda_j}{\partial \theta_k}, \tag{11.43}$$

where the first term captures the explicit dependence on the model parameters θ through the pricing step and the second term captures the implicit dependence via the calibration step.

The computation of the calibration component of the price sensitivities with standard bump and reval approaches is particularly onerous because it involves repeating the calibration step for each perturbation. Especially for portfolio of simple credit derivatives, like CDS, this can easily represent the bulk of the computational burden. In addition, finite size perturbations of credit spreads, recovery, or interest rates often correspond to inputs that do not admit an arbitrage-free representation in terms of a nonnegative hazard rate curve, thus making the robust and stable computation of sensitivities challenging.

11.5.3 Adjoint calculation of risk

Both the computational costs and stability of the calculation of credit risk can be effectively addressed by means of the AAD implementation of the chain rule 11.43. In particular, the adjoint of the algorithm consisting of the *Calibration Step* and *Pricing Step*, described earlier reads

$\overline{\textit{Pricing Step}}$:

$$\bar{\theta}_k = \bar{V} \frac{\partial V}{\partial \theta_k} \quad \bar{\lambda}_j = \bar{V} \frac{\partial V}{\partial \lambda_j}, \tag{11.44}$$

$\overline{\textit{Calibration Step}}$:

$$\bar{\theta}_k = \bar{\theta}_k + \sum_{j=1}^{M} \bar{\lambda}_j \frac{\partial \lambda_j}{\partial \theta_k}. \tag{11.45}$$

Although in the following we will give explicit examples of the adjoint of the pricing step for portfolios of CDS and credit default index swaptions, here we focus our discussion on the adjoint of the calibration step in Equation 11.45 which is a time-consuming and numerically challenging step common to all pricing applications within the hazard rate framework.

11.5.4 Implicit function theorem

The adjoint of the calibration step $\theta \rightarrow \lambda(\theta)$ can be produced following the general rules of AAD. The associated computational cost can be generally expected to be of the order of the cost of performing the bootstrap algorithm a few times (but approximately less than 4 according to the general result of AAD quoted above). This in itself is generally a very significant improvement with respect to bump and reval approaches, involving repeating the bootstrap algorithm as many times as sensitivities required. However, following the suggestions of [20,21], a much better performance can be obtained by exploiting the so-called *implicit function theorem*, as described later.

By differentiating with respect to θ the calibration identity 11.38, we get

$$\frac{\partial G_i}{\partial \theta_k} + \sum_{j=1}^{M} \frac{\partial G_i}{\partial \lambda_j}\frac{\partial \lambda_j}{\partial \theta_k} = 0,$$

for $i = 1, \ldots, M$, and $k = 1, \ldots, N_\theta$, or equivalently

$$\frac{\partial \lambda_i}{\partial \theta_k} = -\left[\left(\frac{\partial G}{\partial \lambda}\right)^{-1} \frac{\partial G}{\partial \theta}\right]_{ik}. \tag{11.46}$$

This relation allows the computation of the sensitivities of $\lambda(\theta)$, locally defined in an implicit fashion by Equations 11.38 and 11.39, in terms of the sensitivities of the function 11.39.

In the specific case, when $\theta_k \neq s_j$ for $j = 1, \ldots, M$, that is, when considering sensitivities with respect to market risk factors other than the credit spreads, Equation 11.46 can be expressed in turn as

$$\frac{\partial \lambda_i}{\partial \theta_k} = -\left[\left(\frac{\partial s(\lambda,\theta)}{\partial \lambda}\right)^{-1} \frac{\partial s(\lambda,\theta)}{\partial \theta}\right]_{ik}. \tag{11.47}$$

Here we have used that θ_k is not a credit spread so that $\partial G/\partial\theta_k = -\partial s(\lambda,\theta)/\partial\theta_k$, where the par spread *functions*

$$\begin{aligned} s(\lambda,\theta) &= (s_1(\lambda,\theta), \ldots, s_M(\lambda,\theta)), \\ s_j(\lambda,\theta) &= \frac{\mathcal{L}(t, T_j; \lambda, \theta)}{\mathcal{A}(t, T_j; \lambda, \theta)} \end{aligned} \tag{11.48}$$

are defined by Equations 11.38 and 11.39.

In the case of credit spread sensitivities, $\theta_k = s_k$, Equation 11.46 simplifies as follows:

$$\frac{\partial \lambda_i}{\partial s_k} = \sum_{j=1}^{M} \left[\frac{\partial s(\lambda,\theta)}{\partial \lambda}\right]^{-1}_{ij} \frac{\partial G_j}{\partial s_k} = \sum_{j=1}^{M} \left[\frac{\partial s(\lambda,\theta)}{\partial \lambda}\right]^{-1}_{ij} \frac{\partial s_j}{\partial s_k} = \left[\frac{\partial s(\lambda,\theta)}{\partial \lambda}\right]^{-1}_{ik}, \tag{11.49}$$

where we have used that the par spread functions do not explicitly depend on the credit spreads s_k.

Equations 11.47 and 11.49 express the implicit function theorem in the context of hazard rate calibration. These allow the computation of the sensitivities $\partial\lambda_i/\partial\theta_k$ by (i) evaluating the sensitivities of the par spread functions with respect to the model parameters, $\partial s_j(\lambda, \theta)/\partial \theta_k$, and the hazard rates, $\partial s_k(\lambda, \theta)/\partial \lambda_i$, and (ii) solving a linear system, for example, by Gaussian elimination. This method is significantly more stable and efficient than the naïve approach of calculating the derivatives of the implicit functions $\theta \to \lambda(\theta)$ by differentiating directly the calibration step either by bump and reval or by applying AAD to the calibration step. This is because $s(\lambda, \theta)$ in Equation 11.48 are *explicit* functions of the hazard rate and the model parameters that are easy to compute and differentiate.

Combining the implicit function theorem with adjoint methods results in extremely efficient risk computations, as we will demonstrate later.

11.5.5 Adjoint of the calibration step

All the sensitivities necessary to compute Equations 11.47 and 11.49 can be obtained through the adjoint of the function

$$s_j = s_j(\lambda, \theta)$$

defined by Equation 11.48, namely, using the definitions 11.1 and 11.6,

$$(\bar{\lambda}, \bar{\theta}) = \bar{s}_j(\lambda, \theta, \bar{s}_j),$$

where the scalar $\bar{s}_j$ is the adjoint of the jth par spread with $j = 1, \ldots, M$. By applying the rules of AAD, this can be implemented as

$$\begin{aligned}
\overline{\mathcal{A}}_j &= -\bar{s}_j \frac{\mathcal{L}(t, T_j; \lambda, \theta)}{\mathcal{A}(t, T_j; \lambda, \theta)^2} \\
\overline{\mathcal{L}}_j &= \bar{s}_j \frac{1}{\mathcal{A}(t, T_j; \lambda, \theta)} \\
(\bar{\lambda}, \bar{\theta}) &\mathrel{+}= \overline{\mathcal{A}}(t, T_j; \lambda, \theta, \overline{\mathcal{A}}_j) \\
(\bar{\lambda}, \bar{\theta}) &\mathrel{+}= \overline{\mathcal{L}}(t, T_j; \lambda, \theta, \overline{\mathcal{L}}_j),
\end{aligned}$$

where $\overline{\mathcal{A}}(t, T_j; \lambda, \theta, \overline{\mathcal{A}}_j)$ and $\overline{\mathcal{L}}(t, T_j; \lambda, \theta, \overline{\mathcal{L}}_j)$ are the adjoints of $\mathcal{A}(t, T_j; \lambda, \theta)$ and $\mathcal{L}(t, T_j; \lambda, \theta)$, respectively.

Combining AAD and the implicit function theorem results therefore in the following algorithm for the adjoint of the calibration routine, $\bar{\theta} = \bar{\lambda}(\theta, \bar{\lambda})$:

1. Execute $(\bar{\lambda}, \bar{\theta}) = \bar{s}_j(\lambda, \theta, \bar{s}_j)$ with $\bar{s}_j = 1$ for $j = 1, \ldots, M$. This gives the derivatives:

   $$\bar{\lambda}_{ij} = \frac{\partial s_j}{\partial \lambda_i} \quad \bar{\theta}_{kj} = \frac{\partial s_j}{\partial \theta_k},$$

   for $i = 1, \ldots, M$ and $k = 1, \ldots, N_\theta$.

2. Find the matrix $\partial \lambda / \partial \theta$ by solving the linear system

   $$\frac{\partial s}{\partial \lambda} \frac{\partial \lambda}{\partial \theta} = -\frac{\partial s}{\partial \theta}.$$

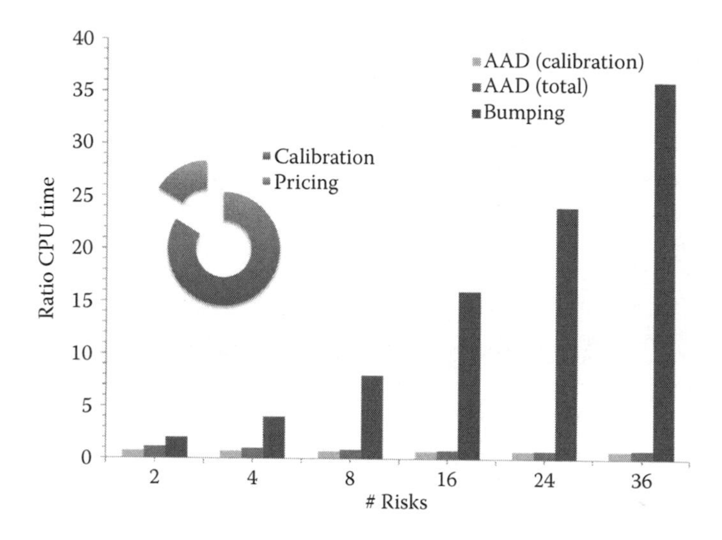

FIGURE 11.5: Cost of computing the sensitivities with respect to the credit spreads and interest rate instruments—relative to the cost of a single valuation—as a function of the number of sensitivities.

3. Return:

$$\bar{\theta}_k = \sum_{i=1}^{M} \bar{\lambda}_i \frac{\partial \lambda_i}{\partial \theta_k},$$

for $k = 1, \ldots, M$.

The adjoint of the calibration algorithm described earlier is extremely efficient. Indeed, as illustrated in Figure 11.5, the sensitivities of the hazard rate with respect to the credit spreads, and interest rate instruments can be computed in ~25% less time than performing a single bootstrap.

11.5.6 Results

11.5.6.1 Credit default swaps

As a first example, we consider the calculation of price sensitivities for a (portfolio of) CDS. In this case, the adjoint of the pricing step simply reads, from Equation 11.42,

$$\begin{aligned}
\overline{\mathcal{L}} &= \bar{V} \\
\overline{\mathcal{A}} &= -\bar{V}c \\
(\bar{\lambda}, \bar{\theta}) &= \overline{\mathcal{A}}(t, T; \lambda, \theta, \overline{\mathcal{A}}) \\
(\bar{\lambda}, \bar{\theta}) + &= \overline{\mathcal{L}}(t, T; \lambda, \theta, \overline{\mathcal{L}}),
\end{aligned}$$

where the risky annuity and expected loss (and their adjoint counterparts) are those of the CDS in the portfolio. In this case, as illustrated in Figure 11.5, the cost of the pricing step is a small portion (~10%) of the overall cost of computing the sensitivities which is instead dominated by the cost of the calibration step. As a result, all the sensitivities can be obtained by means of AAD for ~15% less than the cost of performing a single valuation. In typical applications, where computing sensitivities with respect to 18 spread tenors and interest rate instruments is commonplace, this results in a reduction of the computational cost by a factor of 50 or more.

11.5.6.2 Credit default index swaptions

As a second example, also of significant practical relevance, we consider credit default index swaptions. The value of these instruments at time t is given by

$$V_t = Z(t, T_E; \theta)\mathbb{E}_t\Big(\max\Big(\zeta\Big[V_{\mathrm{iCDS}}(T_S, T_E) + L(T_E) - P_E\Big], 0\Big)\Big) \qquad (11.50)$$

where $\zeta = 1$ for a payer and $\zeta = -1$ for a receiver option, $V_{\mathrm{iCDS}}(T_E, T_M)$ is the value at time T_E of the underlying credit default index swap (long protection) with standard coupon rate and maturity T_M, P_E is the exercise fee, and $L(T_E)$ is the value at time T_E of the loss given default associated to the names that have defaulted before expiry,

$$L(T_E) = \sum_{i=1}^{N} I(\tau^i < T_E) N^i (1 - R^i_\tau),$$

where N is the number of names in the index, I is the indicator function, and N^i, τ^i, and R^i_u are the notional, default time, and recovery function of the ith name in the portfolio.[7]

According to the *de facto* market standard model [22], the value at time T_E of the random quantity given by the sum of the loss amount, $L(T_E)$, and the value of the credit default index swap, $V_{\mathrm{iCDS}}(T_E, T_M)$, are modeled in terms of a single state variable, the default adjusted forward spread s_{T_E}, as

$$V_{\mathrm{iCDS}}(T_E, T_M) + L(T_E) = N_{\mathrm{tot}} \mathcal{A}_{\mathrm{isda}}\left(s_{T_E}, T_E, T_M\right)\left(s_{T_E} - c\right), \qquad (11.51)$$

where c is the fixed rate in the underlying credit default index swap and $N_{\mathrm{tot}} = \sum_{i=1}^{N} N^i$ is the total notional of the index. Here $\mathcal{A}_{\mathrm{isda}}(s, t, T)$ is the standardized risky annuity of Equation 11.41 calculated assuming a flat term structure of the credit spread s, according to the standard ISDA conventions [19]. In the simplest setting, the default adjusted forward spread is assumed lognormally distributed,

$$s_{T_E} = F_{T_E} \exp\left[-\frac{1}{2}\sigma^2_{T_E}(T_E - t) + \sigma_{T_E}\sqrt{T_E - t}\,\tilde{Z}\right], \qquad (11.52)$$

where σ_{T_E} is the volatility of the default adjusted forward spread, $\tilde{Z}$ is a standard normal random variable and the forward F_{T_E}, can be determined by

[7] Here for simplicity of exposition, we assume that no names in the index have defaulted at valuation time.

taking the expectation of both sides of Equation 11.51 giving

$$G_F(F_{T_E}, \boldsymbol{\lambda}, \theta) \equiv V^{\text{adj}}_{\text{iCDS}}(T_E, T_M; \boldsymbol{\lambda}, \theta) - V^{\text{isda}}_{\text{iCDS}}(T_E, T_M; F_{T_E}, \theta) = 0. \quad (11.53)$$

The first term in the equation above,

$$V^{\text{adj}}_{\text{iCDS}}(T_E, T_M; \boldsymbol{\lambda}, \theta) = \mathbb{E}_t\Big[V_{\text{iCDS}}(T_E, T_M) + L(T_E)\Big],$$

can be computed according to the standard hazard rate model using the time t default and recovery curves of the index constituents:

$$\begin{aligned} V^{\text{adj}}_{\text{iCDS}}(T_E, T_M; \boldsymbol{\lambda}, \theta) =& \tilde{\mathcal{L}}(t, T_E; \boldsymbol{\lambda}, \theta) + Z(t, T_E; \theta) \\ & \times \sum_{i=1}^{N} N^i \left(\mathcal{L}(T_E, T_M; \lambda^i, \theta) - c\,\mathcal{A}(T_E, T_M; \lambda^i, \theta)\right), \end{aligned} \quad (11.54)$$

with

$$\tilde{\mathcal{L}}(t, T_E; \boldsymbol{\lambda}, \theta) = Z(t, T_E; \theta) \sum_{i=1}^{N} N^i \tilde{\mathcal{L}}^i(t, T_E; \lambda^i, \theta),$$

where $\tilde{\mathcal{L}}^i(t, T; \lambda^i, \theta)$ is defined by setting in Equation 11.40 $Z(t, u; \theta) \to 1$ to reflect that the loss amounts occurred before option expiry are settled at T_E. The second term can be computed instead by numerical integration over the distribution of s_{T_E}, Equation 11.52,

$$V^{\text{isda}}_{\text{iCDS}}(T_E, T_M; F_{T_E}, \theta) = \mathbb{E}_t\Big[N_{\text{tot}} \mathcal{A}_{\text{isda}}\left(s_{T_E}, T_E, T_M\right)\left(s_{T_E} - c\right)\Big]. \quad (11.55)$$

The calibration Equation 11.53 defines implicitly the loss adjusted forward spread, F_{T_E}, as a function of its volatility σ_{T_E}, the hazard rates and expected recoveries of the index constituents, and the risk parameters of the discount curve, in short

$$F_{T_E} = F_{T_E}(\boldsymbol{\lambda}; \theta). \quad (11.56)$$

For a given set of input parameters θ and the calibrated hazard rates for the index constituents $\boldsymbol{\lambda}$, the pricing algorithm consists of the following steps:

Step 1 Calibrate the forward by solving the calibration Equation 11.53. This involves computing Equation 11.54 using the hazard rate model and Equation 11.55 by numerical integration for each trial value of F_{T_E}.

Step 2 Compute the option value 11.50 using Equation 11.51, for example, using Gaussian quadrature

$$V_t = Z(t, T_E) \sum_{k=1}^{L} w_k \phi(x_k; F_{T_E}, \theta) P_k, \tag{11.57}$$

where $\phi(x_k; F_{T_E}, \theta)$ is the probability density function of s_{T_E},

$$P_k = \left(\zeta \left[N \mathcal{A}_{\text{isda}}(x_k, T_E, T_M)(x_k - c) - P_E\right]\right)^+, \tag{11.58}$$

L is the number of quadrature points, and w_k the quadrature weights.

The adjoint of the implicit forward function 11.56,

$$(\bar{\boldsymbol{\lambda}}, \bar{\theta}) = \bar{F}_{T_E}(\boldsymbol{\lambda}, \theta, \bar{F}_{T_E}), \tag{11.59}$$

can be computed by means of the implicit function theorem, similarly to what we described for the adjoint of the hazard rate calibration. More explicitly, one first computes the adjoint of the calibration function 11.53

$$(\bar{F}_{T_E}, \bar{\boldsymbol{\lambda}}, \bar{\theta}) = \bar{G}_F(F_{T_E}, \boldsymbol{\lambda}, \theta, \bar{G}_F)$$

with

$$\bar{G}_F = \overline{V}^{\text{adj}}_{\text{iCDS}}(T_E, T_M; \boldsymbol{\lambda}, \theta, \bar{G}_F) - \overline{V}^{\text{isda}}_{\text{iCDS}}(T_E, T_M; F_{T_E}, \theta, \bar{G}_F). \tag{11.60}$$

Here

$$(\bar{\boldsymbol{\lambda}}, \bar{\theta}) = \overline{V}^{\text{adj}}_{\text{Idx}}(T_E, T_M; \boldsymbol{\lambda}, \theta, \bar{V}^{\text{adj}}_{\text{iCDS}})$$

and

$$(\bar{F}_{T_E}, \bar{\theta}) = \overline{V}^{\text{isda}}_{\text{iCDS}}(T_E, T_M; F_{T_E}, \theta, \bar{V}^{\text{isda}}_{\text{iCDS}})$$

are the adjoints of Equations 11.54 and 11.55, respectively. For $\bar{G}_F = 1$, Equation 11.60 gives $\bar{F}_{T_E} = \partial G_F / \partial F_{T_E}$, $\bar{\lambda}^i_j = \partial G_F / \partial \lambda^i_j$, and $\bar{\theta}_k = \partial G_F / \partial \theta_k$, for $i = 1, \ldots, N$, $j = 1, \ldots, M$, $k = 1, \ldots, N_\theta$. Applying the implicit function theorem to the function G_F, one finally obtains the outputs of the function in Equation 11.59:

$$\bar{\lambda}^i_j = \bar{F}_{T_E} \frac{\partial F_{T_E}}{\partial \lambda^i_j} = -\left(\frac{\partial G_F}{\partial F_{T_E}}\right)^{-1} \frac{\partial G_F}{\partial \lambda^i_j},$$

$$\bar{\theta}_k = \bar{F}_{T_E} \frac{\partial F_{T_E}}{\partial \theta_k} = -\left(\frac{\partial G_F}{\partial F_{T_E}}\right)^{-1} \frac{\partial G_F}{\partial \theta_k}.$$

The adjoint of the pricing algorithm consists therefore of the following steps:

Step $\bar{2}$ Set:

$$\bar{Z} = \bar{V} \frac{V_t}{Z(t, T_E; \theta)}$$

and

$$\bar{\theta} = \bar{Z}(t, T_E; \theta, \bar{Z}),$$

where $\bar{Z}(t, T; \theta, \bar{Z})$ is the adjoint of the discount function. Then compute the adjoint of the Gaussian quadrature Equations 11.57 and 11.58, namely set $\bar{F}_{T_E} = 0$, and

$$\bar{\phi}_k = \bar{V} Z(t, T_E; \theta) w_k P_k,$$
$$(\bar{F}_{T_E}, \bar{\theta}) \mathrel{+}= \bar{\phi}(x_k; F_{T_E}, \theta, \bar{\phi}_k),$$

for $k = 1, \ldots, L$, where $\bar{\phi}(x_i; F_{T_E}, \theta, \bar{\phi}_i)$ is the adjoint of the probability density function. Note that due to the linearity of the adjoint function with respect to the adjoint input, these instructions can be re-expressed in terms of a numerical integration of the form

$$(\bar{F}_{T_E}, \bar{\theta}) = Z(t, T_E; \theta) \sum_{k=1}^{L} w_k \bar{\phi}(x_k; F_{T_E}, \theta, \bar{V}) P_k,$$

that is, the adjoint of a Gaussian quadrature can be expressed in terms of the quadrature of the adjoint of the integrand.

Step $\bar{1}$ Set $\bar{\boldsymbol{\lambda}} = 0$ and execute the adjoint of the implicit forward function 11.56,

$$(\bar{\boldsymbol{\lambda}}, \bar{\theta}) \mathrel{+}= \overline{F}_{T_E}(\boldsymbol{\lambda}, \theta, \bar{F}_{T_E}),$$

computed as described earlier. Note that the adjoint function in Equation 11.55 can also be expressed in terms of a Gaussian quadrature.

Steps $\bar{2}$ and $\bar{1}$ provide the outputs of the adjoint of the pricing step in Equation 11.44. Performing the adjoint of the calibration step 11.45 as previously described generates the full set of sensitivities.

The remarkable computational efficiency achievable for swaptions is illustrated in Figure 11.6. Here we plot the cost of computing the sensitivities with respect to the volatility, the constituents' credit spreads and interest rate instruments—relative to the cost of performing a single valuation—for different numbers of index constituents, ranging from 10 (e.g., for iTraxx SOVX Asia Pacific) to 125 (e.g., for iTraxx Europe or CDX.NA.IG). Combining AAD with the implicit function theorem allows the computation of interest rate and (constituents) credit spread risk in 20% less than the cost of computing the option value, resulting in up to 3 orders of magnitude savings (note the logarithmic scale) in computational time.

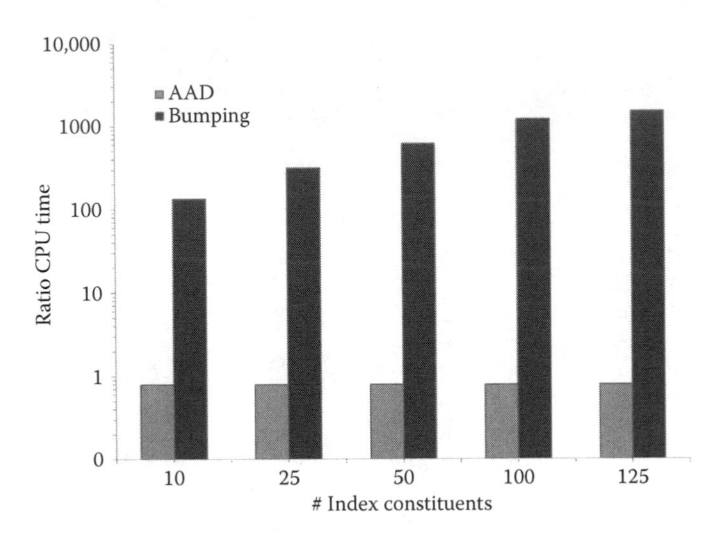

FIGURE 11.6: Cost of computing the sensitivities with respect to the volatility, the constituents' credit spreads and interest rate instruments—relative to the cost of performing a single valuation—as a function of the number of index constituents.

11.6 Conclusion

In conclusion, we have shown how AAD is extremely beneficial for the risk management of financial derivatives by discussing three examples: (i) interest rate products; (ii) counterparty credit risk management; and (iii) flow credit products. These examples illustrate how AAD is effective in speeding up, by several orders of magnitude, the computation of price sensitivities both in the context of MC applications and for applications involving faster numerical methods. In particular, we have shown how by combining adjoint ideas with the implicit function theorem one can avoid the necessity of repeating multiple times the calibration step of financial model which, especially for flow products, often represents the bottle neck in the computation of risk. A recent publication [23] illustrates the application of these ideas to the calculation of risk for Partial Differential Equation application.

These examples illustrate how AAD allows one to perform in minutes risk runs that would take otherwise several hours or could not even be performed overnight without large parallel computers. AAD therefore makes possible real-time risk management on an industrial scale without onerous investments in calculation infrastructure, allowing investment firms to hedge their positions more effectively, actively manage their capital allocation, reduce their infrastructure costs, and ultimately attract more business.

The opinions and views expressed in this chapter are uniquely those of the authors and do not necessarily represent those of Credit Suisse.

References

1. Giles, M. and Glasserman, P. Smoking adjoints: Fast Monte Carlo greeks. *Risk*, 19:88–92, 2006.

2. Capriotti, L. Fast greeks by algorithmic differentiation. *Journal of Computational Finance*, 3:3–35, 2011.

3. Capriotti, L. and Giles, M. Algorithmic differentiation: Adjoint greeks made easy. *Risk*, 25:92–98, 2010.

4. Capriotti, L. and Giles, M. Fast correlation risk by adjoint algorithmic differentiation. *Risk*, 23:79–85, 2010.

5. Broadie, M. and Glasserman, P. Estimating security price derivatives using simulation. *Management Science*, 42:269–285, 1996.

6. Griewank, A. *Evaluating Derivatives: Principles and Techniques of Algorithmic Differentiation*. Frontiers in Applied Mathematics, Philadelphia, 2000.

7. Sorella, S. and Capriotti, L. Algorithmic differentiation and the calculation of forces by quantum Monte Carlo. *Journal of Chemical Physics*, 133:234111, 2010.

8. Brace, A., Gatarek, D., and Musiela, M. The market model of interest rate dynamics. *Mathematical Finance*, 7:127–155, 1997.

9. Glasserman, P. *Monte Carlo Methods in Financial Engineering*. Springer, New York, 2004.

10. Denson, N. and Joshi, M. Fast and accurate greeks for the Libor Market Model. *Journal of Computational Finance*, 14:115–125, 2011.

11. Leclerc, Q. M. and Schneider, I. Fast Monte Carlo Bermudan greeks. *Risk*, 22:84–88, 2009.

12. Capriotti, L., Peacock, M., and Lee, J. Real time counterparty credit risk management in Monte Carlo. *Risk*, 24:86–90, 2011.

13. Brigo, D. and Capponi, A. Bilateral counterparty risk with application to CDSS. *Risk*, 22:85–90, 2010.

14. Jarrow, R., Lando, D., and Turnbull, S. A Markov model for the term structure of credit risk spreads. *Review of Financial Studies*, 10:481–523, 1997.

15. Schonbucher, P. *Credit Derivatives Pricing Models: Models, Pricing, Implementation*. Wiley Finance, London, 2003.

16. Joshi, M. and Kainth, D. Rapid computation of prices and deltas of nth to default swaps in the li model. *Quantitative Finance*, 4:266–275, 2004.

17. O'Kane, D. *Modelling Single-name and Multi-name Credit Derivatives*, volume 573. John Wiley & Sons, 2011.

18. Capriotti, L. and Lee, J. Adjoint credit risk management. *Risk*, 27:90–96, 2014.

19. ISDA. ISDA CDS standard model. *Lehman Brothers Quantitative Credit Research*, 2003.

20. Christianson, B. Reverse accumulation and implicit functions. *Optimization Methods and Software*, 9:307–322, 1998.

21. Henrard, M. Adjoint algorithmic differentiation: Calibration and implicit function theorem. *Open Gamma Quantitative Research*, 2011.

22. Pedersen, C. M. Valuation of portfolio credit default swaptions. *Lehman Brothers Quantitative Credit Research*, 2003.

23. Capriotti, L., Jiang, Y., and Macrina, A. Real-time risk management: An AAD–PDE approach. *International Journal of Financial Engineering*, 2:1550039, 2015.

Chapter 12

Tackling Reinsurance Contract Optimization by Means of Evolutionary Algorithms and HPC

Omar Andres Carmona Cortes and Andrew Rau-Chaplin

CONTENTS

12.1 Introduction

Risk hedging strategies are at the heart of prudent risk management. Individuals often hedge risks to their property, particularly from infrequent but expensive events such as fires, floods, and robberies, by entering into risk transfer contracts with insurance companies. Insurance companies collect premiums from those individuals with the expectation that at the end of the year they will have taken in more money than they have had to pay out in losses and overhead, and therefore remain profitable or at least solvent. Perhaps not surprisingly, insurance companies themselves try to hedge their risks, particularly from the potentially enormous losses often associated with natural catastrophes such as earthquakes, hurricanes, and floods. Much of this hedging is facilitated by the global "property cat" reinsurance market [1], where reinsurance companies insure primary insurance companies against the massive

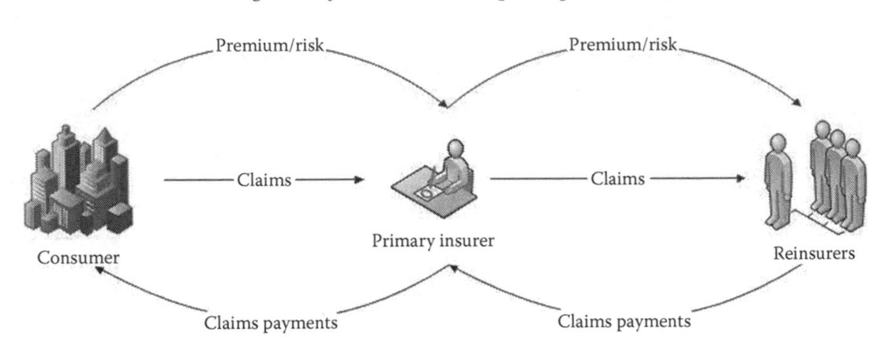

FIGURE 12.1: Risk and premium flow.

claims that can occur due to natural catastrophes. Figure 12.1 illustrates how this flow works.

Analytics in the reinsurance market is becoming increasingly complex for at least three reasons. First, factors like climate change are skewing the data in ways that are not fully understood, making experience less useful for decision making. Second, the global distribution of economic activity is changing rapidly with key supply chains now having significant presence in parts of the world where catastrophic risk is not as well understood. For example, few in 2011 understood that a Thailand flood event could cost US$47 billion in property losses and cause a global shortage of hard disk drives that lasted throughout 2012. Lastly, there is a tendency for risk transfer contracts to become ever more complex, in large part by increasing the number of subcontracts (called layers) that make up a contract. This in turn makes it increasingly important to have good computational tools that can help underwriters understand the interaction between layers and to decide on placement percentages—placement percentages involve choosing layers to buy and how large a share or percentage of them to buy in order to maximize the risk hedging and the expected return.

From the perspective of an insurance company, the problem is known as *Reinsurance Contract Optimization* (RCO). In this problem, we can identify a reinsurance contract consisting of a fixed number of layers and a set of expected loss distributions (one per layer) as produced by a Catastrophe Model [2], plus a model of current costs in the global reinsurance market. The main difficulty in this problem is to identify the optimal combinations of placements (percent shares of subcontracts).

In order to solve the RCO problem, an enumeration method can be used; however, this approach presents two main problems: (i) it has to be discretized, demanding some changes in numerical algorithms and (ii) it is only applicable in small problem instances ranging from two to four layers, whereas real instances of the RCO problem can have seven or more layers. For instance, a seven-layered problem can take several weeks to be solved with a 5% level of discretization on the search space using the enumeration method as presented in Reference 3. Thus we discard the enumerative search in favor of an evolutionary heuristic search.

12.2 Modeling the RCO Problem

As previously stated, insurance organizations, with the help of the global reinsurance market, look to hedge their risk against potentially large claims or losses. This transfer of risk is done in a manner similar to how a consumer cedes part of the risk associated with their private holdings. Unlike the case of the consumer, who is usually given options as to the type of insurance structures to choose from, the insurer has the ability to set its own structures and offers them to the reinsurance market. Involved in this process are decisions around what type of and the magnitude of financial structures, such as deductibles and limits, as well as the amount of risk the insurer wishes to maintain. The deductible describes the amount of loss that the insurer must incur before being able to claim a loss to the reinsurance contract. The limit describes the maximum amount in excess of the deductible that is claimable. Lastly, the placement describes the percentage of the claimed loss that will be covered by the reinsurer.

Typically, companies try to hedge their risk by placing multiple layers at once as illustrated in Figure 12.2, that is, they may have multiple sets of limit and deductible combinations. These different layers may also have different placement amounts associated with them. At the same time, insurers are price takers in terms of the compensation paid to reinsurers for assuming risk. This compensation, or premium, depends on both the amount of risk associated with a layer and the placement amount of the layer. For this reason, it is important for insurers to choose placements when seeking to buy multiple layers. This optimal placement ensures that the insurer is able to maximize their returns on reinsurance contracts for potentially large future events.

In order to simplify the problem description, we focus on the primary contractual terms. Secondary terms such as the contractual costs associated with brokerage fee and contractual expenses, as well as provisions such as reinstatement premiums, are straightforward to add. As is typically done in reinsurance markets, contracts are assumed to be enforced for a 1-year period.

12.2.1 Reinsurance costs

The basic cost of reinsurance to an insurer comes in the form of premium payments. As mentioned previously, the amount of premium paid for a layer can vary with the amount of the layer being placed in the market. In general, premiums are stated per unit of limit, also known as a *rate on line*. The cost of the reinsurance layer can then be expressed as

$$p = \pi\mu(\pi, l, d) \times l \tag{12.1}$$

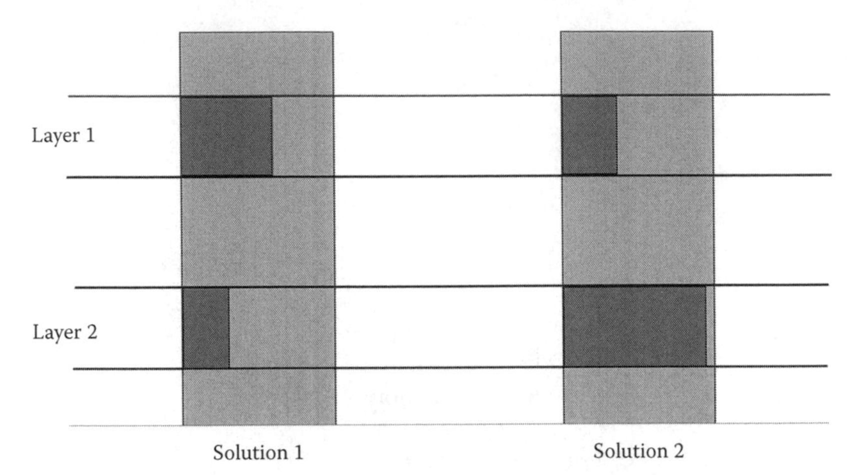

Solution	Layer	Limit	Deductible	Placement	Max recovery	Premium
1	Layer 1	200	300	60%	120	$40
	Layer 2	200	100	20%	40	$40
2	Layer 1	200	300	30%	60	$25
	Layer 2	200	100	95%	190	$100

FIGURE 12.2: An example two-layer reinsurance contract optimization problem with two sample solutions.

where p is the monetary value of the premium, μ is the rate on line, π is the placement, d is the deductible, and l is the limit.

For contracts with multiple layers, Equation 12.1 can be generalized to Equation 12.2,

$$p = \vec{\mu}^{\mathrm{T}}\mathbf{L}\vec{\pi} \tag{12.2}$$

where $\mathbf{L}$ is an $n \times n$ diagonal matrix of limits, $\vec{\mu}$ is an $n \times 1$ vector of rate on lines, $\vec{\pi}$ is an $n \times 1$ vector of placements, and n is the number of layers being placed. This matrix defines a model of expected reinsurance costs.

12.2.2 Reinsurance recoveries

Losses affecting an insurer can be defined as a random variable X, such that

$$X \sim f_X(x) \tag{12.3}$$

where f_X is some distribution that represents severity of X. These losses, once claimed, are subject to the financial terms associated with the contract they are being claimed against. Any one instance of X, x_i, then results in a claim of

$$c_i = max\{0, min\{l, x_i - d\}\}\pi \tag{12.4}$$

where c_i is the value of the claim for ith instance of X. Equation 12.4 can then be extended to contracts with multiple layers as follows:

$$c_i = \sum_{j=1}^{n} max\{0, min\{l_j, x_i - d_j\}\}\pi_j \tag{12.5}$$

where l_j, d_j, and π_j are the limit, deductible, and placement of the jth layer, respectively. In addition to this, many contracts allow for multiple claims in any given contractual year. The yearly contractual loss is then, assuming no financial terms that impose a maximum amount claimable, simply the sum of all individual claims in a given contractual year

$$y_j = \sum_{i=1}^{n} c_{ij} \tag{12.6}$$

where y_j is the annual amount claimed for the jth layer. The annual return for reinsurance contract is then defined as

$$\begin{aligned} r &= \vec{y}^{\mathrm{T}}\vec{\pi} - p \\ &= \vec{y}^{\mathrm{T}}\vec{\pi} - \vec{\mu}^{\mathrm{T}}\mathbf{L}\vec{\pi} \\ &= (\vec{y}^{\mathrm{T}} - \vec{\mu}^{\mathrm{T}}\mathbf{L})\vec{\pi} \end{aligned} \tag{12.7}$$

where $\vec{y}$ is an $n \times 1$ vector of annual claims for each layer.

12.2.3 The risk value and optimization problem

Given a fixed number of layers and loss distributions, the insurer is then faced with selecting an optimal combination of placements. As with most financial structures, the problem faced involves selecting an optimal proportion, or placement, of each layer such that, for a given expected return on the contracts, the associated risk is minimized. This is generally done by using a risk value such as a variance, Value at Risk (VaR), or a Tail-Value at Risk (TVaR). The TVaR is also referred to as a conditional Value at Risk (CVaR) or a conditional tail expectation (CTE). Unlike the traditional finance portfolio problem, in the insurance context a claim made, or loss, to the contract is income to the buyer of the contract. This means, from the perspective of the insurer, there is a desire to maximize the amount claimable for a given risk value. In doing so, they minimize amount of loss the insurer may face in a year.

Equation 12.7 can be rewritten in matrix format such that

$$\mathbf{R} = (\mathbf{Y} - \mathbf{ML})\vec{\pi} \tag{12.8}$$

where $\mathbf{R}$ is an $m \times 1$ vector of recoveries, $\mathbf{Y}$ is an $m \times n$ matrix of annual claims, and $\mathbf{M}$ is an $m \times n$ matrix of rates on line (ROL), which is a layers times share percentage matrix of rates-on-line values, that is, a model of the cost of placing risk in the marketplace. Since the same year is being simulated each row in matrix $\mathbf{M}$ is the same. This formulation leads to this optimization problem:

$$\begin{array}{ll} maximize & VaR_\alpha(\mathbf{R}(\pi)) \\ s.t. & E(\mathbf{R}(\pi)) = a \end{array} \tag{12.9}$$

Given that the expected return a is not specified, Equation 12.9 can be rewritten as a Pareto Frontier problem such that

$$maximize \qquad VaR_\alpha(\mathbf{R}(\pi)) - qE(\mathbf{R}(\pi)) \tag{12.10}$$

where q is a risk tolerance factor greater than zero.

This problem can be approached using a number of different methods. Mistry et al. [4] use an enumeration approach by discretizing the search space for each layer's placements. The discretization of the placements may be desirable for practical reasons (i.e., a placement with more than two decimal places may be invalid in negotiations) and the full enumeration method lends itself well to parallel computation. However, the computational time to evaluate all possible combinations increases exponentially as the number of layers and the resolution of the discretization increases. This renders the enumeration approach infeasible for many practically sized problems.

Mitschele et al. introduced the use of heuristic methods for addressing reinsurance optimization problems [5]. They show the power of two multiobjective evolutionary algorithms (EAs) in finding nondominated combinations, in comparison to the true nondominated set of points. However, their work is done exclusively in continuous space and focus on algorithms that change the limit and deductible aspects of a reinsurance contract. Their methods are therefore not directly applicable here.

12.3 Evolutionary Algorithms

EAs keep track of the most fit or optimal solution for a specific problem by employing a population of individuals where each individual represents a possible solution. The population has to undergo genetic operators on many iterations until some stop criteria is reached, either: (i) after a certain amount of iterations, (ii) when there is no more evolution, or (iii) when the algorithm cannot acquire a more optimal solution. Figure 12.3 shows this process.

In the first step of an EA, the population is initialized at random, normally using a uniform distribution. The population is then evaluated in order of its members to determine how each individual scores in fitness as a solution. The better an individual's fitness, the stronger the individual is within that population. Subsequently, the probability of an individual to be selected for genetic operators or to go to the next iteration (generation) is higher for stronger members of the population.

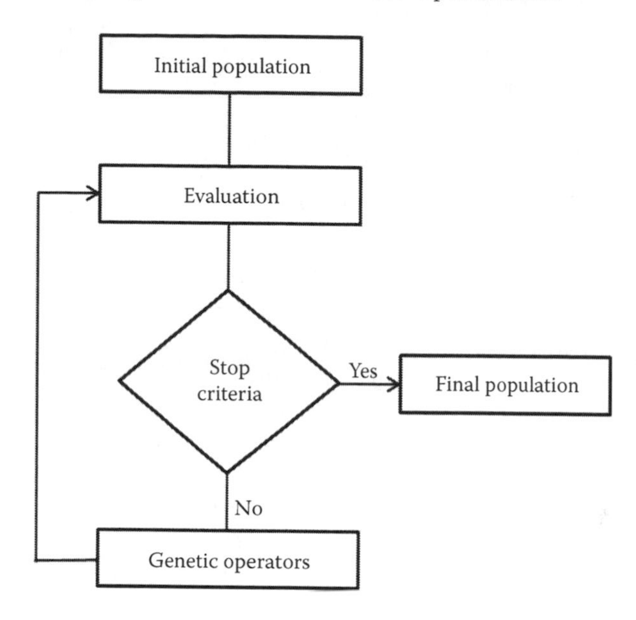

FIGURE 12.3: Structure of an EA.

Typically, the genetic operators are: selection, cross-over, and mutation. Selection is the process of choosing an individual to undergo genetic operators or to go to the next generation. In cross-over operators, parents exchange information (genes) between themselves in order to create one or more offspring. Ideally, when two strong individuals exchange their genes, the offspring tends to be stronger than its parents [6] and then spreads its genes to further generations. On the other hand, this behavior can lead to a premature convergence of the solution because the population can be trapped in a local optima. The mutation operator has the purpose of avoiding the premature convergence by applying modifications to one or more genes. In other words, the process of using genetic operators tends to improve the solution quality as new generations carry on [7].

Taking these operators into account, we can observe that several EAs share similar features. For instance, genetic algorithms and evolutionary strategies may use all of these mentioned genetic operations. However, the sequence that these operators are used can differ. Evolutionary strategy applies other genetic operators before selection operators, whereas genetic algorithms select the individuals and then perform those genetic operators. Moreover, evolutionary strategies need two vectors for representing an individual instead of only one as is typical for genetic algorithms.

12.3.1 Population-based incremental learning (PBIL)

PBIL was first proposed by Baluja [8]. The algorithm's populations are encoded using binary vectors and an associated probability vector, which was then updated based on the best members of a population. Unlike other EAs, which transform the current population into new populations, a new population is generated at random using an updated probability vector on each generation. Baluja describes his algorithm as a "combination of evolutionary optimization and hill-climbing" [8].

Since Baluja's work, extensions to the algorithm have been proposed for continuous and base-n represented search spaces [9–11]. The extension to continuous search spaces using histograms (PBIL_H) and real-code (RPBIL) suggests splitting the search space into intervals, each with their own probability [9,11]. For multivariate cases, the probability vector is then substituted for a probability matrix, such that each row or column of the matrix represents a probability vector for any given independent variable. While those methods support continuous spaces, a similar idea extended PBIL to a discrete approach in Reference 3 in order to deal with the reinsurance problem constraints, that is, we substitute the intervals for equidistant increments in the lower and upper bounds of the search space.

Algorithm 12.1: DiPBIL

Initialization: $P_{ij} = 1/I$, LR_N, NLR_N, $\mathbf{x}_G^{best}$={}, $\mathbf{x}_i^{best}$={}

for $i = 1$ to N_G **do**

  Generate a population $\mathbf{X}$ of size n from P_{ij}

  Evaluate $\mathbf{f} = \text{fun}(\mathbf{X})$

  Find $\mathbf{x}_G^{best}$ from the current and previous populations

  Find $\mathbf{x}_i^{best}$ for top $q-1$ members of the current population

  Update P_{ij} based on $\mathbf{x}_G^{best} \cup \mathbf{x}_i^{best}$ using LR_N and NLR_N

end

In the same spirit as canonical PBIL, the probability matrix is initialized with all increments having equal probability. This matrix is then updated after every generation with the best combinations member. The updating of each vector in the matrix, however, is done using the base-n method, with an adjusted learning rate and updating function [10]. RPBIL suggests its own updating function which exponentially increases the probabilities as you move toward intervals that are closer to the best individuals [9]. The base-n updating method, however, has the side effect of slightly increasing the probability of increments further from the best individuals in the search space and may allow for a chance at more population diversity [10]. To ensure more population diversity from across generations, the probability matrix is updated with best member from previous generations as well as the top q members from the current generation. This modifies the updating equation slightly to the one presented in Equation 12.11

$$p_{ij}^{NEW} = \sum_{k=1}^{q} p_{ij}^{OLD} \frac{LF_{ijk}}{q} \tag{12.11}$$

where LF_{ijk} is the ith learning factor, as described in Reference 10, for the kth best result for the jth variable.

The main drawback of the discrete PBIL is that it requires all of the objective functions get transformed into a single function. In order to address this issue, and compute a better Pareto frontier, a true multiobjective optimization approach, called MOPBIL, is presented in Algorithm 12.2.

Algorithm 12.2: MOPBIL (Sketch)

Input: N_G = number of generations; n_{best} = number of best individuals;
while *(N_G not reached)* **do**
    Create the population using probability matrix;
    Evaluate objective function on each member of the population;
    Merge archive and the new population;
    Determine the nondominated set;
    Cluster the nondominated set into k clusters;
    Select k representative individuals;
    Insert the k individuals into the best population;
    Update and mutate the probability matrix;
end
Determine the final Pareto frontier;

MOPBIL creates a random population by using the probability matrix as it is proposed in Reference 3. Then, nondominated solutions are found and the resulting set is clustered. Finally, k representative solutions are chosen to update the probability matrix. The clustering process is introduced to help to keep the diversity along the generations, that is, the algorithm tries to choose representatives from kth most different solutions keeping them in the population and also updating the probability matrix. The individual which presents the best risk value is chosen to represent a cluster. In other words, we are trying to drive the search toward the optimal risk values and retain the diversity in the population at the same time. When the number of generations is reached, the last results are combined with the archive in order to get the final Pareto frontier. Figure 12.4 graphically illustrates a high-level structure of this process.

12.3.2 Differential evolution (DE)

The DE algorithm was proposed by Storn and Price [12] in 1995 where it is based on the difference between two individuals that is summed up to a third one. The process is sketched in Algorithm 12.3. As in Particle Swarm Optimization, the first step is to create a population at random. Then, while the stop criteria is not reached, the vector of differences is calculated according to the equation $v = P_{idx_3} + F * (Pop^k_{idx_1} - Pop^k_{idx_2})$, where idx_i is a vector with three individual randomly chosen, and F is a multiplication factor normally between 0 and 1. This strategy is called $DE/Rand/1$ because P_{idx_3} is

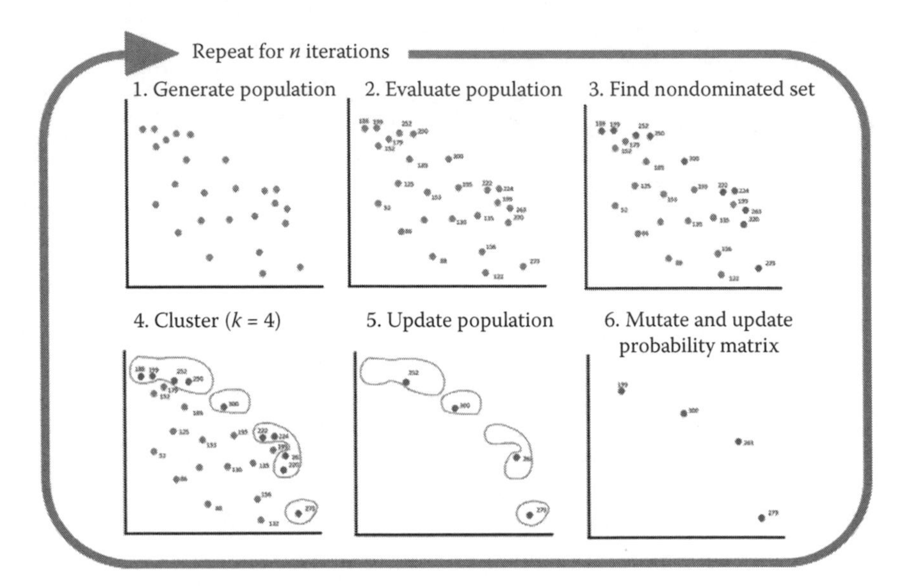

FIGURE 12.4: How the MOPBIL creates points and updates the probability matrix.

randomly chosen. When P_{idx_3} is the best individual in the population, the strategy is called $DE/Best/1$.

The process of computing v is called mutation. Afterwards, a new individual is created in a similar way as the discrete cross-over of genetic algorithms, that is, for each dimension d, a gene is chosen from the vector of differences v with a probability of CR, or from the target individual i with a probability of $1 - CR$. Finally, if the fitness of the new individual is better than the fitness of the target one, then the new individual replaces the individual i.

The canonical DE presents the same problem of PBIL single objective, that is, it has to aggregate all functions into only one evaluation function. Thus, a multiobjective version, called DEMO, is shown in Algorithm 12.4, where we can observe that it is similar to the canonical version of DE whose strategy is $DE/Rand/1$ [13]. The differences start in line 16 when the new population is selected for the next iteration. Thus if a new individual ($indiv$) dominates the target one (Pop_i), then the new one is added into a new population; if the target individual dominates the new one, then the target element is added into the new population; otherwise, both individuals go to the new population. The dominance process builds a new population whose size ranges from pop_size to $2 \times pop\_size$. Finally, if the size of the new population is larger than pop_size, then the new individuals which go to next iteration are selected by crowding distance ($select\_cdistance$ function).

The main drawback of the original DEMO was not to maintain an archive thereby loosing good solutions when the number of nondominated points overcome the size of the population. Taking this into account, we changed

Algorithm 12.3: DE

```
Pop ← generate_pop(n,d)
fit ← evaluate_(Pop^k)
while (Stop Criteria is FALSE) do
    for i = 1 to #pop_size do
        idx ← select_indiv(3)
        v ← Pop_idx3 + F * (Pop^k_idx1 − Pop^k_idx2)
        for j = 1 to dimension do
            nj = rand()
            if (nj < CR) then
                pop'← v_j
            else
                pop'← pop_i j
            end
        end
        fit'_i ← evaluate_(P_i)
        if fit'_i < fit_i then
            pop_i ← pop'_i
            fit_i ← fit'_i
        end
    end
end
```

the original algorithm into two parts. First, we introduce an archive in the algorithm (after line 31, i.e., it is done on each iteration) in order to not lose nondominated solutions from one iteration to another due to the crowding distance algorithm in line 30. Doing so, we are able to use different mutation operators such as those presented in Equations 12.12 through 12.15. These strategies are called DE/ND/Rand/1, DE/ND/Rand/RF/1, DE/Arch/Rand/1, and DE/Arch/Rand/RF/1, respectively. Equation 12.12 uses an random individual from the set of nondominated ones. In order to do so, it is necessary to compute the nondominated set between lines 3 and 4, that is, before starting the loop which deals with the population. Equation 12.13 is similar to the previous one; however, F is a random number between 0 and 1. Then, both Equations 12.14 and 12.15 use the first individual chosen from the archive. The difference between them is the use of F which is randomly chosen in Equation 12.15.

$$v \leftarrow non\_dominated_{idx} + F \times (Pop_{idx_1} - Pop_{idx_2}) \tag{12.12}$$

$$v \leftarrow non\_dominated_{idx} + Rand() \times (Pop_{idx_1} - Pop_{idx_2}) \tag{12.13}$$

$$v \leftarrow archive_{idx} + F \times (Pop_{idx_1} - Pop_{idx_2}) \tag{12.14}$$

$$v \leftarrow archive_{idx} + Rand() \times (Pop_{idx_1} - Pop_{idx_2}) \tag{12.15}$$

Algorithm 12.4: DEMO

```
Pop ← generate_pop(n, d)
fit ← evaluate(Pop)
while (Stop Criteria is FALSE) do
    for i = 1 to #pop_size do
        idx ← select_indiv(3)
        v ← Pop_idx3 + F * (Pop_idx1 − Pop_idx2)
        for j = 1 to dimension do
            nj = rand()
            if (nj < CR) then
                indiv← v_j
            else
                indiv← Pop_i j
            end
        end
        fit' ← evaluate(indiv)
        if (fit' dominates fit_i) then
            pop' ← indiv
            nf ← fit'
        else if (fit_i dominates fit') then
            pop' ← Pop_i
            nf ← fit_i
        else
            add indiv and Pop_i into pop'
            add fit and fit' into nf
        end
    end
    if (nrow(pop') == nrow(Pop_i))) then
        Pop ← pop'
    else
        [Pop, fit] ← select_cdistance(pop', nf)
    end
end
```

12.4 Case Study

12.4.1 Metrics

In this section, we discuss the experimental evaluation of the MODE algorithm. First, the average number of nondominated points (number of solutions) found in the Pareto frontier was determined. Second, the average hypervolume, which is the volume of the dominated portion of the objective space as presented in Equation 12.16, was measured, where for each solution $i \in Q$ a hypercube v_i is constructed. Having each v_i, we calculated the final hypervolume by the union of all v_i. The final number of solutions after all trials is showed as well.

$$hv = volume\left(\bigcup_{i=1}^{|Q|} v_i\right) \tag{12.16}$$

Third, the dominance relationship between Pareto frontiers obtained with different mutation operators was calculated as depicted in Equation 12.17. Roughly speaking, $C(A, B)$ is the percentage of solutions in B that are dominated by at least 1 solution in A [14]. Therefore, if $C(A, B) = 1$, then all solutions in A dominate B; $C(A, B) = 0$ would indicate that all solutions in B dominate A. It is important to notice that this metric is neither complementary by itself nor symmetric, that is, $C(A, B) \neq 1 - C(B, A)$ and $C(A, B) \neq C(B, A)$ making it important to compute in both directions: $C(A, B)$ and $C(B, A)$.

$$C(A, B) = \frac{|\{b \in B | \exists a \in A : a \preceq b\}|}{|B|} \tag{12.17}$$

Finally, the resulting frontiers can be reviewed by experts for reasonability. For further details about the use of these metrics, see Reference 15.

In terms of parallelism, we calculated the speedup according to Equation 12.18,

$$speedup = \frac{T_{\text{s}}}{T_{\text{p}}} \tag{12.18}$$

where T_{s} is the execution time considering one thread and T_{p} represents the time in parallel using p threads. This kind of metric is called weak speedup and it was suggested in Reference 16 because the code is exactly the same regardless the number of threads. Thus it is not necessary to guarantee that the serial version is the best one.

12.4.2 Results

All tests were conducted using R version 3.2.1 on a Red Hat Linux 64-bit operating system with an Intel Xeon processor comprised of two Xeon processors E5-2650 running at 2.0 GHz with 8 cores, hyperthreading and 256 GB of memory. Considering 250 and 500 with a population size equals to 50. The following parameters were used for MOPBIL:

- Population size = 50
- Slice size = 0.05
- Number of generations = 250, 500
- Best population = 3

- learn.rate = 0.1
- neg.learn.rate = 0.075
- mut.prob = 0.02
- mut.shift = 0.05

In terms of DEMO, the following parameters were used:

- Population size = 50
- Strategy = DE/Arch/Rand/RF/1

Table 12.1 shows metrics for 250, 500, and 1000 iterations (Figure 12.5). As expected, as we increase iterations results tend to be better. On the other hand, we can see DEMO is the best of the two executed algorithms in terms of both number of solutions and hypervolume.

12.4.2.1 Parallel version

Because DEMO tends to present better results, we only included it in the parallel tests. So, in order to parallelize the code we used the Snow [17] package from R, which is a package for automatic parallelization. We parallelized the iteration loop that presents better results than parallelizing the population loop. The mutation operator we used is the M5 with 1000 iterations because it presented better results than the other ones.

TABLE 12.1: Metrics for 7 layers as we increase iterations

	NS	**HV**	**Time**
250 it			
MOPBIL	119.7666667	1.79E+15	162.5251333
	20.24451111	2.66E+14	2.625853618
DEMO	235.1	2.17E+15	164.4676
	12.38004512	2.81E+14	1.362389321
500 it			
MOPBIL	122.6	1.91E+15	325.8486333
	16.31500262	2.03E+14	6.554302917
DEMO	289.4666667	2.15E+15	318.5643
	11.34941388	2.80E+14	1.866340697
1000 it			
MOPBIL	136.8	1.95871E+15	671.1731333
	18.55541561	2.44412E+14	99.93907566
DEMO	337.3666667	2.25E+15	626.3866667
	12.60673967	2.14E+14	3.61856901

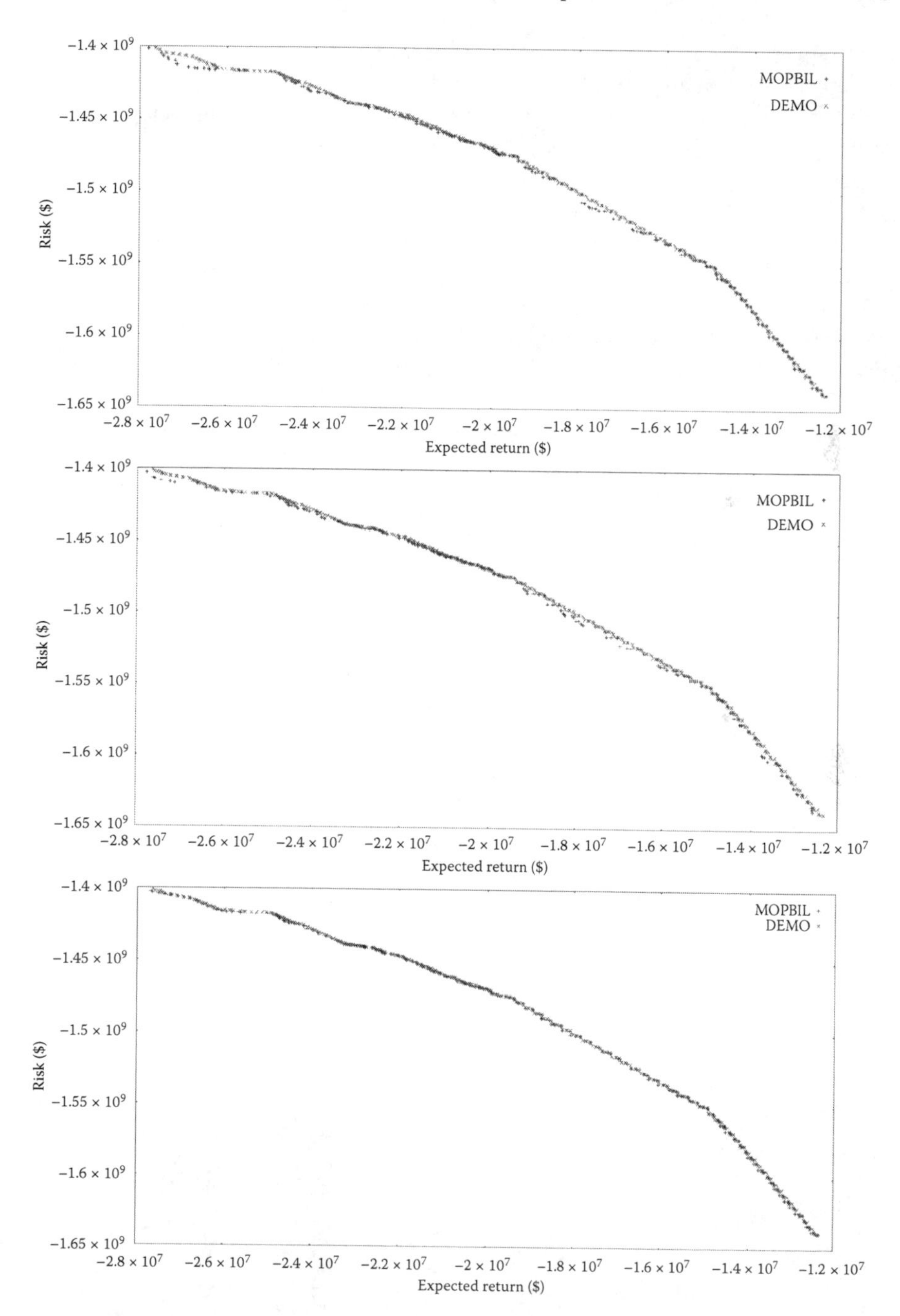

FIGURE 12.5: Final Pareto frontier for DEMO and MOPBIL using 7 layers, 250, 500, and 1000 iterations.

Figures 12.6 and 12.7 show the time and speedup reached in the Xeon architecture variating the thread count. Regardless the number of layers, the best efficiency is reached using two threads representing an efficiency of 96.7% and 98.2%, respectively. In terms of speedup, it is almost linear up to four threads. Then, the best one is reached using 32 threads representing 9.38 and 8.33 for 7 and 15 layers, respectively; however, the use of 32 threads represents an efficiency of 29.3% and 26% for 7 and 15 layers. Moreover, the best speedups are reached by 7 layers saturating in approximately 16 threads.

Figure 12.8 presents the Pareto frontier obtained by varying the thread count for 1000 iterations and 7 layers, where we can observe that visually all Pareto frontiers seem to be the same. Table 12.2 depicts the average in term of metrics. Even though, the number of solutions decrease as we increase the number of threads, the final number of solutions is not affected. Moreover,

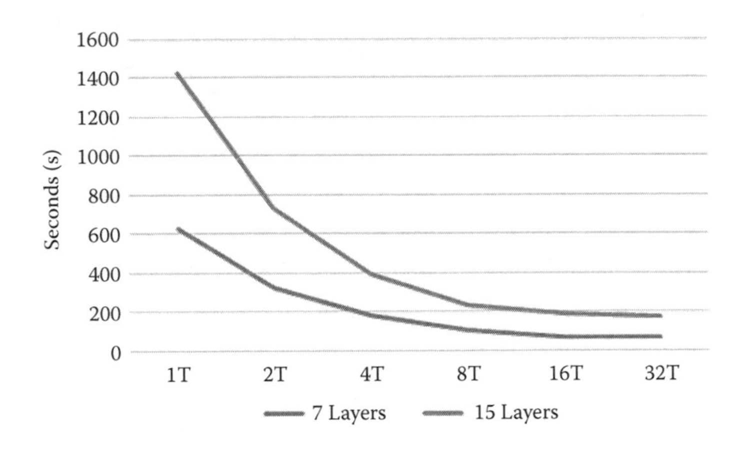

FIGURE 12.6: Time for 7 and 15 layers and 1000 iterations on Xeon.

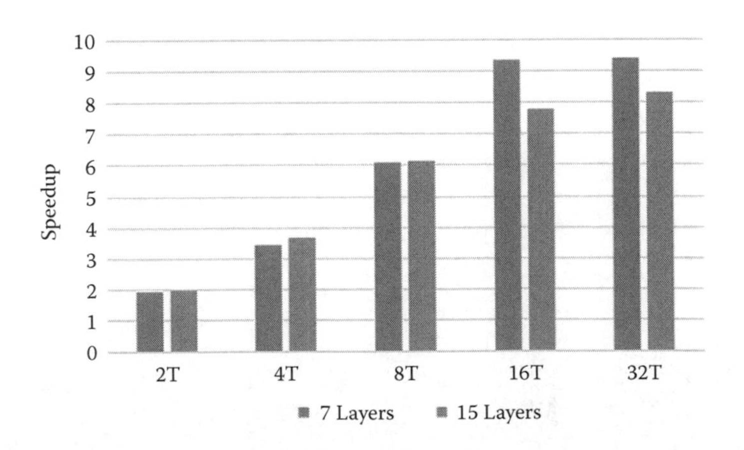

FIGURE 12.7: Speedup for 7 and 15 layers and 1000 iterations on Xeon.

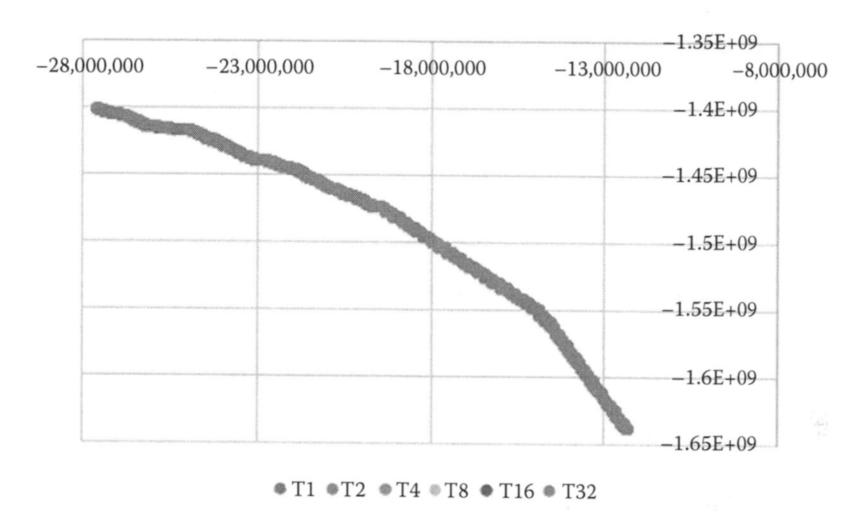

FIGURE 12.8: Pareto frontier varying thread count for 1000 iterations and 7 layers.

TABLE 12.2: Metrics for 7 layers and 1000 iterations

	#NS	Hypervolume	Time	#NS final
1T	337.3666667	2.25E+15	626.3866667	403
	12.60673967	2.14E+14	3.61856901	
2T	336.6333333	2.30E+15	323.8888	398
	11.60999371	1.59E+14	1.890259066	
4T	329.6333333	2.34E+15	181.6333667	390
	10.49296425	1.05E+14	0.808078286	
8T	315.8666667	2.35E+15	103.0467333	406
	12.01359383	1.28E+13	0.340365719	
16T	288.9666667	2.35E+15	67.123	403
	9.86628999	3.00E+13	0.711079801	
32T	246.9	2.35E+15	66.7494	390
	13.47628824	1.45E+13	2.711422067	

the hypervolume is quite stable between threads, therefore, the faster the execution the better. In fact, the small numbers in Table 12.3, which represent the coverage, mean that the Pareto frontiers are very similar regardless the number of threads.

Figure 12.9 shows the Pareto frontier obtained by varying the thread count for 1000 iterations and 15 layers, where we can observe that, visually, the difference between Pareto frontiers obtained by different counting of threads is not meaningful. On the other hand, Table 12.4 presents how the number of solutions decrease as we increase the number of threads; nonetheless, the hypervolume indicates that this decrement is worth up to eight threads

TABLE 12.3: Coverage for 7 layers and 1000 iterations

	T1	T2	T4	T8	T16	T32
T1	–	0.028	0.03	0.08	0.16	0.20
T2	0.04	–	0.04	0.09	0.16	0.215
T4	0.03	0.03	–	0.07	0.14	0.20
T8	0.030	0.035	0.028	–	0.14	0.17
T16	0.019	0.015	0.015	0.057	–	0.16
T32	0.017	0.022	0.026	0.02	0.086	–

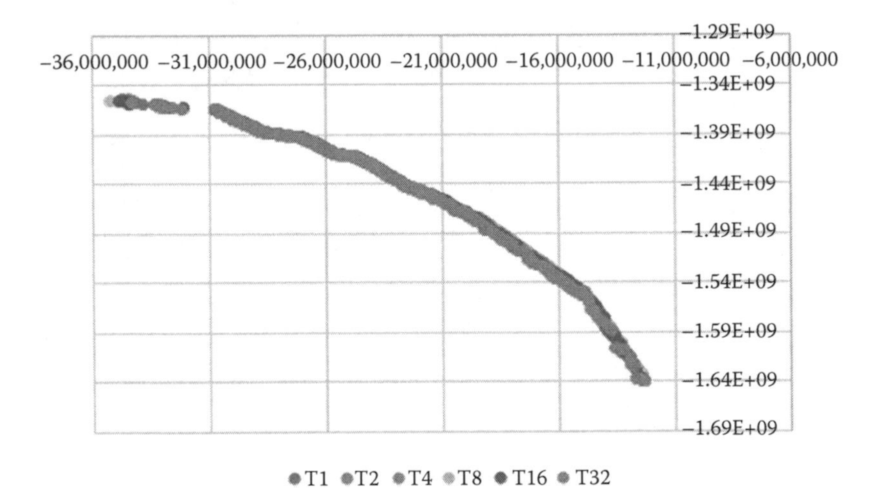

FIGURE 12.9: Pareto frontier varying thread count for 1000 iterations and 15 layers.

TABLE 12.4: Metrics for 15 layers and 1000 iterations

	#NS	Hypervolume	Time	#NS final
1T	290.73	3.39E+15	1426.90	517
	22.97	8.93E+14	6.33	
2T	296.03	3.82E+15	726.32	515
	15.83	7.22E+14	2.16	
4T	280.37	4.03E+15	387.39	470
	12.70	5.74E+14	1.13	
8T	237.40	4.22E+15	232.96	450
	13.63	3.72E+14	1.12	
16T	201.00	4.12E+15	182.87	378
	13.43	3.30E+14	9.25	
32T	164.00	3.99E+15	171.36	336
	12.71	4.05E+14	16.34	

TABLE 12.5: Coverage for 15 layers and 1000 iterations

	T1	**T2**	**T4**	**T8**	**T16**	**T32**
T1	–	0.50	0.57	0.68	0.79	0.87
T2	0.40	–	0.32	0.54	0.65	0.77
T4	0.30	0.14	–	0.46	0.61	0.75
T8	0.20	0.09	0.15	–	0.52	0.70
T16	0.13	0.05	0.10	0.16	NA	0.55
T32	0.059	0.017	0.04	0.08	0.20	NA

because it is larger. Actually, the time saved using eight threads is also of great note. Table 12.5 reinforces that the quality of using eight threads is an attractive option because it dominates 52% and 70% of solutions from 16 and 32 threads, respectively.

References

1. Cai, J., Tan, K. S., Weng, C., and Zhang, Y. Optimal reinsurance under VAR and CTE risk measures. *Insurance: Mathematics and Economics*, 43(1):185–196, 2008.

2. Grossi, P. and Kunreuther, H. *Catastrophe Modeling: A New Approach to Managing Risk, volume 25 of Catastrophe Modeling.* Springer, US, 2005.

3. Cortes, A. C., Rau-Chaplin, A., Wilson, D., Cook, I., and Gaiser-Porter, J. Efficient optimization of reinsurance contracts using discretized PBIL. In *The Third International Conference on Data Analytics*, pp. 18–24, Porto, Portugal, 2013.

4. Mistry, S., Gaiser-Porter, J., McSharry, P., and Armour, T. Parallel computation of reinsurance models, 2012. (Unpublished)

5. Mitschele, A., Oesterreicher, I., Schlottmann, F., and Seese, D. Heuristic optimization of reinsurance programs and implications for reinsurance buyers. In *Operations Research Proceedings*, pp. 287–292, 2006.

6. Herrera, F., Lozano, M., and Verdegay, J. L. Tackling real-coded genetic algorithms: Operators and tools for behavioural analysis. *Artificial Intelligence Review*, 12(4):265–319, 1998.

7. Michalewicz, Z. *Genetic Algorithms + Data Structure = Evolution Programs*, 3rd edition, Springer-Verlag, Berlin, Heidelberg, New York, 1999.

8. Baluja, S. Population based incremental learning. Technical report, Carnegie Mellon University, Pittsburgh, Pennsylvanian, 1994.

9. Bureerat, S. *Improved Population-Based Incremental Learning in Continuous Spaces*, pp. 77–86. Number 96. Springer, Berlin, Heidelberg, 2011.

10. Servais, M.P., de Jager, G., and Greene, J. R. Function optimisation using multiple-base population based incremental learning. In *The Eighth Annual South African Workshop on Pattern Recognition*, Rhodes University, Grahamstown, South Africa, 1997.

11. Yuan, B. and Gallagher, M. Playing in continuous, some analysis and extension of population-based incremental learning. In *IEEE Congress on Evolutionary Computation, IEEE*, pp. 443–450, Canberra, Australia, 2003.

12. Storn, R. and Price, K. Differential evolution: A simple and efficient heuristic for global optimization over continuous spaces. *Journal of Global Optimization*, 12(4):341–359, 1997.

13. Qin, A. K., Huang, V. L., and Suganthan, P. N. Differential evolution algorithm with strategy adaptation for global numerical optimization. *Transaction on Evolutionary Computation*, 13(2):398–417, 2009.

14. Zhang, Q. and Li, H. MOEA/D: A multiobjective evolutionary algorithm based on decomposition. *IEEE Transactions on Evolutionary Computation*, 11(6): 712–731, 2007.

15. Deb, K. *Multi-Objective Optimization Using Evolutionary Algorithms*. John Wiley & Sons LTDA, 2001.

16. Alba, E. Parallel evolutionary algorithms can achieve super-linear performance. *Information Processing Letters*, 82(1):7–13, 2002.

17. Tierney, L., Rossini, A. J., Li, N., and Sevcikova, H. Snow, https://cran.r-project.org/web/packages/snow/snow.pdf, 2017.

Chapter 13

Evaluating Blockchain Implementation of Clearing and Settlement at the IATA Clearing House

Sergey Ivliev, Yulia Mizgireva, and Juan Ivan Martin

CONTENTS

13.1 ICH (IATA Clearing House)'s Current Clearing Procedure and Potential for Improving Its Efficiency

IATA (The International Air Transport Association) is a trade association in the airline industry. In 2016, it had more than 260 airline members, which together represent about 83% of total air traffic.

IATA defines its mission as "to be the force for value creation and innovation driving a safe, secure, and profitable air transport industry that sustainably connects and enriches our world." In other words, IATA intends to represent, lead, and serve the airline industry. It aims to simplify processes by developing the commercial standards for the airline business, to increase passenger convenience, and to reduce costs by improving efficiency.

The ICH is an organization serving the air transport industry which provides clearing and settlement services for its members. ICH was founded in 1947, 2 years after the foundation of IATA. Now it has about 275 airline members (both IATA and non-IATA members) and 75 nonairline members.

The ICH members are in close collaboration with each other, and typically they have two-way billings. By applying the principles of set-off/netting, it reduces significantly the amount of cash required to settle these billings.

13.1.1 ICH's clearance procedure

The ICH takes responsibility for a clearing and settlement procedure which is held on a weekly basis and lasts for 2 weeks. It means that during a week before the **closure day**, each member sends their invoices to ICH (the invoice contains the information about the payer, the sum of money, the currency of payment, and so on). After the **closure day**, ICH does the offset of the invoices and on the **advice day**, it sends confirmation messages of the final balances to the members. A week after this, on the **call day**, the net debtors must pay their balance to ICH and 2 days later (on the **settlement day**) the net creditors will receive their payments. A clearance cycle covers a single week with a year of 48 cycles. So, the payments are received about 2 weeks after the day when the last invoice of current clearance cycle came to ICH.

A schematical illustration of the ICH's current clearing process is represented in Figure 13.1. Such a process of clearing has its advantages and disadvantages. The main advantages are the following:

- IATA guarantees that all invoices will be submitted for payment and this means that ICH's clearing and settlement procedure reduces the credit risk.

- About a 64% offset ratio (the relation between the volume of billings and the amount of cash required to settle them). This means that from

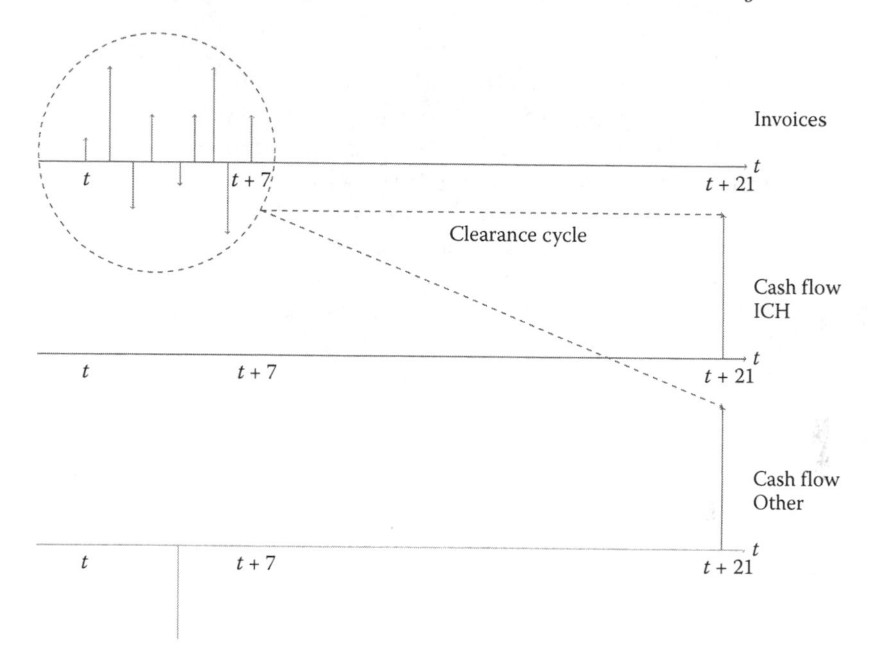

FIGURE 13.1: Example of an airline member's cash flow (1 clearing cycle)—how it is now.

a total volume of billings which, for example, is \$54.3 B, the cash-out is about \$19.8 B.

There are also disadvantages for both sides—for ICH and its members. They are:

- Slow payment procedure (a huge delay between sending the invoices and receiving the money)
- High banking fees
- Potentially not an optimal offset ratio

These disadvantages can be overcome with the help of a blockchain technology implementation.

13.1.2 Improving ICH's clearance procedure with blockchain technology

Blockchain is a novel development in the world of computing science that has potentially far-reaching implications for many aspects of global commerce. Over the coming decade, this technology is expected to fundamentally disrupt established modes of operation in many sectors, including banking, logistics, and government, to name just a few. Similarly, blockchain has the potential to be a positive transformative force within the airline industry.

A distributed ledger is essentially an asset database that is shared across a network of multiple sites, geographies, or institutions. The assets can be financial, legal, physical, or virtual in nature. Selected participants within the network have copies of the ledger and any changes to the ledger are reflected in all copies within minutes, or in some cases, seconds. The security and accuracy of the transactions stored in the ledger are ensured cryptographically through the use of keys and signatures to control who can do what within the shared ledger. Entries can also be updated by one, some, or all of the participants according to rules agreed by the network. Distributed ledger technologies (DLTs) hold the potential to redefine a number of industries and various aspects of society. From reducing the transaction costs experienced by large companies to providing greater possibilities for distributed economies and business models, they hold great disruptive potential. DLT can bring multiple benefits to its users, including but not limited to:

- High availability due to structural multiple redundancy
- Cryptographically secured records
- Multiple layers of cryptographically enforced access control
- Configurable trust levels depending on the degree of decentralization

A distributed ledger allows untrusting parties with common interests to cocreate a permanent, unchangeable, and transparent record of exchange and processing. This aim aligns well with the need of the airline industry to have some means of cross-organizational value exchange. The degree to which a high-throughput low-cost solution can be realized in the context of any given use case depends primarily on the openness of the system in terms of who is permitted to submit, validate, and process transactions. A cryptocurrency solution will help streamline the conversion of funds by allowing fast, secure, low-cost movement of value. Cryptocurrencies do not require long clearing times, while their cryptographic mechanisms provide heightened protection from fraudulent transactions [1].

By creating a new industry cryptocurrency, IATA gives a chance for members to submit invoices immediately and exclude them from the clearing process. Potentially it would help to overcome the disadvantages of the current clearing procedure and

- Accelerate cash flows (the payments will become significantly faster with settlement and confirmation on a distributed ledger)
- Reduce transaction costs (the amount of SWIFT transfers will decrease)
- Increase the offset ratio

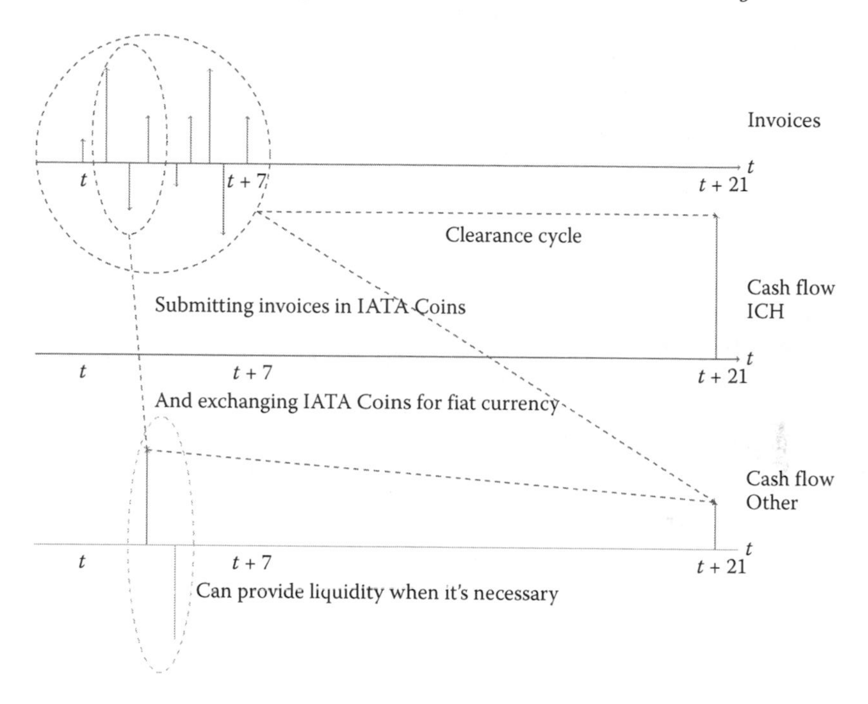

FIGURE 13.2: Example of an airline member's cash flow (1 clearing cycle)—how it will be.

- Provide additional liquidity for members inside the current clearance cycle

IATA would be the liquidity provider, guaranteeing the retro-convertibility of IATA Coins into fiat currency. Airlines will be able to present the coins to IATA to cash out in fiat currencies.

A schematical illustration of the ICH's clearing process with the opportunity to submit the invoices in IATA Coins is represented in Figure 13.2.

13.2 Data Simulation Description and Model Assumptions

For analysis of the potential benefits of blockchain technology implementation for IATA, a simulation model will be developed. Initially, the model will be focused on only one company. The main assumptions which will be necessary to take into account during the building of the model are as follows.

1. There are two main agents whose interests are not the same:
   - ICH, which tries to increase the offset ratio and to accelerate the payment procedure

- The company, which has its own view (and its own criteria of optimality) in choosing the dates when invoices should be submitted

2. The company is acting rationally and trying to reduce transaction costs.
3. ICH current clearance cycle is still working, but all invoices can be submitted in IATA Coins and consequently excluded from ICH's clearing procedure.
4. When the company sends an invoice to ICH, it can mark whether it wants to receive the money in IATA Coins or in fiat currency (as the company is a rational agent let us assume that for all sent invoices, submission in IATA Coins is desirable, because this can help to decrease transaction costs).
5. The company can choose the times when it wants to exchange IATA Coins for fiat money (according to the liquidity profile of the company).
6. On the settlement day of the clearance cycle, the payments are effected not in IATA Coins, but in fiat currency.
7. For calculating the offset ratio, only fiat currency payments will be taken into account.
8. Due to the absence of historical data, the sample for analysis will be generated according to the following distributional assumptions:
   - The number of invoices per day for a company will be considered as a random variable with a Poisson distribution (with an arbitrary value of the parameter)
   - The volume of invoices will be considered as a random variable with a lognormal distribution (with arbitrary values of the parameters)
9. For simplicity we assume:
   - Only one fiat currency
   - The exchange rate is not taken into account
   - The company has enough IATA Coins to cover the invoices and enough money in fiat currency to cover the after-clearing payments (borrowing activities are not taken into account)

So, the sample for constructing the model in the simplest case will contain the following attributes:

- Date
- Value (positive value—paid by the company, negative—paid to the company)

- Dummy variable for indicating if this particular observation is an invoice sent to IATA or a payment which does not go through IATA

13.3 Mathematical Formulation of the Model

As was said above, ICH and every company counterparty have their own goals and their own opinions about the dates when the invoices should be submitted. From ICH's point of view, the main criterion is the higher offset ratio, meanwhile a company intends to minimize its transaction costs and to improve its liquidity profile. In fact, each company in the network, if it is a rational agent, will have its own strategy for choosing the invoices to submit in IATA Coins.

The situation can be considered from the game theoretic point of view, wherein the decision of each company will influence the strategy chosen by every other company. Due to the complexity of this financial system (there are more than 260 members in ICH), it is almost impossible to formalize the model as a game model and it seems that there would not be a clear equilibrium. But the problem can also be considered as involving an optimization model from the point of view of one agent of the system—either ICH or a typical company.

In this chapter, the second model form, namely the optimization model, will be considered. In fact, the only decision makers in this financial system are the companies, and ICH cannot influence the decisions that will be made. That is why we will focus on the construction of the model from a typical company's point of view.

It also should be mentioned that the model will take as an input not the whole clearing year, but only the current clearing cycle. This means that in each clearing cycle a company is trying to find an optimal strategy disregarding the information from other clearing cycles.

The mathematical formulation of the model will be represented in the following sections.

13.3.1 Discrete-time optimal control model

The notation and main equations will correspond to Figures 13.1 and 13.2. We will consider the clearance cycle for the time period $[t, t+7]$.

Let the volume of the invoice paid to (or received from) the company A be denoted by Inv_i, where i is the invoice number in the period $[t, t+7]$ and $i = \overline{1, N}$. The fiat currency payment which does not go through IATA will be denoted by R_j, where j is the payment number in the period $[t, t+7]$ and $j = \overline{1, M}$.

Then the after-clearing sum of money for the period $[t, t+7]$ which the company A should pay to (or receive from) IATA can be calculated from Equation 13.1:

$$Cl_{t+21} = \sum_{i=1}^{N} Inv_i \left(1 - u_i\right), \tag{13.1}$$

where

Cl_{t+21} is the after-clearing sum of money for company A and the time period $[t, t+7]$;

Inv_i is the volume of the invoice i of company A;

u_i is the control variable defined by Equation 13.2 as

$$u_i := \begin{cases} 0, & \text{if the company decides to include } Inv_i \text{ into clearing;} \\ 1, & \text{if the company decides to submit } Inv_i \text{ in IATA Coins} \\ & \text{and exclude it from clearing.} \end{cases} \tag{13.2}$$

The company has the opportunity to exchange IATA Coins for fiat currency. Let the proportion of IATA Coins exchanged for fiat currency be denoted by a. The sum of money for the period $[t, t+7]$ which will be paid to (or received from) the company A in fiat currency can be calculated by Equation 13.3:

$$P_{t+7} = a \sum_{i=1}^{N} Inv_i u_i + \sum_{j=1}^{M} R_j, \tag{13.3}$$

where

P_{t+7} is the total volume of payments of company A in the time period $[t, t+7]$ which are not included into clearing and are submitted in fiat currency;

R_j is the volume of fiat currency payment j which does not go through IATA.

The control variable a can take on the values $[0, 1]$. If $a = 1$, the sum of all the volumes of invoices submitted in IATA Coins will be exchanged for fiat money. If $0 < a < 1$, only some IATA Coins will be exchanged and the others will be equal to $(1 - a) \sum_{i=1}^{N} Inv_i u_i$.

It is easy to see that

$$Cl_{t+21} + P_{t+7} + (1 - a) \sum_{i=1}^{N} Inv_i u_i = \sum_{i=1}^{N} Inv_i + \sum_{j=1}^{M} R_j. \tag{13.4}$$

This means that the factual cash flow (in fiat currency) can be reduced by the sum $(1 - a) \sum_{i=1}^{N} Inv_i u_i$ which will be held in IATA Coins.

The following section presents the possible variants of the objective functions.

13.3.2 The variants of the objective function

The objective function from the company's point of view can be formulated in two ways:

- Reducing the transaction costs
- Improving the liquidity profile

13.3.2.1 Reducing the transaction costs

This objective function can be formulated in terms of decision variables $a, u_1, \ldots, u_N$ as follows:

$$\frac{|\sum_{i=1}^{N} Inv_i(1-u_i)| + \sum_{j=1}^{M} |R_j| + |Cl_{t+7}|}{\sum_{i=1}^{N} |Inv_i|(1-u_i) + \sum_{j=1}^{M} |R_j| + |Cl_{t+7}|} \rightarrow min. \tag{13.5}$$

Expression (13.5) is the equivalent of the offset ratio maximization. This rational multiextremal function is nonlinear and nonconvex, and its global solution can only be found by exhaustive search. It may be approximated by a suitable global optimization procedure.

13.3.2.2 Improving the liquidity profile

This objective function can be formulated as follows:

$$|P_{t+7} + Cl_{t+7}| = \left| a \sum_{i=1}^{N} Inv_i u_i + \sum_{j=1}^{M} R_j + Cl_{t+7} \right| \rightarrow min. \tag{13.6}$$

In other words, the company tries to decrease the absolute value of the sum of all the fiat currency payments on the time period $[t, t+7]$. This potentially can help to smooth the liquidity profile by receiving additional liquidity from exchanging IATA Coins for fiat currency.

In order to give this bijective model, only a single optimization criterion, the second objective (13.6) will be implemented as a constraint:

$$\left| a \sum_{i=1}^{N} Inv_i u_i + \sum_{i=1}^{M} R_j + Cl_{t+7} \right| \leq \left| \sum_{i=1}^{M} R_j + Cl_{t+7} \right|. \tag{13.7}$$

13.3.3 How to choose the values of the control variables?

In this section, some recommendations for a company on how to find an optimal strategy will be presented. Each control variable will be considered separately.

13.3.3.1 The control variable $a \in [0, 1]$

There are four main situations that may influence the decision on how to choose a. They depend on the aggregated volume of the invoices in IATA Coins $\left(\sum_{i=1}^{N} Inv_i\right)$ and the total volume of fiat currency payments $\left(\sum_{j=1}^{M} R_j + Cl_{t+7}\right)$ in the current clearance cycle.

1. $\sum_{i=1}^{N} Inv_i > 0$ and $\sum_{j=1}^{M} R_j + Cl_{t+7} > 0$: It does not make sense to exchange IATA Coins for fiat currency $\Rightarrow a = 0$.
2. $\sum_{i=1}^{N} Inv_i > 0$ and $\sum_{i=1}^{M} R_j + Cl_{t+7} < 0$: It makes sense to exchange IATA Coins for fiat currency $\Rightarrow 0 < a \leq 1$.

   There are two cases which may appear:

   a. $\sum_{i=1}^{N} Inv_i u_i > -\left(\sum_{j=1}^{M} R_j + Cl_{t+7}\right)$: The sum of IATA Coins is greater than the value of the negative fiat currency balance. In this case, the company will intend only to cover the value of the negative balance: $a \sum_{i=1}^{N} Inv_i u_i = -\left(\sum_{j=1}^{M} R_j + Cl_{t+7}\right) \Rightarrow 0 < a < 1$.

   b. $\sum_{i=1}^{N} Inv_i u_i \leq -\left(\sum_{j=1}^{M} R_j + Cl_{t+7}\right)$: The sum of IATA Coins is less or equal to the value of the negative fiat currency balance. In this case, the company will intend to exchange all IATA Coins that it possesses $\Rightarrow a = 1$.
3. $\sum_{i=1}^{N} Inv_i < 0$ and $\sum_{j=1}^{M} R_j + Cl_{t+7} > 0$: The balance in IATA Coins is negative, there is nothing to exchange $\Rightarrow a = 0$.

   It does not make sense to exchange fiat currency for IATA Coins, because it seems to be more rational to include the invoices in clearing and later to pay for them in fiat currency.
4. $\sum_{i=1}^{N} Inv_i < 0$ and $\sum_{j=1}^{M} R_j + Cl_{t+7} < 0$: The balance in IATA Coins is negative, there is nothing to exchange $\Rightarrow a = 0$.

13.3.3.2 The control variables $u_i \in \{0; 1\}, i = \overline{1, N}$

The choosing of u_i can be considered from both the company's and from ICH's point of view.

1. *The company's opinion*: Using IATA Coins, the company will try to decrease the cash amount which is circulated in the network. This can be formulated as the following constraints:

   a. Total cash payments after IATA Coin adoption should be not greater than before it:

$$\left|\sum_{i=1}^{N} Inv_i\,(1-u_i)\right| + \left|a\sum_{i=1}^{N} Inv_i u_i\right| + \sum_{j=1}^{M} |R_j| + |Cl_{t+7}| \leq \left|\sum_{i=1}^{N} Inv_i\right| + \sum_{j=1}^{M} |R_j| + |Cl_{t+7}| \tag{13.8}$$

   or

$$\left|\sum_{i=1}^{N} Inv_i\,(1-u_i)\right| + \left|a\sum_{i=1}^{N} Inv_i u_i\right| \leq \left|\sum_{i=1}^{N} Inv_i\right| \tag{13.9}$$

   b. The same inequality for the absolute volumes of invoices:

$$\sum_{i=1}^{N} |Inv_i|\,(1-u_i) + \left|a\sum_{i=1}^{N} Inv_i u_i\right| \leq \sum_{i=1}^{N} |Inv_i| \tag{13.10}$$

   c. Liquidity profile is improved by using IATA Coins:

$$\left|a\sum_{i=1}^{N} Inv_i u_i + \sum_{j=1}^{M} R_j + Cl_{t+7}\right| \leq \left|\sum_{j=1}^{M} R_j + Cl_{t+7}\right|. \tag{13.11}$$

2. *Counterparty's opinion*: It should be mentioned that the decision about the submission of a particular invoice in IATA Coins also depends on the opinion of the company's counterparty. Let us assume that both sides of the invoice have the same criterion for choosing the invoices that will be submitted in IATA Coins. And let us also suppose that the invoice will be submitted in IATA Coins only when the both sides have made this decision.

3. *ICH's criterion*: As was said above, ICH's main goal is to increase the offset ratio. In fact, ICH is not a decision maker in the model, so it only can hope that the decisions made by the companies will improve the offset ratio. Actually, it can be taken for granted (according to the formulation of the model) that no matter what decisions are made by the companies (of course, if they are rational), they will not decrease the initial value of the offset ratio.

13.3.4 The final mathematical formulation of the model

Combining all above mentioned, the final optimization model becomes

$$
\begin{aligned}
&\frac{\left|\sum_{i=1}^{N} Inv_i\,(1-u_i)\right| + \sum_{j=1}^{M}|R_j| + |Cl_{t+7}|}{\sum_{i=1}^{N}|Inv_i|\,(1-u_i) + \sum_{j=1}^{M}|R_j| + |Cl_{t+7}|} \to min,\\
&\left|\sum_{i=1}^{N} Inv_i\,(1-u_i)\right| + \left|a\sum_{i=1}^{N} Inv_i u_i\right| \le \left|\sum_{i=1}^{N} Inv_i\right|,\\
&\sum_{i=1}^{N}|Inv_i|\,(1-u_i) + \left|a\sum_{i=1}^{N} Inv_i u_i\right| \le \sum_{i=1}^{N}|Inv_i|,\\
&\left|a\sum_{i=1}^{N} Inv_i u_i + \sum_{j=1}^{M} R_j + Cl_{t+7}\right| \le \left|\sum_{j=1}^{M} R_j + Cl_{t+7}\right|,\\
&a \in [0,1]\,,\ u_i \in \{0;1\},\ i=\overline{1,N}.
\end{aligned}
\tag{13.12}
$$

There are three variants of simplification depending on the value of the control variable a:

1. For the situation where $a = 0$, the model will be

$$
\begin{aligned}
&\frac{\left|\sum_{i=1}^{N} Inv_i\,(1-u_i)\right| + \sum_{j=1}^{M}|R_j| + |Cl_{t+7}|}{\sum_{i=1}^{N}|Inv_i|\,(1-u_i) + \sum_{j=1}^{M}|R_j| + |Cl_{t+7}|} \to min,\\
&\left|\sum_{i=1}^{N} Inv_i\,(1-u_i)\right| \le \left|\sum_{i=1}^{N} Inv_i\right|,\\
&\sum_{i=1}^{N}|Inv_i|\,(1-u_i) \le \sum_{i=1}^{N}|Inv_i|,\\
&a \in [0,1]\,,\ u_i \in \{0;1\},\ i=\overline{1,N}.
\end{aligned}
\tag{13.13}
$$

2. For the situation where $0 < a < 1$, the model will be

$$
\begin{aligned}
&\frac{\left|\sum_{i=1}^{N} Inv_i\,(1-u_i)\right| + \sum_{j=1}^{M}|R_j| + |Cl_{t+7}|}{\sum_{i=1}^{N}|Inv_i|\,(1-u_i) + \sum_{j=1}^{M}|R_j| + |Cl_{t+7}|} \to min,\\
&\left|\sum_{i=1}^{N} Inv_i\,(1-u_i)\right| + \left|\sum_{j=1}^{M} R_j + Cl_{t+7}\right| \le \left|\sum_{i=1}^{N} Inv_i\right|,\\
&\sum_{i=1}^{N}|Inv_i|\,(1-u_i) + \left|\sum_{j=1}^{M} R_j + Cl_{t+7}\right| \le \sum_{i=1}^{N}|Inv_i|,\\
&a \in [0,1]\,,\ u_i \in \{0;1\},\ i=\overline{1,N}.
\end{aligned}
\tag{13.14}
$$

3. For the situation where $a = 1$, the model will be

$$\begin{aligned}
&\frac{\left|\sum_{i=1}^{N} Inv_i\,(1-u_i)\right| + \sum_{j=1}^{M} |R_j| + |Cl_{t+7}|}{\sum_{i=1}^{N} |Inv_i|\,(1-u_i) + \sum_{j=1}^{M} |R_j| + |Cl_{t+7}|} \to min,\\
&\left|\sum_{i=1}^{N} Inv_i\,(1-u_i)\right| + \left|\sum_{i=1}^{N} Inv_i u_i\right| \le \left|\sum_{i=1}^{N} Inv_i\right|,\\
&\sum_{i=1}^{N} |Inv_i|\,(1-u_i) + \left|\sum_{i=1}^{N} Inv_i u_i\right| \le \sum_{i=1}^{N} |Inv_i|,\\
&\left|\sum_{i=1}^{N} Inv_i u_i + \sum_{j=1}^{M} R_j + Cl_{t+7}\right| \le \left|\sum_{j=1}^{M} R_j + Cl_{t+7}\right|,\\
&a \in [0,1]\,,\ u_i \in \{0;1\},\ i = \overline{1,N}.
\end{aligned} \tag{13.15}$$

In the following section, the above described optimization model will be evaluated experimentally.

13.4 Practical Implementation of the Model

Before presenting the results of the model, some words should be said about how the results of the clearing procedure can be represented.

13.4.1 Results representation

All the payments which are included to the clearing cycle can be represented as a payment matrix in the forms that are shown in the following sections.

13.4.1.1 Before IATA Coin adoption

In the current clearing procedure, all the invoices can be combined into one matrix:

Total

	1	2	...	n	$\sum$	Offset
1		a_{12}	...	a_{1n}	$a_{1\cdot}$	$a_1 = a_{1\cdot} - a_{\cdot 1}$
2	a_{21}		...	a_{2n}	$a_{2\cdot}$	$a_2 = a_{2\cdot} - a_{\cdot 2}$
...	...	...		...	...	...
n	a_{n1}	a_{n2}	...		$a_{n\cdot}$	$a_n = a_{n\cdot} - a_{\cdot n}$
$\sum$	$a_{\cdot 1}$	$a_{\cdot 2}$	...	$a_{\cdot n}$		

where

a_{ij} is the sum of money, that airline i needs to pay to airline j in current clearing cycle (the sum of invoices from airline i to airline j during the clearing cycle, where N_{ij} is the number of invoices from airline i to airline j):

$$a_{ij} = \sum_{k=1}^{N_{ij}} Inv_{ijk}, \tag{13.16}$$

$a_{i\cdot}$ is the sum of money, that airline i needs to pay to all of the members of ICH:

$$a_{i\cdot} = \sum_{j=1}^{n} a_{ij}, \tag{13.17}$$

$a_{\cdot j}$ is the sum of money, that airline j should receive from all of the members of ICH:

$$a_{\cdot j} = \sum_{i=1}^{n} a_{ij}. \tag{13.18}$$

Then offset ratio will be

$$\mathit{Offset}_1 = 1 - \frac{\sum_{i=1}^{n} |a_i|}{\sum_{i=1}^{n} \sum_{j=1}^{n} \sum_{k=1}^{N_{ij}} |Inv_{ijk}|}. \tag{13.19}$$

13.4.1.2 After IATA Coin adoption

After the adoption of IATA Coins, it will be necessary to consider separately the payments in IATA Coins and in fiat currency and, instead of one payment matrix, there will be two:

In IATA Coins

	1	2	...	n	$\sum$
1		b_{12}	...	b_{1n}	$b_{1\cdot}$
2	b_{21}		...	b_{2n}	$b_{2\cdot}$
...	...	...		...	...
n	b_{n1}	b_{n2}	...		$b_{n\cdot}$
$\sum$	$b_{\cdot 1}$	$b_{\cdot 2}$	...	$b_{\cdot n}$	

In Fiat Currency

	1	2	...	n	$\sum$	Offset
1		$\tilde{a}_{12}$	...	$\tilde{a}_{1n}$	$\tilde{a}_{1\cdot}$	$\tilde{a}_1 = \tilde{a}_{1\cdot} - \tilde{a}_{\cdot 1}$
2	$\tilde{a}_{21}$		...	$\tilde{a}_{2n}$	$\tilde{a}_{2\cdot}$	$\tilde{a}_2 = \tilde{a}_{2\cdot} - \tilde{a}_{\cdot 2}$
...	...	...		...	...	...
n	$\tilde{a}_{n1}$	$\tilde{a}_{n2}$	...		$\tilde{a}_{n\cdot}$	$\tilde{a}_n = \tilde{a}_{n\cdot} - \tilde{a}_{\cdot n}$
$\sum$	$\tilde{a}_{\cdot 1}$	$\tilde{a}_{\cdot 2}$	...	$\tilde{a}_{\cdot n}$		

where

b_{ij} is the sum of payments in IATA Coins, that airline i will pay to airline j.

$\tilde{a}_{ij}$ is the sum of cash, that airline i needs to pay to airline j in the current clearing cycle (invoices submitted in IATA Coins are excluded).

Generally, the sum of billings both in cash and IATA Coins is equal to the volume of billings in the case when airlines do not use IATA Coins:

$$b_{ij} + \tilde{a}_{ij} = a_{ij}. \tag{13.20}$$

The offset ratio for fiat payments can be calculated in a similar way as in Equation 13.19. But actually, for invoices in IATA Coins if we calculate the offset (in the same meaning as for fiat currency invoices), it will be equal to 0, because the invoices submitted in IATA Coins are not aggregated.

The total offset ratio for the system:

$$\mathit{Offset}_2 = 1 - \frac{\sum_{i=1}^{n} |\tilde{a}_i| + \sum_{i=1}^{n} \sum_{j=1}^{n} b_{ij}}{\sum_{i=1}^{n} \sum_{j=1}^{n} \sum_{k=1}^{N_{ij}} |Inv_{ijk}|}. \tag{13.21}$$

But, if we say that offset ratio is the relation between the volume of billings and the amount of cash required to settle them and, in fact, IATA Coins are not the cash, in that sense it can be said that the offset ratio for the second case will be

$$\mathit{Offset}_2^* = 1 - \frac{\sum_{i=1}^{n} |\tilde{a}_i|}{\sum_{i=1}^{n} \sum_{j=1}^{n} \sum_{k=1}^{N_{ij}} |Inv_{ijk}|}. \tag{13.22}$$

13.4.2 The model results

The final version of the model was practically implemented on invoice samples simulated according to the assumptions that

- The number of invoices per day for a company was considered as a random variable with Poisson distribution. The parameter λ was fitted on a sample of the invoices between 2 airlines which was provided by IATA. It is approximately equal to 10, meaning that for each pair of counterparties on each day there are 10 invoices on average. The model was also implemented on sample invoices with other values of λ.

- Other payments which do not go through IATA are also included into the sample (for simplicity, there is only one counterparty nonmember of ICH and λ is set the same, equal to 10).

- The volumes of invoices (and other payments) were a random variable with a lognormal distribution. The parameters were fitted on the sample of the invoices between two airlines which was provided by IATA. So, $\hat{\mu} = 8.190488$ and $\hat{\sigma} = 2.118524$.

The solution of the optimization problem was found via a genetic algorithm [2–4]. It is rather a resource-intensive optimization method which explains why the number of simulations was only 1000 and the number of the companies in the sample was equal to 3 in order to decrease the time needed for calculations.

The above described model is implemented for one clearing cycle and for each company. This means that every company chooses its optimal strategy independently from the others. Then for each invoice in the sample, the opinions of both counterparties (the payer and the payee) are compared. If both companies decide that the submission of the invoice in IATA Coins is more profitable for them, then the invoice will be submitted in IATA Coins, otherwise it will be included in the clearing procedure.

13.4.2.1 Example of one simulation

The results of the implementation of the model calculated on one simulated sample are given in the tables.

1. Total cash flow before the IATA Coins adoption

	1	2	3	$\sum$	Offset
1		1,269,504	1,636,788	2,906,292	1,149,624
2	779,289		920,626	1,699,915	−725,878
3	977,379	1,156,289		2,133,668	−423,746
$\sum$	1,756,668	2,425,793	2,557,414		**2,299,248**

2. Cash flow in IATA Coins

	1	2	3	$\sum$
1		226,544	13,102	239,646
2	0		0	0
3	0	0		0
$\sum$	0	226,544	13,102	**239,646**

3. Cash flow in fiat currency

	1	2	3	$\sum$	Offset
1		1,042,960	1,623,686	2,666,646	909,978
2	779,289		920,626	1,699,915	−499,334
3	977,379	1,156,289		2,133,668	−410,644
$\sum$	1,756,668	2,199,249	2,544,312		**1,819,956**

We see that using IATA Coins can potentially help to reduce the cash flow:

- Total cash flow without IATA Coins is equal to 2,299,248
- Total cash flow using IATA Coins is equal to 239,646 + 1,819,956 = 2,059,602

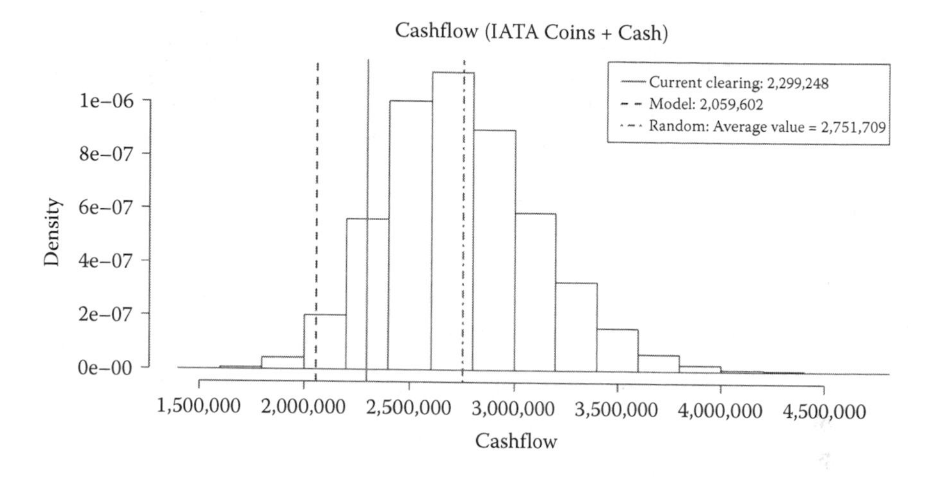

FIGURE 13.3: Example of model implementation.

In order to check the quality of the model, the results of the model were compared with random behavior in which invoices that should be submitted in IATA Coins were selected in a random way.

The results of the comparison are shown in Figure 13.3. For this example, a large number (namely, 100,000) of scenarios of random invoice selection were generated and for each scenario the value of after-clearing net cash flow was calculated. Then these values were represented in the form of histogram, where also three values are marked with vertical lines:

- The value of net cash flow without using IATA Coins (as in the current clearing procedure)
- The modeled value of net cash flow
- The average value of the cash flow in the randomly generated scenarios

We see that the current clearing procedure gives a better result than the average of the random scenarios, but the model can help to improve this result.

In Table 13.1, there are main metrics calculated for these three cases: the values of cash flow (going through ICH and externally), the sum of the invoices, and the offset ratio. In this example, the offset ratio has increased from 65.9% to 69.4%, an increase of 5.3%.

13.4.2.2 Increasing the number of simulations

In order to check whether the results are sensitive to the number of simulations or not, the model was also run on 1000 simulations. The average values of all the metrics are given in Table 13.2.

TABLE 13.1: Metrics calculated on one example simulation

	Current clearing procedure	Model	Random (averaged over 100,000 simulations)
1. Cash flow	20,899,899	20,660,253	21,352,360
1.1. Total (ICH)	2,299,248	2,059,602	2,751,709
1.1.1. In fiat	2,299,248	1,819,956	2,078,041
1.1.2. In IATA Coins	0	239,646	673,668
1.2. External	18,600,651	18,600,651	18,600,651
2. Sum of invoices	6,739,875	6,739,875	6,739,875
3. Offset (ICH)	65.9%	69.4%	59.2%
4. Offset (including external cash flow)	17.5%	18.5%	15.7%

TABLE 13.2: Metrics averaged over 1000 example simulations

	Current clearing procedure	Model	Random (averaged over 100,000 simulations)
1. Cash flow	20,720,171	20,382,077	21,726,810
1.1. Total (ICH)	6,270,014	5,931,921	7,276,653
1.1.1. In fiat	6,270,014	5,245,561	5,909,896
1.1.2. In IATA Coins	0	686,360	1,366,757
1.2. External	14,450,157	14,450,157	14,450,157
2. Sum of invoices	13,662,411	13,662,411	13,662,411
3. Offset (ICH)	56.8%	59.5%	49.2%
4. Offset (including external cash flow)	27.3%	28.6%	23.6%

Due to the nature of a small sample from the lognormal invoice volume distribution, the averaged offset ratio is lower than in the example from previous subsection, but the model result is better than the current procedure result by 4.8%.

13.5 Summary

This chapter covers the issues of the implementation of blockchain technologies in the clearing and settlement procedure of the ICH. We have developed an approach to estimate the industry level benefits of adoption of the blockchain-based industry money (IATA Coin) for clearing and settlement. The potential system-wide offsetting benefits are mainly driven by the following factors:

- Immediate settlement of the value on the blockchain
- Shortened cycle of expense recognition

- More flexible liquidity management enabled by *ad hoc* retroconvertibility of IATA Coin

- Increased adoption of the IATA Coin as a means of settlement for the supply chain, as well as for institutional and retail customers

To evaluate the magnitude of these benefits, we have developed a model that allows simulation of the current cycle and of the proposed innovations and their comparison in terms of offset ratio.

The main result gives practical evidence that using IATA Coins can help to increase the offset ratio, reduce the transaction costs, and improve the liquidity profile for a company, assuming that all members of the financial network are acting rationally and are intending to decrease their transactions costs and to improve their liquidity profile.

References

1. Committee on Payments and Market Infrastructures (CPMI), 2017. *Distributed Ledger Technology in Payment, Clearing and Settlement: An Analytical Framework*. Basel. http://www.bis.org/cpmi/publ/d157.pdf

2. Lucasius, C.B. and Kateman, G., 1993. Understanding and using genetic algorithms—Part 1. Concepts, properties and context. *Chemometrics and Intelligent Laboratory Systems*, 19:1–33.

3. Lucasius, C.B. and Kateman, G., 1994. Understanding and using genetic algorithms—Part 2. Representation, configuration and hybridization. *Chemometrics and Intelligent Laboratory Systems*, 25:99–145.

4. Willighagen, E. and Ballings M., 2015. *Package "genalg": R Based Genetic Algorithm*. https://cran.r-project.org/web/packages/genalg/

Part III

HPC Systems: Hardware, Software, and Data with Financial Applications

Chapter 14

Supercomputers

Peter Schober

CONTENTS

14.1 Introduction

Generally speaking, a supercomputer is a computer which is one of the fastest computers of its time. Usually, the computer's performance is measured in Floating Point Operations Per Second (FLOPS) when running a certain benchmark program. Since 1993, the TOP500 list [1] ranks the top 500 supercomputers worldwide according to their maximal performance achieved when running the LINPACK benchmark.[1] As of November 2016, the fastest supercomputer in the world is the *Sunway TaihuLight* at the National Supercomputing Center in Wuxi, China. Its maximal performance is at 93 Peta FLOPS, that is, 93×10^{15} FLOPS.

All supercomputers on the current TOP500 list are massively parallel computers designed either in a computer cluster or in a Massively Parallel Processing (MPP) architecture. A computer cluster consists of many interconnected, independent computers that are set up in a way that they are virtually a single system. The computers in the cluster (often called compute nodes) usually are standalone systems and are configured to work on a parallel job, which is distributed over the nodes and controlled by specific job scheduling software. In contrast, an MPP computer consists of many individual compute nodes that do not qualify as a standalone computer and are connected by a custom network. For example, in IBM's Blue Gene series, the compute nodes comprise multiple CPUs and a shared RAM, but no hard disk, and are interconnected via a multidimensional torus network. In MPP architecture, often times the complexity of a single CPU (e.g., the number of the transistors or the clock frequency) is reduced to allow for a higher number of parallel cores. Both architectures are frequently combined with certain hardware acceleration methods such as coprocessors or general-purpose GPUs (GPGPUs).

An example of a supercomputer with MPP architecture is the Blue Gene/Q supercomputer *JUQUEEN*, currently ranked 19 on the TOP500 list. It combines compute nodes which have 16 cores with 1.6 GHz clock frequency per node sharing 16 GB RAM and run the lightweight Compute Node Kernel (CNK). These are stacked together in 28 racks, share a main memory, and are interconnected by infiniband with five-dimensional torus structure.[2] The supercomputer can be accessed via the SSH protocol on separate log-in nodes that run Linux. Jobs can be sent to the compute nodes via a central job manager called LoadLeveler. A simple example for a computer cluster is a compilation of conventional servers; each with multiple cores, a high amount of RAM and its own hard drive running on a Windows HPC Server. These servers are put in one rack and are connected via Gigabit Ethernet. A central software on a dedicated head node, the HPC Cluster Manager, schedules the jobs to the compute nodes.

---

[1] Theoretical maximum performance can also be reported and is calculated as follows: maximum number of floating point operations per clock cycle × maximum frequency of one core × number of cores of the computer.

[2] The full specifications can be found in Reference 2.

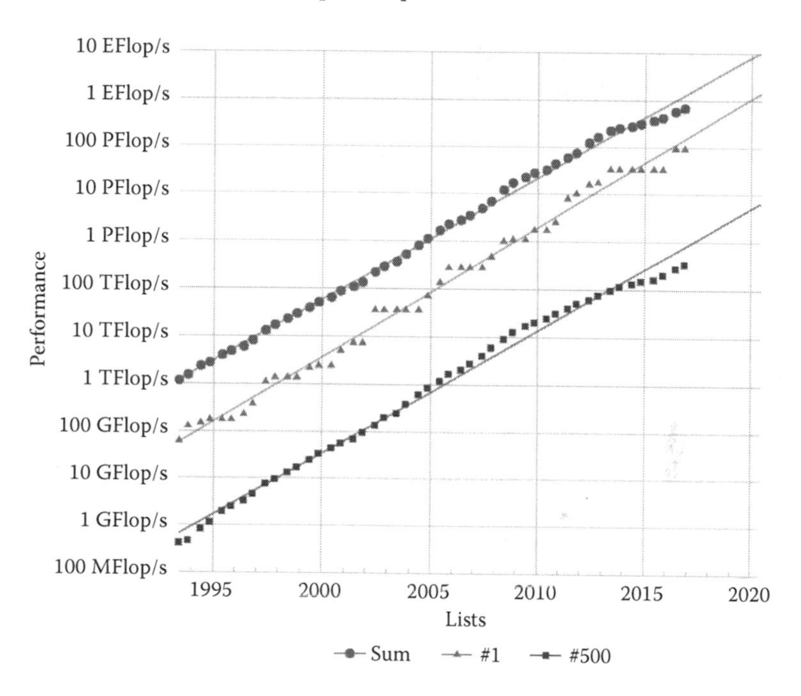

FIGURE 14.1: Exponential performance development and projection. (Adapted from the Top500 homepage. Available: http://top500.org/, 2017.)

A reason for the development of massively parallel computers is that Moore's Law of doubling the number of transistors in an integrated circuit (and by this also the speed of CPUs) roughly every 2 years is considered to be outdated due to the physical limits in transistor size and difficulties with heat dissipation. Today's supercomputers cluster a large number of interconnected CPUs with multiple cores each, and code has to be run in parallel on these machines. The *Sunway TaihuLight*, for example, sources its computing power from 10,649,600 cores with only 1.45 GHz clock frequency.[3] Supercomputers at the end of the TOP500 list still have several thousand cores and around 350 Tera FLOPS LINPACK performance. The tendency for parallelization can also be observed in personal computer and embedded systems development: the total number of cores per device now roughly doubles every 2 years. This way, the performance of (super)computers in terms of FLOPS is still steadily growing exponentially, compare Figure 14.1.

One of the main advantages of working with supercomputers is that applications can be programmed using a standard programming language (such as C++) combined with a standardized parallel programming interface (such as

[3]For comparison: today, common desktop computers comprise multicore CPUs with two, four, or eight cores with clock frequencies around 3 GHz.

MPI). This leads to easy portability of code as well as independence of the application from the actual hardware. Essentially, this opens the door to run the same application developed for a multicore desktop computer to be run on small size supercomputers or even on the top supercomputers worldwide. In addition, if exponential growth of computing power persists, the same applications as implemented for supercomputers today may be suitable to run on personal computers in a few years time. Having said this, the following sections can be applied to any hardware that is similar to the architectural approaches of today's TOP500 supercomputers—even though the reader's "supercomputer" might have far less cores and "only" a performance of a few Tera FLOPS.

14.2 History, Current Landscape, and Upcoming Trends

In 1837, computer pioneer Charles Babbage proposes the *Analytical Engine*, which can be considered as the first concept of a mechanical general-purpose computer function in the same manner as modern computers. The *Analytical Engine* stays a concept and is never actually built by Babbage. Around one century later, in 1941, Konrad Zuse unveils the *Zuse Z3*, the first universally programmable fully functional digital computer operating at 5.3 Operations Per Second (OPS). Another 38 years later, in 1979, the *Cray-1*[4] at the Los Alamos National Laboratory, New Mexico, USA, leads the first compiled LINPACK performance list with ca. 250 Mega FLOPS. Jack Dongarra's LINPACK software library [4] to solve a dense system of linear equations soon becomes established as an important benchmark for the measurement of supercomputer performance. At the 8th Mannheim Supercomputer Seminar in 1993, the first TOP500 list is published. At its top, Thinking Machines Corporation's *CM-5*, also located at the Los Alamos National Laboratory, performs at 59.7 Giga FLOPS when running the LINPACK benchmark on 1024 cores. In 2008, IBM's *Roadrunner* is the first supercomputer to break the barrier of one Peta FLOPS, which is more than twice as fast as the number one system on the TOP500 list at that time. Since 2010, hybrid systems have also made it to the top of the TOP500 list. First, the *Tianhe-1A* at the National Supercomputing Center, Tianjin, China, with 14,336 CPUs and 7168 GPUs used as accelerators makes it to the top. In 2012, when the 40th TOP500 list is published, the *Titan* supercomputer at Oak Ridge National Laboratory, Tennessee, USA, combines GPUs and traditional CPUs to become the world's fastest supercomputer. *Titan*'s performance of 17.6 Peta FLOPS on the LINPACK benchmark is facilitated by 18,688 nodes that contain a GPU

[4]The *Cray-1* was built by Cray Research, Inc., founded by Seymour Cray, who is broadly considered the "father of supercomputing" [3].

and a CPU with 16 cores, which totals 560,640 cores. Lately, China matches up with the USA (171 systems each in the TOP500 list as of November 2016, Germany is third with 32 systems) and manifests its dominance at the top of the list. In June 2016, the Chinese *Sunway TaihuLight*, developed by the National Research Center of Parallel Computer Engineering & Technology, with 93 Peta FLOPS and over 10 million cores, took the position as the world's fastest supercomputer from the likewise Chinese *Tianhe-2* (32.9 Peta FLOPS). Thus the top two Chinese systems alone account for about 127 Peta FLOPS of the total aggregate of 672 Peta FLOPS on the current TOP500 list.

The focus on speed as the only measure for performance leads to increasing power consumption of the top supercomputers during the last decade. Half of the costs of a conventional cluster can be accounted for cooling of the facilities, the other half for the actual computations. In 2006, an awareness of social responsibility regarding climate change and global warming, but also cost efficiency within the High Performance Computing community, initiates the announcement of the Green500 list [5]: the Green500 aims at ranking the most energy efficient supercomputers worldwide (measured by Mega FLOPS per Watt). Since then, architecture develops toward hybrid supercomputers combining CPUs with GPGPU and coprocessor accelerators, which are connected with low latency network technologies like Infiniband. Besides the good suitability of GPGPUs and coprocessors for data flow tasks, a reason for the increased use of these technologies is that their power and cost efficiency is about 10 times higher than for conventional CPUs. In November 2014, the *L-CSC* at the GSI Helmholtz Center claims the number one position in the Green500 list with 5271 Mega FLOPS per Watt. It is developed at the Frankfurt Institute for Advanced Studies (FIAS) [6], a public-private partnership of the Goethe University Frankfurt and private sponsors. *L-CSC* reaches its top position by massive use of GPU acceleration and an especially efficient cooling system. Since November 2016, the in-house supercomputer *DGX SaturnV* of the GPU manufacturer NVIDIA based in Santa Clara, Calif., USA, is at the top of the Green500 list. It operates at 9462 Mega FLOPS per Watt and is closely followed by the Swiss *Piz Daint* (7453 Mega FLOPS per Watt), which also claims a rank of 8 in the TOP500 list. Recently, in the advent of energy efficient supercomputers, the European Mont-Blanc project [7] aims at building power efficient supercomputers based on low-power embedded system technology as used in smart phones or tablet computers. Within this project funded by the European Commission, Systems on Chip (SoC) are stacked together in blades and are interconnected by 10 Gigabit Ethernet. The first *Mont Blanc* prototype has a total of 2160 CPUs and 1080 GPUs and runs parallel applications from physics and engineering using conventional programming languages and parallelization approaches.

Whichever technology is going to prevail in the future, if computing power doubles every 2 years, supercomputers will be operating at Exa scale FLOPS by 2020.

14.3 Programming Languages and Parallelization Interfaces

When developing a parallel application, the following two questions have to be answered:

1. Which programming language should be used? A compiled programming language, such as Fortran, C or C++, or an interpreted language, such as MATLAB or Python?
2. Which parallelization interface is appropriate? Shared memory parallelization, such as OpenMP or POSIX threads (Pthreads), or distributed memory parallelization with the Message Passing Interface (MPI)?

Table 14.1 depicts features of programming languages that are widely used by scientific researchers and practitioners for writing applications for supercomputers. Of course, this overview has no claim to completion, and various other programming languages exist that can be used on supercomputers in combination with parallelization. Table 14.2 shortly summarizes the characteristics of the most important parallelization interfaces. Usually, these are extended by task-specific (e.g., I/O of data) or hardware-specific parallelization interfaces (such as CUDA for the use of GPUs).

Still widely used programming languages in scientific computing are Fortran and C followed by C++, recently more often combined with Python.

TABLE 14.1: Overview on programming languages

Programming language	Features
Compiled	
C	General-purpose, imperative programming language that is available on all computer systems. C supports structured programming and has a static type system
C++	Object-oriented extension of C
Fortran (and its variants)	General-purpose, imperative programming language that is especially suited to numeric computation and scientific computing. Fortran 2003 interoperates with C and C++
Interpreted	
MATLAB	High-level, interpreted, interactive programming language that supports object orientation. Commercial software
Python	Interpreted, interactive, object-oriented, extensible programming language. Open source

TABLE 14.2: Overview on parallelization interfaces

Parallelization interface	Features
MPI	Standardized interface for message passing between computing units on distributed (and shared) memory (super)computers. Parallelization is organized by MPI processes that communicate with each other using interface commands. Parallelizing is the duty of the programmer
Pthreads	Thread-level parallelism using the POSIX library. Only limited support for data-parallel operations
OpenMP	Library and set of compiler directives for Fortran, C, and C++. The compiler handles the parallel thread creation, which makes OpenMP far more simple than MPI or Pthreads programming

Parallelization of code is mostly done with MPI, because it can handle the distributed memory of the systems, but is often combined with shared memory parallelization using OpenMP, Pthreads, or more recently, GPUs. Table 14.3 provides an overview on programming languages and parallelization interfaces used by codes that run on all 458,752 cores of the supercomputer *JUQUEEN*. Double counting is possible as nearly all codes use combinations of multiple programming languages and parallelization interfaces. However, all codes use MPI for parallelization, mostly in combination with Fortran, closely followed by C and C++. OpenMP is frequently used on node level, because all cores on one node share the node's memory. Pthreads are rarely used. While interpreted programming languages as Python or MATLAB play a minor role in

TABLE 14.3: Overview on programming languages and parallelization interfaces of codes that run on all 458,752 cores of the supercomputer *JUQUEEN*—the so-called *high-Q club*

	C	C++	Fortran	CUDA/openCL
MPI	16	10	18	2
OpenMP	13	9	14	–
Pthreads	4	–	3	–
Other (mostly for I/O)	6	3	9	–

Source: High-Q Club—Highest Scaling Codes on JUQUEEN. [Online]. Available: http://www.fz-juelich.de/ias/jsc/EN/Expertise/High-Q-Club/_node.html, 2017.

scientific computing,[5] practitioners often use interpreted, interactive programming languages which already include parallel capabilities.

As pointed out in the introduction, an application that can be implemented on your desktop computer is—when sticking with standard programming languages and parallelization interfaces—easily portable to a supercomputer. Usually, the trade off in choosing a combination of programming language and parallelization interface resides in programming flexibility and application performance against implementation time and ease of use. For example, C++ in combination with MPI, that is, a parallelization approach that can handle distributed memory, can be run on multicore desktops, smaller supercomputers, and TOP500 supercomputers with nearly no code changes (although recompiling might be required). On the downside, to program C++ a deeper understanding of memory management, compilation and features of high-level programming languages (such as object orientation) is necessary. Using MPI as the parallelization interface is flexible but comes at the cost of higher programming effort to go from a serial to a parallel version of the application. Additionally, debugging is a cumbersome task.[6]

In contrast, using an interpreted interactive programming language such as MATLAB together with MATLAB's Parallel Computing Toolbox provides a convenient way to parallelize independent computations. MATLAB wraps the MPI library and abstracts basic MPI functionalities to high-level, user-friendly functionalities, such as parallel for-loops or distributed arrays. In addition, certain MPI functions that enable communication are exposed to the user. MATLAB can easily be configured to use the multicore architecture from desktop computers and even supercomputers (which is a more difficult task). On the downside, MATLAB requires licenses for as many cores as are used by the computation—which is costly[7]—and MATLAB has to be installed in the appropriate version on the supercomputer. Besides that, the MATLAB's user-friendliness comes at the cost of loss of control over the parallelization, meaning that it is hardly possible for experienced programmers to overcome shortcomings of the high-level parallel interfaces MATLAB provides. This is further aggravated by MATLAB being closed source. Python, with its many easy-to-use frameworks for parallel computing, provides capabilities very similar to MATLAB's parallel for-loops and beyond. Also, Python supports native thread level parallelism with its multiprocessing module, and MPI extensions exist.[8] However, oftentimes, Python is only used as a wrapper for parallel, high-performance modules, which are in turn programmed in a compiled programming language like Fortran or C.

---

[5]Python is sometimes used to write wrappers for pre- and postprocessing of the results or similar tasks.

[6]There is (commercial) software that facilitates parallel debugging with MPI.

[7]However, these licenses are part of a special license for a computer clusters called Distributed Computing Server and are cheaper than full MATLAB licenses.

[8]Most MPI extensions for Python, similarly to MATLAB, provide (a subset) of the MPI functionalities of the wrapped MPI library.

14.4 Advantages and Disadvantages

As already addressed, an important advantage of the use of supercomputers is the extensive hardware independence and easy portability of applications. Another advantage is that parallelization using standard programming languages and parallelization interfaces not only scales well for data flow problems but also for control flow problems. That is, for example, not only can floating point operations on vectors of data be parallelized easily, but also complex function calls including conditional branches. This makes supercomputers especially suitable for coarse-grain parallelization, where large parts of the application can be run independently and, optimally, without communication between the cores. Of course, this makes proper load balancing of utmost importance, as the overall runtime of a parallel application is governed by the runtime of the slowest independently run part.

Coincidentally, this is also one of the major disadvantages of supercomputers: no or only little problem-specific optimization can be done. For example, the use of GPUs for data flow problems can be highly favorable over CPUs. A contemporary GPU has several thousand cores at a lower frequency (around 500 MHz) for stream processing of data, whereas a high-end CPU has only eight cores at about 3 GHz.[9] In addition, GPUs are more power efficient in terms of FLOPS per Watt. Other hardware that can be tailored to specific financial applications (such as Field Programmable Gate Arrays) promise an even better performance. However, there is a high awareness for this disadvantage within the supercomputing community and recently more and more supercomputers are complemented with problem-specific hardware such as GPGPUs and coprocessors. The application programmer can then decide which calculations are suitable to be run on a GPGPU—potentially using GPU-specific code parts—and which bigger building blocks are parallelized on a CPU level using standard programming languages and parallelization interfaces. Modern coprocessors constitute a hybrid solution as they also have a large number of cores and are especially suitable for data flow problems, such as GPUs, but the CPU automatically decides which calculation tasks it offloads to the coprocessor. Besides configuration of the application's runtime environment, there is no additional problem-specific programming necessary. In addition, the openCL standard further aids in overcoming the obstacles of programming for hybrid hardware approaches as it intends to provide a unified framework for programming applications that run on heterogeneous platforms consisting of CPUs, GPUs, and coprocessors.

Another disadvantage of the use of supercomputers is the ease of access and associated costs. Computing time of many supercomputers cannot simply be bought. Also, many of the supercomputers are only accessible for research

[9]Usually, GPUs have many more cores and threads with a lower clock frequency than standard CPUs.

purposes.[10] Setting up, operating, and maintaining a proprietary supercomputer is costly due to high initial costs, high energy costs—especially for appropriate cooling—and high costs for qualified staff.[11]

14.5 Supercomputers for Financial Applications

Besides many financial research institutions that lease computing time on supercomputers from the TOP500 list to efficiently solve finance problems using parallelization, there are several (undisclosed) institutions from the finance industry operating supercomputers that are ranked in several TOP500 lists. When considering how costly it is to set up and maintain such supercomputers, there obviously is demand within the financial industry for supercomputing.

14.5.1 Suitable financial applications

There is a vast set of financial applications from various fields: pricing of financial products, risk management, computation of solvency capital requirements, portfolio optimization, and many more. However, the different numerical methods employed can be roughly clustered:

- Monte Carlo simulations
- Numerical integration
- Finite differences and finite element methods

Some of these numerical methods are inherently parallel, such as Monte Carlo simulations; others are inherently coupled, like finite difference methods.

The problems that are inherently parallel are often called *embarrassingly parallel* problems. This term usually applies if one big problem can be disassembled into a set of decoupled problems. Examples include:

- Simulation of multiple paths in a Monte Carlo simulation for determining solvency capital requirements of an insurer
- Variance decompositions of partial differential equations (PDEs), which can be used for the pricing of derivatives using finite difference methods (this is only possible under certain assumptions)
- Portfolio optimization over the life cycle with time discrete dynamic programming

[10]Due to the high costs of implementing and maintaining, supercomputers are often run by government financed nonprofit organizations, huge corporations, or industry consortiums.

[11]An exemplary cost calculation is presented in Section 14.5.

The latter two are discussed as case studies in Sections 14.6 and 14.7. For embarrassingly parallel problems, parallelization is straightforward and, using the right combination of programming language and parallelization interface, often a "quick win" when using supercomputers.

In the case of coupled problems, parallelization is usually not so straightforward and harder to implement, for example, for finite differences or finite element methods for general pricing of derivatives.

14.5.2 Performance measurement

To assess how well a parallel application speeds up when using an increasing number of computing units (that is, CPUs, cores, processes or threads, and so on), scaling efficiency can be measured. Depending on the application at hand, there are two ways to measure the parallel performance: strong and weak scaling.

When measuring strong scaling, usually the overall runtime of the application is the limiting factor. In this case, the problem size (e.g., the number of paths in a Monte Carlo simulation) stays fixed and the number of computing units is increased until the whole problem is solved in parallel (e.g., every path is simulated on a single core). Let the time needed to solve the problem on the minimum number of available computing units P_{min} be t_{max} (in reference to the maximum runtime of the application). When scaling strongly, an application is considered to scale linearly if the speedup $S_{\text{strong}} = t_{\text{max}}/t$ is proportional to the number of computing units P used: $S_{\text{strong}} \sim P$. The optimally achievable speedup is thus $S_{\text{strong}}^{\text{opt}} = P/P_{\text{min}}$. A measure of overall parallel efficiency is then defined by

$$E_{\text{strong}} = \frac{S_{\text{strong}}}{S_{\text{strong}}^{\text{opt}}}. \tag{14.1}$$

For example, if the runtime on one computing unit, $P_{\text{min}} = 1$, is $t_{\text{max}} = 80$ seconds, then one expects the runtime on $P = 8$ cores to be $t = 10$ seconds and the application would scale linearly with speedup $S_{\text{strong}} = 80/10 = 8 = S_{\text{strong}}^{\text{opt}}$. Parallel efficiency is $E_{\text{strong}} = 1$. Embarrassingly parallel problems often times scale strongly, but, in general, it is harder to achieve high strong scaling efficiencies at larger number of computing units since, depending on the parallelization approach, the communication overhead for many applications also increases proportionally in the number of computing units used. In addition, ideally the problems should be distributed in a way that the variance of the runtime distribution of the atomic components (e.g., the paths) is minimal. Otherwise, the longest running computing unit governs the overall solution time. As already indicated, an example is a Monte Carlo simulation where the development of a single path is decoupled from the development of all other paths and hence all paths can be computed in parallel. If a fraction of the code cannot run in parallel, the realized speedup S_{strong} cannot achieve the optimal speedup $S_{\text{strong}}^{\text{opt}}$. Consider the following example: the sequential part of the code takes 1 second and the parallelizable part of the code takes 4 seconds. With four computing units available, the optimal speedup would be 4, that is, a runtime of 1.25 seconds. However, since there is a part of the code with runtime of 1 second that is not parallelizable, the best achievable runtime is 2 seconds, that is, at best a speedup of 2.5. *Amdahl's law* connects the fraction α of the code that can run in parallel with the theoretically achievable speedup given this fraction by

$$S_{\text{strong}}^{\text{theo}} = \frac{1}{(1-\alpha) + \frac{\alpha}{S_{\text{strong}}^{\text{opt}}}} \tag{14.2}$$

and the theoretically achievable parallel efficiency given Amdahl's law is

$$E_{\text{strong}}^{\text{theo}} = \frac{S_{\text{strong}}}{S_{\text{strong}}^{\text{theo}}}. \tag{14.3}$$

For an application that scales weakly, usually the system or node level resources like the RAM are the limiting resource. That means, the problem size per computing unit stays constant and additional computing units are used to solve an altogether bigger problem. For example, when parallelizing the solution routine to a PDE using a finite difference grid, the grid is partitioned into pieces and each computing unit works on a piece of the grid that just fits in its memory. If the total number of grid points is to be increased, more computing units have to be used. In the case of weak scaling, optimal scaling is achieved if the run time stays constant while the problem size is increased proportionally to the number of computing units, that is, $S_{\text{weak}}^{\text{opt}} = 1$. The realized speedup is then given by $S_{\text{weak}} = t_{\max}/t$ and parallel efficiency is measured by

$$E_{\text{weak}} = \frac{S_{\text{weak}}}{S_{\text{weak}}^{\text{opt}}} = S_{\text{weak}}. \tag{14.4}$$

For example, if the runtime on one computing unit, $P_{\min} = 1$, is $t_{\max} = 10$ seconds, then one expects the runtime of an eight times larger problem on $P = 8$ cores to be $t = 10$ seconds and the application would scale with speedup $S_{\text{weak}} = 10/10 = 1 = S_{\text{weak}}^{\text{opt}}$. In contrast to embarrassingly parallel problems, coupled problems often scale weakly. Most applications scale well to larger numbers of computing units as they typically employ nearest-neighbor communication. That is, in this example the pieces of the grid only need to know the values at the grid points of their direct neighbors. Thus the communication overhead is constant regardless of the number of computing units used. For weak scaling, *Gustafson's law* describes the theoretically possible speedup in presence of inherently serial code parts:

$$S_{\text{weak}}^{\text{theo}} = \frac{(1-\alpha)P_{\min}}{P} + \alpha. \tag{14.5}$$

The theoretically achievable parallel efficiency given Gustafson's law is

$$E_{\text{weak}}^{\text{theo}} = \frac{S_{\text{weak}}}{S_{\text{weak}}^{\text{theo}}}. \tag{14.6}$$

The following best practice rules should be taken into consideration when measuring parallel code performance:

1. Optimally, start the scaling efficiency measurement at one computing unit. Then always double the number of computing units (and the problem size, when measuring weak scaling efficiency). Keep in mind that often times computing resources can only be allocated node or even rack wise, meaning that a certain minimum number of computing units is always allocated and allocation step size is fixed.[12]
2. Report average runtimes and standard deviations to account for load imbalances of the network or other supercomputer-specific architectural shortcomings. The architectures of large supercomputers usually set great store on homogeneity of the compute nodes and their interconnects, which leads to low standard deviations of runtimes when running the same code multiple times. However that might not hold true for simple supercomputers, especially when heterogeneous servers are connected via standard Ethernet to form a computer cluster.
3. Calculate the speedup for all scaling stages on basis of the average runtimes and compare it to the respective ideal linear speedup in your setting.
4. Calculate the nonparallel fraction of your code and report the theoretically achievable speedup given by Amdahl's or Gustafson's law, respectively.
5. Use an application configuration that reflects your production run of the application.

14.5.3 Access to and costs of supercomputers

Due to the high costs and special knowledge needed for developing, setting up, and maintaining a supercomputer, access to the largest supercomputers in the world is mostly open for governmental or research facilities and granted via centrally organized application processes. However, there are supercomputers also open for commercial use. For example, the High Performance Computing Center (HLRS) in Stuttgart, Germany, which operates *Hazel Hen* (ranked 14

---

[12]The allocation step size is often times a power of 2.

TABLE 14.4: Total costs of a small computer cluster consisting of 8 servers with two 3.30 GHz CPUs and 192 GB RAM each

Total costs of a small computer cluster (128 cores)		
Estimated purchase price	85,000	EUR
Per year		
Staff	30,000	EUR
Power	3687	EUR
Cooling	5091	EUR
Linear depreciation	21,250	EUR
Total costs over lifetime (4 years)	240,111	EUR

Source: Own calculations and Fujitsu, Fujitsu PRIMERGY Servers Performance Report PRIMERGY RX200 S8, Fujitsu K.K., Tech. Rep., 2013.
Note: The systems total up to 128 cores and approximately 6.4 Tera FLOPS LINPACK performance.

on the current TOP500 list) offers access and a variety of services to the industry and small- and medium-sized enterprises. In Germany, computing time at the tier 0 supercomputers *Hazel Hen*, *JUQUEEN* (currently ranked 19), and *SuperMUC* (rank 36) can be applied for at the Gauss Centre for Supercomputing. In Europe, the Partnership for Advanced Computing in Europe (PRACE)—an international nonprofit association consisting of 25 member countries—provides access to supercomputers for large-scale scientific and engineering applications. Though access to the supercomputing infrastructure can be granted to the industry, it is foremost for research purposes such as simulations in the automotive or aerospace sector. Thereby, the costs commercial institutions have to bear are not publicly disseminated. Moreover, computing time is distributed among the users using a job queuing system. This means, that the job execution start time is conditional on the queue priority, which is determined by the total runtime of the job (the so-called *walltime*) and the availability of the requested computing resources, that is, the number of cores and potentially special accelerators. Altogether, having access to a shared supercomputer is not sufficient for common operational tasks in financial institutions, such as overnight portfolio valuation or pricing of derivatives at the trading desk.

Proprietary supercomputers, run and maintained by an institution or shared over different entities in a bigger institution or consortium, pose an often chosen but costly alternative to shared access to the largest supercomputers of the world. Table 14.4 depicts the costs associated with setting up and maintaining a small proprietary computer cluster of 128 cores with an estimated LINPACK performance of 6.4 Tera FLOPS. A medium-sized supercomputer with roughly 3200 cores has total costs of approximately 700,000 EUR.[13]

[13]Based on calculations presented by Bull Atos Technologies at the ICS 2015 [9].

14.6 Case Study: Pricing Basket Options Using C++ and MPI

This case study demonstrates how to efficiently calculate the price of an arithmetic average basket put option on the S&P 500 index using the supercomputer *JUQUEEN*. Since every constituent of the S&P 500 contributes to one dimension in the PDE representation of the pricing problem, the solution to the problem suffers from the *curse of dimensionality*, which refers to the exponential growth of the resulting linear equation system to be solved depending on the dimensionality of the PDE. However, supercomputers can be utilized to calculate solutions despite the curse of dimensionality. Problem, approach, and results of this case study have been extensively studied in, and this section is closely based on [11].

14.6.1 Problem

In higher dimensions, the development of d correlated stocks over time can be described by a vector process $x = (x_1, \ldots, x_d) \in \mathbb{R}^d$, where the component i follows a geometric Brownian motion with drift

$$\mathrm{d}x_i(t) = x_i(t)\mu_i \,\mathrm{d}t + x_i(t) \sum_{j=1}^{n} \sigma_{ij} \,\mathrm{d}W_j(t), \quad t \geq 0, x_i(0) = x_i^0. \tag{14.7}$$

Here, μ is a d-dimensional mean vector, $\sigma_{ij} = \sigma_i \sigma_j \rho_{ij}$ with correlation ρ_{ij} and standard deviation $\sigma \in \mathbb{R}^d$ is an entry of the $d \times d$ covariance matrix $\{\sigma\}_{i,j}$, and W is a d-dimensional vector of independent Wiener processes. Using the Feynman–Kac formula, the arbitrage free price can be expressed as the solution u to the multidimensional Black–Scholes equation

$$\frac{\partial u}{\partial t} - r \sum_{i=1}^{d} x_i \frac{\partial u}{\partial x_i} - \frac{1}{2} \sum_{i,j=1}^{d} \sigma_i \sigma_j \rho_{ij} x_j x_j \frac{\partial^2 u}{\partial x_i \partial x_j} + ru = 0, \tag{14.8}$$

where r denotes the risk-free rate. Subject of this case study is a European put option with strike K on the S&P 500 index with an arithmetic average payoff. Hence, the initial condition for the PDE 14.8 is

$$u(x, 0) = \left(K - \sum_{i=1}^{500} \gamma_i x_i \right)_+ \quad \forall x \in \mathbb{R}_+^d. \tag{14.9}$$

The sample basket comprises of the S&P 500 index constituents as of June 2013 with uniformly distributed weights γ_i and estimated volatilities as well as correlations on basis of daily logarithmic increments of the last 360 days.[14]

14.6.2 Approach

Firstly, a decomposition technique is employed to decompose the high-dimensional PDE into a linear combination of low-dimensional PDEs. For these low-dimensional PDEs, the curse of dimensionality is then broken by use of the so-called *combination technique* on sparse grids, which in turn allows for straightforward parallelization. This parallelization approach significantly reduces the overall runtime of the solution routine for the decomposed high-dimensional PDE.

14.6.2.1 Decomposition

The *Taylor-like ANOVA* decomposition is a special form of an anchored ANOVA-type decomposition of the solution $u(x)$, $x \in \mathbb{R}^d$, to a d-dimensional PDE. In addition to the anchor point a, all contributing terms also depend on the first $r \geq 1$ coordinates. With $s \leq d - r$, the solution can be written as

$$u(x) = u_0^{(r)}(a; x_1, \ldots, x_r) + \sum_{p=1}^{s} \sum_{\substack{\{i_1,\ldots,i_p\} \\ \subseteq \{r+1,\ldots,d\}}} u_{i_1,\ldots,i_p}^{(r)}(a; x_1, \ldots, x_r; x_{i_1}, \ldots, x_{i_p}). \tag{14.10}$$

For u being a function of the eigenvalues $(\lambda_1, \ldots, \lambda_d)$ of the covariance matrix of the vector process x, the first-order approximation, $s = 1$, $r = 1$, is given by

$$\begin{aligned} u(x) &\approx u_0^{(1)}(a; x_1) + \sum_{j=2}^{d} \left(u_j^{(1)}(a; x_1; x_j) - u_0^{(1)}(a; x_1) \right) \\ &\approx \sum_{j=2}^{d} u_j^{(1)}(a; x_1; x_j) - (d-2) u_0^{(1)}(a; x_1), \end{aligned} \tag{14.11}$$

where $u_{1,j}^{(1)}(a; x_1; x_j)$ is a solution to the heat equation

$$\frac{\partial u}{\partial t} - \frac{1}{2}\lambda_1 \frac{\partial^2 u}{\partial x_1^2} - \frac{1}{2}\lambda_j \frac{\partial^2 u}{\partial x_j^2} = 0 \quad \forall (t, x_1, x_j) \in [0, T] \times \mathbb{R} \times \mathbb{R} \tag{14.12}$$

with appropriate initial conditions.

14.6.2.2 Combination technique

For a given level $n \in \mathbb{N}$, the combination technique combines solutions $u_l(x)$ to the PDE on multiple, mostly anisotropic full grids with mesh widths

[14] As there were seven stocks without correlations, actually the solution to a basket option on the "S&P 493" is calculated.

2^{-l} for multi-index $l \in \mathbb{N}^d$ by the combination formula:

$$u_n(x) = \sum_{n \leq |l|_1 \leq n+d-1} (-1)^{n+d-|l|_1-1} \binom{d-1}{|l|_1 - n} u_l(x). \tag{14.13}$$

Here, $|l|_1 = \sum_{i=1}^{d} l_d$.

Computing the approximated solution $u_n(x)$ to the (high-dimensional) PDE thus boils down to computing the solution to each of the $\mathcal{O}(dn^{d-1})$ full grid solutions $u_l(x)$ and each of these grids has only $\mathcal{O}(2^n)$ grid points. This solution approach breaks the curse of dimensionality as it only involves $\mathcal{O}(2^n dn^{d-1})$ grid points compared to $\mathcal{O}(2^{nd})$ grid points of the full grid solution with mesh width 2^{-n}. It can be shown that the error of the approximation is only slightly deteriorated by a logarithmic factor, if u fulfills certain smoothness conditions. In addition, the combination technique facilitates straightforward parallelization as the up to $\mathcal{O}(dn^{d-1})$ full grid solutions can be computed in parallel.

14.6.2.3 Parallelization

The Taylor-like ANOVA decomposition 14.11 generates a set of $d = 493$ independent problems (1 one-dimensional, 492 two-dimensional). To solve the two-dimensional PDEs in parallel, computing units are grouped and each group of computing units solves one (or more) of the two-dimensional PDEs on sparse grids using the combination technique. In dependence of the level n used in the combination technique, let $P_{\max}$ be the maximum number of computing units for a fully parallel solution. When having $P_{\max}$ computing units available, every computing unit calculates the solution to exactly one full grid. If there were more than $P_{\max}$ computing units available, the additional computing units would be idle. Assuming there are less than $P_{\max}$ computing units available, it is favorable to split the available computing units into as many groups as independent problems exist. Figure 14.2 sketches the idea.

14.6.2.4 Implementation

The calculations are run on *JUQUEEN*. For the combination technique, the PDEs are solved on the full grids using a Crank–Nicolson time stepping scheme combined with a finite difference discretization in space. The resulting set of linear equations is solved using a BiCGSTAB solver with a precision of the minimum residual norm of 10^{-12}. The implementation of the parallelization approach is done with C++ and MPI and the code is compiled using the IBM XL compiler.

14.6.3 Results

For an at-the-money basket put option with strike $K = 1$, the price of the basket option computes to $u(x_0, 0) = 0.0276$ using the Taylor-like ANOVA

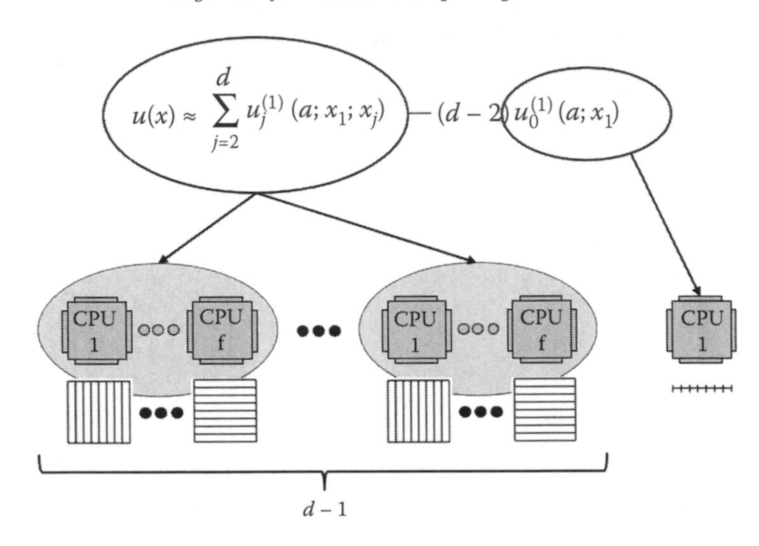

FIGURE 14.2: Schematic overview on solving multiple Taylor-like ANOVA terms in parallel using sparse grid parallelization. (Adapted from Schober, P., Schröder, P., and Wittum, G., *In revision at the Journal of Computational Finance*, available at SSRN 2591254, 2015.)

decomposition of first order 14.11 and the combination technique 14.13 on level $n = 10$. Since there is no analytical solution, a benchmark result was calculated using a Monte Carlo simulation with 100,000 paths ($u(x_0, 0) = 0.0281$). To test for strong scaling, the number of computing units is repeatedly doubled until $P_{\max} = 13{,}285$ computing units are reached, that is, full parallelization when utilizing 13,296 computing units, because of round lot allocation on *JUQUEEN*. The code runs seven times for every scaling stage and average runtimes μ and standard deviations σ are reported to account for jitter due to the job scheduler of the cluster and the cluster's network. Table 14.5 depicts realized mean runtimes, standard deviations, and speedups. Figure 14.3 plots

TABLE 14.5: S&P 500 realized mean runtime μ, standard deviation σ, and speedup S_{strong} as well as $S_{\text{strong}}^{\text{opt}}$

P	μ [s]	σ [s]	S_{strong}	$S_{\text{strong}}^{\text{opt}}$
1024	1471.71	0.44	1.00	1.00
2048	778.63	1.89	1.89	2.00
4096	433.48	1.92	3.40	4.00
8192	265.88	3.55	5.54	8.00
13,296	170.97	3.77	8.61	12.98

Source: Schober, P., Schröder, P., and Wittum, G., *In revision at the Journal of Computational Finance*, available at SSRN 2591254, 2015.

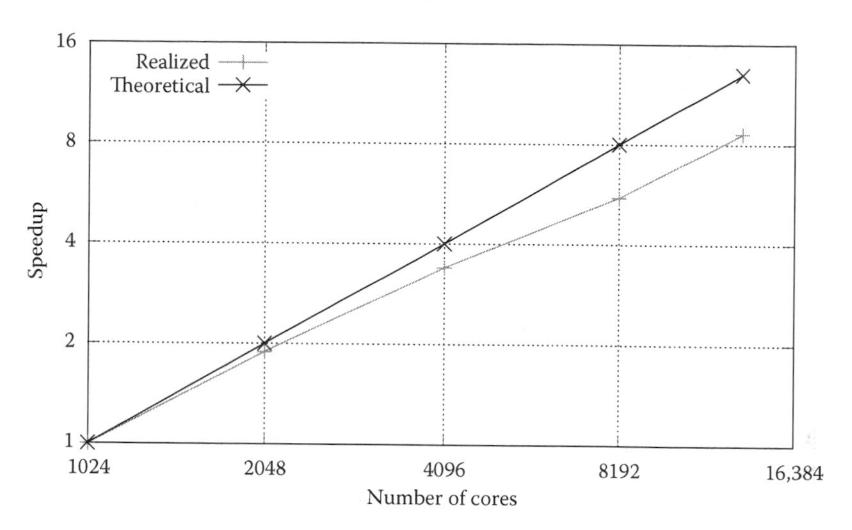

FIGURE 14.3: Strong scaling results for the arithmetic basket option on the S&P 500. (Adapted from Schober, P., Schröder, P., and Wittum, G., *In revision at the Journal of Computational Finance*, available at SSRN 2591254, 2015.)

the realized speedup against the theoretically achievable speedup. The parallel efficiency is $E_{\text{strong}} = 63.85\%$ at 13,296 computing units with respect to 1024 computing units. Looking at the realized mean runtime μ, the basket option was priced within 3 minutes using massive parallelization.

14.7 Case Study: Optimizing Life Cycle Investment Decisions Using MATLAB

How an individual can maximize her utility from consumption over the lifetime is the subject of this case study. Therefore, the individual can dynamically decide (i.e., adapt her decisions as time evolves) on how much to consume momentarily and how to allocate her remaining wealth in certain financial products to finance future consumption. This section demonstrates that the resulting optimization problem can be solved efficiently by parallelizing a discrete time dynamic programming approach for an exemplary dynamic portfolio choice problem. A small proprietary computer cluster with 97 cores and standard MATLAB is used to reduce the total overall runtime of the numerical solution routine from roughly 12.5 hours on a single core to about 8 minutes. References 12 and 13 serve as references for all details left out here for the sake of brevity.

14.7.1 Problem

Formally, the problem of maximizing the individual's expected utility from her choices $p_t \in \mathbb{R}^k$ over all time periods $t \in \{0, \ldots, T\}$ needs to be solved:

$$\max_{p_t} \sum_{t=0}^{T} \rho^t \mathbb{E}_0 \left[u \left(p_t \right) \right]. \tag{14.14}$$

Here, u denotes a Constant Relative Risk Aversion utility function and $\rho < 1$ is the time discount factor. At every point in time t, her choices are how much to consume, C_t, to invest in stocks S_t yielding a risky return $r_{t+1}^S \sim LN(\mu_S, \sigma_S^2)$, to invest in bonds B_t yielding a risk-free return r_B, and how much annuity claims to buy giving a yearly, lifelong income stream of $1/\ddot{a}_t$ dollar; $p_t = (C_t, S_t, B_t, A_t)^T$. Here, $\ddot{a}_t$ is the age-dependent actuarial premium charged by an insurance company in exchange for a lifelong payment of 1 dollar yearly to an individual in time period t starting in $t+1$, that is, the annuity factor. Her decision is conditional on her state $s_t \in \mathbb{R}^d$, where current wealth W_t, multiples of average permanent labor income P_t, and amount of yearly annuity payments L_t are tracked; $s_t = (W_t, P_t, L_t)^T$. The individual stays alive with probability π_t. Uncertainty in the labor income is introduced via a permanent risk factor ν and transitory risk factor ϑ that are uncorrelated and iid lognormally distributed with $\mathbb{E}[\nu_{t+1}] = \mathbb{E}[\vartheta_{t+1}] = 1$. That is, $\nu \sim LN(-\sigma_\nu^2/2, \sigma_\nu^2)$ and $\vartheta \sim LN(-\sigma_\vartheta^2/2, \sigma_\vartheta^2)$. The deterministic component of labor income is given by G_t, which reflects the dollar amount associated with $P_t = 1$. After retirement, income is deterministic and a fraction λ of the last income $P_t G_t$. Since the stochastic risk factors in the model are all lognormal, the random variable $\omega_{t+1} = (r_{t+1}^S, \nu_{t+1}, \vartheta_{t+1})^T \sim LN\left((\mu_S, -\sigma_\nu^2/2, -\sigma_\vartheta^2/2)^T, (\sigma_S^2, \sigma_\nu^2, \sigma_\vartheta^2)^T\right)$ is multidimensionally lognormally distributed. With these definitions, the state space dynamics from t to $t+1$, which is also random variable $f_{t+1} \colon \mathbb{R}^k \times \mathbb{R}^d \times \Omega \mapsto \mathbb{R}^d$, can be defined as

$$\begin{aligned} f_{t+1}^1 &:= W_{t+1} = S_t r_{t+1}^S + B_t r_B + \mathbb{1}_{\{t<t_R\}} G_t P_t \nu_{t+1} \vartheta_{t+1} + \mathbb{1}_{\{t \geq t_R\}} \lambda G_t P_t \\ f_{t+1}^2 &:= P_{t+1} = \mathbb{1}_{\{t<t_R\}} P_t \nu_{t+1} + \mathbb{1}_{\{t \geq t_R\}} P_t \\ f_{t+1}^3 &:= L_{t+1} = L_t + \frac{1}{\ddot{a}_t} \end{aligned} \tag{14.15}$$

with $\mathbb{1}$ being the indicator function and t_R the retirement age.

Problem 14.14 can be solved using a dynamic programming approach by setting up the Bellman equation for the value function j_t at discrete points in time $t = 0, \ldots, T-1$:

$$j_t(s_t) = \max_{p_t} \left\{ u(p_t) + \rho \pi_t \mathbb{E}_t \left[j_{t+1} \left(f_{t+1}(p_t, s_t, \omega_{t+1}) \right) \right] \right\} \tag{14.16}$$

$$j_T(s_T) = v(s_T) \tag{14.17}$$

subject to

$$S_t, B_t, A_t \geq 0 \tag{14.18}$$
$$C_t - \varepsilon \geq 0 \tag{14.19}$$
$$W_t - C_t - S_t - B_t - A_t = 0. \tag{14.20}$$

Here, v is a known function of the final state and a minimal consumption of $\varepsilon > 0$ is assumed.

14.7.2 Approach

A general way to solve problem 14.16–14.20 is discrete time dynamic programming stepping backwards in time, which allows for a simple parallelization easily implemented in MATLAB.

14.7.2.1 Discrete time dynamic programming

The state space is discretized by a mesh of a finite number of equidistant grid points $n \in \mathbb{N}^d$ with mesh size $h = ((u_1 - l_1)/n_1, \ldots, (u_d - l_d)/n_d)^T \in \mathbb{R}^d$, where $l \in \mathbb{R}^d$ is the lower and $u \in \mathbb{R}^d$ the upper boundary for every dimension of the state space. A grid point s_t^i can then be represented by a multi-index $i \in I = \{i \in \mathbb{N}^d \,|\, i \leq n \text{ element-wise}\}$ and is addressed by $s_t^i = (l_1 + i_1 h_1, \ldots, l_d + i_d h_d)^T$. On this grid, let the interpolating function be

$$j_t(s_t) \approx \sum_{i \in I} c_t^i \phi^i(s_t), \tag{14.21}$$

where the basis functions ϕ^i can be global polynomials or Ansatz functions with local support (such as B-splines). The same grid is chosen for every time step t and the coefficients c_t^i are determined in such way that the approximation fits the known function values at all grid points at a given time. The last period's optimal value function 14.17 is given by $j_T(s_T^i) = v(s_T^i)$, $\forall i \in I$. To determine the optimal solution of the next-to-last period $j_{T-1}(s_t^i)$ and all earlier periods $t \in \{0, \ldots, T-2\}$ at all grid points s_t^i, $i \in I$, an optimization routine which maximizes 14.16 over the real-valued vector p_t is used (subject to the boundary and budget constraints 14.18–14.20). Since within the optimization routine f_{t+1} does not generally correspond to a grid point, the expectation is approximated by a Gaussian quadrature rule with $q = 1, \ldots, m$ nodes ω^q and weights w^q:

$$\mathbb{E}_t \left[j_{t+1} \left(f_{t+1}(p_t, s_t^i, \omega_{t+1}) \right) \right] \approx \sum_{q=1}^{m} \sum_{i \in I} c_t^i \phi^i (f_{t+1}(p_t, s_t^i, \omega_{t+1}^q)) \, w_{t+1}^q. \tag{14.22}$$

14.7.2.2 Parallelization

The dynamic programming approach yields a set of independent optimization problems in every period t: for every grid point in the state space grid the maximum of the continuation value of the Bellman equation has to be computed. As the number of grid points is fixed, the maximum number of computing units that can be employed equals the number of grid points $\#I$ (i.e., the cardinality of the multi-index set I):

$$P_{\max} = \#I. \tag{14.23}$$

Note that this is a fixed-size parallelization problem that should scale strongly until $P_{\max}$ is reached.

For the parallelization, a master–worker pattern is employed: one of the P computing units is the designated master, the other $P - 1$ computing units become the workers. The master firstly generates a (possibly random) execution order over all multi-indexes i from the index set I, and then assigns an optimization problem associated to the grid point indexed by i to all workers. Every time any computing unit idles, the master dispatches the next problem from the predefined order to this idle computing unit until all problems are solved. Figure 14.4 contains a schematic representation of the approach. If there is no *a priori* knowledge about the runtime of the single problems that are dispatched in parallel, the master–worker pattern is a simple load-balancing solution that in practice achieves very high speedups.

14.7.2.3 Implementation

The computations are performed using MATLAB version R2012b on a proprietary computer cluster consisting of heterogeneous high-performance servers with CPU speed varying from 2 GHz to 3.3 GHz, number of cores from 12 to 16 and RAM size from 48 GB to 128 GB per server. All servers are stacked in one air-cooled rack and are connected via Gigabit Ethernet. The servers run on Windows HPC Server 2008 with Microsoft HPC Pack as a job scheduler.

The parallelization is done using MATLAB's `parfor` command that implements a master–worker pattern. For the optimization, the gradient-based constrained optimization solver `fmincon` from the MATLAB Optimization Toolbox is used. Besides standard MATLAB, the Parallel Computing Toolbox and MATLAB Distributed Computing Server are required to use the `parfor` construct for parallelization across MATLAB sessions running on multiple servers.

14.7.3 Results

To evaluate the optimal choices obtained from the optimization, a Monte Carlo simulation over the life cycle from age 20 to age 99 for 100,000 paths is performed. In every path and every time step, the optimal choices from the numerical solution are evaluated to determine the path's evolution. Finally, labor income, stock and bond investment, consumption, and annuity purchase profiles are generated by averaging over all paths, see Figure 14.5.

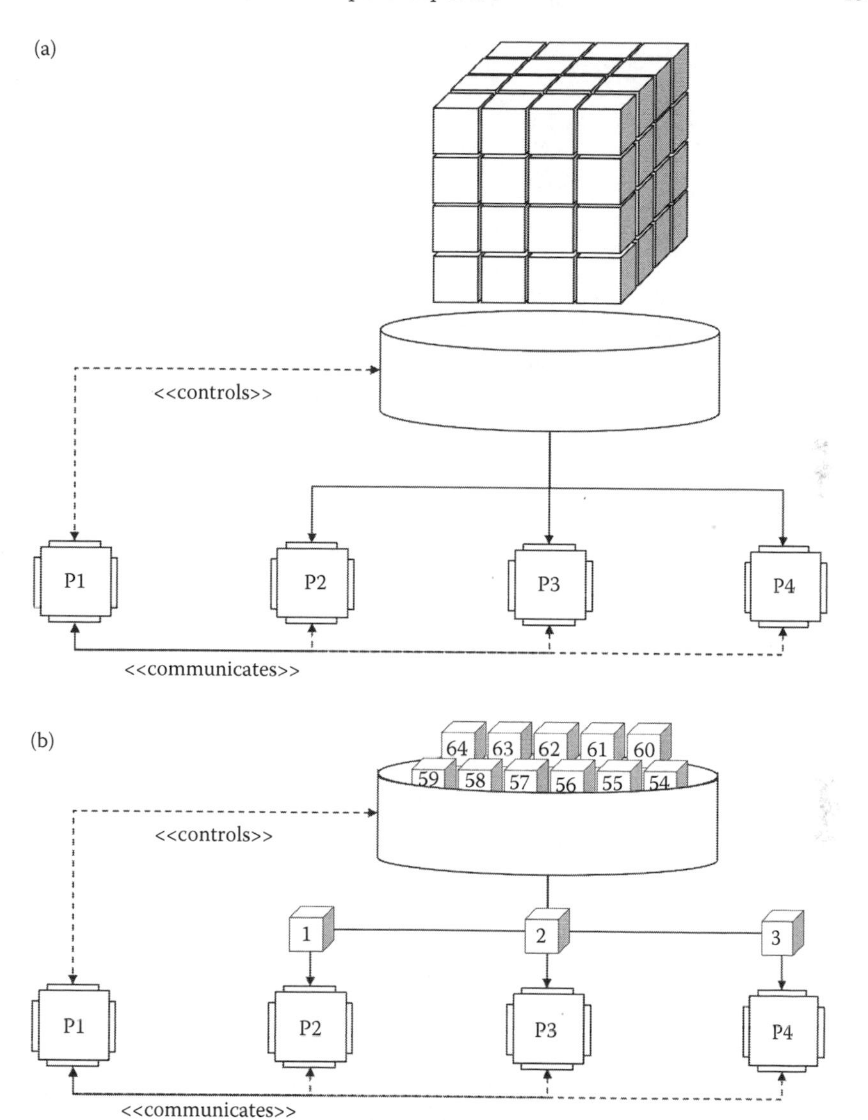

FIGURE 14.4: Figurative state space decomposition and parallel solution of decomposed, independent optimization problems: (a) state space decomposition and (b) distribution of independent optimization problems. (Adapted from Horneff, V., Maurer, R., and Schober, P., *Efficient parallel solution methods for dynamic portfolio choice models in discrete time*, available: SSRN 2665031, 2016.)

To test for strong scalability, the number of computing units is doubled repeatedly until the limits of the cluster are reached ($P = 97$). The code is run seven times for every scaling stage to account for jitter due to the job scheduler of the cluster and the cluster's network. Table 14.6 and Figure 14.6

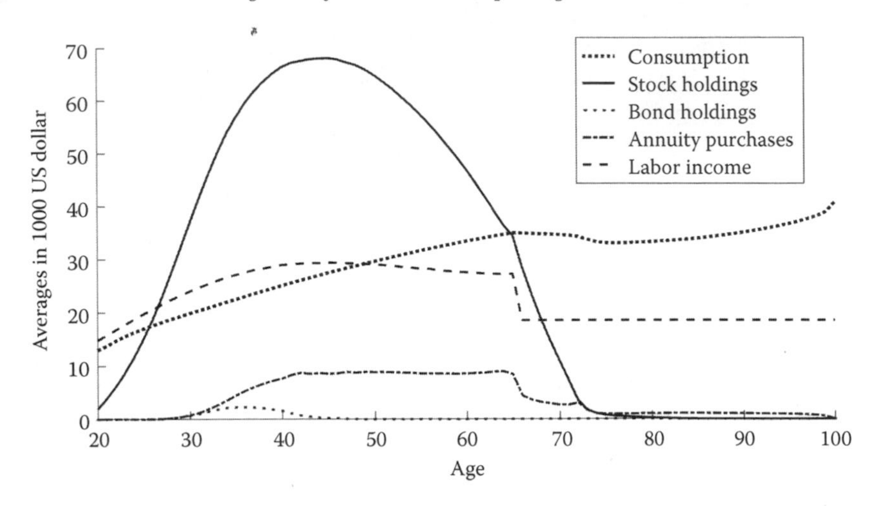

FIGURE 14.5: Average life cycle profile for consumption, stock and bond holdings, annuity purchases, and labor income in total US dollar values.

TABLE 14.6: Realized mean runtime μ, standard deviation σ, and speedup S_{strong} as well as $S_{\text{strong}}^{\text{opt}}$

P	$\mu[s]$	$\sigma[s]$	S_{strong}	$S_{\text{strong}}^{\text{opt}}$
8	5659.60	21.55	1.00	1.00
16	3283.04	5.71	1.72	2.00
32	1321.91	7.28	4.28	4.00
64	669.96	10.22	8.45	8.00
97	481.90	11.30	11.74	12.13

depict realized mean runtimes, standard deviations, and realized speedup as well as optimal speedup. Parallel efficiency is nearly optimal at 96.86%. Note that the realized speedup crosses the optimal speedup upwards at 32 cores. This is due to the heterogeneity of the cluster: the first 16 allocated cores only operate at 2 GHz clock speed, all allocated cores afterwards operate at 3 GHz and more.

Further Reading

Section 14.1 [1] contains the current TOP500 list, various statistics, historical data, and an extensive FAQ regarding the TOP500. In Reference 4, the LINPACK benchmark is published and available for download. For detailed information on the LINPACK benchmark, see also [14]. Supercomputers as a platform for heavily parallel applications are discussed in Reference 15.

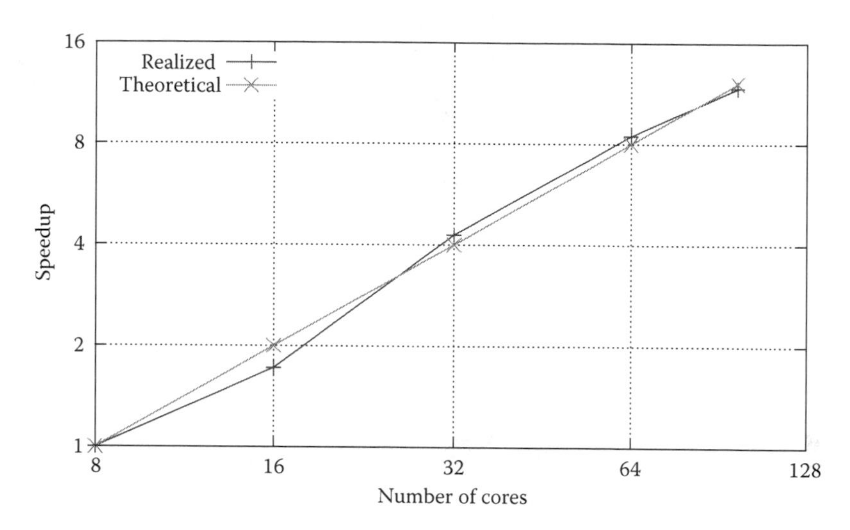

FIGURE 14.6: Strong scaling results for the parallelization approach.

Section 14.2 [16] contains an overview on the evolution of supercomputers. The homepage of the TOP500 [1] has a timeline depicting the milestones of supercomputing. The homepage of the Green500 [5] provides the current Green500 list and information as well as historical data regarding the Green500. Details on the construction of *L-CSC* can be found in Reference 17. About the use of SoCs in supercomputing, the homepage of the Mont-Blanc project [7] has various information about the project itself and applications. The accompanying article [18] compares the use and performance of SoC with similar trends in the 1990s.

Section 14.3 [15] gives an alternative overview on parallel programming languages (and, additionally, parallel compilers). A comprehensive discussion of MATLAB's parallel programming features, design aspects, and implemented parallelization paradigms is given in Reference 19. Reference 20 is one recommendable library (of many) to include MPI in Python. The MPI standard itself is documented in Reference 21. A good compilation of pros and cons for the usage of OpenMP and MPI can be found in Reference 22.

Section 14.4 advantages and disadvantages of supercomputers and the associated programming and parallelization approaches are also discussed in Reference 15.

Section 14.5 [23] provides a comprehensive overview on quantitative models for finance and common numerical methods. Various numerical applications in finance, including dynamic stochastic programming and Quasi Monte Carlo methods, are covered in Reference 15. Reference 24 introduces into strong and weak scaling measurement. Amdahl's law bases on his conference contribution [25] and [26] contains Gustafson's response. The cost example was estimated on basis of the technical report [10].

Section 14.6 [11] is the basis for the case study. Reference 27 covers the special class of ANOA decomposition used. Reference 28 is a comprehensive reference for sparse grids. Reference 29 proposes the parallelization approach applied.

Section 14.7 [13] comprises and extends this case study for the same life cycle model, which is taken from Reference 12.

References

1. Top500 homepage. Available: http://top500.org/, 2017.

2. About *JUQUEEN*—Jülich Blue Gene/Q Supercomputer. [Online]. Available: http://www.fz-juelich.de/ias/jsc/EN/Expertise/Supercomputers/JUQUEEN/Configuration/Configuration_node.html, 2017.

3. IEEE Computer Society, Tribute to Seymour Cray. [Online]. Available: https://www.computer.org/web/awards/about-cray, 1996.

4. Petitet, A., Whaley, C., Dongarra, J., and Cleary, A., LINPACK. [Online]. Available: http://www.netlib.org/benchmark/hpl/ 2008.

5. Green500 homepage. [Online]. Available: http://www.green500.org/, 2017.

6. Frankfurt Institute for Advanced Studies. [Online]. Available: https://fias.uni-frankfurt.de/en/, 2017.

7. Mont-Blanc project homepage. [Online]. Available: http://www.montblanc-project.eu/, 2017.

8. High-Q Club—Highest Scaling Codes on JUQUEEN. [Online]. Available: http://www.fz-juelich.de/ias/jsc/EN/Expertise/High-Q-Club/_node.html, 2017.

9. Warschko T. M., Advanced cluster computing, In *Presentation at the ISC Conference, Frankfurt*, 2015.

10. Fujitsu, Fujitsu PRIMERGY servers performance report PRIMERGY RX200 S8, Fujitsu K.K., Tech. Rep., 2013.

11. Schober, P., Schröder, P., and Wittum, G., Efficient parallel solution methods for high-dimensional option pricing problems, *In revision at the Journal of Computational Finance, available at SSRN 2591254*, 2015.

12. Horneff, W. J., Maurer, R. H., and Stamos, M. Z., Life-cycle asset allocation with annuity markets, *Journal of Economic Dynamics and Control*, vol. 32, no. 11, pp. 3590–3612, 2008.

13. Horneff, V., Maurer, R., and Schober, P., Efficient parallel solution methods for dynamic portfolio choice models in discrete time, Available: SSRN 2665031, 2016.

14. Dongarra, J., Luszczek, P., and Petitet, A., The LINPACK benchmark: Past, present and future, University of Tennessee, Tech. Rep., 2002.

15. Vajteršic, M., Zinterhof, P., and Trobec, R., *Overview—Parallel Computing: Numerics, Applications, and Trends.* Springer, London, United Kingdom, 2009.

16. Kaufmann, W. J. and Smarr, L. L., *Supercomputing and the Transformation of Science.* WH Freeman & Co., New York, NY, 1992.

17. Lindenstruth, V., The L-CSC construction and its applications, In *Presentation at the ISC Conference, Frankfurt*, 2015.

18. Rajovic, N., Carpenter, P. M., Gelado, I., Puzovic, N., Ramirez, A., and Valero, M., Supercomputing with commodity CPUs: Are mobile SoCs ready for HPC? In *High Performance Computing, Networking, Storage and Analysis (SC), 2013 International Conference for IEEE*, pp. 1–12, 2013.

19. Sharma, G. and Martin, J., Matlab®: A language for parallel computing, *International Journal of Parallel Programming*, vol. 37, pp. 3–36, 2009.

20. Dalcin, L. MPI for Python. [Online]. Available: http://mpi4py.scipy.org/, 2017.

21. Message Passing Interface Forum, MPI: A Message passing interface standard, Version 2.2. [Online]. Available: http://mpi-forum.org/docs/mpi-2.2/mpi22-report.pdf, 2009.

22. Pros and Cons of OpenMP/MPI. [Online]. Available: https://www.dartmouth.edu/rc/classes/intro_mpi/parallel_prog_compare.html, 2011.

23. Wilmott, P., *On Quantitative Finance*, 2nd ed. John Wiley & Sons, Ltd., Chichester, United Kingdom, vol. I–III, 2006.

24. Measuring Parallel Scaling Performance. [Online]. Available: https://www.sharcnet.ca/help/index.php/Measuring_Parallel_Scaling_Performance, 2017.

25. Amdahl, G. M., Validity of the single processor approach to achieving large scale computing capabilities, In *Proceedings of the April 18–20, 1967, Spring Joint Computer Conference.* ACM, pp. 483–485, 1967.

26. Gustafson, J. L., Reevaluating Amdahl's law, *Communications of the ACM*, vol. 31, no. 5, pp. 532–533, 1988.

27. Schröder, P., Gerstner, T., and Wittum, G., Taylor-like ANOVA expansion for high-dimension PDEs in finance, *Working paper*, 2013.

28. Bungartz, H.-J. and Griebel M., Sparse grids, *Acta Numerica*, vol. 13, pp. 147–269, 2004.

29. Schröder, P., Mlynczak, P., and Wittum, G., *Dimension-Wise Decompositions and Their Efficient Parallelization.* World Scientific, ch. 13, pp. 445–472, 2013, [Online]. Available: http://www.worldscientific.com/doi/abs/10.1142/9789814436434_0013.

Chapter 15

Multiscale Dataflow Computing in Finance

Oskar Mencer, Brian Boucher, Gary Robinson, Jon Gregory, and Georgi Gaydadjiev

CONTENTS

15.1 Introduction

Computer technology has become an essential driver for the financial industry in almost all of its areas. Advances in hardware and software technology, novel numerical methods, financial models, and algorithms have made computers a key technology that became essential for all financial institutions. High-performance computing (HPC) systems are widely used to price financial products or to quickly calculate the risk of complex portfolios. Often, the available computational power determines the types of problems that can be practically solved. Being able to handle a more complex problem or to obtain the results faster than all other organizations directly translates into a competitive advantage.

Conventional computer architectures used in many areas of everyday life including mobile devices, desktop computers, and HPC systems generally follow the basic concepts of general-purpose processing [1]. Such processors perform calculations by executing a sequence of instructions that can either carry out arithmetic, control, or input/output (IO) operations. This model of execution is generic and hence extremely flexible; however, it is also inherently sequential. Over many decades, the performance of processors has been improved by increasing the clock rates, and also by extending the basic processor architecture with complex structures to deal with issues like control divergence, main memory access penalties and to recover low-level binary instructions parallelism. Many micro-architectural innovations such as caches, branch prediction, out-of-order execution, and Single Instruction Multiple Data (SIMD) extensions were developed to alleviate the fundamental drawbacks of the general-purpose processor inherently sequential paradigm. This has led to the modern complex processor microarchitectures where only a tiny part of the chip area is dedicated to useful calculations at very high speeds while the rest of the device is used for auxiliary functions such as caching of instructions and data. With the end of clock frequency improvements offered by the CMOS technology scaling, additional performance can now be only obtained through exploiting parallelism. Multithreaded implementations on multiple cores or SIMD extensions are just two examples. However, the individual cores (or threads) still rely on a fundamentally sequential computing principle, that is, performing a sequence of instructions. In addition, legacy applications have to be rewritten, analyzed, and optimized in order to achieve satisfactory performance levels. Attempting to compute larger and larger problems by simply scaling over existing processor technology is no longer practically possible for many current and future HPC applications [2]. Even if performance requirements can be met by using a large number of machines, the cost, area, and power requirements may exceed practical limits.

These limitations have led to an increased interest in special-purpose computing where an algorithm or parts of it are targeted onto a customized architecture, leading to both increased performance and improved power efficiency. A special-purpose architecture can be customized and tailored to the unique requirements of the application, resulting in a combination of increased performance, reduced power, smaller area, and lower economical cost as compared to its general-purpose counterparts. Nowadays, special-purpose units are added to many processors to perform specific, frequently used, demanding tasks such as encryption or digital video decoding. However, these special-purpose units are available only for a limited set of common functionalities, and those are "frozen" during the design of the processor. HPC can also benefit from special-purpose processing, but due to the vast space of possible applications with different characteristics, prefabricated (and hence fixed) accelerators are not practical. Instead, a flexible computing substrate that can be customized on demand by the designer according to the application requirements is required. Reconfigurable devices, such as Field-Programmable Gate Arrays (FPGAs), offer such a substrate technology, and significant speed-ups over conventional computing systems have been reported for a wide range of applications [3,4]. Public Cloud computing has recently started to provide reconfigurable solutions for the high-end computing market. However, the downside of this highly capable technology is often a complex, very low-level programming model that requires highly specialistic knowledge in hardware design.

Maxeler Technologies is pioneering a novel approach to dataflow oriented supercomputers. Maxeler computing systems are a type of special-purpose system that can be customized to the unique requirements of an application. At the heart of a Maxeler system are one or several Dataflow Engines (DFEs) that combine a large and powerful reconfigurable device with significant amount of DRAM memory. DFEs are programmed using a simple dataflow model that enables domain experts to optimize both their algorithms and the underlying architecture simultaneously, cutting through the typical layer approach of custom libraries, standard libraries, operating systems, and hardware organization. This approach has led to orders-of-magnitude higher performance, lower power consumption, and significantly lower data center space as compared to traditional approaches. A wide range of applications ranging from 3D finite-difference partial differential equation solvers [5] to Monte Carlo simulations have been successfully accelerated in commercial products [6]. In addition, speedups can also enable completely new computational models that were previously not feasible under hard timing constraints. For example, computing a 24-hour forecast is not practical if the computation would take 48 hours to complete. If, however, the same computation can be achieved in 2 hours, then running 24-hour forecasts becomes a realistic scenario.

15.2 The Dataflow Paradigm

Maxeler's dataflow oriented computing paradigm fundamentally differs from conventional processors which are control-flow centric. This approach is illustrated in Figure 15.1 and it represents an evolution of dataflow and systolic array concepts [7,8]. A conventional processor operates by reading and decoding an instruction, loading the required data, performing an operation on the data, and returning the result to memory. This process is iterative in nature and requires complex control mechanisms that manage the operation of the processor. The dataflow execution model is greatly simplified in comparison. Data is streamed from memory into the chip where arithmetic operations are performed by chains of functional units (dataflow cores) statically interconnected in a structure corresponding to the implemented functionality. It should be noted that the dataflow structure preforms computations entirely without instructions. Data flows from one functional unit directly to the next without the need of complex control mechanisms. Data simply arrives when it is needed and the final results are streamed back into memory. Each dataflow

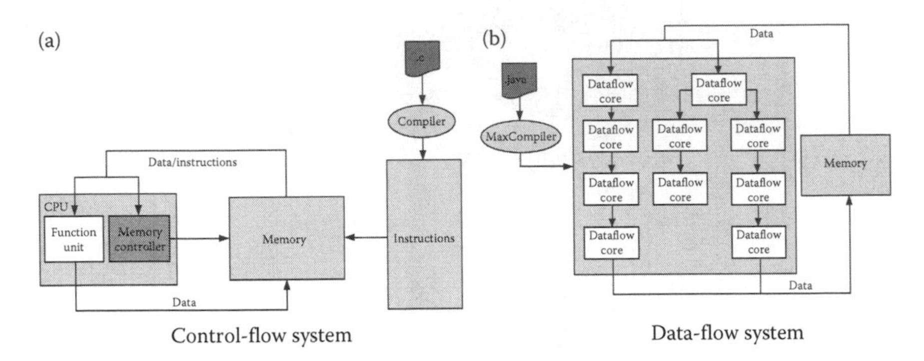

FIGURE 15.1: Conventional control-flow oriented processor (a) as compared to Maxeler's DFE (b).

core performs only a simple operation such as addition or multiplication and hence, thousands of operations can fit on the chip surface characterized by a specific area.

Unlike the control-flow based processor, where operations are computed on a time-shared functional unit ("computing in time"), the complete dataflow computation is laid out in space over the entire chip ("computing in space"). Dependencies in the dataflow are resolved statically at compile time, and because there is no data-dependent behavior present at run time, the entire DFE can be deeply pipelined. Every stage of the pipeline computes in parallel with the dataflow architecture maintaining overall throughput of one result per clock cycle. A simple analogy of this approach is an assembly line in a car factory. The most efficient way to realize large-scale productions (computations) is a specialized assembly line (pipeline) where parts (data) move from storage (memory) to a chain of dedicated workstations (custom functional units) where they are assembled together (data is processed) and moved forward in lock step to produce complete cars (final results) at the end. There are no instructions and hence, instruction decoding logic is not required. Also, a static dataflow model does not require control-flow techniques such as branch prediction and out-of-order execution for obvious reasons. General-purpose caches are equally not necessary and data is always kept on chip with the minimum amount of buffering memory for intermediate results. By eliminating these extraneous functions, all the chip's resources can be dedicated to perform useful computations instead of managing the execution.

Maxeler realizes this dataflow oriented computing approach by mapping an application described in its dataflow model onto a DFE. DFEs are highly efficient for large-scale computations with a static execution model due to the elimination of sequential execution and control, and the optimization of memory access to a simple feed-forward model. However, DFEs are inefficient for small-scale computations with control-dominated dynamic behavior. The key to effective dataflow computing systems is therefore the combination of DFEs with a conventional processor. The DFE carries out the compute-intensive part of the application while host processors are tasked with control-intensive tasks and also with setting up and controlling the computation on the DFE. Depending on the nature of the problem, one can also adopt a combined processing approach where the processor computes the less demanding part of the application while the DFE will target the performance-critical part. This results in a codesign approach where we develop a conventional processor application together with a customized DFE implementation. In the following, we first cover Maxeler dataflow systems, followed by programming principles and custom optimizations.

15.3 Maxeler Dataflow Systems

At the center of Maxeler's dataflow systems sits its proprietary DFE hardware. In state-of-the-art MAX4 generation systems, DFEs are based on large Altera Stratix-V FPGAs that provide the reconfigurable computing substrate for dataflow cores. This device is surrounded by large amounts of DRAM memory (currently between 48 and 96 GB), providing a very high memory capacity enabling large computational problems. This is called Large Memory (*LMem*). In addition, the FPGA itself also provides embedded on-chip memories which are spread throughout the chip's fabric and can be used to hold local values of the computation. These embedded memories are called Fast Memory (*FMem*) as they can be accessed with a total bandwidth of several terabytes/second. This is an important factor for the efficiency of DFE computations because data can be kept locally where it is needed and accessed with very high speeds. This is in contrast to CPU caches where data is kept on a speculative basis, and replicated several times, with only the smallest L1 cache providing very high speed to the computational unit.

As previously mentioned, DFEs are not intended to fully replace conventional CPUs; instead, they are integrated into an HPC system consisting of CPUs, DFEs, storage, and networking. Various system architectures are possible and the overall balance of components can be tailored to the requirements of the user application. As a key feature, DFEs always contain large amounts of DRAM to facilitate the previously described model of dataflow processing. Various configurations between DFEs and CPU, as well as between multiple DFEs are possible. In the following, we give a brief overview of the current Maxeler MPC-C, MPC-X, and MPC-N series.

Maxeler MPC-C systems couple x86 server-grade CPUs with up to 4 DFE cards (see Figure 15.2). Each DFE card contains 48 GB of DRAM as LMem, and each DFE card is connected to the CPUs via a PCI Express (PCIe) bus. DFE cards are also directly connected to each other through a dedicated high-speed, low-latency link called MaxRing. This provides fast communication between neighboring DFEs, enabling larger applications to scale across multiple DFEs without the PCIe link becoming a communication bottleneck.

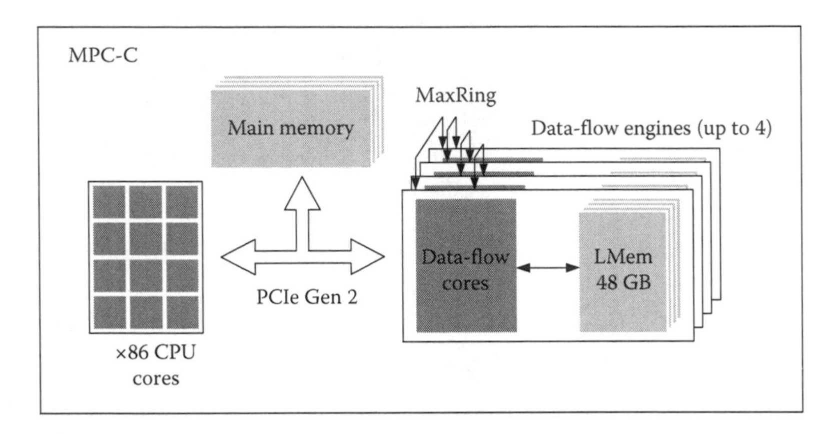

FIGURE 15.2: MPC-C series architecture. A single node contains both ×86 CPUs and 4 DFEs connected via PCIe and MaxRing.

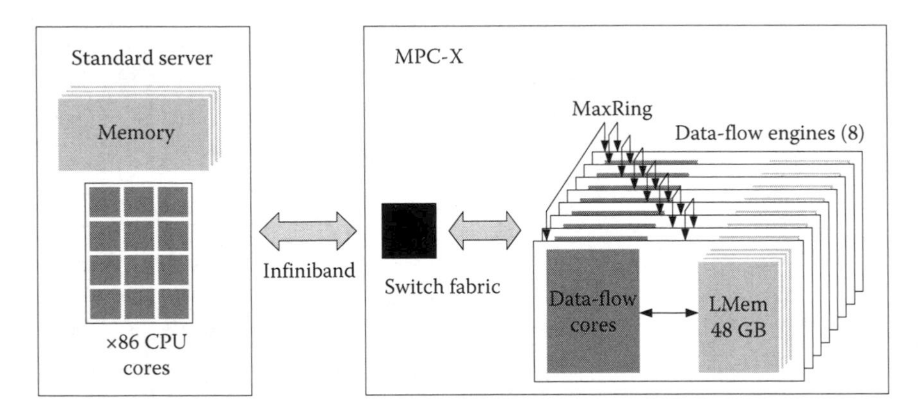

FIGURE 15.3: MPC-X series architecture with eight DFEs inside a node.

The system also includes storage and networking, and it is integrated into a dense 1U[1] industry standard rack unit. Such a system supports simple stand-alone deployment of DFE technology, tightly coupled with high-end CPUs. The architecture is beneficial for high-performance applications that run on a fixed number of CPU cores and continuously use one or multiple DFEs.

The MPC-X series enable a more heterogeneous system architecture supporting dynamic balancing of CPU and DFE resources. MPC-X series systems are pure DFE nodes without any CPUs (see Figure 15.3). An MPC-X system combines 8 DFE cards in a 1U chassis directly connected through MaxRing. DFEs are also linked through Infiniband to a cluster of CPU nodes. The system can dynamically allocate arbitrary (often large) numbers of DFEs,

[1]U is the abbreviation of Rack Unit. It is the standard unit of measure for the vertical usable data center space, or height defined as 1.75 inches (44.45 mm). A typical full-size rack cage is 42U high, composed by multiple 1U, 2U, or 4U boxes.

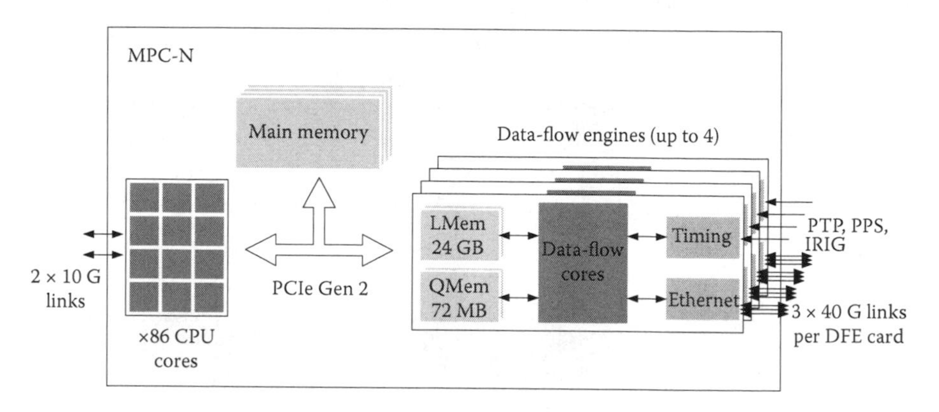

FIGURE 15.4: MPC-N series architecture with four DFEs inside a node.

providing good scalability and flexibility for applications with changing behavior, for example, when the computation has several stages that vary in their characteristics. The CPUs to DFEs ratio can be matched to user application requirements.

Maxeler's MPC-N series systems (see Figure 15.4) are a network-oriented platform that provides Ethernet connections directly to the DFEs, supporting ultra low-latency line-rate processing of multiple 10–40 Gbit data streams. A single MPC-N node contains up to 4 DFE cards similar to the MPC-C series architecture. However, each DFE card also supports up the 3 QSFP+ 40 Gbit Ethernet connections where each 40 Gbit port can be split into 4 × 10 Gbit ports. Providing fast Ethernet connections directly to the DFE enables network processing with minimal latency. The memory architecture in DFE also differs from the two previous system architectures: in addition to 24 GB DRAM available as LMem, the DFE also integrates 72 MB of QDR SRAM (QMem) supporting very low latency off-chip data access. The system contains additional 10 Gbit connections to the CPU. MPC-N series systems are well suited for a range of networking applications including gateways, aggregators, or endpoints.

Maxeler systems are provided with a compilation and simulation environment (called MaxCompiler) for application development, and the MaxelerOS system management environment. MaxelerOS coordinates the use of DFE resources at run time, and manages the scheduling and data movement within Maxeler systems. MaxCompiler provides a high-level programming environment to express dataflow structures, and produces the necessary binaries for CPU and DFE binaries.

15.3.1 DFEs in the cloud

Public cloud computing has been taking the world by storm, with over 32,000 attendees at the Amazon AWS re:Invent in Las Vegas in November

2016. Why is cloud computing becoming so popular? A public cloud appeals to both senior management and to IT departments of many organizations. For senior management, a public cloud creates a second source for IT services, removing the lock-in with large internal IT teams and expensive infrastructure. For the new generation of IT professionals, the public cloud offers the power to manage a massive amount of resources with much less effort and much lower expertise is required.

In Finance, the public cloud had a hard youth. Data confidentiality issues, the negotiation power of current IT infrastructure owners, as well as service levels (e.g., liability contract clauses) of public cloud providers made it prohibitive for many years for senior management of banks to use public cloud services. This is now rapidly changing, with cloud providers offering service guarantees, and addressing critical business continuity services, which are desperately needed by many financial institutions. Cloud providers are also starting to also look into high-performance services: 2016 has seen the integration of Altera into Intel, on the back of a large deployment of Altera products at Microsoft; and more recently Amazon announced the new AWS EC2 F1 instance as their top end performance instance offering [9].

This Amazon EC2 F1 instance (not to be mistaken for a Formula 1 event) offers raw computer hardware which is fully compatible with the MAX5 generation of DFEs. F1 instances come in single and eight-way configurations with eight DFE compatible units, each with 64 Gigabytes of DRAM. Each MAX5 DFE compatible F1 unit has about 3× the compute performance of a MAX4 card. The instances come with 480 GB and 960 GB of SSD[2] storage, respectively. The instances are therefore similar to the Maxeler single DFE MPC-C appliances and the MPC-X 8-way appliance.

With the availability of Amazon F1 instances, DFEs now have a high volume second source, as well as the Amazon Marketplace creating a commercial platform for vendors to build various applications on top. For large-scale deployment, a combined solution of on-premise private DFE cloud plus the elasticity to the Amazon F1 instances further enhances the efficiency and operational cost reduction provided by DFEs while significantly reducing the operational risks associated with mixed technology stacks. So a bank would use DFEs on premise based on predicted average load requirements and elastically expand into the cloud during peak times, as well as being able to build a solid business continuity solution, where if a bank data center goes down, the Amazon cloud can pick up seamlessly without service interruption.

On top of all Cloud advantages, the Multiscale Dataflow programming ecosystem enables a simple software development process and a debug and optimization process that has been fine tuned for over a decade to enable scientists and domain experts to achieve Maximum Performance Computing for ultra-large scale software packages and workloads. Previously this was only available to a few top computing experts and typically worked only for

[2]Solid-State Drive: A type of mass storage device similar to a hard disk drive (HDD) with no moving parts and hence faster and more predictable read and write times.

tiny compute examples such as simple matrix multiply. Moreover, Multiscale Dataflow programming has successfully proven through its five generations to be truly independent of the underlying technology with built in by design performance portability towards the next DFE generation. These two properties are both considered of extreme value in the Cloud deployment context. With all the above, DFEs offer a manageable migration path from standard solutions to cloud-enhanced ultra high-end computing.

15.4 Dataflow Programming Principles

In the following, we outline the dataflow oriented programming model that is used in Maxeler systems. As described in the previous section, Maxeler dataflow systems are based on a combination of DFEs and CPUs. The basic logical architecture of such a system is illustrated in Figure 15.5. The CPU is responsible for setting up and controlling the computation on the DFE. The DFE contains one or multiple dataflow kernels that perform the accelerated arithmetic and logical computations. Each DFE also contains a manager that is responsible for the connections between kernels, DFE memory, and the various interconnects such as PCIe, Infiniband, and MaxRing.

Separating computation and communication into kernels and managers is beneficial because it allows the data path inside the kernels to be deeply pipelined without any synchronization issues. When developing the kernel, a designer would simply focus on achieving high degrees of pipelining

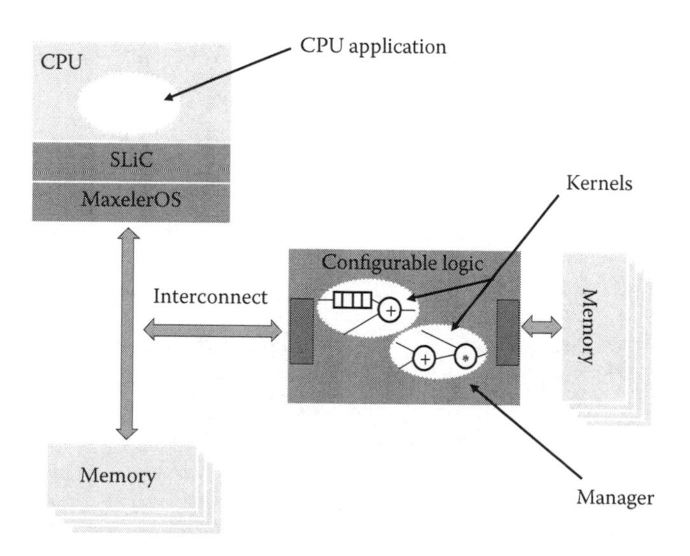

FIGURE 15.5: Logical architecture of a dataflow computing system with one CPU and one DFE.

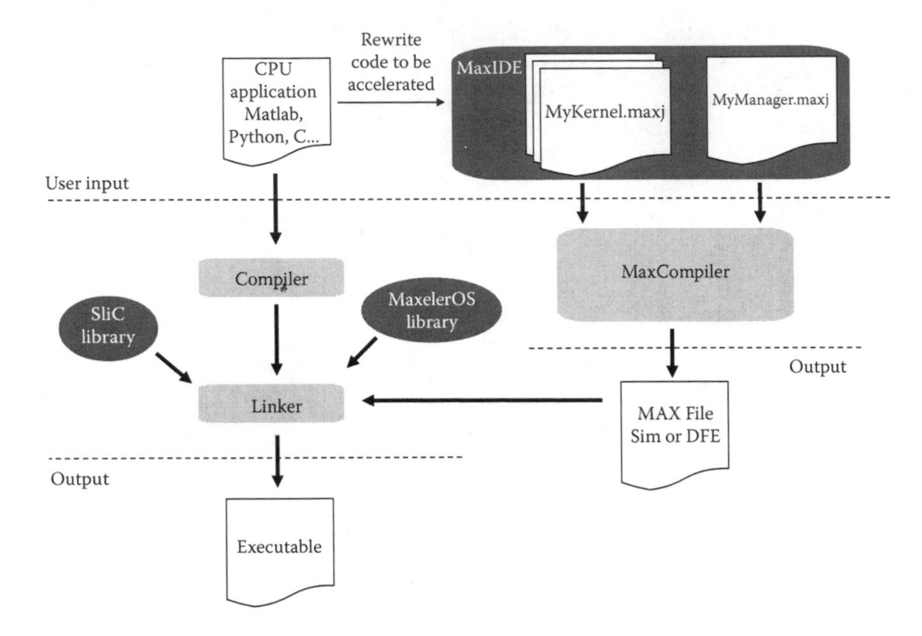

FIGURE 15.6: Compiling a dataflow application with MaxCompiler.

and parallelism without worrying about scheduling or synchronization. The scheduling of operations inside the kernel will be performed automatically by the compiler. The manager code describes how kernels are connected to memory and other IO interfaces, and the necessary synchronization logic will also be generated by the compiler.

Developing an application for a DFE-based system therefore includes three parts:

1. A CPU application typically written in C/C++, Matlab, Python, or FORTRAN
2. One or multiple dataflow kernels written in extended Java[3]
3. A manager configuration, also written in extended Java[3]

The compilation flow of a Maxeler dataflow design is illustrated in Figure 15.6. The design typically starts with a CPU application where a performance-critical part needs to be accelerated. This part of the application will be targeted on a DFE. Designing a DFE application involves describing one or multiple kernels and a manager in MaxJ. MaxJ is a Java-based meta-language that describes dataflow. It is important to note that executing the MaxJ program will not perform the computations described within the program. Instead, it will trigger the generation of a configuration file for the DFE

[3]Maxeler provides extensions to the Java language, referred to as MaxJ.

(the so-called .max file). The computation will later be performed by loading the .max configuration file into the DFE and streaming the data through it. Before we can do this, we need to modify the CPU application to invoke the DFE. To simplify this process, MaxCompiler will generate the necessary function prototypes and header files. The CPU code is then compiled as usual and is linked with the .max file and Maxeler's Simple Live CPU (SLiC) interface library. The result of this is a single executable file that contains all the binary code to run on both the conventional CPUs and the DFEs in a system.

Let us focus on the principles of dataflow programming in MaxJ. As mentioned previously, MaxJ is a metalanguage that describes dataflow computing structures; it uses Java syntax but is in principle different from regular Java programming (or other imperative programming paradigms that describe computations by changing state). The most important principle in MaxJ is that we describe a fixed spatial dataflow structure that can perform computations by simply streaming through data, and not a sequence of instructions to be executed on a traditional processor.

To illustrate these principles, we show how a simple loop computation can be transformed into a dataflow description using MaxJ. Let us assume we want to calculate $y = x^2 + 3x + 17$ over a data set. Even though there is nothing inherently sequential in this computation, a conventional C program would require a for loop. This is illustrated in Figure 15.7. The calculation is repeated for the number of data elements in a loop. Within the loop body, all operations also run sequentially.

In contrast, a dataflow implementation would focus on identifying the core part of the computation and creating a data path for it. Figure 15.8 illustrates such a dataflow implementation. The same computation that is described inside the loop body can be performed by a fixed data path that contains two multipliers and two adders. It is one of the key features of dataflow computing having several operators present at the same time and running concurrently, instead of using a time-shared functional unit inside a processor. A practical dataflow implementation can have thousands of operators in a data path all running concurrently. Another important principle is the absence of control and instructions. The data path is fixed and the computation is performed by streaming data from memory directly into the data path.

Figure 15.9 depicts the MaxJ kernel description that can generate the data path shown in Figure 15.8. The MaxJ descriptions begin by extending the `kernel` class (line 1). The kernel class is part of the Maxeler Java extensions

```
1 for (i = 0; i < numDataElements; i++) {
2     float x = input[i];
3     float y = x * x + 3 * x + 17;
4     output[i] = y;
5 }
```

FIGURE 15.7: C code of a simple computation inside a loop.

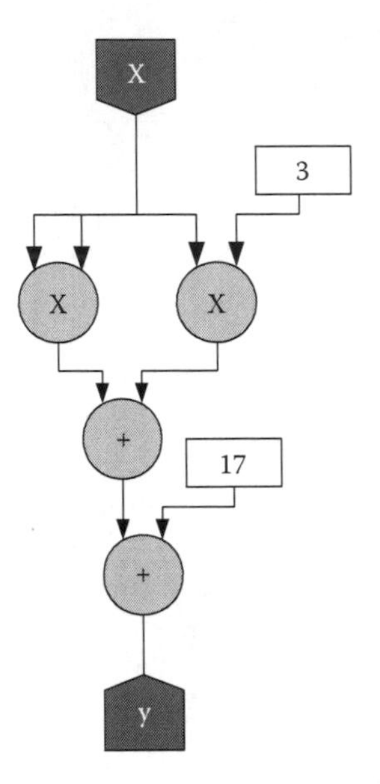

FIGURE 15.8: A dataflow implementation for the computation inside the loop body.

and the user develops their own kernels by using inheritance. Next, we define a constructor for `SimpleCalc` class (line 2). It is important to remember that this MaxJ program will only run once to build the DFE configuration; the constructor will facilitate building the dataflow implementation. To create the streaming inputs and outputs for the kernel, the methods `io.input` (line 3) and `io.output` (line 5) are used. Streaming inputs and outputs replace the `for` loop in the original C code that iterates over data. The input method takes two arguments: the name on the input that will be used by the manager to connect the kernel and the data type of the input. In this case, we use a standard single precision floating point format (8-bit exponent and a 24-bit mantissa), but MaxJ also supports custom data types that can be defined by the user. This is useful when optimizing the numerical behavior and performance, which will

```
1  class SimpleCalc extends Kernel {
2      SimpleCalc() {
3          DFEVar x = io.input("x", dfeFloat(8,24));
4          DFEVar y = x * x + 3 * x + 17;
5          io.output("y", y, dfeFloat(8,24));
6      }
7  }
```

FIGURE 15.9: A MaxJ description that generates the dataflow implementation shown in Figure 15.8.

```
1 int x = 10;
2 DFEVar y;
3 DFEVar z;
4 y = x; // ok, assign constant to run-time variable
5 x = y; // compiler error, cannot read run-time variable into
      compile-time Java variable
6 z = y; // ok, both handle run-time data
```

FIGURE 15.10: DFEVars handle run-time data, Java constants are evaluated only at compile time.

be covered later. The output method uses three arguments: the name of the output to be used by the manager, the variable to connect to the output, and the data format. The computation itself (see Figure 15.9) is expressed in a very similar way as in the original C code (line 4).

In MaxJ, the `DFEVar` object is used to handle run-time data. Since MaxJ describes a dataflow graph rather than a procedure, we have to distinguish between run-time values and compile-time values. Regular Java variables such as `int` will be evaluated and fixed at compile time. Such variables can be used as constants for improved code readability or to control the build of the dataflow graph. The values of DFEVars are known only at run time when data is streamed through the kernel. This means assigning a Java variable to a DFEVar will result in a constant. However, it is not possible to read a DFEVar and assign its value to a Java variable (see Figure 15.10).

This principle described above means that we can use Java variables and control constructs to shape the structure of our dataflow graph. Let us consider an example of a nested loop as shown in Figure 15.11. We observe that the outer `for` loop performs an iteration over data, while the inner `for` loop describes a computation with a cyclic dependency of `v` from one loop iteration to another.

This example can be effectively transformed into a dataflow description as illustrated in Figure 15.12. Again, the outer loop is replaced by streaming inputs and outputs. The inner loop is described with the same for `for` loop statement in Java, but the compilation of this loop will result in an unrolled implementation of the loop body in space, as depicted in Figure 15.13. Unlike

```
1 for (i = 0; i < numDataElements; i++) {
2     float d = input[i];
3     float v = 2.91 - 2.0 * d;
4     for (iteration = 0; iteration < 4; iteration++) {
5         v = v * (2.0 - d * v);
6     }
7     output[i] = v;
8 }
```

FIGURE 15.11: C code of a nested loop with dependency.

```
1  class Loop extends Kernel {
2      Loop() {
3          DFEVar d = io.input("d", dfeFloat(8,24));
4          DFEVar v = 2.91 - 2.0 * d;
5          for (int iteration = 0; iteration < 4; iteration +=
               1) {
6              v = v * (2.0 - d * v);
7          }
8          io.output("output", v, dfeFloat(8,24));
9      }
10 }
```

FIGURE 15.12: A MaxJ implementation of the inner loop will be statically evaluated resulting in spatial replication.

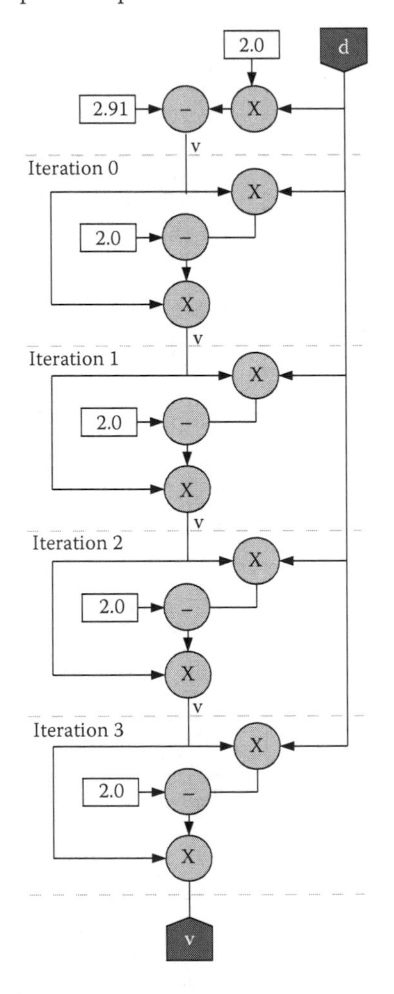

FIGURE 15.13: The result of the MaxJ loop is an unrolled and pipelined data path.

```
1 DFEVar x = io.input("x", dfeFloat(9,31));
2 DFEVar a = io.input("y", dfeFloat(9,31));
3
4 DFEVar y1 = x * 5;
5 DFEVar y2 = x - 7;
6
7 DFEVar y = a > 3 ? y1 : y2;
8
9 io.output("y", y, dfeFloat(9,31));
```

FIGURE 15.14: Data-dependent control with the ternary operator and use of a custom number format.

the original loop in C, the `for` loop in MaxJ does not carry out four iterations at run time. Instead, the compiler can resolve the dependency of `v` from one loop iteration to another and construct an unrolled, acyclic data path where the calculation inside the loop body is replicated four times, and each `v` is connected to the result from the previous iteration.

The previous example has shown how a Java `for` loop can be used to control the replication of statements inside the loop body into an unrolled data path. Likewise, it is possible to use Java conditionals such as `if` or `case` to control the construction of the dataflow graph. The Java `if` condition is evaluated at compile time, and the block of code inside the conditional statement will be added into the dataflow graph only if the condition is evaluated as true.

However, we cannot use a Java conditional on DFEVars because their value will be only known at run time. As previously mentioned, run-time dependent behavior is undesirable as it is against the principles of static dataflow computing. If a data-dependent decision needs to be made, then this can be expressed using the ternary operator `? :` (see Figure 15.14). This example results in data-dependent control, but in the data path, both `y1` and `y2` will be computed concurrently. At the output, we simply select one of the two results, depending on the value of `a`. This switching will be very fast and will not delay or stall the stream processing. However, it also means that we require resources for both computations on the DFE chip even though only one of the two outputs will be used at any time. This makes this type of control effective for fast, small-scale switching. For switching between larger blocks of computation, it might be more effective to implement separate DFE kernels and handle the switching and control from the CPU host.

Figure 15.14 also illustrates that custom number formats other than conventional single or double-precision floating point can be used. In this example, we use a 9-bit exponent and a 31-bit mantissa, which offers better scaling and precision than single precision (8, 24 bit) but less than double precision (11, 53 bit). Likewise, it is possible to use any arbitrary fixed-point or integer format. The application developer can use such custom number formats to

```
1 DFEVar x = io.input("x", dfeFloat(8,24));
2 DFEVar prev = stream.offset(x,-1);
3 DFEVar next = stream.offset(x,1);
4 DFEVar y = (prev + x + next)/3;
5 io.output("y", y, dfeFloat(8,24));
```

FIGURE 15.15: Using stream offsets to access values with relative offsets in the stream.

tailor the implementation to the numerical requirements of the application, and using such custom formats will yield better resource utilization and performance than relying on the next larger standard format.

All previous examples have considered operations where the output is a function of inputs with the same array index within the stream, for example:

$$z_i = 5x_i + y_i, \quad z_{i+1} = 5x_{i+1} + y_{i+1}, \quad \ldots \tag{15.1}$$

However, in some cases we need to access values that are ahead or behind the current element in the data stream. For example, in a moving average filter we need to compute:

$$y_i = \frac{x_{i-1} + x_i + x_{i+1}}{3} \tag{15.2}$$

In dataflow computing, `x` is a stream rather than an indexed array, and we need a way of accessing elements of the same stream with other indices than the current one. This can be achieved with the `stream.offset` method that accesses values with a relative offset from the current value in the stream. In the moving average example (see Figure 15.15), we need the previous value (-1) and the next value $(+1)$:

Figure 15.16 illustrates how a DFE application interacts with the CPU host application. On the right side, we see the moving average kernel `MAVKernel` from our last example. As previously mentioned, we also create a manager to describe the connectivity between the kernel and the available DFE interfaces. In Figure 15.16, the kernel is connected directly to the CPU, and all of the communication will be facilitated via PCIe. The manager also makes visible to the CPU application all the names of the kernel streaming inputs and outputs. Compiling the manager and kernel will produce a .max file that can be included in the host application code. In the host application, running the moving average calculation will be performed with a simple function call to `MAVKernel()`. In this example, the host application is written in C but MaxCompiler can also generate bindings for a variety of other languages such as MATLAB or Python.

MaxelerOS and the SLiC library provide a software layer that facilitates the execution and control of the DFE applications. The SLiC Application Programming Interface (API) is used to invoke the DFE and process data on it. In the example in Figure 15.16, we use a simple SLiC interface and

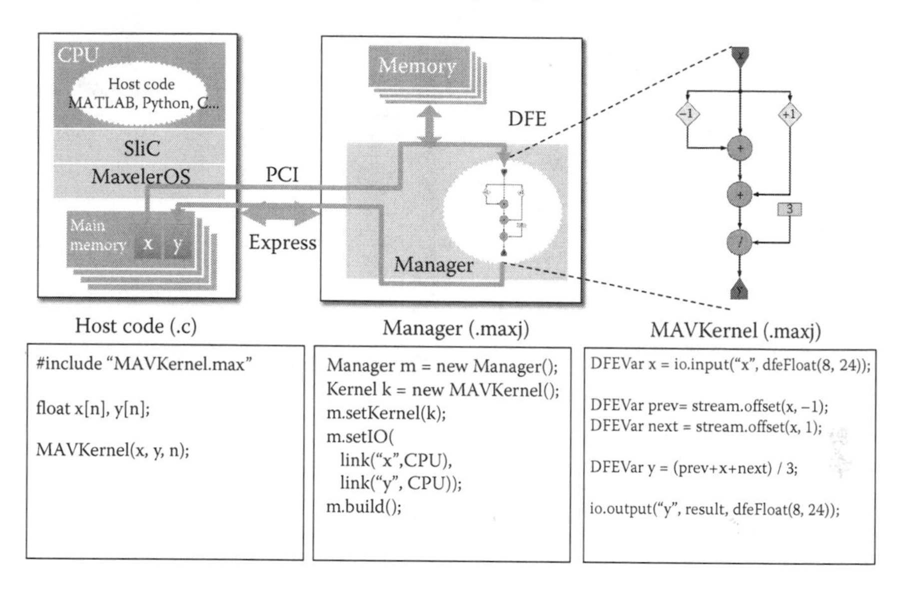

FIGURE 15.16: Interaction among host code, manager, and kernel in a dataflow application.

the simple function call `MAVKernel()` will carry out all DFE control functions such as loading the binary configuration file and streaming data in and out over PCIe. More advanced SLiC interfaces are also available that provide the user with additional control over the DFE behavior. For example, in many cases it is beneficial to transfer the data to DFE memory (LMem) first and then start the computation. This is one of many performance optimizations, which we will briefly cover in the following section.

15.5 Development Process and Design Optimization

In the previous section, we have introduced the principles of dataflow programming. We now outline how to develop dataflow applications in practice and how to improve their performance. In traditional software design, a developer usually targets a given platform and optimizes the application based on available libraries that reflect the capabilities and architectural characteristics of the targeted platform. Developing a dataflow implementation fundamentally differs in that we *codesign* the application and architecture. Instead of mapping a problem to preexisting APIs and data types, we enable domain experts, for example, physicists, mathematicians, and engineers, to create a solution all the way from the formulation of the computational problem down to design of the best possible dataflow architecture. A developer would therefore optimize the scientific algorithm to match the capabilities of the dataflow

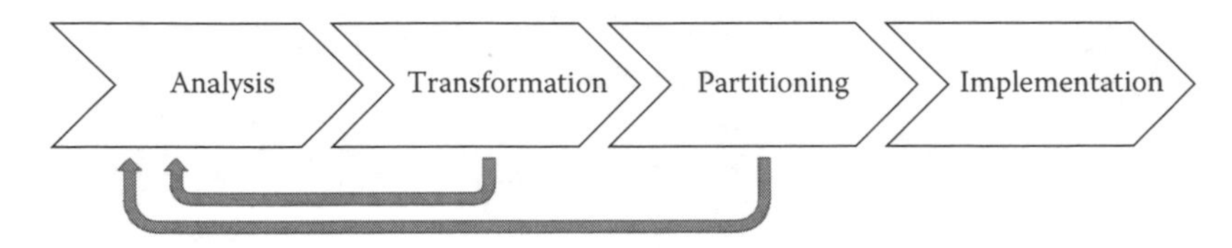

FIGURE 15.17: Process for developing and optimizing dataflow applications.

architecture while at the same time optimizing the dataflow structure to match the requirements of the algorithm. Another key difference to traditional software design is the implementation and optimization cycle. In software design, a developer would typically implement a design, go on to profile and evaluate the performance of the current implementation, and then tweak the implementation. In dataflow design, we adopt a different approach where the design is optimized before it is implemented. The behavior inside a DFE is very predictable and we can therefore plan and precisely predict the performance of a possible solution without even implementing it. This means the design will be analyzed and optimized with simple spreadsheet calculations before we create the final implementation.

This development process is illustrated in Figure 15.17. The first step consists of an application analysis phase. The purpose of this step is to establish an understanding of the application, the data set, the algorithms used, and the potential performance-critical parts. Since we will codesign an algorithm and its dataflow architecture, this analysis should cover all parts of the computational problem, from the mathematical formulation and algorithm to the architecture and implementation details. Typical considerations are the type and regularity of the computation, the ratio between computation and memory accesses, the ratio of computation to disk IO or network communication, and the balance between recomputation and storage of precomputed results. All these aspects can have a significant impact on the performance of the final implementation. If, for instance, an application is limited by the speed at which data can be read from disk, then optimizing the throughput of the compute kernel beyond that limit will have no benefit.

The second step involves algorithmic transformations. A designer could attempt to choose a different algorithm to solve the problem, or transform the code, data access patterns or number representations. A typical example of an algorithmic transformations is to change the number format: Choosing a smaller number representation can support more IO bandwidth, and higher computational performance, but the numerical effects of the algorithm have to be well understood. The reconfigurable technology used inside the DFEs support far greater flexibility in the available number formats than all conventional processors. Instead of choosing from single or double precision floating point, a design can exploit a custom format with arbitrary bit-widths of its exponent and mantissa. Another common optimization is the reordering of data access patterns to support better dataflow. The impact of algorithmic transformations has to be evaluated through iterative analysis of the design.

The third step is to partition the application between the CPU and the DFE. This partitioning covers program code as well as data. For the program code, we can choose whether the code should run on the CPU or the DFE. Large-scale applications typically involve multiple DFEs and this also involves partitioning DFE code over multiple DFEs. Furthermore, it is often beneficial to follow a coprocessing approach where the CPU and DFE work on different parts of the computation at the same time. For instance, the CPU can perform lightweight precalculations or more control-intensive parts of the application. For this purpose, the SLiC library provides nonblocking functions to control the DFEs. Another consideration is the partitioning of data. The example in Figure 15.17 showed DFE data being streamed from main CPU memory. For processing larger data sets, it is usually beneficial to locate the data in the large DFE memory (LMem). Coefficients or frequently accessed values can be kept inside the DFE reconfigurable substrate in fast memory (FMem).

A high-level performance model is used to evaluate the design as it undergoes various transformations, code, and data partitionings. The process of analysis and optimization is repeated iteratively as additional possibilities are explored. Only when the design is fully optimized, will the designer proceed to step 4: the implementation of the design.

15.6 A Case Study: Correlation

To illustrate the principles of Section 15.5, we consider the following real-world example: given a collection of price time series $\{S_t^i\}$ for times t and stocks $i = 1, \ldots, N$, with N significantly large, we wish to compute the correlations $\rho_t^{i,j}$ for each of the $N(N-1)/2$ pairs $0 \leq i < j \leq N$ over a sliding time interval from $t - T$ to t. These correlation values are important in portfolio management and trading strategies such as statistical arbitrage, where a statistical relationship between a pair of stocks can be used to predict price movements in one from the other. Typical values for this example are $N = 6000$ and $T = 100$.

The formula for correlation is

$$\rho_t^{i,j} = \frac{T \sum_\tau r_\tau^i r_\tau^j - \sum_\tau r_\tau^i \sum_\tau r_\tau^j}{\sqrt{T \sum_\tau (r_\tau^i)^2 - (\sum_\tau r_\tau^i)^2} \cdot \sqrt{T \sum_\tau (r_\tau^j)^2 - (\sum_\tau r_\tau^j)^2}} \tag{15.3}$$

where $t - T \leq \tau < t$ and $r_t^i = \ln(S_{t+1}^i / S_t^i)$ is the return on stock i in period t. We begin our *analysis* step by writing a CPU program to perform this calculation, and evaluating it for performance bottlenecks using well-known code analysis tools.

We can *transform* the problem by noting that our goal is to calculate the sequences $\{\rho_t\}$ over time as new price data $\{S_t\}$ becomes available. In a single time step increment from t to $t+1$, we can update each of the sums in Equation 15.3 by adding a single term for $\tau = t$ and subtracting a term for $\tau = t - T$—in particular, this greatly reduces the number of multiplications needed to calculate the $\sum r^i r^j$ terms from $N(N-1)(T/2)$ to $N(N-1)$. This type of "streaming" calculation is particularly well-suited to DFEs, which consume data at a consistent and predictable rate.

To *partition* the problem between CPUs and DFEs, we note that the most computationally intensive components are the pairwise product terms $\sum r^i r^j$, which must be calculated for nearly 18 million pairs in our example, while the other terms need only be calculated once for each of the 6000 stocks and the values reused. Accordingly, we first calculate the single stock terms on the CPU and store the values in LMem to be consumed by the DFE later—only the pairwise multiplications are implemented on the DFE. We also add logic to the DFE to calculate the final correlation value and store the pairs with the highest correlation.

Turning to the *implementation* stage, we need to determine the number of multiplications we will perform simultaneously on the DFE. This is a product of the number of multiplier units, or *pipes*, that can physically fit on the chip, as well as the number of clock cycles required for each multiplication

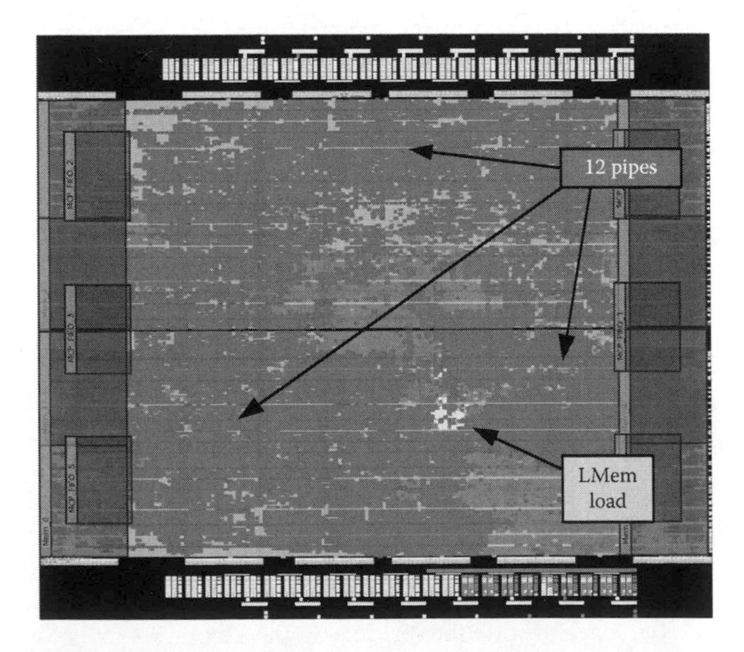

FIGURE 15.18: Layout of the correlation calculation.

(the *pipeline depth*). These values will vary between DFE generations and can only be truly determined through performing place-and-route using the reconfigurable chip vendor tools. The latter takes a significant amount of compute time (typically 10s of hours). Figure 15.18 shows the layout of the DFE correlation with 12 parallel pipes each with a pipeline depth of 12 stages, so that 144 data elements can be simultaneously processed with a throughput of 12 data elements per clock cycle. Even when assuming a low clock frequency, this design is able to perform all 36 million required multiplications in less than 30 ms.

15.7 Financial Application Examples

Maxeler dataflow technology has been deployed in a number of areas including finance [6,10], oil and gas exploration [11,12], atmospheric modelling [13], and DNA sequence alignment [14]. The range of applications includes Monte Carlo, finite difference, and irregular tree-based partial differential equations, to name a few. Maxeler provides a number of products and solutions in the financial domain, including financial analytics and trading applications, particularly for low-latency/high frequency electronic trading on organized exchanges.

15.7.1 Maxeler RiskAnalytics platform

Maxeler RiskAnalytics is a financial valuation and risk management platform designed from the ground-up, where the core analytic algorithms are accelerated on Maxeler dataflow systems. The purpose of the platform is to go beyond simply providing highly efficient computational finance capabilities, but rather the aim is to provide a complete, vertically integrated application stack that provides a platform containing all the necessary components for streamlined front-to-back portfolio risk management, including:

- Front-end, pretrade valuation, and risk checking
- Exchange-based, electronic trade execution, portfolio valuation, and risk management
- Front-end trade booking, portfolio management, model and risk reporting and analysis
- Posttrade model and risk metric selection and verification
- Rapid and flexible transaction analysis and reporting
- Application layer in software for quick and flexible functional reconfiguration

- Large memory to enable rapid and flexible in-memory portfolio risk analysis
- Regulatory reporting for Basel III, EMIR, Dodd-Frank, Volker-rule, Solvency II, and so on
- Adaptive load balancing
- Database integration

All core RiskAnalytics components have been implemented in both software and on Maxeler DFE-based systems, requiring integration of the DFE technology with expertise of quantitative analysts with extensive investment banking experience. The platform has been designed in a modular fashion to maximize flexibility and performance. Each module realizes a core analytics component, such as curve bootstrapping or Monte Carlo path generation. To support flexible hardware/software coprocessing and to enable ease of integration with existing systems, each module is available as both a CPU and DFE library component. As outlined in Section 15.5, achieving an efficient implementation depends on the overall system composition, architecture, and application structure. Making use of preexisting CPU and DFE library components greatly simplifies this process. In the following, we show the use of Maxeler's RiskAnalytics library in several commercial use cases.

15.7.2 Interest rate swap pricing

An interest rate swap is a financial derivative with high liquidity that is commonly used for hedging. Such a swap involves exchanging interest rate cashflows based on a specified notional amount from one interest rate to another, for example, exchanging fixed interest-rate flows for floating interest-rate flows. Figure 15.19 illustrates a typical module configuration for pricing

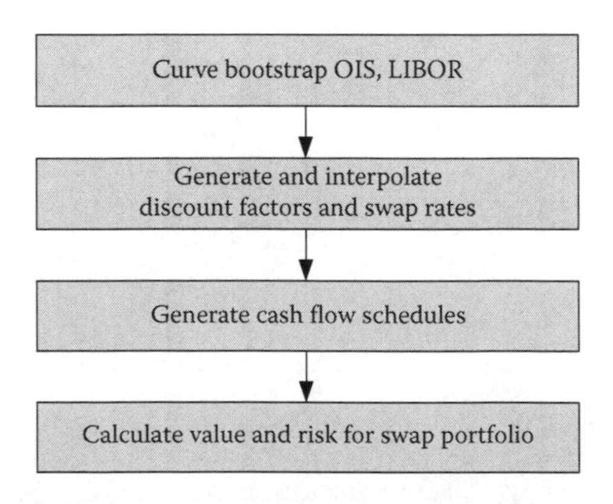

FIGURE 15.19: A typical swap pricing pipeline.

TABLE 15.1: Possible configurations for swap pricing pipeline

Application characteristics	OIS	LIBOR	Cashflow	Pricing
Many curves, few swaps	DFE	DFE	CPU	CPU
Few curves, many swaps	CPU	CPU	DFE	DFE

interest rate swaps, involving bootstrapping the Overnight Index Swap (OIS) curve and the London Interbank Offered Rate (LIBOR) curve, followed by generating swap cashflow schedules, valuing swaps, and calculating swap portfolio risk. Each stage is available as either a CPU or a DFE library component and can be accessed via number of convenient APIs. The implementation provides construction of and access to all intermediate and final objects.

Depending on the characteristics of the swap pricing application, DFE acceleration can be beneficial at one or more stages of the computation. Table 15.1 illustrates two possible module configurations where the performance-critical DFE acceleration can be carried out at different stages of the pipeline. Modular design of Maxeler's RiskAnalytics allows the user application to dynamically load balance between CPUs and DFEs, and to target heavy compute load to DFEs, leaving CPUs to support application logic and lighter compute loads. DFE functionality can be switched in real time by using MaxelerOS SLiC API functions. Fully pipelined, a Maxeler DFE-equipped 1U MPC-X node can value a portfolio of 10-year interest rate swaps at a rate of over 2 billion per second—including bootstrapping of the underlying interest rate curves.

15.7.3 Value-at-risk

Value-at-risk, or VaR, a measure widely used to evaluate the risk of loss on a portfolio over a given time period. VaR defines the loss amount that a portfolio is not expected to exceed for a specified level of confidence over a given time frame. VaR can be calculated in a number of ways (e.g., using fixed historical scenarios, or using arbitrarily specified scenarios, a delta-based approach, or using Monte Carlo generated scenarios). Irrespective of the method chosen, the VaR computation involves evaluating many possible market scenarios, a technique that is computationally very demanding. Regardless of the chosen approach, the computation of VaR using conventional technology is frequently slow and often inaccurate, as well as being unstable in the tail of the loss distribution, resulting in uncertainty in risk attribution and difficulty in optimizing against portfolio VaR targets. This is illustrated in Figure 15.20a, where the tail of the loss distribution for a mixed portfolio of interest rate swaps exhibits a stepwise profile, making it extremely difficult to accurately manage portfolio VaR.

Mitigating these problems requires massively increased number of scenarios, in order to provide higher resolution in the tail of the loss distribution, in order to significantly improve stability for risk attribution and/or provide

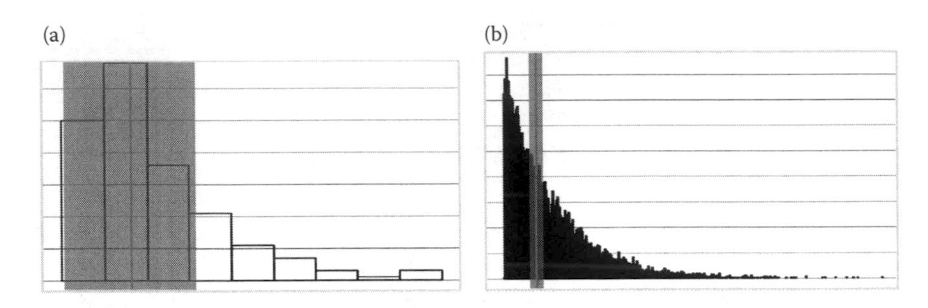

FIGURE 15.20: Value-at-Risk with 10,000 scenarios (a) and 500,000 scenarios (b).

greater visibility of the impact of market and portfolio changes. This is clearly illustrated when comparing Figure 15.20a and b. In the second case, the number of Monte Carlo scenarios is increased by a factor of 50, resulting in far greater granularity in the tail of the loss distribution leading to improved accuracy of portfolio risk management. Fully pipelined, a Maxeler DFE-equipped 1U MPC-X node can compute full revaluation VaR on a portfolio of 250,000 10-year interest rate swaps (equivalent to a rate of over 2 billion swaps per second)—including bootstrapping of the underlying interest rate curves, as well as scenario construction.

Increasing the number of Monte Carlo scenarios as suggested above obviously increases the computational requirements, but with DFE acceleration, the extra scenarios can be easily and practically achieved. When the accuracy of computation is increased, several new approaches to VaR can become feasible:

- Prehorizon cashflow generation and dynamic portfolio hedging
- Sensitivity metrics for enhanced risk explanation and attribution
- Stable and efficient portfolio optimization

15.7.4 Exotic interest rate pricing

Another use case for DFEs is pricing an exotic product such as a Bermudan swaption, which is an option to enter into an interest rate swap on any one of a number of predetermined dates. One of the industry standard approaches to this pricing problem is to use the LIBOR market model (LMM) which employs a high-dimensional Monte Carlo simulation with complex dynamics and a large state space. Pricing involves a multistage algorithm with forward and backward cross-sectional (Longstaff–Schwartz) computations across the full path space. Here, the challenge is to manage large path data sets, typically several gigabytes, across multiple stages.

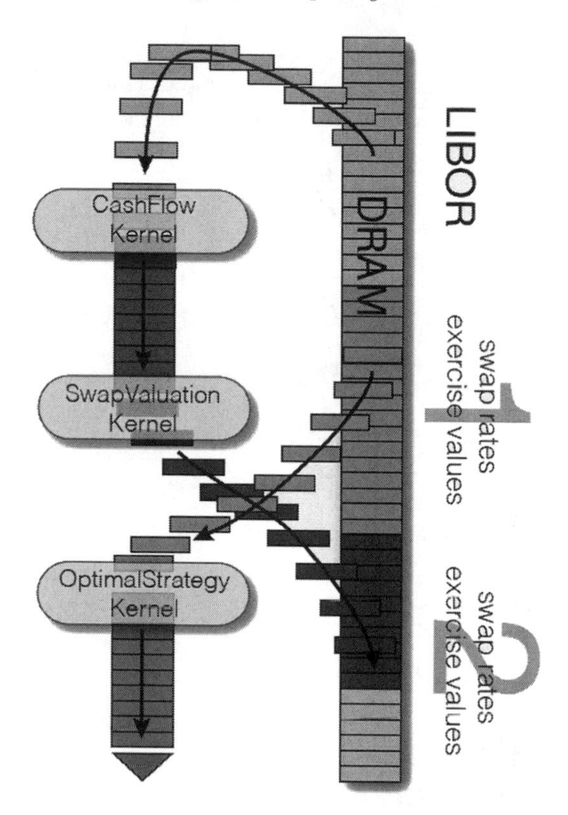

FIGURE 15.21: Bermudan swaptions computation on a DFE.

Figure 15.21 illustrates the RiskAnalytics DFE implementation, including cashflow generation and Longstaff–Schwartz backward regression. By closely coordinating between multiple DFE stages and DRAM memory, 6666 quarterly 30-year Bermudan swaptions can be priced per second on a Maxeler 1U MPC-X node. This represents an 23× improvement over an 1U CPU node.

Table 15.2 provides a comparison of different instruments priced per second for a range of instrument types supported in RiskAnalytics. As it can be seen,

TABLE 15.2: Performance comparison of 1U CPU and DFE nodes

Instruments priced per second	**Conventional 1U CPU-node**	**Maxeler 1U MPC-X node**	**Comparison**
European swaptions	848,000	35,544,000	42×
American options	38,400,000	720,000,000	19×
European options	32,000,000	7,080,000,000	221×
Bermudan swaptions	296	6666	23×
CDS	432,000	13,904,000	32×
CDS bootstrap	14,000	872,000	62×

a single 1U MPC-X node can replace between 19 and 221 conventional CPU-based units. The power efficiency advantage due to the dataflow nature of the implementation also ranges between 1 and 2 orders of magnitude.

15.7.5 Credit value adjustment capital

Nowadays, banks are ever more focused on returns measured against their regulatory capital requirements for OTC (over-the-counter) derivatives. This has even led to the introduction of a valuation adjustment to represent such costs, namely capital value adjustment (KVA) alongside other similar adjustments such as credit value adjustment (CVA) and funding value adjustment (FVA).

Future regulatory changes will increase the amount of regulatory capital banks, required to hold against counterparty credit risk (CCR) and CVA. Such changes will make OTC derivative businesses yet more expensive and make return on capital targets even harder to achieve. Many banks may decide to minimize their activity in derivative businesses, only participating in key business areas and those that service the needs of key clients (see Figure 15.22).

However, an alternative route to downsizing is to manage CVA more actively. By doing this, banks can save capital and thus increase profitability in two ways. First, by using more risk sensitive capital methodologies, it will be possible to benefit from smaller regulatory capital charges. Second, by actively managing CVA, further capital savings will be possible thanks to the capital reducing benefit of CVA-related hedges allowed under these more risk sensitive approaches. There will likely also be further nontangible benefits of taking a more proactive approach, such as improving the risk culture within a firm. Banks pursuing this active management route will not only enhance

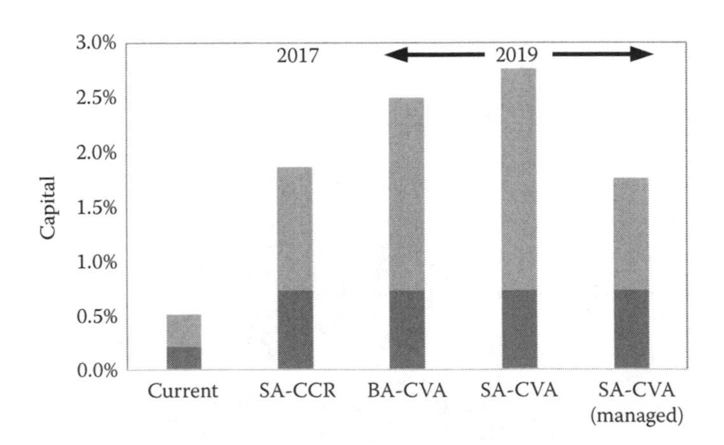

FIGURE 15.22: Estimated impact of future regulatory change on the capital requirements for an uncollateralized 10-year interest rate swap denominated in Euros.

their return on capital but should also be able to expand their OTC derivative franchises relative to competitors who are more capital (and leverage ratio) constrained.

A limiting factor for market participants is the computational complexity of the CVA calculation, which requires an entire subportfolio of derivative assets to be projected through multiple scenarios to determine the potential exposure to counterparty default. Since large portfolios of trades between two parties can extend to tens of thousands of assets, each asset must be projected until its maturity date (which can run as long as 30 years) and a large number of scenarios must be used to appropriately capture tail risk, this quickly becomes a problem too large in scale for all traditional CPU-based implementations (including state-of-the art and overclocked systems).

Utilizing the DFE-accelerated components of the Maxeler RiskAnalytics library, CVA calculations can be made practical at enterprise scale for large financial institutions. Additionally, banks are also becoming more rigorous about pricing capital via KVA. This requires the simulation of future capital requirements rather than just the calculation of spot capital. Under the SA-CVA[4] approach, for example, this is particularly time-consuming because this methodology is sensitivity based and so these sensitivities would need to be calculated in all possible future states. Nevertheless, due to the convexity of capital requirements (i.e., capital can go up more than it can go down) such a calculation is extremely important to accurately quantify KVA. Figure 15.23 shows such a capital projection run on Maxeler DFEs. The bold red line is the expected value, often known as the expected capital profile (ECP).

15.7.6 Standard initial margin model

Prior to the 2008 financial crisis, it was common for many derivative contracts to be traded bilaterally without significant margin or collateral requirements. Since the crisis, new regulations such as Dodd–Frank in the United States and EMIR in Europe require many trades to be cleared through a central clearing house, and impose significant margin requirements on uncleared trades. The methodologies for calculating these margin requirements have not yet been fully standardized, which has led to disagreements between counterparties on exactly how much margin to post. In response, the International Swaps & Derivatives Association (ISDA) has created the Standard Initial Margin Model (SIMM), with the first draft released in September 2016.

The SIMM framework is based on first-order greeks to make it more computationally tractable than methodologies such as Expected Historical VaR. Trades are broken down into risk categories, and the sensitivities of each trade

[4]Standardized approach for credit valuation adjustment.

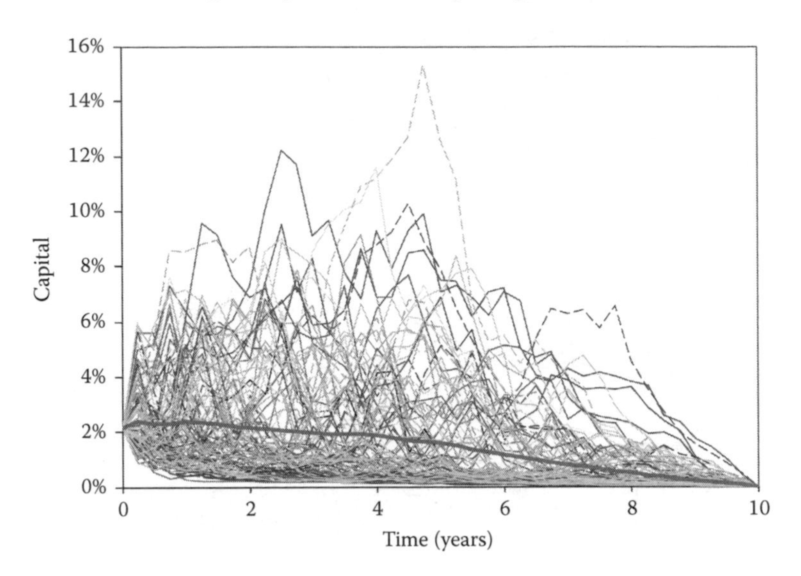

FIGURE 15.23: Capital evolution scenarios.

to risk factors, such as interest rates or equity prices, are computed and then aggregated using a proscribed set of weights and correlations. One key drawback of the framework as currently drafted is that the methodology to calculate these sensitivities is not similarly proscribed, so counterparties may not agree on the inputs to the model. Exposures may be netted within a single counterparty portfolio, but are grossed up between counterparties, so firms have a strong incentive to reduce initial margin requirements through careful matching of trades and portfolio compression.

In order to calculate the initial margin requirement for a new trade, every trade already existing between two counterparties must be revalued under both a base scenario and a number of shocked scenarios, so that any risk offsets between the new trade and the existing portfolio are taken into account. If this trading portfolio is already large, a significant performance cost can be incurred simply transferring it from disk storage to memory—in a DFE-accelerated solution, the portfolio can instead be stored in LMem and accessed directly as needed. Combining this memory model with high-performance pricing engines like that described in Section 15.7.2 allows for margin calculations to occur at the speed of trading.

Maxeler's SIMM calculation product on Amazon Web Services (AWS) combines the ease of deployment of a cloud service with an industry-proven risk analytics infrastructure. Our SIMM product splits naturally into the calculation of sensitivities, and the application of risk weights and aggregation. The Maxeler RiskAnalytics library provides the framework for calculating greeks on CPUs as well as on Maxeler DFEs. Initial margin requirements can be calculated directly from a portfolio of trades supplied in FpML format, or indirectly by supplying sensitivity values from external models (e.g., a liability exposure to be hedged).

The additional performance provided by DFEs also enables more complex use cases, such as:

- Leveraging the cloud to allow a pair of counterparties to view the same SIMM calculation before agreeing to post margin
- For a given trade, automatically selecting the counterparty with the most offsetting risk for minimal margin requirements
- Identifying trades for portfolio compression
- Evaluating "what–if" trades and hedging strategies for margin impacts

The Maxeler competitive advantage is basically twofold: (i) performance advantage for complex risk calculations of 10–50× for real-time results before trading and (ii) transparent methodology, with the option of a source code license to participating clients.

15.8 Conclusion

Cutting-edge applications in computational finance depend on highly capable computational systems while scaling over current computing technology is becoming increasingly problematic. Maxeler has pioneered a novel vertically integrated, dataflow oriented approach that can deliver orders-of-magnitude improvement in performance, data center space, and power consumption for a wide range of real highly demanding applications. DFEs provide a highly efficient computational model for data- and compute-intensive parts of an application. In addition, DFE-based systems can be balanced with all other computing system resources, for example, CPUs and storage, according to the specific requirements of the application. Elastic scaling of these systems to the public cloud to cope with peak performance demands is another key advantage to be mentioned. Maxeler supports a high-level programming model that allows application experts to harness the computational power of dataflow systems and focus on optimizing their applications all the way from the formulation of the algorithm down to the design of the best possible dataflow architecture for its solution. This dataflow technology is key to many finance applications where a more complex model, larger data sets, or more frequent recomputation often directly translates into monetisable, competitive advantage. A number of DFE-based products for analytics and trading are available from Maxeler, and we described several practical application scenarios that could not be achieved with conventional, general-purpose computing technology.

References

1. Godfrey, M., Hendry, D.: The computer as von Neumann planned it. *Annals of the History of Computing, IEEE* 15(1), 11–21, 1993.

2. Pell, O., Mencer, O.: Surviving the end of frequency scaling with reconfigurable dataflow computing. *SIGARCH Computer Architecture News* 39(4), 60–65, 2011.

3. Chau, T.C.P., Niu, X., Eele, A., Maciejowski, J., Cheung, P.Y.K., Luk, W.: Mapping adaptive particle filters to heterogeneous reconfigurable systems. *ACM Transactions on Reconfigurable Technology and Systems* 7(4), 36:1–36:17, 2014.

4. Thomas, D.B., Luk, W.: Multiplierless algorithm for multivariate gaussian random number generation in FPGAs. *IEEE Transactions on VLSI Systems* 21(12), 2193–2205, 2013.

5. Lindtjorn, O., Clapp, R., Pell, O., Fu, H., Flynn, M., Mencer, O.: Beyond traditional microprocessors for geoscience high-performance computing applications. *IEEE Micro* 31(2), 41–49, 2011.

6. Weston, S., Spooner, J., Racanière, S., Mencer, O.: Rapid computation of value and risk for derivatives portfolios. *Concurrency and Computation: Practice and Experience* 24(8), 880–894, 2012.

7. Dennis, J.B.: Data flow supercomputers. *Computer* 13(11), 48–56, 1980.

8. Kung, H.T.: Why systolic architectures? *Computer* 15(1), 37–46, 1982.

9. Amazon Web Services: https://aws.amazon.com/ec2/instance-types/f1/

10. Jin, Q., Dong, D., Tse, A.H.T., Chow, G.C.T., Thomas, D.B., Luk, W., Weston, S.: Multi-level customisation framework for curve based Monte Carlo financial simulations. In: *Reconfigurable Computing: Architectures, Tools and Applications—8th International Symposium, ARC*, pp. 187–201. Springer, 2012.

11. Fu, H., Gan, L., Clapp, R.G., Ruan, H., Pell, O., Mencer, O., Flynn, M.J., Huang, X., Yang, G.: Scaling reverse time migration performance through reconfigurable dataflow engines. *IEEE Micro* 34(1), 30–40, 2014.

12. Pell, O., Bower, J., Dimond, R., Mencer, O., Flynn, M.J.: Finite-difference wave propagation modeling on special-purpose dataflow machines. *IEEE Transactions on Parallel Distributed Systems* 24(5), 906–915, 2013.

13. Gan, L., Fu, H., Yang, C., Luk, W., Xue, W., Mencer, O., Huang, X., Yang, G.: A highly-efficient and green data flow engine for solving Euler atmospheric equations. In: *24th International Conference on Field Programmable Logic and Applications, FPL 2014*, Munich, Germany, September 2–4, 2014, pp. 1–6. IEEE, 2014.

14. Arram, J., Luk, W., Jiang, P.: Ramethy: Reconfigurable acceleration of bisulfite sequence alignment. In: *Proceedings of the International Symposium on Field-Programmable Gate Arrays (FPGA)*, pp. 250–259. ACM, 2015.

Chapter 16

Manycore Parallel Computation

John Ashley and Mark Joshi

CONTENTS

16.1 Introduction

In this chapter, we will argue that there is a parallelism imperative: quants must learn to write effective parallel code in order to take advantage of future computing hardware. We provide a grounding in the basic computer science and hardware considerations needed to explore effective parallelism in more depth. These are, fortunately, far simpler than the typical financial mathematics encountered in computational finance. We spend the bulk of the chapter applying parts of these foundational points in working a detailed example of coding a nontrivial early exercise LMM problem on the GPU, which highlights the key issues of writing highly parallel code. The results clearly highlight the parallelism imperative—quants who leverage parallel execution well can gain a significant advantage over competitors who do not.

16.1.1 What a practitioner needs to know

A practitioner of the art of computational finance (more usually known as a quantitative analyst, financial engineer, or quant), once could have counted on a solid knowledge of mathematical finance and a cursory knowledge of programming to be successful. In today's environment, however, very few practitioners will thrive with anything less than solid programming skills. In the future, the ability to efficiently use the compute resources available will be a critical factor in judging the performance of practitioners—and so the perspectives the practitioner needs to adopt will expand, to include that of the computer engineer, the computer scientist, and the computational (traditional supercomputing or technical computing) scientist. The future of computational finance will be written by professionals with a solid grounding in mathematical finance, who also have the complete programming toolkit of the modern HPC specialist at their fingertips.

16.1.2 Outline of the chapter

This chapter is divided into background and practice. It is possible to dive directly into the practice section, but we hope that even those with a strong existing background in parallel computing will still find some benefits in the earlier sections. As background, we first present our argument, in Section 16.2, based on current computer architecture and manufacturing trends, for the existence of a parallelism imperative. We look at computer architecture in

Section 16.3. In Sections 16.4 and 16.5, we start to fill our parallel computing design considerations toolbox. We switch gears in Section 16.6, and begin to work an actual example, which we detail in Section 16.7.

16.2 The Parallelism Imperative

From a computer hardware engineering perspective, parallelism is the future of computing. Power and its always present companion heat currently conspire to limit practical power onto a single chip in commodity hardware to around 300 W. Current materials, design, and manufacturing are beginning to run into the physical limits of the materials and processes as we understand them today. We will soon run into the theoretical physical limits of what can be done with electrons. Engineers have been unable to break through the power and frequency walls; instead they have circumvented them, lowering or holding constant clock rates but adding more parallelism to build newer and better chips. Quants need to understand the trends in the computer hardware industry to design code that will be performant for more than one product generation.

16.2.1 Moore's Law

Moore's Law was coined by Gordon Moore (1965) and was, for many years, taken to mean that because of underlying advances in hardware, software would run twice as fast every 18–24 months. The true point of Moore's Law, however, is that the number of transistors on a single die will roughly double every 18 months. There was a long period of time where both statements were effectively equivalent—but that time has now passed.

Moore's Law was primarily predicated on process improvements in the manufacture of transistors. As the size of the building blocks of transistors shrinks (process shrink), the number of transistors per unit area increases. Additional process improvements have resulted, over time, in our being able to economically produce larger silicon dies. Taken together these two manufacturing improvements have given us Moore's Law.

Quantum effects dictate that there is a lower bound to the size of a transistor. Current public vendor roadmaps run out to features in the 7–10 nm range; the quantum limit is normally quoted as being around 1 nm—this is on the order of the scale of the wavelength of an electron. Process shrink, while not yet over, cannot continue forever.

While transistor counts are still following Moore's Law, the performance of most computer programs is increasing much more modestly—unless those programs can exploit the new levels of parallelism being introduced with each chip generation to keep pace with Moore's Law. Why is that?

16.2.2 Dennard Scaling

Another benefit of smaller feature sizes is that smaller features inherently have lower electrical capacitance. Lower capacitance helps allow higher frequencies at the same power level. Additionally, improvements in materials science and transistor design, coupled with the ability to fit more complex circuits in the same area, meant that chip operating voltages could also be lowered. Again, lower voltage equals lower power dissipation, which can be used to enable higher frequency (Dennard and Gaensslen, 1974).

This Dennard Scaling, coupled with Moore's Law, meant that for a long time, programs just got faster with every new microprocessor generation. Tuning code to be efficient was frequently not required, as it would simply run twice as fast next year anyway.

Unfortunately, physics giveth, and physics taketh away. Materials science, fabrication technology, and transistor design have reached a point where significant voltage reduction isn't possible. This is due to leakage current—no transistor is perfect—and to mainatin correct operations in the face of leakage current requires a certain minimum voltage plus some safety margin. Without voltage scaling, the energy efficiency of the improvements in capacitance caused by a process shrink don't outweigh the increase in energy dissipation per unit area caused by having more transistors. To maintain total overall power usage, frequency needed to be reduced (or at least could not be increased). Code suddenly ran SLOWER on the newest hardware. What could the industry do?

16.2.3 Performance then and now

In the single-core, single-threaded era of computing, the benefits of Moore's Law and Dennard Scaling were generally spent in three ways—increasing frequency, increasing cache sizes, and adding new hardware functions.

Frequency increases had obvious benefits for any code that was spending most of its time executing instructions. This was the rising tide that lifted all boats.

Increasing the number and size of caches was important—as core compute frequencies increased, memory, disk, and IO subsystems couldn't keep up. Caches enabled the CPU to hide some of these growing imbalances in IO and continue to compute.

Adding new hardware functions had dramatic impacts on codes that had previously used software emulation or multiple software steps to perform the same task. Floating point math is an obvious example—originally, there was no HW math acceleration on x86 architecture chips—then the math coprocessor was introduced. This was originally on a separate chip—but Moore's Law meant that those transistors could be moved onto the main CPU die. Today, video encode and decode, cryptographic, and graphics features are all routinely included as new hardware functions.

When frequency could no longer scale, the obvious things to do were to make bigger caches and add more functions. But beyond a certain point, caches give diminishing performance returns for many codes; and unless special function hardware is doing something ubiquitous, the additional benefits of new features apply to smaller and smaller portions of a smaller number of codes—again, diminishing returns.

The answer was twofold: provide more general-purpose cores and by using vector instructions, make those cores "wider" for some math and logic operations. CPUs become dual core, then quad core. SSE and AVX evolved to allow SIMD style operations on adjacent data in a vector (hence vector units). Current CPUs can have a dozen or more cores with vector units up to eight double precision values wide (eight lanes); current GPUs have thousands of cores (which are a bit like a cross between a core and a vector lane). Cache sizes continue to grow but more slowly when measured on a per core basis.

Today, achieving maximum performance on a modern architecture requires careful use of parallelism and caches to ensure that all the cores are being fed data as fast as they can process it. And as every new generation becomes more parallel at a hardware level, for software to see a performance gain means it must be written with more parallelism in mind than what today's hardware is delivering or it will be obsolete in one or two hardware generations.

16.3 Systems Architecture

From a computer system engineer's perspective, the early days of the computing industry were dominated by bespoke hardware; with advances in hardware and software, more homogenous architectures and systems became dominant. Today, driven by the commodity availability of compute accelerators and generally increasing levels of parallelism, the architectures that designers need to be able to exploit are again becoming more varied and more parallel. Quants need to know enough about the organization of the hardware to ensure they are using it effectively.

16.3.1 Building blocks

There are many courses in computer architecture, and many levels of detail can be used to analyze various products. This section presents a simplified set of components that can be used to help us think about performance on multiple architectures.

Logically, we will switch between the viewpoint of a program, a task, and a thread as it suits us. For our purposes, a program, abstractly, is a collection of partially ordered tasks that transform data. A task takes a collection of input data and maps it to output data—for our purposes, a task doesn't share intermediate results unless it does so explicitly (via message or shared memory region). A task can be implemented by one or more threads. We will see that CPUs and GPUs map this logical model to hardware differently.

Threads are executed on a core, which may have one or more vector lanes—which can all perform the same mathematical or logical operation on a single operand of a multioperand vector. Cores may be organized into groups that share some resources (like caches).

We define a number of properties of data storage—proximity to compute resources (latency), rate of data movement (bandwidth), organization of data movement, and size. Typically, modern systems will have layers of storage that increase in latency and size, while decreasing in bandwidth, and sharing a common data movement organization through the majority of the layers. The layer nearest to the cores is usually called the registers and the layer furthest from the cores is called system memory. Layers in between are called caches and are usually distinguished by a number indicating distance in layers from the cores (2–4 layers are typical today). A useful rule of thumb is to expect nearer layers to have an order of magnitude less latency and a factor of 1–5 more bandwidth than further layers. The relationship of size of a layer to location relative to the cores is architecture specific. In most CPUs today it is a strict pyramid, whereas in many GPUs it is an hourglass—there can as much or more memory in the register file than in the nearest cache layer (with lower latency and more bandwidth).

Until it moves into registers, data is usually requested and moves through the system in chunks of memory called cache lines—these are generally multiple operands (many bytes) in size. To simplify the circuitry involved in memory management, cache lines always start on some aligned memory boundary—for example, if a cache line is 128 bytes long, the starting address of all cache lines will be an integer multiple of 128. This "chunkiness" of memory access can be exploited to gain more performance.

Specialized hardware determines if a memory request can be served from a nearby cache (cache hit) or has to come from more distant memory (cache miss). The memory subsystem is coherent (all CPU and some GPU memory) if multiple threads always see consistent values for a single memory address.

Disk and network IO can be thought of as additional layers beyond system memory with multiple orders of magnitude more latency and 1 or 2 orders of magnitude lower bandwidth. They aren't terribly relevant to this discussion.

16.3.2 Basic CPU architecture

A typical Intel CPU can have a dozen or more cores per each chip (socket), and the server will have two or sometimes four sockets. Each core has a small local set of registers (capable of supporting one or two active threads at a time) and a local L1 cache. L2 caches are shared between a subset of the cores, and there is a chip wide shared L3 cache. These caches will be coherent—the memory subsystem ensures consistency across all layers of the cache and between sockets. Each core will have a dedicated set of 2/4 or 4/8 single/double precision vector lanes.

CPUs based on other microarchitectures such as ARM or IBM Power will have different characteristics, but these are currently less commonly used in finance and are out of the scope of this chapter, as are systems based on AMD's x86 chips.

16.3.3 Basic GPU architecture

From a compute perspective, the majority of GPUs deployed are from NVIDIA, and so our discussion will focus on NVIDIA hardware. The details of hardware from AMD and other vendors would be quite different in detail but many of the concepts would remain the same.

Each core has a large local set of registers (capable of supporting 16 or more active threads per core). Each group of 32 cores shares some instruction control hardware. While the number varies by hardware generation, the Maxwell (2014+) generation of NVIDIA processors will have sets of 128 cores sharing a private, scratchpad memory space, and a coherent L1 cache; from one to several dozen of these groups (called Streaming Multiprocessors or SMs) form a single GPU and share L2 cache and system memory. Each GPU card has a global memory area called device memory, usually between 4 and 12 GB in size.

16.4 Parallelism and Performance

A computer scientist's perspective has evolved with computer hardware. Increasing performance meant that some performance could be traded off against other goals like system portability and code maintainability. Eventually, the abstractions and isolation of the programmer from the hardware meant that a significant amount of performance could be regained by writing more performance efficient code. This code can maintain much of the abstraction, portability, and maintainability of existing code but fundamentally needs to be better adapted to the increasing parallelism of systems and ensure that it more effectively uses any hardware features that are present. A quant needs to be able to think about how the organization of code and data will interact with hardware and other layers of software, today and tomorrow, to build durable code that performs well.

16.4.1 Amdahl's Law

Amdahl's Law (Amdahl, 1967) is effectively the law of diminishing returns for parallelization of code. If the parallelizable percentage of the execution time of the code is P, and the total execution time is T_{old}, then if you can infinitely accelerate the parallel portion of the code, or equivalently, drive the execution time of the parallel portion of the code to zero, then the new execution time is given by

$$T_{\text{new}} = (1 - P) * T_{\text{old}},$$

and the maximum achievable parallel speedup is

$$T_{\text{old}}/T_{\text{new}} = 1/(1 - P).$$

This is closely related to the concept of strong scaling, which looks at the ratio of improvement in compute time to the number of processing units engaged in the computation. Normally, there is a point where strong scaling breaks down, as communications and coordination costs overwhelm the actual computation. The strong scaling properties of a code on a system limit our ability to achieve the theoretical Amdahl limit. Combined with Gustafson's Law, below, this is extremely useful in defining exceptions for performance on existing and future hardware.

From a parallel speedup perspective, if a code spends 90 seconds in potentially parallel sections and 30 seconds in serial sections, a 3× speedup of the parallel code reduces execution time from 2 minutes to 1 minute. If in a second phase of the performance tuning the parallel section is now accelerated by 9×, the overall elapsed time goes to 40 seconds. This highlights the law of diminishing returns from a performance perspective.

From a strong scaling perspective, if the problem at hand is to evaluate 32,000 paths, and the parallelism of the problem is at the path level, then adding threads up to 32,000 could result in some performance improvement—any threads/cores added after that have no useful work to do and contribute no additional performance. Strong scaling performance cliffs are not always this dramatic.

16.4.2 Gustafson's Law

Gustafson's Law (Gustafson, 1988) is related to Amdahl's Law, but whereas Amdahl's Law focuses on the limit of acceleration of a problem of constant size, Gustafson's Law addresses how much larger a problem can be solved in the same unit of time via parallelization. This is related to the concept of weak scaling—how much more work can we do with more parallel processing units. Eventually, communications and coordination overhead will overwhelm any increase in efficiency of a code on a system and prevent us from reaching the Gustafson limit.

From a theoretical perspective, evolution of Monte Carlo paths has perfect weak scaling in the number of cores (more cores, more paths) until some other system limit is hit (memory or disk space or bandwidth) that prevents full utilization of additional cores.

16.4.3 Little's Law

Little's Law (Little, 1961) comes from queuing theory. Our interest lies in maximizing our use of bandwidth (or throughput)—if a memory system has a request (cache line) size of C bytes, latency to respond to a request of L seconds, and a usable bandwidth W in B/s, then the number of memory requests outstanding to maximize bandwidth used is WL/C. Relative to CPU memory, GPUs tend to have higher values of W, similar or larger values of C, and longer (higher) latency L—meaning that more threads need to have more outstanding memory requests at any point in time to saturate the memory system. Fortunately, GPUs support many more active threads than CPUs and so this is usually not a problem. This same law can apply to any pipelined compute element.

For example, assume a math pipeline 4 clock cycles deep. If the code only has one math operation to do, it submits it to the math processor at time $t = 0$, and gets a result back at time $t = 3$, using only 1/4 of the available pipeline. If, on the other hand, it can submit math operations at time $t = 0, 1, 2, 3, \ldots, n$ then it will get results at $t = 3, 4, 5, 6, \ldots, n + 4$—getting full usage of the pipeline from $t3, \ldots, n$.

16.4.4 Task parallel

In our definitions above, it is obvious that two programs can be run independently and in parallel—we ignore for now higher level data flow dependencies between jobs. We also define a task in such a way that multiple threads may coordinate to execute the task, with only explicit coordination required between those threads. Any section where multiple threads can execute large sections of code between any required communication or coordination can be thought of as task parallel. This is the most commonly used multicore CPU programming paradigm, but as the number of tasks grows, managing the coordination and communications between threads can become either very complicated or must be hidden behind abstractions that result in lower overall performance. Map and reduce style algorithms, for example, identify task parallel steps and then synchronize all threads expensively on disk (as in Hadoop) or in memory (as in Spark). Traditional HPC codes might use a Message Passing Interface (MPI) to coordinate many hundreds or thousands of threads, but finance code rarely requires such large processor sets to cooperate so closely on a single task. Indeed, many production finance codes are single threaded today.

16.4.5 Instruction parallel

Instruction Level Parallelism (ILP) is very common in today's computing environment, but is rarely directly seen by programmers, as it is usually exploited by compilers and dedicated hardware functions. When the compiler builds an internal representation of the data and instruction dependencies of a section of code, it can identify generated instructions that can be performed in any order so long as they are performed prior to a certain point. This can result in the hardware being able to execute multiple instructions in different units simultaneously or may allow scheduling instructions more tightly into pipelined instruction units.

16.4.6 Data parallel

Data Parallelism can be exploited when the same instruction sequence needs to be applied to a large amount of data with limited or well-structured dependencies between output data elements. For example, calculating the sum of two long vectors, the product of two dense matrices, or the membership function of the Mandelbrot set all exhibit data parallelism. The additional structure of a data parallel problem can be exploited, allowing many more processing units to contribute to the work with a reduction in communication and coordination overheads. Pure data parallelism leads to "embarrassingly parallel" problems which generally have good strong and weak scaling properties. In finance, Monte Carlo codes are the poster child for embarrassing parallelism, but PDE and even adjoint methods can be implemented in a highly data parallel way.

16.5 Parallelism and Execution

In the traditional supercomputing space, the computational scientist has an evolving role that sits between a pure domain scientist and a pure programmer. The computational scientist's perspective includes an understanding of how the discrete mathematics of computers interacts with the pure mathematics of models, and also understanding how algorithms and code are mapped to hardware and executed. Both these perspectives are used to tune programs to study some of the largest scientific and engineering questions. A quant needs to understand issues of algorithmic and discrete mathematical accuracy, stability, and performance, just like a computational scientist.

16.5.1 Logical threading models

The logical threading model on the CPU is likely familiar to readers—but we present a broad summary here in order to contrast it with the model on a GPU. A CPU thread may be independent or may be part of a group of several threads under a single process. Generally, a CPU thread has a relatively large startup and teardown cost. It can be actively resident on a core in the register set, or it can be swapped out, with its state stored in main memory. Swapping in and out from main memory is expensive, swapping between threads resident in the register file is fast. Because of the limit on the number of resident threads, the compiler and the hardware need to ensure that a given thread completes as much work as possible, and keeps the CPU core as busy as possible, while it is active. Logically, threads have great flexibility in passing messages and sharing memory, with a few more options for threads within a given process than for threads between processes. CPU threads thrive on task parallelism, have dedicated hardware to take advantage of ILP, and if written to take advantage of the vector units, have some ability to leverage them for data parallelism as well. Highly performant code must take advantage of all three types of parallelism.

A logical thread on a GPU is quite different. A GPU thread is always part of a hierarchy that includes a local thread block and a set of thread blocks called a grid (which in our definitions here is a task). A GPU thread is a very lightweight construct, with the majority of startup and teardown managed in hardware. GPU threads have three main states—waiting to be active, active and in the register set of a core, and completed. All the threads in a local thread block are in the same main state; and threads within an active thread block can communicate and synchronize through a variety of mechanisms. Threads that are not in the same thread block can only communicate through the main device memory. GPUs excel at data parallelism, although they can take limited advantage of ILP and, particularly with more recent GPUs, can exploit task parallelism as well. The logical threading model for the GPU does not allow the programmer to assume or force any particular ordering for the execution of thread blocks.

Whereas efficient CPU code will normally have one or two active threads per core, efficient GPU code will have many active threads per core—and with many more cores. Cooperative CPU codes may use dozens of threads, but GPUs will not function efficiently without thousands of threads, and routinely are used with millions of logical threads. This is very much due to differences in the complexity of the large CPU cores versus the smaller but more numerous GPU cores (bringing Amdahl's and Gustafson's Laws into play) and the latency and bandwidth of the memory subsystems (bringing in Little's Law).

16.5.2 Physical execution models

On a CPU, the physical execution model of a thread is familiar—a process is launched, it spawns whatever threads it needs, and each thread starts to do its work. It swaps in and out of the core, and may execute on many different cores over its lifetime (unless limited by the operating system). It will swap out whenever it is bumped by another thread that is now entitled to a timeslice on the core either by virtue of priority or elapsed time. Hardware and software will aggressively attempt to use branch prediction, speculative execution, and cache prefetching to avoid allowing idle time while the thread is resident on a core.

When a core needs data that isn't already in the local register file, it will normally begin a cascade of checks down the cache hierarchy until the memory management subsystem finds the data it is looking for; this is then hoisted through the various cache layers into the local register file where it can be used as an operand in instructions. Writes that spill out of the register file or are directed at memory proceed to invalidate caches in other cores until they reach system memory, causing cache synchronization traffic.

On a GPU, things are less familiar but in many ways simpler. When a task (kernel) is launched onto the GPU by a CPU side thread, a grid of GPU threads is created. The threads are grouped into thread blocks, and each thread block is independently scheduled onto a single SM. Depending on the hardware and resources required, several thread blocks may be resident on an SM at a given time and many SMs can be servicing thread blocks from a grid at the same time. Once a thread block is active, it will remain active on that SM until all its threads have exited, having processed their section of the overall dataset. As a programmer, you have no control over or ability to assume what order thread blocks will execute in nor can you communicate, coordinate, or synchronize with other thread blocks except through global device memory.

Memory management on the GPU presents several more options that the programmer can consider. Reads and writes to and from a device global memory behave similarly to CPU reads and writes. Alternatively, however, a thread block can declare a small area of memory to be a shared memory. This creates a private allocation of memory, physically drawn from the same pool as the L1 cache, that is visible to all the threads in the thread block and to no threads outside that thread block. Because this data is very near to the cores in the memory hierarchy, it is extremely fast. And because it is not coherent with any other memory in the system, it pays no synchronization or more distant memory management costs. It is effectively a private working space for the threads in a thread block to share data.

Drawing on their graphics heritage, GPUs also have additional paths from global device memory to the cores that bypass some or all of the conventional caches—the most commonly used are the constant cache and texture cache. The constant cache is now well leveraged by the compiler and, as its name suggests, is well suited to delivering a very small number of values to all the threads in the SM. The texture cache is another path to memory that has slightly different caching behavior from the standard caches. The usefulness of the texture cache to the programmer changes from generation to generation on the GPU. Some algorithms and data access patterns can make good use of the texture cache.

16.5.3 Data structures

Earlier in Section 16.5 we discussed cache lines and alignment, and noted that there are potential performance advantages to be gained by being aware of these points. On a CPU, these considerations are most important when using the vector units but also can help performance in general; on a GPU they are critical to gaining maximum performance because of the data parallel nature of the device.

There are two interesting and common cases where this can be observed. In the first case, a container or array of objects or structures is stored. In a simple example, assume a trade has fields of symbol (4 bytes), trade date (8 bytes), trade time (8 bytes), exchange (2 bytes), quantity (4 bytes), price (8 bytes), fees (8 bytes)—a total of 42 bytes. Assume we have a large number of trades, and a cache line size of 64 bytes. We need to get the net total position and value by symbol by day, and we assign each thread to a symbol. The first thread requests the first record, and assuming the array of structures started on a cache line boundary (is 64 bytes aligned), it gets the first record plus 22 bytes of the second record delivered to it. Of the bytes delivered in the first cache line, it uses the symbol, trade date, quantity, price, and fees—32 bytes. If 64 bytes were delivered and only 32 bytes were used, the effective bandwidth is 1/2 what could have been achieved with a more efficient data structure and assignment of work to threads.

If there were many threads in parallel sharing the data request, as there are in the GPU, then the second record's bytes would be partially used and the efficiency might rise to 32/42 on average. Arranging the data as a structure of arrays—an array of symbols, then an array of trade dates, and so forth—means that in the case where there are parallel threads, the unused fields are never read and every byte moved is used by a thread—maximally efficient.

Consider another case: two- or higher dimensional arrays are commonly used in financial mathematics. For simplicity, assume we have a 2D path array, V, indexed by path and time—$V[p, t]$. If we have paths A, B, C and times 0, 1, 2, in C compatible languages these are stored in row major order in memory. So in memory, we would see A0, A1, A2, B0, B1, B2, C0, C1, C2. If each element is 16 bytes and the cache line size is 64 bytes, then the first cache line contains A0, A1, A2, B0, the second contains B1, B2, C0, C1, and the third contains C2, junk, junk, junk.

If each thread is processing one path, for instance in path generation, then the first thread needs the first cache line, the second thread needs the first and second cache lines, and the third thread needs the second and third cache lines—3 threads generate 5 requests to memory and only use 0.75, 0.25, 0.5, 0.5, 0.25 of the returned data.

If each thread is processing a time step, for instance doing a potential future exposure calculation, then the first thread needs the first and second cache lines, the second thread needs the first and second cache lines, and the third thread needs the first, second, and third cache lines. This is 7 requests to memory, using 0.5, 0.25, 0.25, 0.5, 0.25, 0.25, 0.25 of the returned data.

With array padding, rather than declaring $V[3, 3]$, you could declare it to be $V[3, 3 + 1]$. What happens then? We get memory of A0, A1, A2, pad, B0, B1, B2, pad, C0, C1, C2, pad. From our first example, each thread requests a cache line and uses 3/4 of it. Effective BW goes from 0.45 to 0.75 of peak—a huge improvement. In the second example, each thread needs all three cache lines—9 requests to memory, with an efficiency of 0.25 as opposed to 0.32, which is only slightly worse.

Good data structure design needs to balance usability and performance. As these toy examples show, data layout can have a huge impact on how efficient a code can be on any given hardware.

16.6 Putting It All Together

The advent of multicore graphics cards places the possibility of a huge amount of computing power in the desktop or server room at a reasonably low cost. Such cards are known as graphics processing units or GPUs. However, they can be used for many purposes other than graphics. In particular, they are very naturally adapted to Monte Carlo simulations since their design is naturally highly parallel in nature. They work best when threads are performing the same operations with varying initial data (data parallelism).

To gain some idea of the level of computing power, consider one of NVIDIA's recent GPUs, the Tesla K20. This has a peak performance of 3.52 teraflops, that is, it can perform 3.52 trillion single precision floating point operations per second. However, possession of raw computing power is not enough. Different architectures require different coding designs and the question is therefore what level of performance can one achieve for realistic practical problems? For example, Aldrich, Fernandez-Villaverde, Gallant, and Rubio-Ramirez (2011) demonstrate the effectiveness of GPUs for solving dynamic equilibrium problems in economics using iterative methods.

The pricing of nonearly exercisable derivatives using GPUs is straightforward and large speedups can be obtained. The case of Asian options was studied by Joshi (2010). An early piece of work on pricing exotic interest derivatives using the LIBOR market model was produced by Giles and Xiaoke (2008) where a 120 times speedup was achieved. Whilst this was an interesting pilot, the case studied was a little simple in that a one-factor model was used.

The questions we address here are:

- Can a complex multifactor displaced diffusion LIBOR market model be implemented in such a way as to achieve large speedups whilst maintaining genericity?
- Is it possible to implement an effective and fast Monte Carlo pricer for early exercisable Bermudan derivatives using the GPU?

We answer both questions in the affirmative and achieve over a hundred times speedups over single-threaded CPU C++ code. The second problem is much tougher than the first in that the pricing of early exercisable derivatives via Monte Carlo simulation is much more complex than pricing path-dependent derivatives lacking this feature. The fundamental reason is that an exercise strategy has to be developed and the development of the strategy requires interaction between paths. Thus one cannot set many threads going, let each one handle a different path and accumulate results at the end. Unlike many Monte Carlo simulation problems, the method is not "embarrassingly parallel," and it is a challenge to design an effective implementation.

The problem of pricing Bermudan (or American) derivatives by Monte Carlo simulation is well known and used to be regarded as very hard. However, much progress has been made in recent years. We focus on lower bounds in this example. The use of regression to develop estimates of continuation values and therefore to decide exercise decisions has, in particular, proven popular. This was introduced by Carrière (1996) and popularized by Longstaff and Schwartz (2001). Whilst these methods work reasonably well for finding lower bounds in many cases, they are not always successful. Continued work has therefore focussed on enhancing these techniques (Beveridge, Joshi, and Tang, 2013; Broadie and Cao, 2008; Kolodko and Schoenmakers, 2006). Whilst these improved techniques have proven effective, they are very much designed for the virtues and constraints of a CPU. For example, early termination of a subsimulation if an inaccurate number is sufficient makes sense on the CPU, but running lots of subsimulations with differing numbers of paths is unnatural on a GPU. We therefore introduce a new approach to early exercise based on using a cascade of multiple regressions. This builds on previous work (Broadie and Cao, 2008) on double regressions.

In this portion of this chapter, we present and discuss solutions to the challenges of pricing early exercisable exotic interest derivatives on the GPU using NVIDIA's CUDA language. In particular, we discuss in detail the design choices made in the open-source project Kooderive which contains an example of the pricing of a 40-rate cancellable swap using a five-factor displaced diffusion LIBOR market model. The code is fully available for download under the GNU public licence 3.0 (Joshi, 2014). The project is a collection of C++ and CUDA code targeting the Kepler 3.5 architecture and the K20c NVIDIA graphics card. It will, however, run on other cards with earlier or later architectures. It is designed for use with Windows operating systems and comes with project files for Visual Studio which allows immediate building.

Whilst we focus on the LIBOR market model (LMM), we emphasize that the approaches used are generic and they could equally well be applied to other models. In particular, the multidimensional Black–Scholes model is from the point of view of implementation really just a simplification of the LMM in which drifts and discounting are much easier. The implementation of the early exercise code is done in generic fashion and does not use specific features of the model and product. In fact, a key part of the design is that the early exercise strategy is generated in a different component that only interacts with the path generator via the paths produced and so will be very broadly applicable.

We focus here on cancellable swaps because their pricing is mathematically equivalent to that of Bermudan swaptions and callable fixed rate bonds. These products are the most common exotic interest rate derivatives. However, no special features of the product are used and other products can be handled by modifying the functions defining the coupons with no changes elsewhere.

We note that there has been previous work on the use of GPUs for Bermudan/American options. Dang, Christara, and Jackson (2010) proceed using a PDE approach. Abbas-Turki and Lapeyre (2009) use a least-squares Monte Carlo approach. However, the case they study is four-dimensional and indicative rather than realistic so it is difficult to know how their techniques would translate to the high-dimensional interest rate case. Most other work appears to be focussed on binomial trees and/or the one-dimensional case.

In this example, we focus on lower bounds. However, another part of the Bermudan pricing problem is upper bounds: one cannot be sure a lower bound is good without a nearby upper bound. To do a regression-based method such as that in Joshi and Tang (2014) would not be a particularly hard extension of the work here. First, one would have to compute Deltas along each path using adjoint differentiation techniques. Second, one would have to use similar techniques to those here to compute regression estimates of their value. Third, one would run a hedging simulation using these estimates. To do a method such as Andersen and Broadie (2004) would be more challenging since it involves running many subsimulations. One could put each subsimulation on the GPU as a separate kernel; however, that would result in a very large number of kernel launches and possibly not much of a relative speedup unless large numbers of paths were being used. We defer the problem of designing a smarter implementation to future work.

We review the LIBOR market model in Section 16.7. We discuss its algorithmic implementation in a discretized setting in Section 16.7.1. We develop new ideas for early exercise in Section 16.7.2. We discuss the software and hardware used, in Section 16.7.3. We outline the design of the code in Section 16.7.4. The intricacies of memory use are examined in Section 16.7.5. We study how to evolve the LIBOR market model on the GPU in Section 16.7.6. The specification of products is done in Section 16.7.7. We go into the implementation details of regression on the GPU in Section 16.7.8. Section 16.7.9 covers methodologies for data collection to prepare for least-squares. Pricing is discussed in Section 16.7.10. We present timings and numerical results in Section 16.7.11 and we conclude in Section 16.8.

Mark would like to thank Oh Kang Kwon for his assistance with coding a Brownian bridge and with a skipping Sobol generator. Mark is also grateful to Jacques Du Toit for his comments on an earlier version of this work.

16.7 The LIBOR Market Model

In this section, we briefly review the displaced diffusion LIBOR market model. This is standard material. We refer the reader to standard texts such as Andersen and Piterbarg (2010), Brace (2007), Brigo and Mercurio (2006), and Joshi (2011) for more details.

Since it was given a firm theoretical base in the fundamental papers by Brace, Gatarek, and Musiela (1997), Musiela and Rutkowski (1997), and Jamshidian (1997), the LIBOR market model has become a very popular method for pricing interest rate derivatives. It is based on the idea of evolving the yield curve directly through a set of discrete market observable forward rates, rather than indirectly through the use of a single nonobservable quantity which is assumed to drive the yield curve.

Suppose we have a set of tenor dates, $0 = T_0 < T_1 < \cdots < T_{n+1}$, with corresponding forward rates $f_0, \ldots, f_n$. Let $\delta_j = T_{j+1} - T_j$, and let $P(t, T)$ denote the price at time t of a zero-coupon bond paying one at its maturity, T. Using no-arbitrage arguments,

$$f_j(t) = \frac{\frac{P(t,T_j)}{P(t,T_{j+1})} - 1}{\delta_j},$$

where $f_j(t)$ is said to reset at time T_j, after which point it is assumed that it does not change in value. We work solely in the *spot LIBOR measure*, which corresponds to using the discretely compounded money market account as numeraire, because this has certain practical advantages (Joshi, 2003a). This numeraire is made up of an initial portfolio of one zero-coupon bond expiring at time T_1, with the proceeds received when each bond expires being reinvested in bonds expiring at the next tenor date, up until T_n. More formally, the value of the numeraire portfolio at time t will be

$$N(t) = P\left(t, T_{\eta(t)}\right) \prod_{i=1}^{\eta(t)-1} (1 + \delta_i f_i(T_i)),$$

where $\eta(t)$ is the unique integer satisfying

$$T_{\eta(t)-1} \leq t < T_{\eta(t)},$$

and thus gives the index of the next forward rate to reset.

Under the displaced diffusion LIBOR market model, the forward rates that make up the state variables of the model are assumed to be driven by the following process:

$$\mathrm{d}f_i(t) = \mu_i(f, t)(f_i(t) + \alpha_i)\,\mathrm{d}t + \sigma_i(t)(f_i(t) + \alpha_i)\,\mathrm{d}W_i(t), \tag{16.1}$$

where $\sigma_i(t)$'s are deterministic functions of time, α_i's are constant displacement coefficients, W_i's are standard Brownian motions under the spot LIBOR martingale measure, and $\mu_i' s$ are uniquely determined by no-arbitrage requirements. It is assumed that W_i and W_j have correlation $\rho_{i,j}$ and throughout $\{F_t\}_{t \geq 0}$ will be used to denote the filtration generated by the driving Brownian motions. In addition, all expectations will be taken in the spot LIBOR probability measure. The requirement that the discounted price processes of the fundamental tradable assets, that is the zero-coupon bonds associated to each tenor date, be martingales in the pricing measure, dictates that the drift term is uniquely given by

$$\mu_i(f,t) = \sum_{j=\eta(t)}^{i} \frac{(f_j(t)+\alpha_j)\delta_j}{1+f_j(t)\delta_j}\sigma_i(t)\sigma_j(t)\rho_{i,j};$$

see Brigo and Mercurio (2001).

Displaced diffusion is used as a simple way to allow for the skews seen in implied caplet volatilities that have long persisted in interest rate markets (Joshi, 2003a). In particular, the use of displaced diffusion allows for the wealth of results concerning calibrating and evolving rates in the standard LIBOR market model to be carried over with only minor changes. The model presented collapses to the standard LIBOR market model when $\alpha_i = 0$ for all values of i.

16.7.1 The LMM in discrete time

In a computer program, we discretize time into a number of steps. Each step is typically from one reset date to the next. Thus it is a time-discretized version of the LMM that is important when implementing. Our philosophy as expounded in Joshi (2011) is that calibration should be done post discretization. We thus have a finite strictly increasing sequence of reset and payment times for LIBOR rates T_j, and a similar sequence of evolution times as inputs to our calibration. We will take these to be equal in what follows for simplicity, although this is certainly not a necessity. We assume N time steps and an F-factor model.

The effective calibration of the LIBOR market model is the subject of many papers. We will not address it here but instead will assume that a CPU routine has already produced a calibration which we use as an input for our pricing routine. The output of our calibrator is the following:

- The initial value of each forward rate, $f_r(0)$
- The displacement for each of these rates, α_r
- The pseudo-square root, A_{j-1}, of the covariance matrix of the log forward rates for each time step from T_{j-1} to T_j

Note that one could equivalently specify the covariance matrix, C_j, for the time step instead of the pseudo-square root. The pseudo-square root uniquely determines the covariance matrix, of course, and it is the covariance matrix that determines the drifts. However, since we are working with reduced-factor models, a pseudo-square root is a more natural object. See Joshi (2011) for discussion of this approach to calibration. Plus when working with low discrepancy numbers, it is generally believed that working with a spectral pseudo-square root can improve convergence (Giles, Kuo, Sloan, and Waterhouse, 2008), Jäckel (2001), so it is convenient to specify this explicitly.

We use log rates $x_r = \log(f_r + \alpha_r)$ and the rates f_r as convenient. We need to compute drifts; these are state dependent so only the drift at the start of the first step is known in advance. We use a predictor–corrector algorithm so they have to be computed twice per step. The discretized drift of a log forward rate f_r across step j is

$$\mu_{j,r} = -0.5C_{j,rr} + \sum_{l=0}^{r} C_{j,rl} \frac{(f_r(t) + \alpha_r)\delta_r}{1 + f_r(t)\delta_r},$$

where $t = T_{j-1}$ when predicting and T_j when correcting. Whilst this expression is correct, it is inefficient from a computational perspective and an algorithm giving the same numbers with lower computational order is presented in Joshi (2003b) and we use it here.

Our evolution algorithm for the rates on each path is therefore as follows:

1. Draw uncorrelated NF standard normals for a quasi-random generator.
2. Use a Brownian bridge to develop these into F Brownian motion paths.
3. Take the successive difference of these paths to get a vector, Z_j, of F standard normals for each step j.
4. Compute drifts $\mu_{j,r}$ for step j using $(f_r(T_{j-1}))$. (For the initial step, use the stored values instead.)
5. Let
$$(\hat{x}_r(T_j)) = (x_r(T_{j-1})) + A_j Z_j + (\mu_{j,r}).$$
6. Compute drifts $\hat{\mu}_{j,r}$ for step j using $e^{\hat{x}_r(T_j)} - \alpha_j$.
7. For each r, let
$$x_r(T_j) = (\hat{x}_r(T_j)) + \frac{1}{2}(\hat{\mu}_{j,r} - \mu_{j,r}).$$
8. Let $f_r(T_j) = \exp(x_r(T_j)) - \alpha_r$.
9. Unless at end of path go back to 3.

Once the forward rate path has been developed, other ancillary quantities such as discount factors are easy to compute.

16.7.2 Multiple regression

In this section, we discuss a new algorithm for developing the exercise strategy. This algorithm is designed to be simple but requires a large number of paths, making it suited to GPU programming. We first recall the least-squares method for cancellable products (Amin, 2003; Carrière, 1996; Longstaff and Schwartz, 2001). The method generally works in three phases. In the first phase, a set of paths is generated. In the second phase, regression coefficients are estimated, and in the third these are used to develop a lower bound price. We refer the reader to Joshi (2011) for further discussion.

The main choice in the algorithm regards which basis functions are used to regress continuation values against, and this can have a great effect on results (Beveridge, Joshi, and Tang, 2013; Brace, 2007). Here we make the distinction between basis functions and *basis variables.* We make the former polynomials in the latter. The crucial point is that the functions are easily generated from the variables. A variable would typically be a stock price, a forward rate, a swap rate, or a discount bond. We do not investigate the question of basis function choice here but instead refer the reader to the extensive analysis in Beveridge, Joshi, and Tang (2013).

In the second phase, a backwards inductive algorithm is used. First, at the final exercise time, the remaining discounted cash flows for each path are regressed against the basis function values for that point on that path. The regression coefficients then yield an estimate of the continuation value. This estimate is compared to the exercise value for the path and the value of the product at this final exercise time is set to the exercise value if it is greater, and to the discounted value of the remaining cash flows if it is not.

We then step back to the previous exercise time. We discount the value at the succeeding exercise time and add on the discounted value of any cash flows that occur in between on each path. We then regress again and repeat all the way back to zero.

Whilst in simple cases, the method works well, in complicated ones it can do unacceptably poorly or be highly basis function dependent. Many techniques have been developed for improving the method, for example, Beveridge and Joshi (2008), Kolodko and Schoenmakers (2006), Broadie and Cao (2008), and Beveridge, Joshi, and Tang (2013). However, their methodologies tend to be better suited to CPU codes. For example, the use of sub-simulations with varying numbers of paths appears difficult to code effectively on the GPU.

However, we have the advantage that using large numbers of paths is practical. We therefore adapt and extend an idea from Broadie and Cao (2008) and Beveridge, Joshi, and Tang (2013). They suggested using double regression: perform a least-squares regression and then perform a second regression for paths for which the absolute difference between the estimated continuation value and the exercise value is below some threshold such as 3%. The idea is that the second regression yields greater accuracy in the area where it is most needed. However, the approach does implicitly require that the first regression be sufficiently accurate that the exercise boundary is within this truncated domain.

We adapt this approach by picking a fraction $\theta \in (0, 1)$ and a regression depth d equal to say 5. We regress using least-squares, discard the fraction

$1 - \theta$ of paths farthest from the estimated boundary, and then repeat. We do this until we reach the depth d. If we initially have N_1 paths, we finish with

$$N_d = N_1 \theta^d$$

paths. Typically, we would only require the proportion discarded to be approximately correct rather than exact for efficiency. The value θ would be chosen to make N_d of the size required. For example, we might let

$$\theta = 0.1^{1/5}$$

when using 327,680 paths so that the final regression has roughly 32,768 paths. The advantage of this approach is that by only discarding a small fraction of paths that are very far from the money at each stage, we are less likely to be affected by a substantial misestimation of the boundary.

Many authors use second-order polynomials in forward rates and swap rates as basis functions following Piterbarg (2004). We will take the first forward rate, the adjoining coterminal swap rate and the final discount factor as our basis variables. The basis functions are then quadratic polynomials in these with or without cross-terms.

16.7.3 Packages and hardware

In this section, we briefly describe the software and hardware used. The principal software tool is the CUDA 5.5 toolkit available from NVIDIA for free download. This is used in conjunction with Visual Studio Professional Edition 2012 as IDE and as a C++ complier. The Thrust open-source library for developing algorithms on the GPU is used extensively both for algorithms and memory allocation. The Thrust library now ships as part of the CUDA toolkit.

The CUDA code discussed in this chapter is all part of the Kooderive open-source library. Project files for Visual Studio are included and allow immediate building after download. This has a number of components. In particular, "gold" projects are written in C++ and run on the CPU. The main static library project is "kooderive" and the examples discussed here are in the project "kooexample."

The hardware used as a GPU is a single K20c Tesla card.[1] Its programming is purely done using the CUDA language. The compute capability of the device is 3.5 and the code assumes that such a GPU is available. The CUDA code is written in 64 bits.

The CPU used is an Intel(R) Xeon (R) CPU E5-2643 at 3.30 GHz. We use routines from the QuantLib open-source library as a comparison. This is compiled using Visual C++ in 32-bit mode since that is the configuration specified by the current release.

---

[1] We thank NVIDIA for providing this hardware.

16.7.4 Design overview

The pricing of a Bermudan contract by Monte Carlo can be divided into three phases:

1. A number of paths, N_1, are generated and the relevant aspects of these paths for developing an exercise strategy are stored.

2. A backwards induction is performed, generating regression coefficients and updating continuation values at each exercise date.

3. A second Monte Carlo simulation is run using the exercise strategy generated in the second phase using N_2 paths. This is equivalent to pricing a path-dependent derivative with no optionality since the exercise strategy has been fixed.

It is the first phase that requires most care. The reason being that the data generated has to be stored until the end of the second phase. Thus we will require memory proportional to N_1. For our design, we keep all this data on the GPU at all times and so there must be sufficient memory to store it. If the GPU's total global memory is M and we store m bytes per path, we have the immediate constraint

$$N_1 m < M.$$

A Tesla K20c has 4800 megabytes of global memory so if we run 327,680 paths, the maximum storage per path is 15,360 bytes. In practice, since other data must be stored the maximum would be lower. A float takes 4 bytes so we have storage for less than 3840 floats per path.

If our forward rate evolution has N rates and n steps, to store the entire evolution for a path will take Nn rates. If we take $N = n$, we will run out of memory for some $N < 62$. Whilst one could squeeze some more memory by discarding already reset rates, it would complicate accesses and there would be very little left for other computations. In practice, one will often want to develop many pieces of auxiliary data for all paths simultaneously, such as all the implied discount factors, which multiplies the memory requirements.

We therefore adopt an approach based on batching. Thus rather than storing everything about 327,680 paths, we divide into say 10 batches and only store the aspects of the paths that are required for the backwards induction. So what must be stored? First, we note that we only need data at exercise times. We use "exercise step" to mean the step from one exercise date to the next. This may or may not be the same as the step from one reset date to the next.

- The sum of the discounted values of any cash flows generated by the product during the exercise step. This yields 1 float per exercise step.

- The value of the numeraire at the start of the exercise step again yields 1 float.

- The discounted value of any rebate generated on exercise gives 1 more float.
- The basis variables for the exercise time are an input according to choices of basis functions but typically 3 is enough.

Here "discounted" means discounted to the start of the exercise step. We therefore typically have 6 data points per exercise step per path. If we have 40 exercise dates, and 327,680 paths, this requires 300 MB, leaving plenty of space for more dates, paths, or other needs.

Our second step processes the data from the first step and outputs regression coefficients. An interesting feature of the algorithm is that the second step uses no specific features of the model or product other than these outputs. This means that it is flexible and generic. Although developed for cancellable swaps in the LIBOR market model, it could equally well be used for Bermudan max options in a multidimensional Black–Scholes model or, indeed, for any Bermudan derivatives pricing problem priced using martingale techniques.

For the third phase, we templatize the cash-flow generation on the exercise strategy using the coefficients produced during the second phase. This means that the exercise strategy is simply an input that could be changed drastically without changing this phase's design. Thus one could use a parametric strategy instead of a least-squares one whilst only making changes to the second step. Note that the third phase is again batched, and we only need to store a single number from the output of each batch: the mean value of the product. In consequence, there are no memory constraints on how many batches we use for this phase. Thus our only constraint on the size of N_2 is time.

In all phases, we divide the algorithm into a sequence of steps. Each step is performed on all paths in a batch by a single GPU routine. Thus almost everything is done for all these paths before any of them are complete. This is quite different from a typical CPU program such as QuantLib where the first path is complete before anything is done for the second path. Each of these small steps in Kooderive will generally correspond to a single call to the GPU.

16.7.5 Memory use, threads, and blocks

A CUDA program consists of a C++ or C program together with various calls to *kernels* which run on the GPU. Each kernel is configured as a number of blocks (e.g., 64) and each block has the same number of threads (e.g., 512). Threads in a block are divided into groups of 32 called warps. Threads in the same warp are constrained to perform the same instructions, and if the program requires them to do otherwise, idling of some threads occurs as the branches are evaluated serially.

A significant difference between GPU programming and CPU programming is the importance of how memory is used. A GPU program has to explicitly use many different sorts of memory and how this is done can have drastic effects on the speed of a program. The principal sorts of memory are (with the amount on a K20c in parenthesis):

- Host—the computer's ordinary memory that the CPU uses
- Global—the graphics card's main memory (4800 MB), large and plentiful but must be accessed correctly or slowness occurs
- Shared—a small amount of memory shared between threads in the same block (49,352 bytes), very fast but amount is very limited
- Constant—read-only for the GPU but writable by the CPU (65,536 bytes), again fast but use is very constrained
- Textures—a way of placing global memory in a read-only cache, fast for read without constraints, but the memory cannot be changed within a kernel

Memory transfer between host and global memory is typically slow for large data sets and is often the main bottleneck in GPU programs. Kooderive avoids this issue by simply not using host memory after the set-up phase. Thus the model calibration and product specifications are passed to the GPU initially but the only data passed back thereafter is the mean values for paths.

The layout of how data is stored in global memory greatly affects speed. This is a consequence of the fact that threads do not access global memory independently. Each thread has a thread number, t, and a block number, b. We will call the total of number of threads the width, w. The number of threads per block we denote s for size. If the code is written so that in a warp, thread t accesses location $l + t$ for some l, then the access is *coalesced* and occurs quickly. However, if the mapping is more complicated and each thread accesses $f(t)$ for some nontrivial function f such as $f(t) = l + \alpha t$ for some $\alpha > 1$, memory access is slower because more cache lines must be fetched to service all the memory requests. A common approach throughout Kooderive is that thread t in block b is responsible for the path

$$t + bs + kw,$$

for all k such that $t + bs + kw$ is less than the total number of data points (typically the paths in a batch). Data is then stored with the path in the smallest dimension, the time step in the largest dimension, and any other index in the middle. Thus if there are R rates, N steps, and P paths, forward rate r on time step s for path p would be stored at location

$$p + rP + sRP.$$

With this layout, coalescing occurs naturally. This is in contrast to a typical CPU program where all the data for each path would be stored together, and one might use location

$$r + sR + pRN.$$

Coalescing is more important for writes than reads, since textures provide a fast route to memory access provided there are no writes to that part of memory during the kernel. Thus for pieces of data that do not change during the kernel, textures are widely used in Kooderive to speed up memory access and to avoid coalescing constraints. The textures provide a cache that also speeds up access. Note that this cache can be explicitly accessed using the $\_\_ldg()$ function in the Kepler architecture and this is also sometimes done. An alternate approach, used in Kooderive, is to copy constant data into shared memory at the start of a kernel. However, this relies on the amount of data being small and we therefore use this technique only a little. Whilst constant memory provides an additional alternative, the advantages do not seem sufficient to justify its unwieldiness and it is not used in Kooderive.

Shared memory is also useful as a fast workspace and this is done by the main path generation kernel.

16.7.6 Path generation

For both the first and third phases, we want to develop large numbers of LMM paths rapidly. We develop the paths in batches of say 32,768 in size. The batch creation is divided into a number of kernels:

- Generation of Sobol numbers as integers
- Scrambling
- Conversion to normals
- Brownian bridging
- Path generation

The generation of Sobol numbers is done using a modification of the example in the CUDA SDK. The main difference being a skip method has been added to allow the Sobol sequence to start at an arbitrary point, and the return of unsigned integers rather than floats or doubles. The first facilitates batching, each batch simply skips to the end of the previous batch. The passing back of unsigned integers is to allow scrambling. Here we input a fixed vector of unsigned integers for the batch and apply exclusively or to each vector of Sobol draws. If the scrambling vector is drawn randomly, we can view the batch average of quantities as an unbiased estimate of any expectations. This is similar to randomized QMC (Giles, Kuo, Sloan, and Waterhouse, 2008).

To transform to normals, we use the Shaw–Brickman algorithm (Shaw and Brickman, 2009). This is performed using the "transform" algorithm from the Thrust library. We use one call to do both the conversion to uniforms from unsigned ints and to take the inverse cumulative normal.

The Brownian bridge is performed using a two-dimensional grid of blocks. This reflects the fact that the Sobol paths will be of dimension NF with N number of steps and F number of factors. We effectively have to create F paths of N steps from these for each overall path. The x-coordinate of the block and the Thread Id determine overall path. The y-coordinate determines the factor. For each block, we first copy all the auxiliary data needed for the bridge into shared memory. Each thread then develops the bridge for one factor for one path without further interaction with other threads. If the total number of threads across all blocks is less than the total number of paths, then a thread will do multiple paths each separated by the global width. The paths generated by each thread are successively differenced at the end so that the outputs would be independent standard $N(0, 1)$ random variables if the inputs were. We refer the reader who is interested in high performing Brownian bridges to du Toit (2011) for discussion of an alternative approach.

The main kernel that does the forward rate evolution and computes the implied discount factors is:

LMM_evolver_all_steps_pc_kernel_devicelog_discounts.

This implements the Hunter, Jäckel, and Joshi (2001) algorithm for predictor–corrector evolutions. A large fraction (roughly 70%) of the total compute time for phases 1 and 3 is spent in this kernel, and so its efficiency is very important. We therefore discuss it in detail. Its inputs are:

- The pseudo-square roots of the covariance matrices (texture)
- The accruals of the forward rates (texture)
- The displacements (texture)
- The fixed part of the drifts (texture)
- The value of the state-dependent drifts for the first step (texture)
- Indices that denote which rates are not yet reset for each step (texture)
- The initial logs of the rates (texture)
- The quasi-random variates (texture)
- The numbers of paths, rates, and steps (integers)

So all data is passed in as either a texture or an integer. It outputs the full forward rate evolutions, the log displaced rates, and the implied discount factors which are all stored in global memory.

Each path is handled by a single thread. If there are more paths than threads, a thread will handle multiple paths with an index separated by the total width in a serial manner.

The steps are done one at a time. (The first step is handled differently since the drifts are already known.) The use of the fast drift computation algorithm from Joshi (2003b) requires floats equal to the number of factors to store the partially computed sums. We therefore use factors times block-size floats in shared memory for each thread as storage. Given these facts, the evolution for each path is then straightforward and the coding of the algorithm is little different from that of a C program.

For step 0, we multiply the variates by the pseudo-square root and add them to the log rates. We add on the drifts. We then compute the state-dependent drifts at the end of the step. We correct the values of the log rates and the rates. We store their values in the output data and we use this location to retrieve them when needed during this kernel. We then use the rates to compute the discount factors implied by these forward rates for the step. For the other steps, we first compute the drifts at the start of the step, and then do as for step zero.

Whilst the code is robust against variation in block and grid size, the combination of 128 threads per block and 256 blocks proved effective when using 32,768 paths per batch. Note that this implies that each thread does precisely one path. A slight further optimization could be obtained by rewriting the code not to handle other cases, but the gains do not seem sufficient for this to be worthwhile.

16.7.7 Product specification and design

It is not enough just to be able to price a single product using hand-crafted code. One wants a code design that allows flexibility and changes. One approach advocated in Joshi (2008) is to use an object encapsulating the product's termsheet with virtual functions providing the necessary data. However, this does not seem well adapted to working on the GPU with CUDA (although it is supported). We therefore decompose the product into a number of components which can be written independently and then slotted together. First, we regard the product as generating a pair of cash flows at each of a set of evolution times. For each generation, the product is passed three rates as well as the current discount curve and forward rates. The computation or extraction of the three distinguished rates is performed independently and so the product knows nothing of its origin or meaning. Thus they could be swap rates or forward rates or something more complicated. In addition, if one wished to incorporate OIS discounting, then a distinction between LIBOR rates and OIS rates could be made at this point by adding a spread.

Second, the product simply generates two flows but does not specify their timing. Instead, separate payment schedules are specified and these are passed to a discounting routine. This allows changes, for example, from in-advance to in-arrears with minimal changes to the code simply by changing these timings.

The actual cash-flow generation routine which turns the rates into the cash-flow sizes is done using a templatized kernel. The template parameter is the product. It steps through the evolution times passing the rates and discount curves to the product and storing the cash flows as they are generated. The path is terminated when the product indicates to do so. The design is set up so that the product is able to store auxiliary data such a running coupon if necessary which allows the possibility of path dependence. Alternatively, one of the rates passed in could be made path dependent.

Exercise values are generated independently of the product, again allowing maximum flexibility.

16.7.8 Least-squares and multiple regressions on the GPU

The second phase of the least-squares algorithm is to perform regressions on estimated continuation values. Here we enact multiple regressions for each step. The least-squares algorithm works backwards in time. On each step, we first have to find the values of the basis functions. Note here the distinction between basis functions and basis variables. The latter are extracted in phase 1 and would typically be a forward rate, x, the adjoining swap rate, y, and the final discount factor, z. The former are polynomials in these. We focus on quadratic polynomials. We can work with or without cross terms, so we can use

$$1, x, y, z, xy, yz, zx, x^2, y^2, z^2$$

and have 10 basis functions, or use

$$1, x, y, z, x^2, y^2, z^2,$$

and have 7. The code is templatized on the algorithm for turning variables into functions to allow flexibility.

Thus at the start of each step, we first use a kernel to generate the basis functions for the step. Note that we only ever store the full basis functions for one step at a time to reduce memory usage.

Once the basis functions are known, we have to find the minimal least-squares error solution of a highly overdetermined rectangular system

$$Ax = y,$$

where A has N_1 rows. Each row consists of the values of the basis functions for one path for the step. The target y is the discounted future cash flows for the path. Typically, $N_1 = 327{,}680$ and there are 10 basis functions so the system is very overdetermined. We solve in two phases. First, we write

$$(A^{\mathrm{t}}A)x = A^{\mathrm{t}}y,$$

reducing to a 10×10 (or similar) system and then we solve this system.

For the computation of $A^{\mathrm{t}}A$, we use two kernels. The first kernel computes for each j and k with $j \leq k$,

$$\sum a_{ij}a_{ik},$$

with the sum taken over a subset of i. The subsets for different blocks partition the paths, and we use 1024 blocks. So at the end, we have 1024 numbers for each matrix entry still to be summed. These sums are performed by a second kernel which uses a different block for each entry. The computation is then done by copying the data into shared memory and then using a repeated binary summation so that each thread in the first half adds the value for the corresponding thread in the second half. Once a thread reaches the second half it does nothing further. Eventually only thread zero remains and it contains the value of the sum. Note that this approach minimizes the length of the computation chain which is an important consideration when working with floats to avoid round off error. The computation of $A^{t}y$ is done similarly.

For the second part of solving the small system, we copy the problem to the CPU and solve there. The fact that it is only a 10×10 system means that this is fast. The solution of the small system is the vector of regression coefficients.

However, the step is not yet done since we are doing multiple regressions. The regression coefficients yield an implied continuation value for every path. Our multiple regression algorithm requires us to discard the fraction $1 - \theta$ of paths which are furthest from the exercise boundary, that is the ones that yield the largest absolute value for discounted continuation value minus discounted exercise value. First, a cut-off level is found which gives the threshold above which paths are discarded. This is done by repeated bisection with the counting being done using the thrust transform and count algorithms. Second, the data for the remaining paths are moved to be contiguous in memory. This is performed using thrust's scatter-if algorithm. The process is repeated on the remaining data until the preset regression depth is reached (e.g., 5) or too few paths remaining according to a preset cut-off such as 2048. For subsequent estimates of continuation values, cascading through the coefficients is performed until a threshold level is reached or the maximum depth is obtained.

Once the continuation values have been estimated, the next stage of the algorithm is to set the paths' stepwise values to be either the discounted future flows for the path if exercise does not occur, or to the discounted exercise value if it does. This is straightforwardly performed by a simple kernel. The final action for the step is to deflate to the previous exercise time using the ratio of the numeraire values at the two times. This is again straightforward.

We then simply repeat back to step 0. The average value after doing step 0 yields the first pass estimate of the discounted cash flows on or after the first exercise time.

16.7.9 The data collection phase

As discussed earlier, the first phase of the pricing is to collect data for the regression and backward induction. The generation is divided into a number of batches (e.g., 10) and in each one the important data is stored. Each batch is straightforward. The following operations are carried out:

- The paths are generated as in Section 16.7.6.
- The coterminal swap rates and their annuities are computed that is the swap rates with the same final date but the first date varying (c.f. Jamshidian 1997).
- The just reset forward rate at each step is extracted as is the adjoining coterminal swap rate.
- The final discount factor for each step is also extracted.
- The basis variables are computed from the three previous extracted values and stored.
- The rates underlying the product are also extracted.
- The numeraires along the paths are computed.
- The cash flows along the paths are generated and discounted to exercise dates. They are then aggregated to each exercise date and stored.
- The exercise values are discounted and stored.
- The numeraire values at exercise dates are stored.

Each of these is done by a dedicated kernel. We have already discussed the cash-flow generation. The other kernels are straightforward.

16.7.10 The pricing phase

The third and final phase is the actual pricing. At this point, the exercise strategy has been generated during the second phase and so we are pricing conditionally on a set of regression coefficients. Most of the phase is very similar to the first one. The main differences being that there is no need to store data for use after the batch has been processed, and that the cash-flow generation takes the exercise strategy into account.

- The paths are generated as in Section 16.7.6.
- The coterminal swap rates and their annuities are computed that is the swap rates with the same final date but the first date varying (c.f. Jamshidian 1997).
- The just reset forward rate at each step is extracted as is the adjoining coterminal swap rate.
- The final discount factor for each step is also extracted.
- The basis variables are computed from the three previous extracted values and stored.

- The rates underlying the product are also extracted.
- The numeraires along the paths are computed.
- The cash flows along the paths are generated up to the exercise time. If exercise occurs, the exercise value is taken into account.
- The cash flows are deflated.
- The cash flows are summed for each path.
- The pathwise values are averaged using the reduce algorithm from thrust.

For each of these, a simple dedicated kernel is used. The only one of much interest is the cash-flow generation kernel. For maximum flexibility, this is templatized on three parameters: the product, the exercise value computer, and the exercise strategy. The auxiliary data for the strategy is moved into shared memory for rapid access. Once this has been done, the routine is very similar to that for the cash-flow generation in the first phase.

Note that each batch produces one number which is the mean pay-off value. Different batches can be achieved either by making the quasi-random generator skip or by using scrambling.

16.7.11 Speed comparisons and numerical results

We focus on the case of a 40-rate cancellable swap studied in Beveridge, Joshi, and Tang (2013). We choose this as it is challenging enough to be interesting without being esoteric. We also want to study a tough example already in the literature to demonstrate the nonartificiality of our results. The product has 40 underlying rates. The first one starts in 0.5 years. Coupon payments occur at $0.5 + 0.5j$ for $j = 1, 2, \ldots, 40$. It is cancellable at time 1.5 and every 0.5 years thereafter. No rebate is paid on cancellation. The swap pays a fixed rate of 0.04 and receives floating.

The calibration is that we set the forward rate from $0.5j$ to $0.5(j+1)$ to be

$$0.008 + 0.002j \text{ for } j = 0, 1, 2, \ldots, 40.$$

The well-used "abcd" time-dependent volatility structure is used, with

$$\sigma_i(t) = \begin{cases} 0, & t > T_i; \\ (0.05 + 0.09(T_i - t)) \exp(-0.44(T_i - t)) + 0.2, & \text{otherwise}, \end{cases}$$

and the instantaneous correlation between the driving Brownian motions is assumed to be of the form

$$\rho_{i,j} = \exp(-\phi|t_i - t_j|),$$

with $\phi = 2 \times 0.0669$. Displacements for all forward rates are assumed to be equal, with

$$\alpha_j = 1.5\%,$$

for all values of j.

We use a five-factor model. To obtain the reduced pseudo-square root matrices, we proceed as follows:

- Compute the full-factor covariance matrix for the step. This involves integrating the product of the volatility functions and multiplying by the instantaneous correlation for each entry.
- Perform a principal components analysis to obtain eigenvalues and eigenvectors, λ_j and e_j with λ_j decreasing.
- Form a column matrix $B = (\sqrt{\lambda_j} e_j)$.
- Scale the rows of B so that the variances of the log rates are the same as before factor reduction.

Our motivations for using five factors are:

- This seems to be as many as practitioners commonly use in industry.
- It is more than enough to encompass the major modes of deformation of yield curves.
- We wish to study an example already in the literature rather than developing a new one.

We note, however, that there is nothing special about 5 from the implementation perspective and the Kooderive code will function for a differing number of factors.

Beveridge, Joshi, and Tang (2013) achieve a price of 1088 with a standard error of 2.5 using double regression and the exclusion of suboptimal points. They use policy iteration to get an increase of 6.5 with a standard error of 2. Thus their lower bound price is 1094.5 with a standard error of 3.2. Their upper bound has the slightly lower value of 1094 with a standard error of 3.

We present results on timings and price for varying regression depths in Table 16.1. We use 10 batches of 32,768 paths for the first pass and 32 of them for the second. We separate the time actually spent doing regressions from that spent on doing other parts of the exercise strategy building. The price increases substantially when we increase from single regression to double regression. Another slight increase occurs from double to triple and it is stable thereafter. The fraction of paths retained after each regression is given by $0.1^{1/d}$ where d is the total regression depth. The implied price whilst slightly lower than that in Beveridge, Joshi, and Tang (2013) is within one standard error and so can be regarded as accurate.

In Table 16.2, we present timing comparisons for Kooderive versus QuantLib. For simplicity in running the QuantLib (QL) code, we do not consider the first two noncall coupons. The value of these is analytic in any case and a simulation pricing of them is not needed. We see that the first pass of path storage is 179 times faster in Kooderive. Similarly, the second pass pricing

TABLE 16.1: Timings and prices for the 40-rate cancellable swap with varying numbers of regressions

Regression depth	1	2	3	4	5
Time taken for first pass paths	0.207	0.206	0.207	0.206	0.207
Time taken for regression set up	0.206	0.169	0.172	0.179	0.179
Time taken for regression	0.094	0.191	0.254	0.32	0.39
Time taken for second pass	0.697	0.71	0.729	0.747	0.765
Total time	1.204	1.276	1.362	1.452	1.541
Second pass price	0.10740	0.10911	0.10924	0.10927	0.10925

Note: All standard errors are between 0.5 and 0.6 basis points.

TABLE 16.2: Timing comparison for QuantLib versus Kooderive for a 38 rate cancellable swap with 38 call dates

	Time QL	Time Kooderive	Ratio
First pass	34.958	0.195	179.2718
Strategy building	11.037	0.379	29.12137
Second pass	122.013	0.648	188.2917
Total	168.008	1.222	137.4861

Note: Time, seconds; 327,680, first pass paths; 1,048,576, second pass paths; Single regression.

is 188 times faster. Note that despite the fact that QuantLib early terminates path generation when appropriate and Kooderive does not. The timing ratio for the computation of regression coefficients is not quite so impressive as *only* a 29 times speedup is achieved. Note, however, that the division between stages in Kooderive is slightly different from QuantLib. The generation of basis functions from basis variables is done in the first pass in QuantLib and during strategy building in Kooderive. This means that our numbers overstate the speed of the first pass and the slowness of strategy building. The overall ratio, which is what really matters, at 137 is very large. We can run numbers of paths in seconds that would previously have been regarded as silly in a live environment.

Of course, a CPU implementation could also be multithreaded and the first and third parts should scale well since they are embarrassingly parallel. For the second part, one would have to solve similar challenges to that presented for the GPU. We do not explore how to carry out such an implementation here. However, we note that if the CPU used t threads, the best we could hope for is a t-times speed up, and so we would still expect the GPU to $137/t$

times faster. It would take a very large number of CPU cores for the CPU to be competitive against a single GPU.

16.8 Conclusion

We spent the first part of this chapter arguing that:

- There is a parallel performance imperative, driven by the state of computer hardware and manufacturing
- To maintain efficiency and performance, software has to adapt to available parallelism
- Quants need to be able to write efficient parallel code

Given those arguments, we presented a very basic primer in the hardware features that are relevant to quants, on both the more traditional CPU and on modern compute accelerators such as GPUs from NVIDIA. We cover the key concepts in parallel acceleration, hardware, thread, and memory management.

We then shifted gears and presented a realistic example of a challenging computation. The calculations were described in sufficient detail that they can be understood as they are mapped for execution on a GPU. The key implementation choices were explained. Pointers to sample code online were provided.

Ultimately, we have demonstrated that it is possible to develop a full-featured powerful and flexible displaced diffusion LIBOR market model than runs in a highly parallel fashion on the GPU. The speedup obtained is over 100 times versus a comparable nonparallel code, and this is using a single GPU. This speedup is achievable with a basic knowledge of modern hardware architectures and features modern, maintainable code. Such speedups mean that it is now possible to routinely run numbers of paths that would have been regarded as far too large in the past.

A natural extension of the work here would be to consider a more complex model incorporating OIS discounting and smiles. These methods are well covered in the literature and following the example of this code, incorporating them into parallel code is straightforward. Alternatively, following this example, other models can also be adapted for highly parallel execution.

References

Abbas-Turki, L., and Lapeyre, B. 2009. American options pricing on multi-core graphic cards. *2009 International Conference on Business Intelligence and Financial Engineering*, 307, Beijing, China.

Aldrich, E. M., Fernandez-Villaverde, J., Gallant, A. R., and Rubio-Ramirez, J. F. 2011. Tapping the supercomputer under your desk: Solving dynamic equilibrium models with graphics processors. *Journal of Economic Dynamics and Control, 35*(3), 386–393. http://www.sciencedirect.com/science/article/pii/S0165188910002216 doi:10.1016/j.jedc.2010.10.001

Amdahl, G. M. 1967. Validity of the single processor approach to achieving large-scale computing capabilities. *AFIPS Conference Proceedings, 30*, 483–485.

Amin, A. 2003. Multi-factor cross currency Libor market models: Implementation, calibration and examples. *Calibration and Examples (May 1, 2003)*. https://dx.doi.org/10.2139/ssrn.1214042

Andersen, L. and Broadie, M. 2004. A primal–dual simulation algorithm for pricing multi-dimensional American options. *Management Science, 50*, 1222–1234.

Andersen, L. and Piterbarg, V. V. 2010. *Interest rate modelling.* London, New York: Atlantic Financial Press.

Beveridge, C. J. and Joshi, M. S. 2008. Juggling snowballs. *Risk, December*, 100–104.

Beveridge, C. J., Joshi, M. S., and Tang, R. 2013. Practical policy iteration: Generic methods for obtaining rapid and tight bounds for Bermudan exotic derivatives using Monte Carlo simulation. *Journal of Economic Dynamics and Control, 37*, 1342–1361.

Brace, A. 2007. *Engineering BGM.* Sydney: Chapman and Hall.

Brace, A., Gatarek, D., and Musiela, M. 1997. The market model of interest rate dynamics. *Mathematical Finance, 7*, 127–155.

Brigo, D. and Mercurio, F. 2001. *Interest Rate Models: Theory and Practice.* Heidelberg: Springer Verlag.

Brigo, D. and Mercurio, F. 2006. *Interest Rate Models—Theory and Practice: With Smile, Inflation and Credit.* Springer.

Broadie, M. and Cao, M. 2008. Improved lower and upper bound algorithms for pricing American options by simulation. *Quantitative Finance, 8*, 845–861.

Carrière, J. F. 1996. Valuation of the early-exercise price for options using simulation and nonparametric regression. *Insurance: Mathematics and Economics, 19*, 19–30.

Dang, D. M., Christara, C. C., and Jackson, K. R. 2010. Pricing multi-asset American options on graphics processing units using a PDE approach. *2010 IEEE Workshop on High Performance Computational Finance (WHPCF)* (pp. 1–8), New Orleans, Louisiana, USA.

Dennard, R. H., Gaensslen, F. H., Rideout, V. L., Bassous, E., LeBlanc, A. R., and Gaensslen, R. H. 1974. Design of ion-implanted mosfets with very small physical dimensions. *IEEE Journal of Solid-State Circuits, SC-9*(5), 256–268.

du Toit, J. 2011. *A high-performance Brownian bridge for GPUS: Lessons for bandwidth bound applications.*

Giles, M., Kuo, F. Y., Sloan, I. H., and Waterhouse, B. J. 2008. Quasi-Monte Carlo for finance applications. *ANZIAM Journal, 50*, C308–C323.

Giles, M. and Xiaoke, S. 2008. *Notes on using the nVidia 8800 GTX graphics card.* (https://people.maths.ox.ac.uk/gilesm/codes/libor_old/report.pdf)

Gustafson, J. L. 1988. Reevaluating Amdahl's law. *Communications of the ACM, 31*, 532–533.

Hunter, C., Jäckel, P., and Joshi, M. S. 2001. Getting the drift. *Risk, July*, 81–84.

Jäckel, P. 2001. *Monte Carlo Methods in Finance.* New York: John Wiley & Sons Ltd.

Jamshidian, F. 1997. LIBOR and swap market models and measures. *Finance and Stochastics, 1*, 293–330.

Joshi, M. 2003a. *The Concepts and Practice of Mathematical Finance.* London: Cambridge University Press.

Joshi, M. 2003b. Rapid drift computations in the LIBOR market model. *Wilmott Magazine, May*, 84–85.

Joshi, M. 2008. *C++ Design Patterns and Derivatives Pricing* (2nd edition). London: Cambridge University Press.

Joshi, M. 2010. Graphical Asian options. *Wilmott Journal, 2*, 97–107.

Joshi, M. 2011. *More Mathematical Finance.* Melbourne: Pilot Whale Press.

Joshi, M. 2014. *Kooderive version 0.3.* http://kooderive.sourceforge.net

Joshi, M. and Tang, R 2014. Effective sub-simulation-free upper bounds for the Monte Carlo pricing of callable derivatives and various improvements to existing methodologies. *Journal of Economic Dynamics and Control, 40*,25–45 http://www.sciencedirect.com/science/article/pii/S0165188913002376 doi:10.1016/j.jedc.2013.12.001

Kolodko, A. and Schoenmakers, J. 2006. Iterative construction of the optimal Bermudan stopping time. *Finance and Stochastics, 10*, 27–49.

Little, J. D. C. 1961. A proof for the queuing formula: $L = \lambda w$. *Operations Research, 9.3*, 383–387.

Longstaff, F. A. and Schwartz, E. S. 2001. Valuing American options by simulation: A simple least squares approach. *The Review of Financial Studies, 14*, 113–147.

Moore, G. E. 1965. Cramming more components onto integrated circuits. *Electronics, April 19*, 114–117.

Musiela, M. and Rutkowski, M. 1997. Continuous-time term structure models: forward-measure approach. *Finance and Stochastics*, *1*, 261–292.

Piterbarg, V. 2004. A practitioner's guide to pricing and hedging callable LIBOR exotics in forward LIBOR models. *Journal of Computational Finance*, *8*, 65–119.

Shaw, W. T. and Brickman, N. 2009. *Differential equations for Monte Carlo recycling and a GPU-optimized normal quantile* (Tech. Rep.). Citeseer.

Chapter 17

Practitioner's Guide on the Use of Cloud Computing in Finance

Binghuan Lin, Rainer Wehkamp, and Juho Kanniainen

CONTENTS

17.1 What Is Cloud Computing?

It takes mankind centuries to learn how to make use of electricity. In the early age, factories and corporations were powered by on-site small-scale power plants. Maintaining such power plants is expensive due to the additional labor cost. Nowadays with the help of large-scale power plants and efficient transmission networks, electricity powers modern industrial society for transportation, heating, lighting, communications, and so on. Electricity is at everyone's disposal at a reasonable price.

Cloud computing shares many of the similarities with electricity. By connecting end-users via Internet to data centers, where powerful computing hardware is located, cloud computing makes computation available to everyone. The core concept of cloud computing is resource sharing. To harvest the computing power, cloud computing is also a practice of operation research in computing resource optimization. Such technology enables the processing of massive-parallel computations by using shared computing resources. The computing resources usually consist of large numbers of networked computing nodes. The word "cloud" is used to depict such networked computing resources.

The formal definition of cloud computing given by the National Institute of Standards and Technology (NIST) of the U.S. Department of Commerce is

> a model for enabling ubiquitous, convenient, on-demand network access to a shared pool of configurable computing resources (e.g., networks, servers, storage, applications, and services) that can be rapidly provisioned and released with minimal management effort or service provider interaction.
>
> Mell and Grance (2009), NIST

The following five essential characteristics differentiate cloud computing from other computing solutions, such as on-premise servers:

- *On-demand self-service*: A consumer can provision computing capacity as needed without interaction with service provider.

- *Broad network access*: Computing resources are available to the consumer through the network and can be accessed from mobile phones, tablets, laptops, and workstations.

- *Resource pooling*: Resources are dynamically assigned to customers' needs.

- *Rapid elasticity*: Capacities can be reconfigured automatically to scale rapidly in response to the changing demand.
- *Measured service*: The resource utilization is automatically controlled and optimized by the cloud systems. The utilization is monitored, measured, and reported.

17.1.1 Why cloud computing and why now?

The first reason is the *increasing computing demand from industry, especially from financial industry.* International Data Corporation (IDC) Joseph et al. (2014) reported the demand for high-performance computing (HPC) from 13 sectors. They predict an 8.7% yearly growth in the spending of HPC in economics/financial sector from 2013 to 2018, which is among the top 3 of all 13 sectors studied as shown in Figure 17.1. The second reason is the *pervasiveness of cloud computing.* Cloud computing has its deep roots dating back to utility computing in 1960s:

> If computers of the kind I have advocated become the computers of the future, then computing may someday be organized as a public utility just as the telephone system is a public utility... The computer utility could become the basis of a new and important industry.
>
> John McCarthy at MIT Centennial in 1961

The technology has developed since then. To provide an overview of the technology developments, Figure 17.2 shows the advances of cloud computing related technology alongside the innovations in financial engineering. While

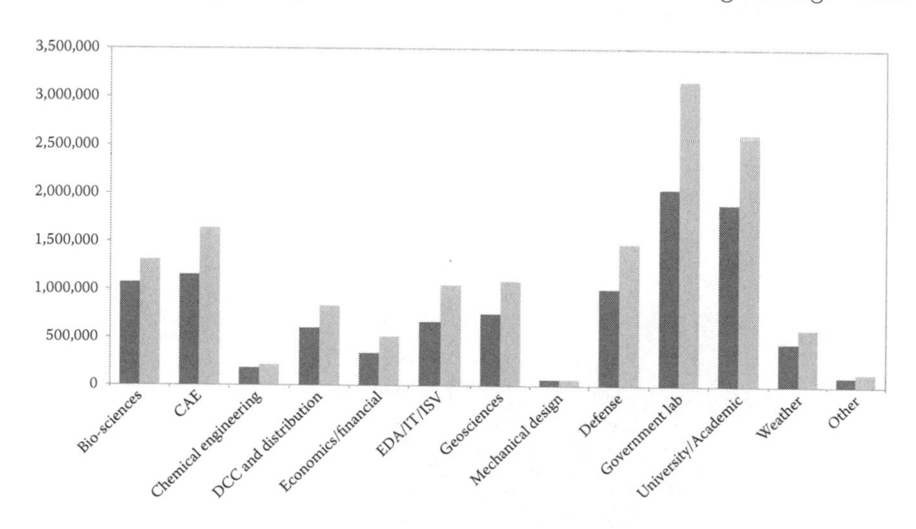

FIGURE 17.1: HPC spending by sector 2013 versus 2018. (Adapted from Joseph, E. et al., 2014. IDC HPC update at ISC'14.)

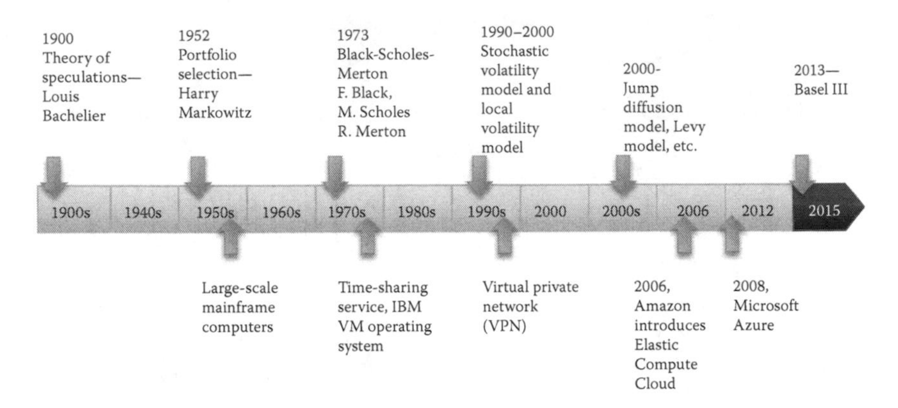

FIGURE 17.2: History of financial engineering and cloud computing.

the innovation of more complex models in financial engineering increases the demand for HPC technology, the supply of HPC technology increases with the development of technologies, such as cloud computing. There is also evidence of increasing awareness of cloud computing from the public. Figure 17.3 shows an increasing search trend for cloud computing using *Google Trends*.

The undergone development and commercialization of cloud computing are significantly boosted by the increase in computation demands in the real world. According to Gartner's report Smith (2008).

> By 2012, 80% of Fortune 1000 enterprises will pay for some cloud computing service and 30% of them will pay for cloud computing infrastructure. Through 2010, more than 80% of enterprise use of

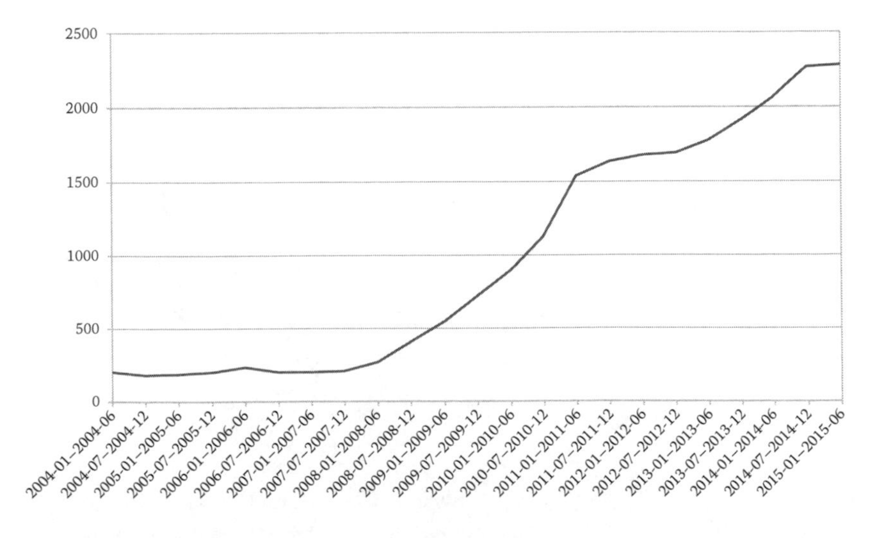

FIGURE 17.3: Google search trend of cloud computing (since 2004).

cloud computing will be devoted to very large data queries, short-term massively parallel workloads, or IT use by startups with little to no IT infrastructure.

Modern day commercial cloud computing has also revolutionized a new level of industrial computing practice, especially in financial industry where the computing need is massive. Our focus is on how to utilize the enormous computing resource from cloud computing to harness the massive computing challenges posed by financial industry. We start by introducing parallel computing problems and massive parallel computing tasks in finance industry. We then compare cloud computing with alternative solutions from the following aspects: performance, cost, and elasticity. To improve understanding about what kind of problems users may face in practice and how they can be solved, we also walk readers through a complete implementation procedure in financial industry with case studies and an implementation of Techila® Middleware Solution.

17.2 Background

17.2.1 The taxonomy of parallel computing

Computer processors process instructions sequentially. Thus traditional computing problems are serial problems by such design. The birth of multi-processors has innovated a new type of computing problem: how to utilize the parallel structure.

Parallel computing problem, in contrast to serial computing problems, refers to the type of computing problems that can be divided into subproblems to be processed simultaneously. Based on the dependency structure of subproblems, it can be further classified into *embarrassingly parallel* and *nonembarrassingly parallel* computing problems. If the processing of one subproblem is independent of other subproblems, then it is an embarrassingly parallel computing problem. It is called nonembarrassingly parallel computing problem otherwise.

The following figure illustrates the structure of embarrassingly parallel and nonembarrassingly parallel problems. There is no communication between jobs in embarrassingly parallel case as in Figure 17.4a, while communication is required in nonembarrassingly parallel case as in Figure 17.4b.

By the nature of underlying problem, it can be classified as *data-parallel* problem and *task-parallel* problem. While data parallelism focuses on distributing data across different processors, task parallelism focuses on distributing execution processes (subtasks) to different processors.

Another important aspect of parallel computing is whether the parallel computing problem is a *scalable* problem. A scalable problem has either a scalable problem size or scalable parallelism. Either the solution time reduces

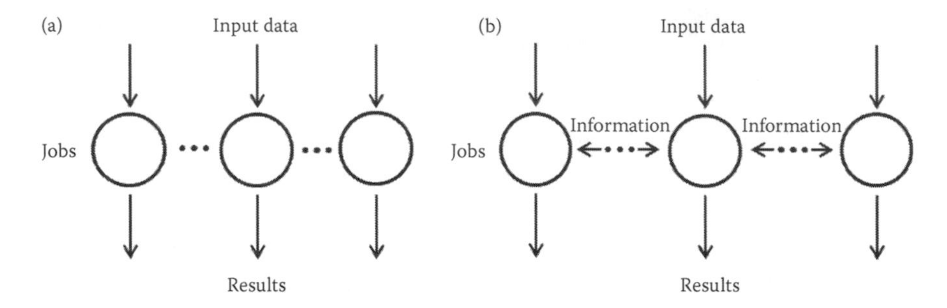

FIGURE 17.4: Parallel computing structure: (a) embarrassingly parallel computing and (b) nonembarrassingly parallel computing.

with the increasing of parallelism or the performance of the solution increases with the problem size. The elasticity of the computing architecture is the key to the success of processing of scalable problems.

Here we provide two examples from finance industry:

Example 17.1: Monte Carlo Option Pricing

Monte Carlo simulation is a typical *embarrassingly parallel* and *task-parallel* computing problem.

By the fundamental theorem of arbitrage pricing, option price is equal to the expected payoff V discounted by a discount factor D. The expectation can be evaluated via Monte Carlo method. The Monte Carlo estimator of option price is given by

$$C_0 = D\frac{1}{N}\sum_{\omega\in\text{sample set}} V(\omega)$$

where N is the number of sample paths.

1. The simulation of each price path is independent of other paths. Thus it is easy to parallel process the simulation of different paths on different computing nodes.

2. It is task-parallel in the sense that simulation of each path is a small task. However, it is not data-parallel since there is no data set to be distributed to different computing nodes.

3. There is benefit in scaling the computation. The accuracy of the price estimator improves with the increase of N. (The error has convergence of $\mathcal{O}(\frac{1}{\sqrt{N}})$.)

To illustrate the difference between task-parallel and data-parallel, we use the following example:

Example 17.2: Backtesting Investment Strategy

Depending on how you implement the computation tasks, backtesting can be either *task-parallel* or *data-parallel.*

Suppose you need to backtest a basket of different investment strategies to identify the optimal strategies. The processing of each investment strategy is independent of each other and can be run simultaneously. Distributing the processing of different strategies to different computing nodes is an *embarrassingly parallel* and *task parallel* implementation.

Algorithm 17.1: Task-Parallel Backtesting

Input: A set of investment strategies, historical data sample;
Output: PnL, Portfolio Attribution, Risk Exposures, etc.;
for $i \leftarrow 1$ **to** $number\_of\_strategies$ **do**
| $PnL \leftarrow$ ProfitandLoss(strategy i, datasample);
| $PA \leftarrow$ PortfolioAttribute(PnL);
| $RE \leftarrow$ RiskExposure(PnL);
end

Backtesting of one strategy can also be implemented as *data-parallel.* By generating subsamples from test data set (e.g., by bootstrapping), strategy can be processed on different subsamples simultaneously. The result on different subsamples is then aggregated to generate the performance and risk report of the strategy.

Algorithm 17.2: Data-Parallel Backtesting

Input: Investment strategy, sub data samples
Output: PnL, Portfolio Attribution, Risk Exposures, etc.;
for $i \leftarrow 1$ **to** $number\_of\_data\_sample$ **do**
| $PnL[i] \leftarrow$ ProfitandLoss(strategy, data_sample_i);
end
PnL_total = aggregate(PnL)

17.2.2 Glossary

- *Computing instance*: refers to a (virtual) server instance that is linked to a computing network to provide computing resources. To offer flexibility to their customers, cloud vendors offer different types of nodes that comprise various combinations of CPU (central processing unit), memory, storage, and networking capacity.[1]

- *Data center*: A data center comprises a large number of computing nodes, the network to connect these nodes and necessary facility to house the computer system.
- *Server–worker nodes*: Server–worker nodes are a typical mechanism to coordinate the computation between different computing nodes. The server nodes assign computing tasks to worker nodes and collect results from worker nodes. Worker nodes receive instructions from server nodes, execute the computations, and send the results back to the server.
- *Middleware*: Middleware is a computer software that "glues" software applications with computing hardware. In cloud computing, middleware is used to enable communications and management of data.
- *Job scheduler and resource manager*: Software to optimize the usage of computing resources based on the resources available and job priority. Commercial solutions usually package job scheduling, resource management with middleware.
- *Virtualization*: Using computer resources to imitate other computer resources. By virtualization, users are not locked with specific operating systems, CPU architecture, and so on. Thus middleware and virtualization are particularly important to ensure on-demand self-service of cloud computing.
- *Cloud bursting*: Cloud computing offers on-demand service. Cloud bursting refers to the process of dynamic deployment of software applications.
- *Elastic computing*: Elastic computing is a computing service which has the ability to scale resources to meet requirements.
- *Public cloud*: Public cloud is the cloud computing service that is available to public and can be accessed through Internet.
- *Private cloud*: Private cloud, in contrast to public cloud, is not available to public. The computing resources are dedicated to select users.
- *Hybrid cloud*: Hybrid cloud is a cloud computing service that combines different types of services, for example, public and private. A hybrid cloud combines public and private clouds and allows workloads move between public and private clouds. The flexibility allows users to optimize the allocations to reduce cost while still having direct control of their environments.

[1] We will provide a list of different computing nodes in Section 17.5.

- *Wall clock time (WCT)*: Wall clock time is the human perception of the passage of time from the start to the completion of a task.

- *CPU time*: The amount of time for which a CPU was used for processing instructions of a computer program or operating system, as opposed to, for example, waiting for input/output (I/O) operations or entering low-power (idle) mode.

- *Workload*: In cloud computing, workload is measured by the amount of CPU time and memory consumption.

- *CPU efficiency*: CPU efficiency measured as the CPU time used for computation divided by the sum of CPU time and I/O time used for data transfer. Thus CPU efficiency measures the overhead of paralleling a computation. A low CPU efficiency, in general, indicates a high overhead.

- *Acceleration factor*: Acceleration factor is measured by wall clock time of running the program locally on the end user's computer divided by the wall clock time of running it on the cloud. In an ideal case, the acceleration factor can be linear in the number of cores used for computation.

- *Total cost of ownership (TCO)*: TCO measures both direct and indirect costs of deploying the solution. In cloud computing and alternative computing solutions, TCO includes the cost of: hardware, software, operating expenses (such as infrastructure, electricity, outage cost, and so on), and long-term expenses (such as replacement, upgrade and scalability expenses, decommissioning, and so on).

17.3 Financial Applications of Cloud Computing

17.3.1 Derivative valuation and pricing

One of the core businesses of the front office is derivative pricing. Even though, the market for exotic derivatives has shrunk after the crisis in 2008–2009, the exoticization of vanilla products has increased the complexity of the valuation process. Moreover, numerical methods are needed with certain models, even with vanilla options, such as nonaffine variance models, infinite-activity jump models, and so on.

The valuation process usually requires high-performance numerical solutions, as well as a high-performance technology platform. The size of the book and time criticalness requires a platform that is both suitable for handling large data and processing massive computing. A recent paper by Kanniainen et al. (2014) evaluates the computation performance of using Monte Carlo methods for option pricing. With the aid of cloud computing and Techila Middleware Solution,[2] the time consumption of valuating option

[2]For more information, please visit: http://www.techilatechnologies.com.

contracts using Monte Carlo methods is comparable with other numerical methods:

> ... valuate once the 32,729 options in Sample A using the Heston–Nandi (HN) model was 25 s with the HN quasi-closed-form solution and 249 s with the Monte Carlo methods. Moreover, with cloud computing with the Techila middleware on 173 Azure extra small virtual machines (173 × 1 GHz CPU, 768 MB RAM) and the task divided into 775 jobs according to 775 sample dates, the overall wall clock time was 55 s and the total CPU time 44 min and 33 s. The Monte Carlo running times were approximately the same for GJR and NGARCH. Substantially shorter wall clock times can be recorded if more workers (virtual machines) are available on the cloud or if the workers are larger (more efficient). Then the wall clock time differs very little between the HN model with the quasi-closed-form solution on a local computer and the HN model or some other GARCH model (such as GJR or NGARCH) with the Monte Carlo methods on a cloud computing platform. Consequently, with modern acceleration technologies closed-form solutions are no longer a critical requirement for option valuation.

Another key postcrisis trend is the populating of XVAs (Fund Valuation Adjustment [FVA], Credit Valuation Adjustment [CVA], and so on). The books of XVAs are usually huge and the time constraint to process the valuation is tight. An industry success is the award-winning in-house system of Danske Bank. The combination of advanced numerical technique and modern computing platform allows real-time pricing of derivative counterparty risk.

17.3.2 Risk management and reporting

The financial crisis also reshaped the business of risk management in financial industry. The implementation of Solvency II for insurance and Basel III for banking, respectively, poses new challenges to financial computing.

First, the computations are highly resource-intensive. Second, the computation needs are dynamic rather than static. Risk report (Solvency II, Basel III, and so on) are required at monthly or quarterly frequency. Computation needs are periodic, where they reach their peak before the reporting deadline. Building and managing a dedicated data center to meet the computation need at its peak will significantly increase the cost. On the other hand, most of the computing resource will be wasted during a relatively less intensive period.

Cloud computing has the advantage of being scalable, which allows it to meet the dynamic computation need from financial industry. Using Google search volume, we find that Google search volume for cloud computing increased rapidly after 2009 and so have the search volumes for Solvency II and Basel III. We are not suggesting any causality between the increasing attention of cloud computing and that of risk regulation. However, such trends show the right timing of popularity of cloud computing as a potential solution for regulation-oriented computation needs.

The financial industry started to embrace cloud solutions, especially when they are integrated to support the need for an effective and timely risk management. IBM's survey (reference) on the implementation of cloud computing for Solvency II in the insurance industry points out the trend of adopting cloud computing as part of the implementation strategy for risk management. Of the 19 firms, 27% either have successfully implemented cloud solutions or are in the process of implementing cloud solutions. Another 23.8% have started considering cloud solutions.

One of the key questions is whether a cloud solution is cost-efficient. Little (2011) from Moody's Analytics analyzes the potential usage of cloud for economic scenario generation and Solvency II in general. They conclude:

> Building a Solvency II platform on the cloud is a realistic and cost-effective option, especially when scenario generation and Asset Liability Modeling are both performed on a cloud.

17.3.3 Quantitative trading

The lowering barrier in the market participation challenges the development of more complicated trading strategies as well as a race of technology. Quantitative trading, especially high frequency trading, requires a quick *time-to-production* as well as a quick *time-to-market.*

Fast R&D of trading strategies and back testing in a timely manner will significantly shorten time-to-market. Firms, by taking advantage of cloud computing, are generating alpha even before trading strategies are actually implemented in market.

The quick prototyping and backtesting of strategies requires close-to-data computing as well as adaptability to the heterogeneity of developer tools, such as different end-user applications, different data storage types, and different programming languages. On the other hand, the adaptability to different end-user software has been one of the key features of the matured cloud computing platform.

17.3.4 Credit scoring

Cloud computing is arguably the solution for big data problems in finance. One typical big data problem in the finance industry is credit scoring. Credit scoring is the procedure for lenders, such as banks and credit card companies, to evaluate the potential risk posed by lending money to consumers. It has been widely treated as a classification problem in machine learning literature, see (Hand and Henley, 1997; West, 2000; Baesens et al., 2003) and many others.

The large number of consumers and the variety of credit report formats create a big data problem. To solve the classification problem over the massive data set of credit history of consumers, an efficient data storage and processing system is required.

As a summary, modern day financial computing requires:

- Adaptability to the heterogeneity of end-user software
- Processing large data and close-to-date computing
- Massive computing
- Data security

In this chapter, we are going to introduce details of how modern cloud computing can help to solve these problems. There are also innovative cloud-supported new business models using the concept of sharing, such as cooperative quantitative strategy development platforms. While we focus on the massive computation part of cloud computing for financial engineering, we refer readers who are interested in those platforms to Internet resources.

17.4 The Nature of Challenges

Integrating massive computing power to existing IT systems may face several challenges. The finance industry poses certain specific requirements to cloud computing.

System needs to be multitenant: The system needs to support multiple users accessing the computing resource at the same time. Meanwhile, the system has to be smart enough to allocate computing resources based on the priority and need of the tasks. The requirements arise from the heterogeneous and dynamic nature of financial computing. Computing from different desks has different priorities and uneven demand for resources.

Compliance requirement and cybersecurity: The finance industry operates with public data/information as well as business critical private data/information. Due to compliance requirements, a hybrid system needs to make computing with sensitive data in-house while allowing utilization of external computing resource with nonconfidential data.

A unified platform for quality assurance: Large financial organizations have teams supporting local business operations across the world. A unified platform will make life of quantitative support and model validation/review teams easier by guaranteeing consistency, and coherency of data and models for users.

IT legacy: Maintaining a monster level of legacy codes is a huge task for IT departments. Any change that needs complete rewrite of the codes will be a nightmare. Thus IT systems have a high level of adaptability to have effortless integrating with existing libraries.

17.5 Implementation and Practices

In the previous section, we reviewed the technical and nontechnical challenges in the integration process of cloud computing in the finance industry. Luckily, with the development of commercial cloud computing services, the complexity of cloud computing is hidden behind user-friendly interfaces.

Aiming to ease the use of cloud computing, many software programs and computing frameworks were developed during the past decade. To mention a few, popular computing framework includes Hadoop+MapReduce, Apache Spark, and so on. Middleware solutions such as Techila Middleware, Sun Grid Engine also help commercial users to distribute computing tasks to computing nodes. In this section, we will introduce using an example of Techila Middleware on how challenges in Section 17.4 can be handled by a cloud computing solution. Then we walk readers through the procedure of implementing cloud computing in practice.

17.5.1 Implementation example: Techila middleware with MATLAB

Techila Middleware Solution, developed by Techila Technologies Ltd,[3] is a commercial software solution aiming to provide user-friendly integration of cloud computing. The service structure of Techila® is shown in Figure 17.5. The specific design of the service structure allows accessing both on-premise and external computing resources to ensure compliance requirements are met when necessary. The system is multitenant, where users can assign different priorities to computing jobs sent to Techila system through secured Gateway. The jobs are scheduled according to the availability of resources and priorities. The solution hides the complexity of integration to heterogeneous end-user software behind a user-friendly interface. To illustrate that, we provide an example of using Techila with MATLAB. For more information on the programming languages and software that Techila supports, please refer to the company's website at: www.techilatechnologies.com.

Before proceeding to use Techila solution, the end user needs some minor configuration. Readers are referred to Techila's online documents for more details (www.techilatechnologies.com).

[3]For more information, please visit: http://www.techilatechnologies.com

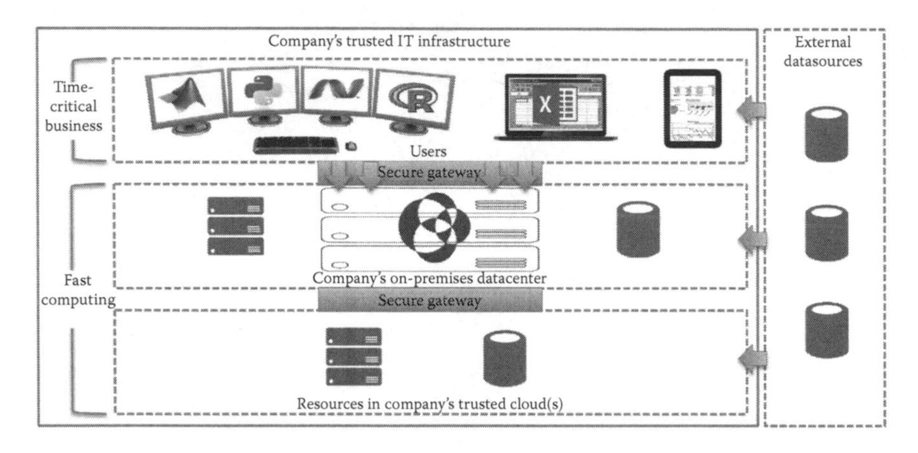

FIGURE 17.5: Techila high-level architecture.

When Techila is successfully installed, using cloud computing with existing codes developed in MATLAB is straightforward.

Suppose an end-user has a code that contains of the following for-loop structure:

```
1  function result = local_loops(loops)
2
3          result = zeros(1,loops);
4          for counter =1:loops
5                  result(counter) = counter*counter;
6          end
```

To parallelize the computation inside the for loop on cloud computing resources, the end user only needs to make a minor change to the code:

```
1  function result = run_loops_dist(loops)
2  % the only change is change for-end to cloudfor-
       cloudend
3          result = zeros(1,loops);
4          cloudfor counter = 1:loops
5                  result(counter) = counter*counter;
6          cloudend
```

17.5.2 Computational needs

Before making a decision to adopt any HPC solution, a key step is to understand your computational need and usage pattern. The best choice of solution depends on the answers to the following questions:

- Do you have a computational bottleneck?
- Where is the computational bottleneck?

17.5.2.1 Do you have a computational bottleneck?

The question may seem to be trivial at first sight. A computational bottleneck exists when the current computing resource cannot process the computing tasks within given time constraints. However, many of the cases, the computational bottleneck arises from another dimension, that is the time it required to upgrade computing resources to meet the increased demand.

The point we would like to emphasize is that the planning of computing needs to be forward-looking. While quants, researchers, and developers are aware of the computational bottleneck, it is usually the IT department's decision whether to expand current IT resources. The procedure may take some time. Thus forward-looking planning is critical in order to ensure an efficient and effective response to the computational need.

The scalability and elasticity of cloud computing may offer an alternative solution by providing computing resource on demand.

17.5.2.2 Where is your computational bottleneck?

A typical computational bottleneck from the finance industry poses one of the following types of challenges:

1. Massive computational time exceeds time constraints
2. Massive memory consumption exceeds limited memory and storage
3. Dynamic usage pattern meets nonscalable computing resource.

There are several solutions for the type 1 challenge. In a production scenario, for example, when implementing a high frequency trading algorithm, hardware accelerators, such as FPGA and GPU, may be better alternatives to cloud computing.[4] The reasons are:

[4] Although in practice, there are firms implementing their algorithms in the cloud to gain benefit for lower latency in connection to exchange when colocation is not possible or too costly to implement.

1. Execution time is critical. Hardware acceleration may be the only solution to accelerate the algorithm.

2. No frequent reconfiguration is needed. The cost to configure and adapt the algorithm for hardware acceleration is less than the profit gain from shorter execution time.

While in an R&D scenario, time-to-market is more important. The easy implementation and massive parallel features provided by cloud computing will enable researchers and developers to quickly prototype and backtest algorithms and models.

Cloud computing is also an economical solution to type 2 and type 3 challenges. A distributed storage and memory consisting of relatively cheaper hardware, compared with expensive local instances that have adequate memory and storage size, reduce significantly the cost to invest in hardware. The elasticity of cloud computing provides users on-demand service. In a type 3 challenge, an investment in computer instances that can process computing demands at their peak time is a waste of resources during periods where computing demands are less intensive.

17.5.3 Solution selection

Both performance and cost should be taken into consideration when choosing a cloud vendor. Among them, Amazon Elastic Compute Cloud (AWS), Google Compute Engine (GCE), and Microsoft Azure (Azure) are three popular cloud computing platforms.

Vendors provide a variety of instance types. For example, the cloud instances used in a recent benchmark by Techila include four different instances from AWS and Azure and two instances from GCE as listed in Table A.1 of (Techila, 2015, Appendix A: Cloud Platform Specifications). These instances are optimized for different purposes: CPU, memory, I/O, cost, storage, and so on.

Vendors adopt different pricing models. The cost of using cloud computing is affected by the pricing model. Table 17.1 provides an overview of pricing models adopted by vendors for data centers based in Europe. The table reports price per instance (PPI), price per CPU core (PPC) and the billing granularity for AWS, Azure, and GCE.[5] Generally speaking, a minute-based (even seconds-based) pricing model offers more flexibility for the utilization of cloud computing. However, the difference among pricing models is relatively insignificant when the computation is massive.

Depending on the vendor, instance types and the operating systems, instances have varying time consumption for configuration and deployment.

[5]GCE machine types are charged a minimum of 10 minutes. After 10 minutes, instances are charged in 1 minute increments, rounded up to the nearest minute.

TABLE 17.1: Pricing model and cost

Cloud platform	Instance type	PPI (USD/h)	PPC (USD/h)	Billing granularity
AWS	c4.4xlarge (Win)	1.667	0.1041875	Hour
AWS	c4.4xlarge (Linux)	1.003	0.0626875	Hour
AWS	c3.8xlarge (Win)	3.008	0.094	Hour
AWS	c3.8xlarge (Linux)	1.912	0.06	Hour
Azure	A11 (Win)	3.5	0.219	Minute
Azure	A11 (Linux)	2.39	0.149	Minute
Azure	D14 (Win)	2.372	0.148	Minute
Azure	D14 (Linux)	1.542	0.096	Minute
GCE	n1-standard-16 (Win)	1.52	0.095	Minute
GCE	n1-standard-16 (Linux)	0.88	0.055	Minute

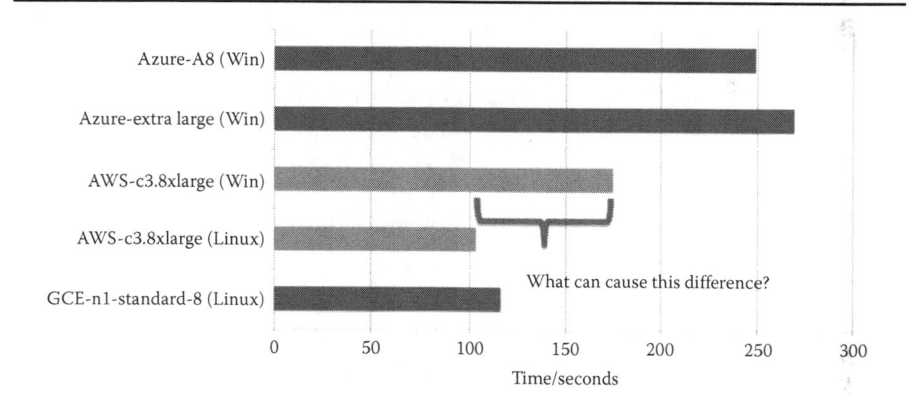

FIGURE 17.6: Configuration time.

According to Techila's benchmark report (Techila, 2014) in 2014, the differences are significant as shown in Figures 17.6 and 17.7. These are nonnegligible factors for the elasticity of cloud computing. As a rule of thumb, the operating systems, instances type, and vendor should be chosen according to the end-user's version of software applications.

We cite the test results from two recent benchmark reports (Techila, 2014, 2015) from Techila to provide readers an impression of the cost of utilizing cloud computing. The test cases simulate real-world financial applications in many areas including: portfolio analytics, machine learning, option pricing, backtesting, model calibration, and so on. However, readers should be aware of the different nature of these applications, whether they are high I/O, high memory consumption, or high CPU consumption.

Figure 17.8 summarizes the cost of computing for different vendors and instances versus the performance of computing. Table 17.2 provides the cost in the portfolio simulation case.[6] The simplified cost provides the cost per unit of

[6] For more information on other user scenarios, please refer to Techila's benchmark report "Cloud HPC in Finance" (Techila, 2015).

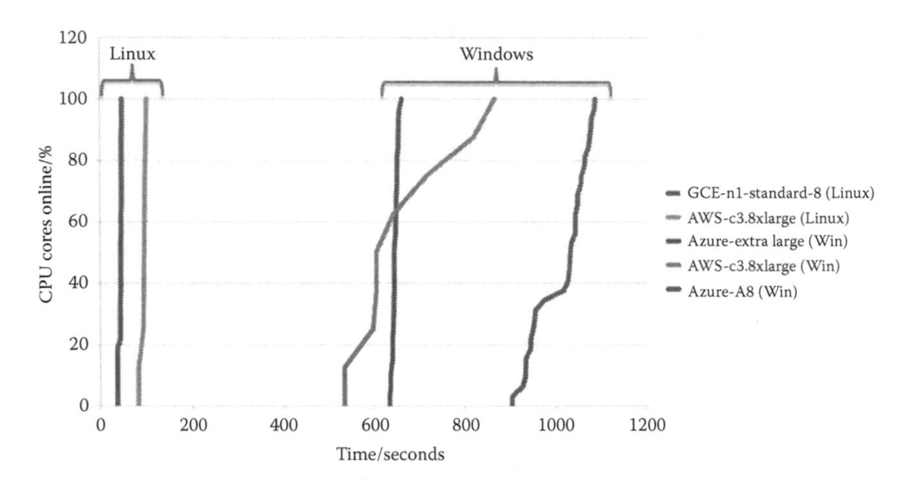

FIGURE 17.7: Deployment time.

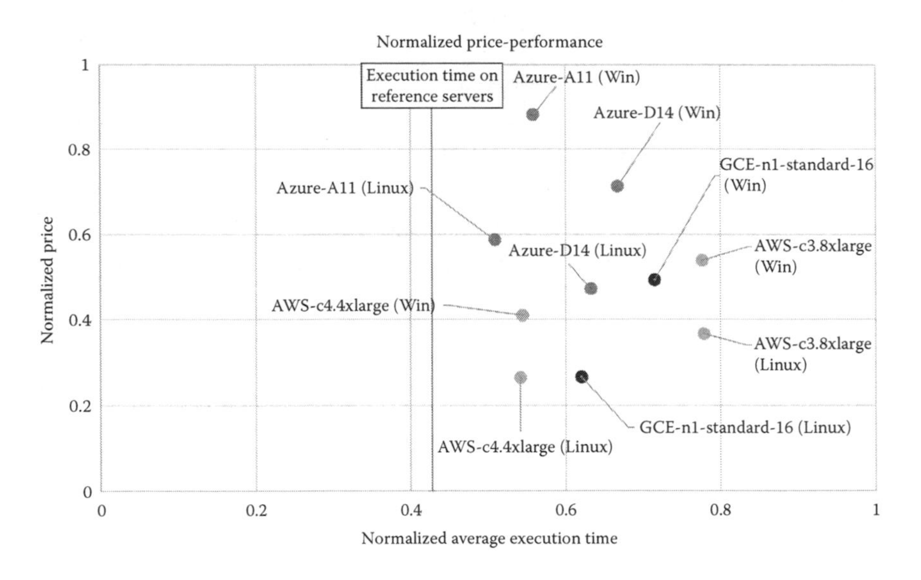

FIGURE 17.8: Cost versus performance.

computation after correcting for the difference in pricing model. For example, for portfolio simulation, the simplified cost ranges from 0.58 USD (GCE with n1-standard-16 instance on Linux Debian 7 operating system) to 1.99 USD (Azure A11 on Windows Server 2012 R2). The difference is significant (about 4 times). However, if we take into consideration the pricing model, the real cost (that is the billing from vendor) differs even more. The cost of using AWS is more than 10 times the cost using Azure or GCE. This is because

TABLE 17.2: Cost of cloud computing case: Portfolio simulation

Cloud platform	Instance type	PPC (USD/hour)	Cost (USD)	Simplified (USD)
AWS	c4.4xlarge (Win)	0.104	26.672	0.926
AWS	c4.4xlarge (Linux)	0.063	16.048	0.566
AWS	c3.8xlarge (Win)	0.094	24.064	1.324
AWS	c3.8xlarge (Linux)	0.060	15.296	0.701
Azure	A11 (Win)	0.219	2.8	1.991
Azure	A11 (Linux)	0.149	1.275	1.073
Azure	D14 (Win)	0.148	1.898	1.550
Azure	D14 (Linux)	0.096	1.234	0.829
GCE	n1-standard-16 (Win)	0.095	4.053	1.486
GCE	n1-standard-16 (Linux)	0.055	2.347	0.583

AWS uses an hour-based pricing model. Users should be able to allocate their computation as units of hours to reduce the cost of computation with AWS.

The report provides valuable insight about the effect of pricing models on the cost of cloud computing. Together with the benchmarks on instance performances, this should provide readers some information on how to choose cloud vendors.

17.5.4 Algorithm design

Designing a well-suited algorithm for a specific problem can significantly boost the performance and benefit from cloud computing.

We use the following simple example to illustrate how the design of algorithm can change the performance.

Example 17.3: Distributed Matrix Multiplication

M is a matrix of size $d \times n$. N is another matrix of size $n \times d$. The matrix multiplication of M and N: $G = MN$ can be done via two schemes.

- Scheme 1: Inner Product. The entry in ith row and jth column of matrix G: $G_{i,j} = \sum_{r=1}^{n} M_{i,r} N_{r,j}$.

```
1 % Scheme 1: via inner product
2 cloudfor i=1:d
3         cloudfor j=1:d
4                 G(i,j) = M(i,:)*N(:,j)
5         cloudend
6 cloudend
```

- Scheme 2: Outer Product. Alternatively, we can return the matrix G as the sum of the outer products between corresponding rows and columns of M and N.

```
1 % Scheme 2: via outer product
2 cloudfor i = 1:n
3         cloudfor j = 1:n
4                 %cf:sum=G
5                 G =M(:,i)*N(j,:)
6         cloudend
7 cloudend
```

%cf:sum=G command is used to sum the return value from each worker node.

The two schemes differ in both storage complexity and computational complexity.

1. To send the data, scheme 1 requires a distributed storage of $O(nd^2)$ while scheme 2 requires $O(nd)$
2. To return the result, scheme 1 requires a local storage of $O(1)$ and total of $O(d^2)$ while scheme 2 requires a local storage of $O(d^2)$ and total of $O(nd^2)$
3. Both schemes require a distributed computation of $O(nd^2)$. Scheme 1 requires a local computation of $O(n)$ on each worker node while scheme 2 requires $O(d^2)$.

Depending on the relative value of n and d, one of the schemes outperforms the other scheme. In terms of computation, outer product scheme parallel computation in the direction of n, thus is preferred when n is large, while the inner product scheme is preferred in large d scenario.

17.6 Case Studies

17.6.1 Portfolio backtesting

In this section, we demonstrate how portfolio backtesting can be accelerated with a distributed computing technique, in particular the Techila Middleware solution. Backtesting is widely used in financial industry to estimate the performance of a trading strategy or a predictive model using historical data. Instead of gauging the performance using the time period forward, which may take many years, traders/portfolio managers can measure the effectiveness of their strategies and understand the potential risks by backtesting on the prior time period using the datasets that are available today. Computer simulation of the strategy/model is the main part of modern backtesting procedure. It might be very time-consuming due to a few computing issues raised during the procedure. Thus it is necessary to seek acceleration using modern techniques and shorten time-to-market in the rapidly changing financial world.

17.6.1.1 Potential computing bottleneck

Backtesting requires three main components: Historical Datasets, Trading Strategy/Model, and Performance Measure. The following computing issues might be raised for each of the components:

1. The datasets used for testing might be huge while the requested output (performance measure) is relatively small. For example, a portfolio consists of N assets. Its historical return data series over the past T time period is $N \times T$. The covariance matrix is of size $N \times N$. In case $N = 75{,}000$, the memory size of the covariance matrix is 450 GB (double precision).

2. The simulation of the strategies/models can be computational-intensive. The intensity of computing is increasing w.r.t the complexity of the strategies/models. Complex logic branching operations can also be involved.

3. The evaluation of the performance measure can also be time-consuming. Measures that are based on the Monte Carlo approach require simulation of thousands and even more paths.

Cloud computing has its natural advantage of processing large data. In general, CPU threads have better performance than GPU threads, especially in handling complex logic branching operations. Thus cloud computing seems to be a suitable technique for accelerating backtesting procedure. To illustrate how to use cloud computing for backtesting, we did some experiments in Microsoft Windows Azure Cloud, as well as a local cluster. The results are presented in the following sections.

17.6.1.2 Computing environment and architecture

By installing Techila SDK on their computer, end users (traders/portfolio managers) can use Techila-enabled computing tools: MATLAB, R, Python, Perl, C/C++, Java, FORTRAN, and so on to access the computing resource managed by Techila Server.

Techila Server works as a resource manager, as well as a job scheduler. Computational jobs are distributed through Techila Server to Techila Workers, which are machines in the Cloud (Azure, Google Computing Engine, Amazon EC2, and so on) or local cluster. When the computation on the worker node finished, the requested results are sent back to the end-user through

Techila Server. In our experiment, we use Techila environment on Windows Azure Cloud. The testing code is written in MATLAB. The computational jobs (optimization and evaluation of each data set) are sent to each of the worker nodes (virtual machine in Azure).

When dealing with large data sets which exceed single machines capability, there are two solutions: (1) Data sets can be stored in a common storage that can be accessed by each of the workers (Blob on Azure for example); (2) Under specific license, workers can easily access data sources such as Bloomberg, Thomson Reuters, and so on.

17.6.1.3 Experiment design and test result

The callback feature of Techila enables streaming results when a computational job is finished. This can also be used to monitor intermediate results of a computational job. This enables us to update the visualization of the result when a computational job is finished. To visualize the result, we plot the time evolution of efficient frontier over the backtesting period as a 3D surface. We also plot the maximum Sharpe Ratio portfolio as a 3D line. In the mean-variance optimization framework of Markowitz, this portfolio is the tangency portfolio. Thus we should expect the line is on the surface[7] as shown in Figure 17.9.

We first perform a small-scale test using 20 stocks. The results based on cloud computing are consistent with the results generated from local run on my own laptop. Using weekly return from 2000 to 2013, we perform several tests using different number of stocks and different length of backtesting period. We set the historical estimation window length equal to 60 weeks, the strategy is re-estimated every 3 weeks. The weekly return data from February 26, 2001, to October 7, 2013, are separated into 220 windows. A straightforward way of distributing the computational load is to treat the backtesting for each window as independent job.

By default, Techila Middleware will automatically distribute the computing project such that each job will have sufficient length to reduce the overhead caused in data transfer. A user can also set the job length (iterations per job) using the job specification parameter.

We ran tests for 50, 100, and 500 stocks. When the number of stocks increased, the optimizer will take a longer time to find the portfolio that maximizes Sharpe ratio. In fact, when the number of stocks is too large, the optimization problem might became an ill-posed problem. However, the performance of the optimizer is not the concern of this report. Compared with simply setting the *stepsperworker= 1*, Techila's default setting significantly improved the CPU efficiency (CPU time/Wall clock time) as shown in Table 17.3.

[7]The visualization code is adapted from Portfolio Demo by Bob Taylor at http://www.mathworks.com/matlabcentral/fileexchange/31290-using-matlab-to-optimize-portfolios-with-financial-toolbox

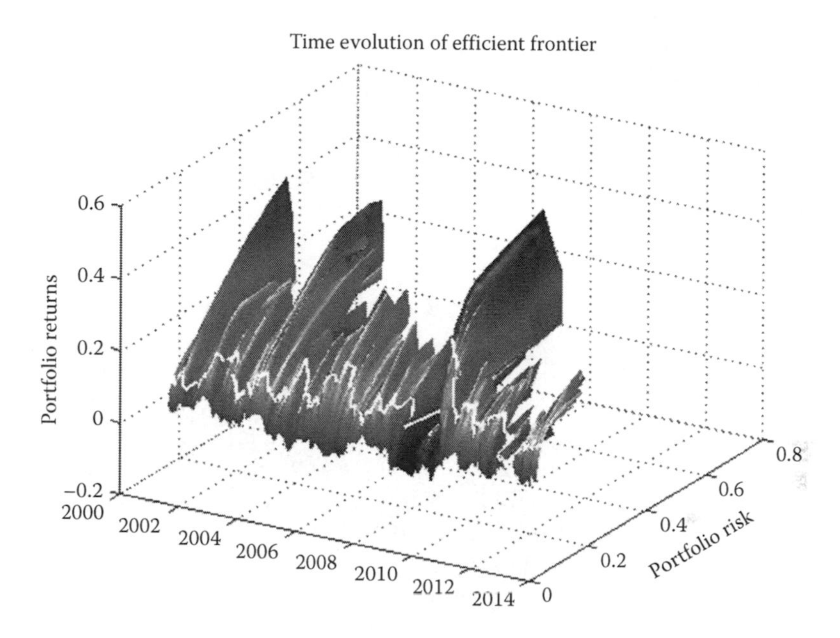

FIGURE 17.9: Time evolution of efficient frontier.

TABLE 17.3: CPU efficiency in portfolio backtesting

NoS	NoJ (step = 1)	ACE (step = 1) (%)	NoJ (Auto)	ACE (Auto) (%)
50	220	88.14	55	96.57
100	220	90.49	74	112.79
500	220	114.13	220	114.13

NoS is number of stocks; *NoJ* is number of jobs; *Auto* refers to Techila's automated job distribution scheme; *step* = 1 refers to assigning 1 step to each job.

17.6.2 Distributed portfolio optimization

In this section, we demonstrate how a large-scale portfolio optimization problem can be solved with a distributed computing technique with specific algorithm design.

17.6.2.1 Challenges in large-scale portfolio construction

Constructing an optimal portfolio consists of two steps. The first step is to construct future belief of the return distribution, which is essentially an inference and prediction problem. The second step is to find the optimal portfolio weights, which is an optimization problem that deals with the trade-off between portfolio risk and portfolio return.

On one hand, this problem is a statistical challenge. Most of the portfolio optimization and risk minimization approaches require estimation of the covariance or its inverse of the return series. When using the sample variance as the expected variance, the estimation error could be large. To achieve a reasonable accuracy, as stated in DeMiguel et al., 2009, an in-sample period of 3000 months is needed for a portfolio of 25 assets to beat naive 1/N strategy. The problem becomes even more significant when the portfolio size is large. As noticed in Fan et al., 2011, estimating the moments of high-dimension distribution is challenging. Among them, one crucial problem is the spurious correlation arise with the curse of dimension.

On the other hand, the problem is also challenging numerically. First, when the degree of freedom is large, finding optimum in high-dimension parameter space is almost impossible to achieve in reasonable time with general optimizers. Additionally, we need to take good care of the property of the matrices to retain feasibility. It is also a data-intensive problem from a hardware perspective. Suppose we are dealing with 75,000 assets (data of the universe), the covariance matrix has 2,812,537,500 parameters. That means, it takes more than 20 GB of memory if we are using double precision. Last but not least, the matrix operation for matrix size of $M \times N$ has a linear computational cost increase with the number of columns.

17.6.2.2 Algorithm design for large-scale mean-variance optimization problem

In the classical Markowitz's mean-variance framework, the portfolio optimization problem is to minimize the variance for given expected return $b = w^{\mathrm{T}}\mu$. The optimum w^* is a solution to

$$\min w^{\mathrm{T}} C w$$

s.t.

$$w^{\mathrm{T}}\mu = b$$
$$w^{\mathrm{T}} 1_N = 1$$

This optimization problem is equivalent to solve:

$$\min E[|\rho - w^{\mathrm{T}} r_t|^2] \tag{17.1}$$

with the same restriction. $\rho = 1_T b$. By replacing the expectation in Equation 17.1 with its sample average, the problem can be considered as a least-square regression.

Regularization methods are introduced targeting to solve the problem that arises with estimation error via shrinkage and achieve either stability and/or sparsity. The regularization can be achieved by adding l^n-penalty term $r(x)$ to the objective function:

$$r(x) = \lambda \left\| x \right\|_n$$

where λ is a constant that scales the penalty term. When $n = 1$, the objective is a LASSO regression. While $n = 2$, it is a ridge regression.

In order to find a solution to this penalized problem, moreover to utilize the modern computing environment–computer cluster/cloud, we solve the problem using distributed optimization technique, namely the alternative direction method of multiplier (ADMM) and block splitting. A detailed introduction of this optimizer can be found in Boyd et al., 2011.

Noticing that we can transform the constraint optimization problem to its consensus form:

$$\min||b1_T - Rw||_2^2 + \lambda||w||_1 + I_C(w) \tag{17.2}$$

where I_C is the indicator function, that is,

$$\begin{aligned} I_C(w) &= 0, \quad \text{if } w \in C \\ I_C(w) &= \infty, \quad \text{if } w \notin C, \end{aligned} \tag{17.3}$$

where C is the constraint set $C = \{w|w^{\mathrm{T}}\mu = b, w^{\mathrm{T}}1_N = 1\}$.

Here we rewrite the problem in ADMM form (denote $b1_T := B$):

$$w_1^{k+1} := \text{argmin}_w \left(\left(\frac{1}{2}\right)||Rw - B||_2^2 + \left(\frac{\rho}{2}\right)||w - z^k + \mu_1^k||_2^2\right) \tag{17.4}$$

$$w_2^{k+1} := \Pi_C(z - \mu_2) \tag{17.5}$$

$$z^{k+1} := \frac{1}{2}\left(S_{\lambda/\rho}\left(w_1^{k+1} + \mu^k\right) + S_{\lambda/\rho}\left(w_2^{k+1} + \mu^k\right)\right) \tag{17.6}$$

$$\mu_1^{k+1} := \mu_1^k + w_1^{k+1} - z^{k+1} \tag{17.7}$$

$$\mu_2^{k+1} := \mu_2^k + w_2^{k+1} - z^{k+1}. \tag{17.8}$$

The update of w_1 is Tikhonov-regularized least squares which have analytical solution:

$$w_1^{k+1} := (R^{\mathrm{T}}R + \rho I)^{-1}(R^{\mathrm{T}}B + \rho(z^k - \mu^k))$$

Via Block Splitting, we can utilize distributed computing environment and solve the problem for small data block of R and B in parallel, depending on the data structure. If the number of assets is larger than the length of historical time series, allocation is preferred. While in the other case, consensus is preferable alternative that is also easier to implement, since it is consistent with the previous decomposition

$$w_i^{k+1} := \left(R_i^{\mathrm{T}}R_i + \rho I\right)^{-1}\left(R_i^{\mathrm{T}}B_i + \rho(z^k - \mu^k)\right).$$

17.7 Cloud Alpha: Economics of Cloud Computing

In their review of cloud computing, Armbrust et al. (2010) proposed a formula to evaluate the economic value of cloud computing by comparing to alternative solutions[8]:

$$UserHours_{cloud} \times (revenue - cost_{cloud}) \geq UserHours_{datacenter} \times \left(revenue - \frac{Cost_{datacenter}}{Utilization}\right).$$

The cost of cloud computing or alternative solutions can be summarized into TCO.

Cloud computing and alternative solutions may have different risks of IT failure. The effect on risk measure should be taken into account when evaluating the potential benefit of cloud computing. Thus we derive the following formula of benefit as the change in revenue plus cost reduction and benefit of risk control:

$$Benefit_{cloud} = \Delta(Revenue) - \Delta(TCO) - \gamma\Delta(Risk) \tag{17.9}$$

where $\Delta(Revenue) = Revenue_{cloud} - Revenue_{alternative}$ is the profit difference from cloud versus alternative solution. $\Delta(TCO) = TCO_{cloud} - TCO_{alternative}$ is the negative of cost reduction. $\Delta(Risk) = Risk_{cloud} - Risk_{alternative}$ measures the change in risk and γ is the risk premium.

The optimal choice of computing solutions is simply the optimum of the following Markowitz-style objective:

$$\max_{s \in \mathbb{S}} Revenue_s - TCO_s - \gamma Risk_s$$

where $\mathbb{S}$ is the set containing all feasible computing solutions.

Quantitative measuring of revenue, cost, and risk is a difficult task and is beyond the scope of this book. Thus, in the following subsections, we only provide qualitative analysis of cost, revenue, and risk to give some intuition of the economics of cloud computing.

17.7.1 Cost analysis

Financial market reduces transaction cost. As an example, asset managers issue ETF and ETN to investors, offering them a lower cost of diversification and exposures to risks and markets that may be costly for an individual investor to access. Cloud computing, by pooling computing resources, offers clients lower TCO and access to up-to-date hardware. Cloud computing may offer cost reduction along one of the following dimensions:

- The first dimension of cost reduction is from lower cost of hardware maintenance and upgrade.
- The second dimension of cost reduction is from elasticity of cloud computing.

[8] Here they compare cloud computing with a dedicated data center.

- The third dimension of cost reduction is from lower cost of human resources.

17.7.2 Risks

The risk of IT system failure is nonnegligible in the finance industry. The following two examples provide some ideas of the importance of having backup IT systems and highly reliable IT systems.

Example 17.4: NYSE and Bloomberg

The New York Stock Exchange crashed at 11:32 am ET, July 8, 2015. The exchange was down for 3 hours and 38 minutes. According to NYSE (reference online document), this was due to a software update to the IT system.

Coincidentally, Bloomberg terminals suffered a widespread outage on April 17, 2015, affecting more than 325,000 terminals worldwide.

IT failure can be costly; however, what would be the best way of risk management for IT systems? *Cloud computing can be viewed as an insurance of IT.* While diversification is a widely accepted concept in the finance industry, cloud computing may be an easy way to diversify the IT failure risk for the finance industry. The distributed file systems, either in-house or in cloud vendors' data centers, protect data from hardware failures. Cloud vendors also offer access to computing to data centers located in various locations around the world. Such a scheme provides constant supply of computing resources in case of catastrophic tail events, such as earthquakes, tsunamis, and so on.

References

Armbrust, M., Fox, A., Griffith, R., Joseph, A.D., Katz, R., Konwinski, A., Lee, G., Patterson, D., Rabkin, A., Stoica, I. et al., 2010. A view of cloud computing. *Communications of the ACM* 53, 50–58.

Baesens, B., Van Gestel, T., Viaene, S., Stepanova, M., Suykens, J., Vanthienen, J., 2003. Benchmarking state-of-the-art classification algorithms for credit scoring. *Journal of the Operational Research Society* 54, 627–635.

Boyd, S., Parikh, N., Chu, E., Peleato, B., Eckstein, J., 2011. Distributed optimization and statistical learning via the alternating direction method of multipliers. *Foundations and Trends in Machine Learning* 3, 1–122.

DeMiguel, V., Garlappi, L., Uppal, R., 2009. Optimal versus naive diversification: How inefficient is the $1/n$ portfolio strategy? *Review of Financial Studies* 22, 1915–1953.

Fan, J., Lv, J. Qi, L., 2011. Sparse high dimensional models in economics. *Annual Review of Economics* 3, 291.

Hand, D.J., Henley, W.E., 1997. Statistical classification methods in consumer credit scoring: A review. *Journal of the Royal Statistical Society. Series A (Statistics in Society)* 160, 523–541.

Joseph, E., Conway, S., Dekate, C., Cohen, L., 2014. IDC HPC update at ISC'14.

Kanniainen, J., Lin, B., Yang, H., 2014. Estimating and using garch models with vix data for option valuation. *Journal of Banking & Finance* 43, 200–211.

Kanniainen, J., Piché, R., 2013. Stock price dynamics and option valuations under volatility feedback effect. *Physica A: Statistical Mechanics and its Applications* 392, 722–740.

Little, M., 2011. ESG and Solvency II in the cloud. *Moody's Analytics Insights.* Published in Barrie+Hibbert (later Moody's Analytics) magazine, see http://docplayer.net/5565696-Esg-and-solvency-ii-in-the-cloud.html.

Mell, P., Grance, T., 2009. The NIST definition of cloud computing. *National Institute of Standards and Technology* 53, 50.

Smith, D.M., 2008. Cloud computing scenario.

Techila, T., 2014. Cloud benchmark—round 1.

Techila, T., 2015. Cloud HPC in finance, cloud benchmark report with real-world use-cases.

West, D., 2000. Neural network credit scoring models. *Computers & Operations Research* 27, 1131–1152.

Yang, H., Kanniainen, J., 2017. Jump and volatility dynamics for the S&P 500: Evidence for infinite-activity jumps with non-affine volatility dynamics from stock and option markets. *Review of Finance* 21, 811–844.

Chapter 18

Blockchains and Distributed Ledgers in Retrospective and Perspective

Alexander Lipton

CONTENTS

PARATOV. The madness of passion soon passes, and what remains are chains and common sense that tells us that these chains are unbreakable. LARISA. Unbreakable chains!

Alexander Ostrovsky
Without a Dowry, A drama in four acts

18.1 Introduction

In this chapter, we discuss blockchains (BCs) and distributed ledgers (DLs) in retrospective and prospective, with an emphasis on their applications to money and banking in the twenty-first century. Additional aspects are discussed in References 1 through 3.

Civilization is not possible without money, and, by extension, banking, and vice versa. Through the ages, money existed in many forms, stretching from the exquisite electrum coins of the Phrygian King Midas, giant stones of Polynesia, cowry shells, the paper money of Khublai Khan and other rulers who came after him, to digital currencies, and everything in between. The meaning of money has preoccupied rulers and their tax collectors, traders, entrepreneurs, laborers, economists, philosophers, writers, stand-up comedians, and ordinary folks alike. It is universally accepted that money has several important functions, such as a store of value, a means of payments in general, and taxes in particular, and a unit of account. The author shares the view of Aristotle formulated in his *Ethics*: "Money exists not by nature but by law" [4]. Thus money is linked to government and government to money. In fact, anything taken in lieu of tax eventually becomes money.

For the last five centuries, money has gradually assumed the form of records in various ledgers. This aspect of money is all important in the modern world. At present, money is nothing more than a sequence of transactions, organized in ledgers maintained by various private banks, and by central banks who provide means (central bank cash) and tools (various money transfer systems) used to reconcile these ledgers. In addition to their ledger-maintaining functions, private banks play a very important role, which central banks are not equipped to perform. They are the system gatekeepers, who provide know your customer (KYC) services, and system policemen, who provide antimoney laundering services (AMLs). We argue that, in addition to the more obvious areas of application of distributed ledger technology (DLT), for instance, digital currencies (DCs), including central bank issued digital currencies (CBDCs), DLT can be used to solve such complex issues as trust and identity, with an emphasis on the KYC and AML aspects [5]. Further, given that all banking activities boil down to maintaining a ledger, judicious applications of DLT can facilitate trading, clearing and settlement triad, payments, trade finance, and so on.

The chapter is organized as follows. We introduce DLs and briefly discuss their different types in Section 18.2. We present historical instances of BCs and DLs in Section 18.3, and describe what happened when they underwent hard forking. Bitcoin, the most popular current application of DLT, is covered in Section 18.4, where a few less well-known facts about bitcoin are presented. Potential applications of DLT to banking are discussed in Section 18.5. As an interesting potential area of applications of BC/DLT, we introduce a modern version of monetary circuit in Section 18.6 and show that it can benefit from the BC/DL framework because money moves in a gigantic circle (or several circles if the world economy as a whole is considered). In addition, in the process of money creation by the banking system as a whole, individual banks become naturally interconnected, so that DLs are particularly suitable to describing their interactions. We discuss topics related to CBDCs in Section 18.7, where we explain the rational for its issuance and discuss practical aspects. In particular, we show that CBDCs can be used to implement the famous Chicago plan [6,7], of moving away from the fractional reserve banking toward the narrow banking. We articulate the differences between Chaum's and Nakamoto's approaches to DCs and consider their respective pros and cons. Conclusions are drawn in Section 18.8.

18.2 Blockchains and Distributed Ledgers

Databases with joint writing access have been known for decades. Several typical examples are worth mentioning: the concurrent versioning system (CVS), Wikipedia, and distributed databases used on board of naval ships [8].

We start with articulating differences between centralized and distributed databases. In a centralized database, storage devices are all connected to a common processor; in a distributed database, they are independent. Furthermore, in a centralized database, writing access is tightly controlled; in a distributed database, many actors have writing privileges. In the latter case, each storage device maintains its own growing list of ordered records, which, for the sake of efficiency, can be organized in blocks, hence, the name Blockchain. To put it differently, in a traditional centralized ledger, the gatekeeper collects, verifies, and performs the write requests of multiple parties, tasks which are distributed in the DL. It should not be taken as fact that these tasks are best distributed: the considerations of efficiency and specialization are relevant as well.

The integrity of the distributed database is cryptographically ensured at two levels. First, only users possessing private keys can make updates to "their" part of the ledger. Second, notaries (also called miners) verify that users' updates are legitimate. Once the updates are notarized, they are broadcast to the whole network, thus ensuring that all copies of the distributed database are in sync. There are several types of distributed databases or ledgers. We list them in increasing order of complexity:

1. Traditional centralized ledger
2. Permissioned private DL (R3 CEV, DAH, and other similar projects)
3. Permissioned public DL (Ripple, and so on)
4. Unpermissioned public DL (Bitcoin, Ethereum and the myriad others)

To control the integrity of DL, a variety of mechanisms can be used—proof of work (PoW), proof of stake (PoS), third party verification, and so on.

Which ledger should be used? It largely depends on the context. If no joint writing access is required, as is the case with most legacy banking applications, a centralized ledger can be used. If participants do need joint writing access, but know each other in advance, have aligned interests, and can be trusted, as is the case in clearing and settlement, a permissioned private DL can be employed. More details are given in Reference 9.

The best known application of BC/DL is the famous bitcoin, which exists on an unpermissioned public DL whose integrity is maintained by anonymous miners via PoW mechanism. BC/DL can be used for issuing CBDCs. However, the sheer scale of the economy precludes unpermissioned public ledger in the spirit of Nakamoto [10], to be used for this purpose, due to the enormous computational effort required for PoW. Resurrecting digicash proposed by Chaum [11] is an exciting possibility.

In many instances, building a DL just to be *au courant* with the times might not be worth the effort.

18.3 Historical Examples of BCs and DLs

18.3.1 Genealogical trees

The idea of a BC is certainly not new. BCs naturally occur whenever power, land, or property change hands. Some of the earliest examples of BC are the genealogical trees of royal (or, more generally, aristocratic or property owning) families. In such a tree (or BC), the transfer of power from one sovereign to the next is governed by well-defined rules and in most cases, occurs without commotion. However, when these rules become ambiguous and open to interpretation the tree can undergo a hard fork.

In addition to being a chain, a genealogical tree is a DL. In order to agree on their respective legitimacy and marriage eligibility, royal houses had to inform each other about births, deaths, marriages, and other life events, thus keeping their versions of BCs in sync. In Figure 18.1, we show the genealogical tree of the House of Habsburg engraved by A. Durer. It was distributed to other royal houses, as well as all imperial cities in the Holy Roman Empire.

Usually, forking of a succession tree is associated with wars and other acts of violence. This is a cautionary tale for proponents of ubiquitous applications of DLs without a possibility of resolving disputes outside of the ledger itself. Here are two (of many) examples.

In Figure 18.2, we show a simplified genealogical tree of the House of Capet. For 10 generations, starting with Hugh Capet, the transfer of power from father to son was smooth. However, the ambiguity occurred when all three sons of Philip IV died without surviving issue, thus creating a power vacuum.

FIGURE 18.1: Albrecht Dürer—The Triumphal Arch of Maximilian *(Ehrenpforte Maximilians I)*—The House of Habsburg generalogical tree by A. Durer. It was distributed to all imperial cities in the Holy Roman Empire. (Albrecht Dürer—National Gallery of Art: online database: entry 1991.200.1. Public Domain.)

In order to resolve it, the peers of France applied the Salic law of Succession, by which persons descended from a previous sovereign only through a woman are not eligible to occupy the throne. The House of Plantagenet did not accept this outcome and started the Hundred Years' War (1337–1453) against the House of Valois, a cadet branch of the Capetian dynasty, which was a dynastic conflict for control of the Kingdom of France. In the end, the Valois established themselves as Kings of France at the expense of the Plantagenets.

Similar conflicts occurred with regularity and for very similar reasons throughout history. For example, the War of the Austrian Succession (1740–1748), which involved all major powers of Europe, was fought to settle the question of the Pragmatic Sanction and to decide whether the Habsburg hereditary possessions could be inherited by a woman. It was finally resolved in favor of Maria Theresa, who became the only female ruler of the Habsburg dominions.

Closer to our times, an interesting example of Ethereum hard forking happened in July of 2016, as a result of fixing a theft of 60 Mil USD worth of Ethereum from DAO. Buterin [12] described the situation as follows:

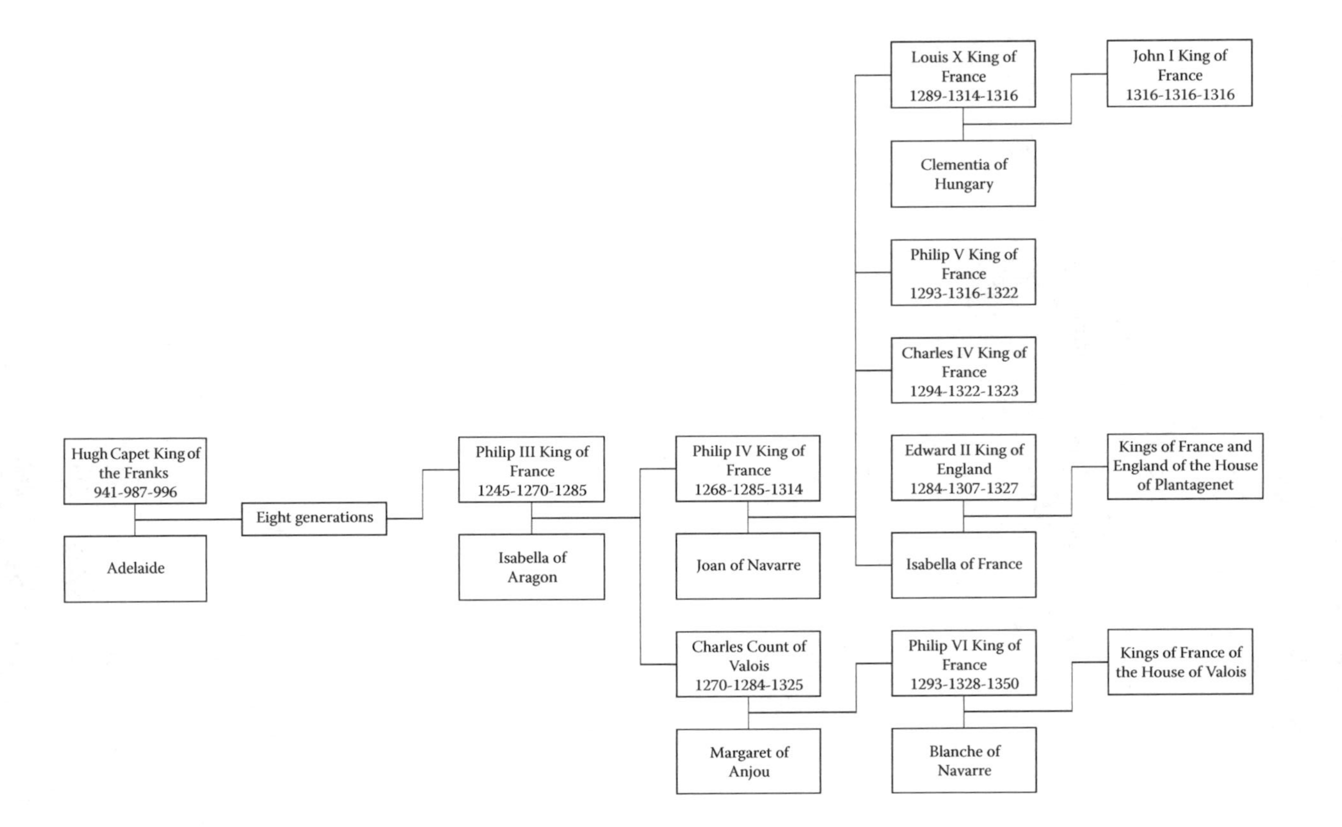

FIGURE 18.2: Genealogical chart (chain) of the House of Capet. Hard fork was resolved in favor of the House of Valois at the expense of the House of Plantagenet by inventing the Salic law. The Hundred Years' War commence as a result. (Adapted from Wikipedia.)

> The foundation has committed to support the community consensus on the admittedly difficult hard fork decision. ... That said, we recognize that the Ethereum code can be used to instantiate other blockchains with the same consensus rules, including testnets, consortium and private chains, clones and spinoffs, and have never been opposed to such instantiations.

Once again, we see that ambiguity within a BC cannot be resolved via its intrinsic mechanisms.

18.3.2 Land titles

In more recent times, land registry title deeds are more relevant examples of BCs. As per Land Registry,

> Title deeds are paper documents showing the chain of ownership for land and property. They can include: conveyances, contracts for sale, wills, mortgages and leases.

It is clear that titles are BCs currently held in a central repository; instead of miners, succession is verified by notaries. Titles are meaningful candidates for being treated on DL. However, there are still some issues which need to be resolved before it can be done. For example, recent lawsuits by Mark Zuckerberg seeking to force hundreds of Hawaiians to sell to him small plots of land located within the external boundaries of his 700-acre beachfront property on the island of Kauai, is a good case in point. It illustrates that in some instances, it is not possible to identify the first owner of land, and then build a chain of ownership from the original owner to the present, resulting in an ambiguous and potentially vulnerable BC.

18.4 The Bitcoin Ecosystem

Bitcoin is not the first digital currency by a long shot, and very likely is not the last major one either. The astute reader will recognize that apart from intriguing technical innovations, bitcoin does not differ that much from the fabled tally sticks used in the Middle Ages. Its precursors include e-cash and digicash invented by D. Chaum [11], and bitgold invented by N. Szabo [13].[1]

[1]There is a heated debate of the true identity of Satoshi Nakamoto. Nick Szabo is often mentioned as a potential inventor of bitcoin. Here is a small piece of evidence, which might be of interest. Nakamoto's initials are SN, while Szabo's are NS. However, Szabo is originally a Hungarian name, where the last name comes first, so his initials would be SN. An interesting coincidence.

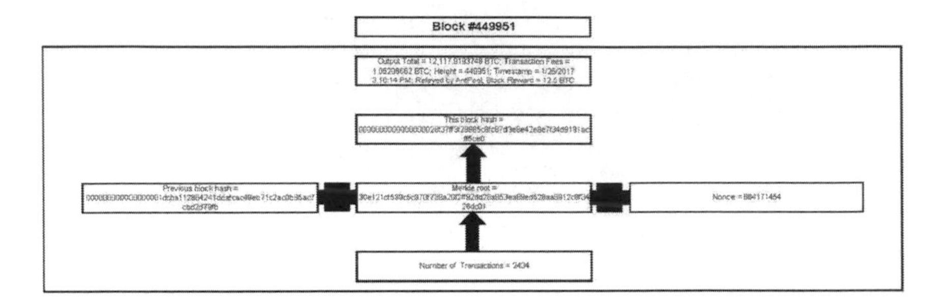

FIGURE 18.3: A typical BTC block. (Adapted from blockchain.info.)

All building blocks of the bitcoin ecosystem have been known for some time, including two of the most important techniques in public-key cryptography, a one-way hash function and the Elliptic Curve Digital Signature Algorithm (ECDSA), (see References 14 through 16). Proof of work, based on cryptographic hash functions, specifically SHA-256, is similar to hashcash invented by Back [17], while Merkle trees were introduced in the seminal paper by Merkle [18].

Ignoring such nuances as wallets, and so on, we can describe the basic setup as follows. Participants of the system are represented by their public/private key pairs. The main control variable is the number of bitcoins belonging to a particular public key. This number is known to all participants at all times (in theory). The owner of a particular public key broadcasts their intent to send a certain quantity of bitcoins to another public key. Miners aggregate individual transactions into blocks, verify them to ensure that there is no double spend by competitively providing proof of work, and receive mining rewards in bitcoins. A transaction is confirmed if there are at least six new blocks built on the top on the block to which it belongs. A typical block is shown in Figure 18.3.

The size of mining rewards is halved at regular intervals so that the total number of bitcoins in circulation converges to 21 Mil. Currently there are about 16 Mil bitcoins in circulation. It is believed that at least one Mil are irretrievably lost or stolen. Some 450,000 blocks have been mined so far; a new block is mined every 10 minutes on average. Due to the fact that mining rewards are paid with *new* bitcoins, transaction costs are claimed to be very low. This is a nifty bit of sleight of hand, however, because the value of existing bitcoins is constantly diluted. Some representative bitcoin statistics is given in Figure 18.4.

Bitcoin promises are grand. Its proponents expect it to become a supranational currency eventually supplanting national currencies, which, in their minds, can be easily manipulated. Many even believe that bitcoin is the modern digital version of gold, due to the effort required for PoW [13]. Whilst bitcoin is clearly an impressive breakthrough, reality is much less grand than

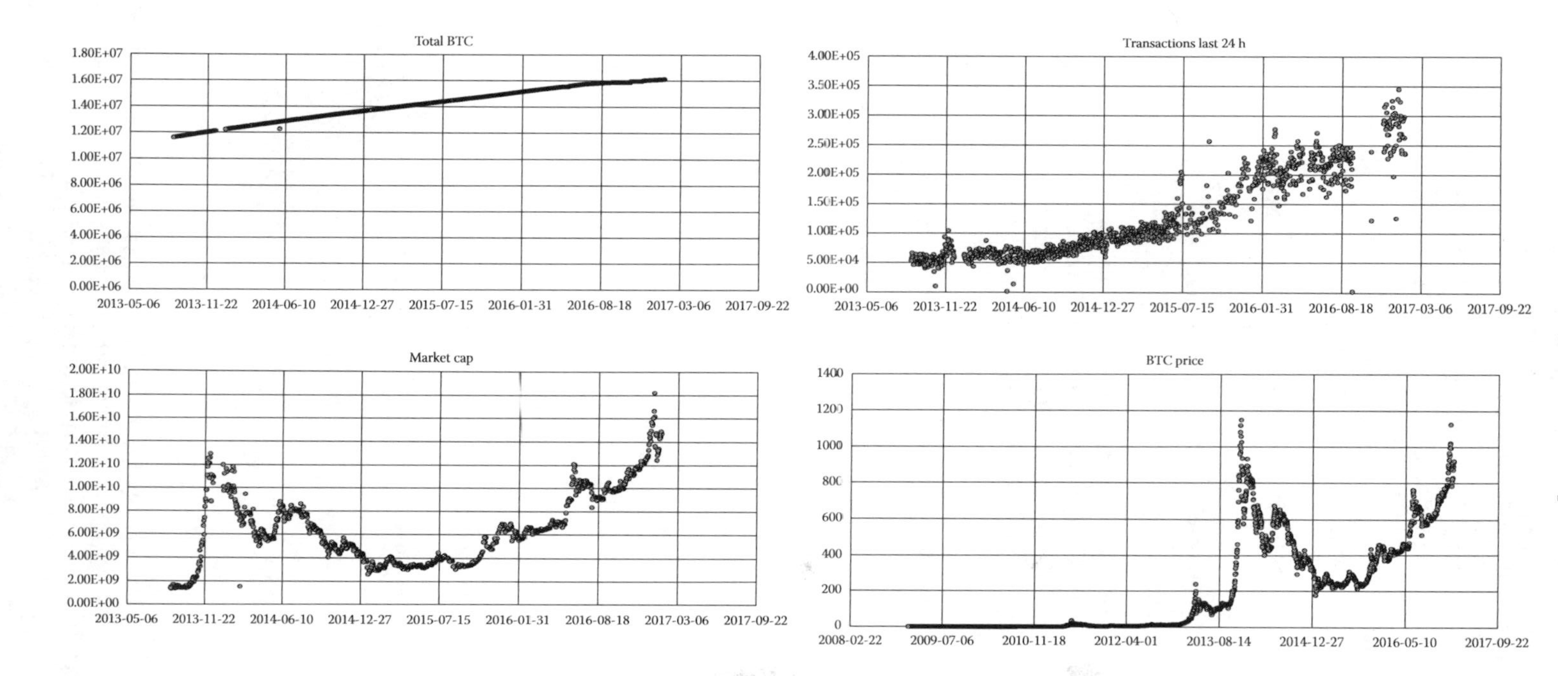

FIGURE 18.4: Some representative bitcoin statistics. (Adapted from blockchain.info.)

perception, and is quite telling (at the time of writing this paper):

1. A new block is created on average every 10 minutes.
2. The number of transactions per second (TpS) is approximately 7, compared to 2000 TpS on average handled by VISA.
3. In monetary terms, the amount of transactions is about 100 Mil USD/day.
4. Current real (not nominal!) transaction costs are 1.5 Mil USD/day, 1.5% of total volume; in 2012 it was whopping 8%, in 2014—6%.
5. Mining is a cost of electricity game. In high energy cost countries miners go bust: Swedish KnCMiner recently declared bankruptcy ahead of halving miner's reward. While exact numbers are not known, it is believed that bitcoin consumes as much electricity as EBay, Facebook, and Google combined.
6. Miners are arranged in gigantic pools (so much for peer to peer (P2P) mining!): AntPool—18.7%, F2Pool—17.7%, BitFury—7.7%, BTCC Pool—7.4%, BW.COM—7.3%. Thus a 51% attack becomes possible! There is a very *high* probability that six consecutive blocks will be mined by the same actor (so much for checks and balances!). Most of all these pools are Chinese, partly due to low electricity cost, partly due to high-tech advances. Not only miners are predominantly Chinese, so are the players—91% CNY, 7% USD, 1% EUR.
7. At the moment, the main purpose of using bitcoin is for speculation and circumvention of capital controls in China.

It is truly amazing to see how miners are prepared to perform socially useless tasks, as long as they are paid for it. A telling historical analogy jumps to mind. During the contest for design of the dome of Santa Maria del Fiore, it was suggested to use dirt mixed with small coins to serve as scaffolding. After the dome's completion, the dirt was to be cleared away for free by the profit-seeking citizens of Florence (proto-miners). It is clear that BC/DL is still awaiting its Brunelleschi who figured out how to build the dome without scaffolding [19].

T. J. Dunning, quoted by Karl Marx in *Das Kapital* [20], put it succinctly:

> With adequate profit, capital is very bold. A certain 10 per cent will ensure its employment anywhere; 20 per cent certain will produce eagerness; 50 per cent, positive audacity; ...

18.5 Potential Usages of DLT in Banking

18.5.1 Banking X-Road

No bank, however big, is an island; banks can only operate as a group. In the process of their day-to-day activities, they become naturally interlinked. Due to these linkages between banks, DLT can provide a useful tool for facilitating, reconciling, and reporting their interactions. Given that internal technology is bank specific, it is impractical to standardize bank infrastructure. However, it is possible to bring them to a common denominator by emulating the success of the Estonian X-Road and creating a DL solution for banking operations, which, by analogy, can be called the e-bank X-Road. In this regard, DL will serve as an adapter, not dissimilar to an electrical adapter.

In 1997, Estonia started to move to digital government. In 2001, Ansper in his master thesis proposed a suitable design [21]. He developed a distributed P2P secure information system called the e-Estonia X-Road based on the idea of an adapter. X-Road is the digital environment which links various heterogeneous public and private databases and enables them to operate in sync. A small company, Cybernetica, implemented this design for around 60 Mil EUR.[2]

Let us describe a possible design for the e-bank X-Road. Given the non-scalable nature of PoW, and unclear security properties of PoS, X-road has to be controlled by trusted notaries or validators. Two financial institutions, represented by their public keys, use their respective adapters to agree on common terms on a deal. They digitally sign and execute a smart contract, hash it, and broadcast the hashed version to the X-Road participants. A quorum of notaries digitally signs the hash ("laminates" it) and reposts the signed hash in the common X-Road layer. Validators are paid for their services, similarly to central securities depositories.[3]

It is worth noting that a BC does not by itself guarantee unambiguous ownership: steps are required to identify and resolve any ambiguities before moving to a BC, and in addition, tools and mechanisms to resolve ambiguities are only discovered when the BC is already well established. Both of these requirements are underemphasized in current discussions of BC/DLT applications.

There are several smaller areas in which DLT can be used to reduce transaction costs and other frictions in the conventional system. Such areas include but are not limited to:

1. Post-trade processing
2. Global payments
3. Trade finance

[2]Other countries tried to follow suit but not all attempts were unqualified successes.

[3]Corda, recently described in a white paper by R3, might be a step in this direction [22].

4. Rehypothecation

5. Syndicated loans

6. Real estate transactions

18.5.2 Trade execution, clearing, settlement

The all-important triad of capital markets is trade execution, clearing, and settlement. While initial public offering of stock is an important rite of passage for a new company, secondary trading is a mechanism for continually reallocating ownership and control in a more or less optimal fashion. In addition to stocks, many other products, such as equity derivatives, interest rate swaps, commodities, and so on, are traded on public exchanges. Moving many over-the-counter (OTC) products to exchanges is an important regulatory imperative [23].

Currently, there are three necessary steps required to trade public securities:

1. Buyers and sellers have to be matched

2. The transaction has to be cleared, that is, novated to a central clearing counterparty (CCP)

3. The transaction has to be settled, that is, delivery versus payment (DvP) has to take place; so that title and money can be transferred as expected.[4]

These steps are characterized by vastly different time scales—trading often takes place in milliseconds, while clearing and settlement take 1–3 days! Although the proverbial T + 2, T + 3 irritate many people, they might be a bit too fast to push for the T + 15″ solution. The actual process is very involved and includes investors, custodial banks, exchanges, brokers (general clearing members of CCPs), CCPs, central securities depositories, regulators, and so on.

It is natural to ask if a different design of exchanges can improve the overall process and make it more stable and less costly. The answer is yes and no. On the pros side, there are several issues which the current set-up solves very well:

1. Counterparty credit risk management

[4]The thriller "Ronin," which is dealing with DvP, is not critically acclaimed [24]. In the author's view, it takes the difficult challenges of transactions among many untrustworthy parties which underlie many great thrillers and brings them to the fore, arguably making "Ronin" arguably one of the greatest of all thrillers ever (Perhaps the ending would have been different had the characters known about DLT).

2. Netting

3. DvP and credit risk more generally, which is addressed by collecting Initial Margin, Variation Margin, and Guarantee Fund contribution from clearing members

4. Anonymity

5. Ability to borrow stocks

On the cons side, numerous issues are rather disconcerting:

1. Cost

2. Speed

3. Need for reconciliation and failures

It is clear that straightforward attempts to apply DLT to clearing and settlement (thankfully, to the best of the author's knowledge, nobody wants to use it in trading *per se*) cannot be successful. The reasons are simple—instantaneous settlement obliterates all the aforementioned advantages of the current system. It increases the money sloshing around by at least an order of magnitude. Thus slow clearing and settlement is not so much a consequence of the technological backwardness of exchanges and CCPs (although they are not always using cutting edge technology), but rather a result of their *modus operandi.*

By using permissioned private ledger(s), one can certainly cut costs, somewhat increase speed of clearing and settlement, and reduce the number of failures and hence the need for reconciliation. In particular, smart contracts, if they can be legally enforced, can solve a *part* of the DvP conundrum, which will require that *both securities and cash* are parts of the same ledger. While smart contracts cannot solve all problems, they represent a step in the right direction. A potential evolution of the trading–clearing–settlement triad is illustrated in Figure 18.5.

18.5.3 Global payments, trade finance, rehypothecation

Global payments is another area where DLT can be potentially useful. It is important to note that, in spite of claims to the contrary, the payment system *is not broken* however, it is rather expensive. For instance, the Real-Time Gross Settlement system works well for domestic transactions but is inefficient and expensive for foreign transactions. Thus some synergies can be gained if a DL, which supports several national currencies at once, is developed to replace the legacy system.

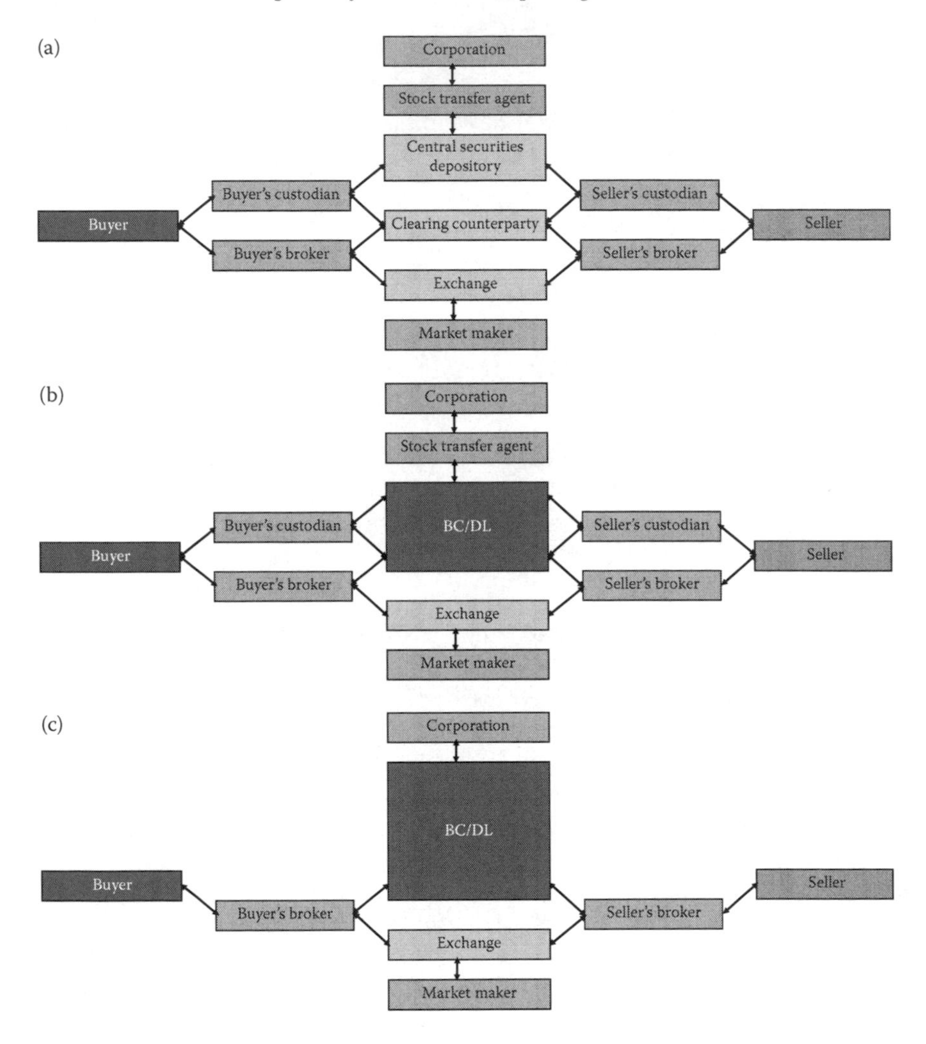

FIGURE 18.5: (a) Current organization of share trading. (b) First improvement of stock trading setup, CSD and CCP are replaced by BC/DL. (c) Second improvement of stock trading setup, in addition to CSD and CCP, custodians and stock transfer agents are replaced by BC/DL.

For trade finance, there is the potential to use BC/DL to simplify the flow of information among all participants and smart contracts to partially solve the DvP problem.

In the rehypothecation setup, it is possible to use BC/DL to untangle the ownership of the collateral. However, this is more of an accounting tool, rather than a comprehensive solution, because in many instances the actual legal ownership of collateral cannot be established with certainty.

18.6 Monetary Circuit and Money Creation

18.6.1 Monetary circuit

For centuries, the origins, properties, and functions of money have been debated in countless expositions. In the fourteenth century, the sagacious French abbot Gilles li Muisis lamented [25]:

> Money and currency are very strange things; They keep on going up and down and no one knows why; If you want to win, you lose, however hard you try.

In the twentieth century, the great British economist John Maynard Keynes shrewdly observed [26]:

> For the importance of money essentially flows from it being a link between the present and the future.

As was mentioned earlier, money is inherently linked with banking, which, over many centuries, gradually evolved from full reserve toward fractional reserve banking. For instance, the Bank of England founded in 1694 already operated as a fractional reserve bank.[5]

In modern societies, commercial banks are almost exclusively fractional and produce money "out of thin air" [27–29]. This important fact is thoroughly misunderstood by the modern macroeconomic thinking, which incorrectly overemphasizes the intermediation aspect of banking and assigns the money creation role to central banks instead of commercial banks. In reality, commercial banks are not constrained by their deposits and can and do issue money at will. At the same time, their ability to do so is restricted by banking regulations, which impose floors on the amount of banks' capital and liquidity, so that money creation cannot go on *ad infinitum*.

To understand the role played by money in the economy, one needs to follow its flow and to account for nonfinancial and financial stocks (cumulative amounts), and flows (changes in these amounts). Here is how Michal Kalecki, the great Polish economist, summarizes the complexity of the issues at hand with his usual flair and penchant for hyperbole [30]:

> Economics is the science of confusing stocks with flows.

[5] The Bank of England was characterized by Marx [20], as follows:

> At their birth the great banks, decorated with national titles, were only associations of private speculators, who placed themselves by the side of governments, and, thanks to the privileges they received, were in a position to advance money to the State. Hence the accumulation of the national debt has no more infallible measure than the successive rise in the stock of these banks, whose full development dates from the founding of the Bank of England in 1694.

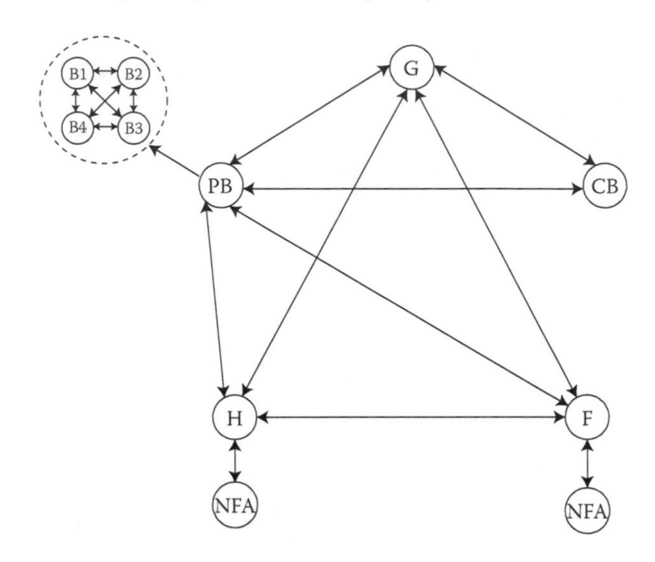

FIGURE 18.6: A sketch of the monetary circuit.

In the author's opinion, the functioning of the economy and the role of money is best described by monetary circuit theory (MCT), which provides a unifying framework for specifying how money lubricates and facilitates production and consumption cycles in the society. MCT describes in the most precise way the dynamics of the economy and explains how and by whom money is created. More specifically, it describes the interactions among the five sectors, including government, central bank, private banks, firms, and households. As part of the monetary circuit, private banks play an outstanding role as credit money creators. In this framework, central banks do not create money directly, but rather accelerate or slow down the process of money creation by private banks by providing a unique universal medium in the form of electronic cash for different banks to control their inventories of assets and liabilities. A schematic representation of the monetary circuit is given in Figure 18.6, which represents money flowing among the above-mentioned five sectors of the economy.

18.6.2 General aspects of money creation

Currently, there are three theories explaining money creation: the credit creation theory, the fractional reserve theory, and the financial theory of intermediation [27–29]. The author firmly believes that only the credit theory advocated by Macleod, Hahn, Wicksell, and Keen among others, correctly reflects the mechanics of linking credit and money creation. Credit creation theory was popular in the nineteenth century, but, unfortunately, gradually lost ground and was overtaken by the fractional reserve theory of banking, which, in turn, was supplanted by the financial theory of intermediation. In the author's view, the latter theory severely underemphasizes the unique and special role of the banking sector in the process of money creation, and cannot rationally explain things like the global financial crisis of 2007–2008 and other similar events, which happen with disconcerting regularity. This aspect is particularly important because currently there is a profound lack of appreciation on the part of the conventional economic paradigm of the special role of banks. For example, banks are excluded from widely used dynamic stochastic general equilibrium models, which are influential in contemporary macroeconomics and popular among central bankers, in spite of the fact that they systematically fail to produce any meaningful results [31]. It is clear that a vibrant financial system cannot operate without banks, and that the banking system is very complex and difficult to regulate because banks become interconnected as a part of their regular lending activities. In addition to their money creation role, banks regulate access to the monetary system, by providing KYC and AML services.

18.6.3 Money creation by individual banks

We start with the simplest situation, and consider a single bank, which lends money to a borrower who immediately deposits it with the same bank. Thus the bank simultaneously creates assets and liabilities. The size of the loan is limited solely by regulations and bank's own risk appetite. The full cycle from money creation to money annihilation is shown in Figure 18.7. Money is pumped into the system (created) when it is lent out by the bank and pumped out (annihilated) when it is repaid. If the borrower repays, the principal is

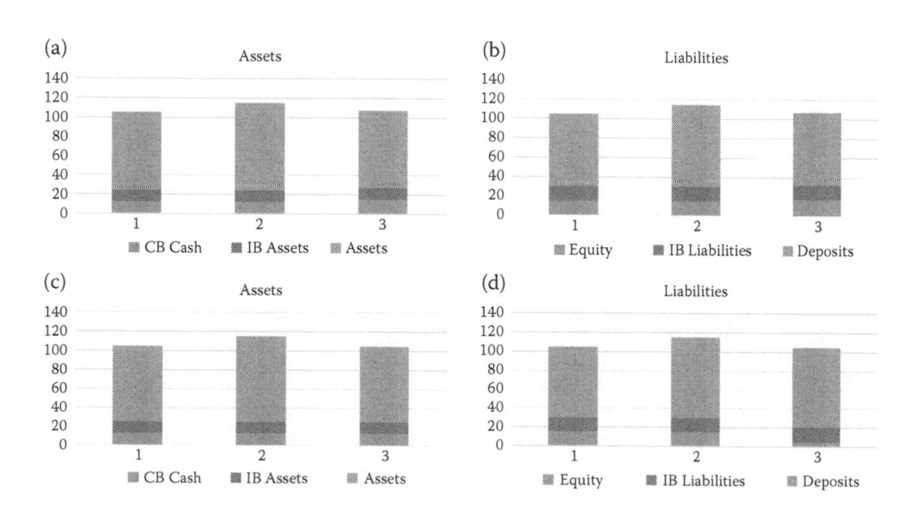

FIGURE 18.7: Money creation by a single bank. (a and b) The case of no borrower's default. (c and d) The case of borrower's default. In the case of no default capital and CB cash increase; in the case of default capital and CB cash decrease.

destroyed, but the interest stays in the system. If the borrower defaults, the money stays in the system indefinitely. The chain of money transfers from one owner to the next is naturally described by a BC, ideally residing on DL.

18.6.4 Money creation by the banking system

A more complex case of asset creation by one bank and liabilities by a second bank is illustrated in Figures 18.8 and 18.9. Linkages between these two banks occur because the first one has to borrow cash from the second, so that their central bank cash holdings reach suitable levels. In this setup, it is clear that central banks do not generate money themselves; instead, they play the role of liquidity providers (if, e.g., the second bank does not want to lend money to the first) and system stabilizers (similar to the Watt's centrifugal governor). Thus central banks are the glue which keeps the financial system together. It is clear that BC is even more relevant in the case in question.

18.6.5 Bank lending versus bitcoin and P2P lending

In view of the above, the key distinction between bank money creation and bitcoin mining, P2P lending, and so on is evident. Banks create money "out of thin air." Since bitcoin transactions are not based on credit, they simply move existing money around. The same is true for P2P transactions—P2P operators are strictly intermediaries, they do not create money at all! Therefore, banks and P2P operators lend on different scales: banks—money they don't have,

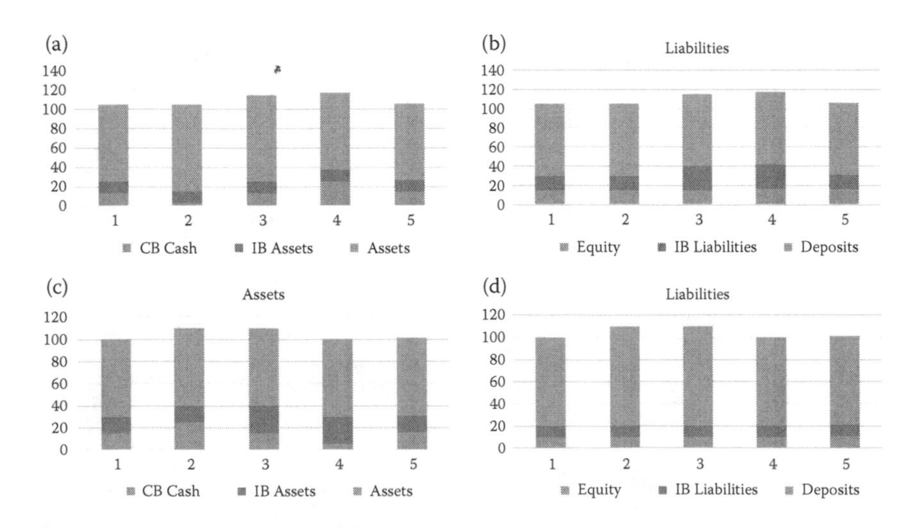

FIGURE 18.8: Money creation by two banks. The case of no borrower's default. (a and b) Assets and liabilities of the first bank. (c and d) Assets and liabilities of the second bank. Capital and CB cash of both banks increase.

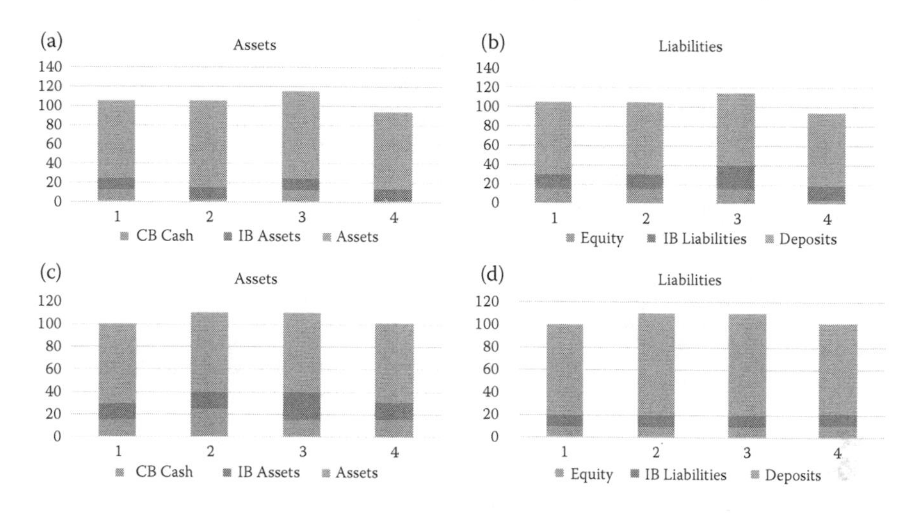

FIGURE 18.9: Money creation by two banks. The case of borrower's default. (a and b) Assets and liabilities of the first bank. (c and d) Assets and liabilities of the second bank. Capital and CB cash of the first bank decrease, while capital and CB cash of the second bank increase.

P2P—only money they have. Hence, the P2P impact on the financial system as a whole is very limited.

18.7 CBDCs and Negative Interest Rates

18.7.1 Why CBDCs?

Can and should central banks issue DCs? Recently, these discussions have been invigorated by the introduction of bitcoin [10], and a persistence of negative interest rates, which plagued Medieval Europe in the form of demurrage, the Brakteaten system, and numerous variations of the same tune for centuries. Recall that demurrage was a tax on monetary wealth and required a massive apparatus of coercion to be imposed efficiently. Today, even the best-in-class economists seem to be unsure of its true nature; for instance, Rogoff equates it with currency debasement, which is a very different mechanism, see Reference 32. The idea of scrip money, that is, money which requires the paying of a periodic tax to stay in circulation, thus emulating demurrage, was proposed by S. Gesell, the German–Argentinian entrepreneur and self-taught economist, in the febrile post-WWI atmosphere [33]. Subsequently, it was regurgitated by Irving Fisher during the Great Depression [34].

In the author's view, it is a sad reflection of the present state of economic affairs, and the level of economic insight, that the current low interest rate environment has prevailed for such a long time, in spite of it being such an ineffective tool. Moreover, in some economies, such as Switzerland and Denmark, interest rates have reached seriously negative levels.[6]

Negative interest rates can be used to simulate inflation; the crucial difference between these two regimes is that physical cash is very valuable under the former, and highly undesirable under the latter. The last line of defense between us and meaningfully negative rates is paper currency. However, in many societies, particularly in Scandinavia, cash is relegated to the far corners of the economy already. It is not hard to imagine that in a few years' time instead of banknotes, we shall have CBDCs [36–38]. Once cash is abolished, interest can be made as negative as desired by central bankers.

18.7.2 How CBDCs can be issued?

Currently, there are two approaches to creating digital currencies on a large scale. The first one, which has gained popularity since the invention of Bitcoin, is based on unpermissioned DL, whose integrity is maintained by notaries (or miners) [39]. Participants in this BC are pseudo-anonymous since they are hidden behind their public keys. However, in principle, they can be identified by various inversion techniques applied to old recorded transactions [40].

An earlier approach was developed by Chaum, who introduced a blind signature procedure for converting bank deposits into anonymous cash [11]. Chaum's approach is much cheaper, faster, and more efficient than the Bitcoin-style one. However, it heavily relies on the integrity of the cash-issuing bank rather than on trustless integrity of Bitcoin secured by computational efforts of miners. Central banks can follow either avenue for issuing digital cash. By doing so, central banks will be indirectly providing access to their balance sheets to the general public. However, in either eventuality, central banks won't be able to perform KYC and AML functions and would still have to rely on commercial banks, directly or indirectly, for doing so.

One possibility is as follows: a central bank issues numbered currency units into DL, whose trust is maintained by designated notaries receiving payments for their services. Thus, at any moment, there is an immutable record showing which public key is the owner of a specific currency unit. Given that notary efforts are significantly cheaper and faster than that of bitcoin miners, this construct is easily scalable to satisfy the needs of the whole economy. Moreover, since the records of transactions are immutable, it is possible to deanonymize transactions thus maintaining AML requirements.

In summary, modern technology makes it possible to abolish paper currency and introduce CBDCs, which can also be used to address some of the societal ills, such as crime, drug trafficking, illegal immigration, and so on, and eliminate costs of handling physical cash, which are of order of 1% of the

[6]One cannot help but notice with a modicum of satisfaction, that critics of the celebrated Vasicek model for interest rates [35], who vigorously attacked him for allowing short rates to become negative, proved to be completely wrong.

country's GDP see, e.g., [38]. It will smooth the motion of the wheels of commerce and help the unbanked to become participants in the digital economy, thus positively affecting the society at large.

18.7.3 How CBDCs can be used to implement the Chicago Plan?

Moreover, CBDCs make the execution of the celebrated Chicago Plan of 1933, originally proposed by Ricardo in 1824, for introducing narrow (full-reserve) banking entirely possible—both firms and ordinary citizens can have accounts directly with central banks, thus negating the need of having deposits with commercial banks [6,7,41–43]. In this case, banks will lose their central position in the economy and become akin to utility providers. They would have to maintain the amount of central bank cash equal to the amount of time deposits. Such narrow banks would in essence become the guardians of the system by providing KYC and AML services and executing simple transactions. In fact, in the wake of the global financial crisis, many central banks massively increased their balance sheets, while commercial banks have chosen to keep enormous quantities of nonmandatory deposits with them. Thus the system *de facto* has moved toward narrow banking.

18.8 Conclusion

While the idea of BC/DLs is not new, modern technology gives it a new lease of life. DLT opens new possibilities for making conventional banking and trading activities less expensive and more efficient by removing unnecessary frictions. Moreover, if built with skill, knowledge, and ambition, it has the potential for restructuring the whole financial system on new principles. We emphasize that achieving this goal requires overcoming not only technical but also political obstacles.

While DLT has numerous applications, it is not entirely clear which financial applications should be handled first. Exchanges, payments, trade finance, rehypothecation, syndicated loans, and other similar areas, where frictions are particularly high, are attractive candidates. DCs, including CBDCs, are another very promising venue.

Currently, many applications of DL and related technology appear to be misguided. In some cases, they are driven by a desire to apply these tools for their own sake, rather than because the result would be clearly superior. In other cases, they are driven by a failure to appreciate that the current systems may not be as they are because of technological reasons, but rather because of business and other considerations.

So far, practical applications of DLT in finance have been limited and a lot remains to be done in order to achieve real breakthroughs.

Acknowledgments

The invaluable help of Marsha Lipton from Numeraire Financial in thinking about and preparing this chapter cannot be overestimated. I am grateful to several colleagues, including Alex Pentland and David Shrier from MIT, Damir Filipovic from EPFL, Matheus Grasselli from McMaster, Julian Phillips from Standard Charter Bank, and Paolo Tasca from UCL for their help and suggestions. As a CEO of StrongHold Labs, I am currently working on a new type of a digital bank, which will be utilizing some of the ideas presented in this chapter. This chapter is reprinted with permission from the *Journal of Risk Finance*, 19(1), 2018.

References

1. Lipton, A., 2016, Banks must embrace their digital destiny, *Risk Magazine*, Vol. 29, No. 8.

2. Lipton, A., Shrier, D., and Pentland, A., 2016, Digital Banking Manifesto: The End of Banks? in *Frontiers of Financial Technology*, Visionary Future, pp. 117–140.

3. Tasca, P., Aste, T., Pelizzon, L., and Perony, N. (Eds.) 2016, *Banking Beyond Banks and Money: A Guide to Banking Services in the Twenty-First Century*, Springer, Switzerland.

4. Aristotle, Aristotle's Nicomachean Ethics, R.C. Bartlett and S.D. Collins (translators), University of Chicago Press, Reprint edition.

5. Zyskind, G., Nathan, O., and Pentland, A., 2015, Enigma: Decentralized computation platform with guaranteed privacy, MIT Working Paper.

6. Allen, W.R., 1993, Irving Fisher and the 100 percent reserve proposal, *The Journal of Law and Economics*, Vol. 36, No. 2, pp. 703–717.

7. Beneš, J. and Kumhof, M., 2012, The Chicago plan revisited, IMF Working Paper.

8. Miller, S.J., 1993, A fully replicated distributed database system, Research Note ERL-0719-RN, Electronics Research Laboratory.

9. Greenspan, G., 2015, Avoiding pointless blockchain project, Working Paper.

10. Nakamoto, S., 2008, Bitcoin: A peer-to-peer electronic cash system, Working Paper.

11. Chaum, D., 1983, Blind signatures for untraceable payments, in *Advances in Cryptology*, Springer, US, pp. 199–203.

12. Buterin, V., 2016, Blog post, https://blog.ethereum.org/2016/07/26/onward_from_the_hard_fork/.

13. Popper, N., 2015, *Digital Gold: The Untold Story of Bitcoin*, Penguin, UK.

14. Diffie, W., and Hellman, M., 1976, New directions in cryptography, *IEEE Transactions on Information Theory*, Vol. 22, No. 6, pp. 644–654.

15. Miller, V.S., 1986, Use of Elliptic Curves in Cryptography. In: Williams H.C. (Eds). *Advances in Cryptology—CRYPTO '85 Proceedings, Lecture Notes in Computer Science*, Vol. 218. Springer, Berlin, Heidelberg, pp. 417–426.

16. Koblitz, N., 1987, Elliptoc curve cryptosystems, *Mathematics of Computation*, Vol. 48, pp. 203–209.

17. Back, A., 2002, Hashcash—A denial of service counter-measure, Working Paper.

18. Merkle, R.C., 1987, A digital signature based on a conventional encryption function, in *Conference on the Theory and Application of Cryptographic Techniques*, Springer, Berlin, Heidelberg, pp. 369–378.

19. King, R., 2013, *Brunelleschi's Dome: How a Renaissance Genius Reinvented Architecture*, Walker & Company, New York, NY.

20. Marx, K., 1867, *Das Kapital: Kritik der Politischen Őkonomie*, Verlag von Otto Meisner, Germany.

21. Ansper, A., Buldas, A., Freudenthal, M., and Willemson, J., 2003, Scalable and efficient PKI for inter-organizational communication, in *Computer Security Applications, Proceedings of 19th Annual Conference*, IEEE, pp. 308–318.

22. Brown, R.G., Carlyle, J., Grigg, I., and Hearn, M., 2016, Corda: An Introduction, R3 CEV Working Paper.

23. Skeel, D., 2010, *The New Financial Deal: Understanding the Dodd–Frank Act and Its (Unintended) Consequences*, John Wiley & Sons, Hoboken, NJ.

24. Turan, K., 2004, *Never Coming to a Theater Near You: A Celebration of a Certain Kind of Movie*, PublicAffairs, New York.

25. Bloch, M., 1953, Mutations monétaires dans l'ancienne France: Première Partie, *Annales Economies, Societes, Civilisations*, Vol. 8, No. 2, pp. 145–158.

26. Keynes, J.M., 1936, *General Theory of Employment, Interest and Money*, Macmillan, London.

27. Keen, S., 2001, *Debunking Economics: The Naked Emperor of the Social Sciences*, Zed Books, London & New York.

28. Werner, R.A., 2014, Can banks individually create money out of nothing?—The theories and the empirical evidence, *International Review of Financial Analysis*, Vol. 36, pp. 1–19.

29. Lipton, A., 2016, Modern monetary circuit theory, stability of interconnected banking network, and balance sheet optimization for individual banks, *International Journal of Theoretical and Applied Finance*, Vol. 19, No. 6, pp. 1650034-1–1650034-57.

30. Robinson, J., 1977, Michal Kalecki on the economics of capitalism, *Oxford Bulletin of Economics and Statistics*, Vol. 39, No. 1, pp. 7–17.

31. Buiter, W.H., 2009, The unfortunate uselessness of most "state of the art" academic monetary economics, MPR A Working Paper.

32. Rogoff, K.S., 2016, *The Curse of Cash*, Princeton University Press, Princeton and Oxford.

33. Ilgmann, C., 2015, Silvio Gesell: "A strange, unduly neglected" monetary theorist, *Journal of Post Keynesian Economics*, Vol. 38, No. 4, pp. 532–564.

34. Fisher, I., Cohrssen, H.R., and Fisher, H.W., 1933, *Stamp Scrip*, Adelphi Company, New York, NY.

35. Vasicek, O., 1977, An equilibrium characterization of the term structure, *Journal of Financial Economics*, Vol. 5, No. 2, pp. 177–188.

36. Barrdear, J. and Kumhof, M. 2016, The macroeconomics of central bank issued digital currencies, Bank of England, Working Paper.

37. Broadbent, B. 2016, Central banks and digital currencies, Speech at London School of Economics.

38. Lipton, A., 2016, The decline of the cash empire, *Risk Magazine*, Vol. 29, No. 11, p. 53.

39. Danezis, G. and Meiklejohn, S., 2015, Centrally banked cryptocurrencies, UCL Working Paper.

40. Reid, F. and Harrigan, M., 2013, An analysis of anonymity in the bitcoin system, in *Security and Privacy in Social Networks*, Springer, New York, pp. 197–223.

41. Baynham-Herd, X., 2016, Banking Balance Sheets and Blockchain: A Path to 100% Digital Money, UBS Discussion Paper.

42. King, M., 2016, *The End of Alchemy: Money, Banking, and the Future of the Global Economy*, WW Norton & Company, New York, NY.

43. Dwyer, J., 2016, Central Bank-Issued Digital Currency: Assessing Central Bank Perspectives of DLT and Implications for Fiat Currency and Policy Stimulus, Celent Working Paper.

Chapter 19

Optimal Feature Selection Using a Quantum Annealer

Andrew Milne, Mark Rounds, and Peter Goddard

CONTENTS

19.1 Introduction

Quantum computing is still in its infancy. Its potential is sensed, but not yet widely applied. Part of this is due to its specialized nature, and the small size of the problems that can currently be handled. However, small does not mean zero, and with the aid of software like the 1QBit Quantum-Ready Software Development Kit, machines like the D-Wave quantum annealer can be used to solve small but useful problems.

The software development kit (SDK) forms an abstraction layer between the quantum hardware and the financial application program. In the specific case of D-Wave, the SDK provides the objects needed to represent the objective function for a quadratic unconstrained binary optimization (QUBO) problem. The SDK also provides tools for translating constrained problems into unconstrained problems, integer problems into binary problems, and so on.

Optimization is a computational paradigm that follows naturally from the physics of quantum annealing. However, other types of hardware can have other paradigms. The abstraction layer is designed to dispatch its high-level problem representations to the appropriate physical solver. 1QBit refers to the SDK as *quantum ready*, and to its overall architecture as *hardware agnostic*.

The practical details of abstracting from multiple-state qubits to conventional ones and zeros are outside the scope of this chapter. Suffice it to say that human beings have been doing experiments with quantum mechanical systems for over a century now. A lot is known about how to distinguish between energy states, and how to accumulate observations until some level of certainty has been reached. Some of this knowledge is encoded into the software made by the hardware manufacturers themselves. However, for the quantitative analyst or software developer working on a business problem, it is easier to work with a consistent set of abstract entities that map more closely onto the problem domain. The goal of the SDK is to provide these.

In the rest of this chapter, we will approach the problem of optimal feature selection for credit scoring and classification as a perfectly ordinary problem from the literature, that we just happen to solve with a quantum computer. Only at a few select points will we pull back the curtain to reveal the hardware being used.

19.2 Credit Scoring and Classification as a Business Problem

Credit scoring and classification is a significant problem. The total amount of money loaned globally is difficult to measure [1,2]. If we focus on the United States, household debt alone is estimated to be around $14 trillion [3]. The Federal Reserve also reports that approximately 2% of these loans are nonperforming [4]. Superficially, this indicates that lenders are making good decisions 98% of the time. However, the U.S. Federal Deposit Insurance Corporation [5] publishes yearly summaries of bank failures, and in the 15 years since 2001, there have been 547 failed institutions. Nonperforming loans have been a major cause.

According to the 2016 Credit Access Survey by the U.S. Federal Reserve Bank of New York [6], approximately 40% of U.S. credit applications are rejected. Moreover, between 20% and 40% of consumers *expect* their applications to be rejected (it depends on the type of credit), and many do not even apply. Yet, among these people, there may well be qualified customers for the right kind of lender.

In a literature survey by Huang [7], the academic approach to credit scoring is typically one of bigger data and bigger models. However, in a recent article in *Forbes* magazine, consumer lending veteran Matt Harris [8] takes a different view:

> Most start-up originators focus on the opportunity to innovate in credit decisioning. The headline appeal of new data sources and new data science seem too good to be true as a way to compete with stodgy old banks. And, in fact, they are indeed too good to be true.
>
> The recipe for success here starts with picking a truly underserved segment. Then figure out some new methods for sifting the gold from what everyone else sees as sand; this will end up being a combination of data sources, data science and hard won credit observations.
>
> ...[P]rogressive and thoughtful traditional lenders like Capital One have mined most of the segments large enough to build a business around. The only way to build a durable competitive moat based on Customer Acquisition is to become *integrated* into a channel that is *proprietary, contextual* and *data-rich.* (original emphasis)

If Harris is correct, the thing to look for in new credit scoring and classification tools will not be their success rate in large-scale applications for which tools like FICO [9] already exist, but in their flexibility and ease of integration into specialized applications.

Feature selection has a natural role in this. More and more data is available all the time, and although there are various complex schemes for using it, the idea of finding a small set of key features is simple and easy to grasp. Ideally, we would want these features to be both influential and independent, but here too there are nuances. People lie. Data can mislead [10]. The "redundant" feature might actually be the *corroborative* feature. The correct balance of influence and independence will depend on the "hard-won credit observations" that Harris sees as crucial, and on the lender's confidence that the model genuinely includes them.

It should be noted that even the largest and most elaborate system for credit scoring has to begin *somewhere*. Moreover, the addition of new credit instruments to an existing system needs its own stage of analysis and validation. Thus the work of a quantitative analyst may involve both small feature sets and large feature sets. The development of new instruments is not so very different from the development of new markets as Harris pictures them.

19.3 Quadratic Unconstrained Binary Optimization as an Established Approach

QUBO has been applied to the credit scoring problem by several researchers, beginning with Demirer in 1998 [11], and taken up more recently by Huang [7] and Waad [12]. For comparative purposes, the technique is usually applied to a widely available data set, such as the "German Credit Data," that is, the Hofmann data in the Statlog Data Set, as published by the Machine Learning Repository at the University of California, Irvine (UCI) [13]. However, the optimization step has often been seen as time-consuming, and the technique has not yet made it into the most popular (i.e., free) software toolkits.

For convenience, we will split the credit scoring problem into two parts: feature selection and classification (i.e., classification using the selected features instead of the full feature set). Sometimes, these are "wrapped" together so that the parameters for the selection algorithm and the parameters for the classifier can be optimized holistically for the best overall accuracy scores. However, in order to focus on the feature selection part, we will keep them separate and use the same classifier throughout.

In order to compare QUBO Feature Selection with other techniques, we will also show the accuracy scores from feature subsets obtained from recursive feature elimination (RFE), a widely used technique available in packages such as scikit-learn [14]. We will look at two variants:

- Stand-alone RFE, where the desired number of features is set explicitly. The least influential features are eliminated one at a time until the desired number of features is left.

- RFE wrapped with cross-validation (RFECV), where the program evaluates the performance of its classifier after every feature removal and terminates when the highest accuracy score has been found. Cross-validation is discussed later in Section 19.9. Broadly speaking, it involves using a part of the data set for training, and the remaining part for evaluation. The roles are then switched so that every data point is used an equal number of times in each role.

The main difference between QUBO Feature Selection and RFE lies in how aggressively each tries to reduce the number of feature variables in the feature subset. QUBO Feature Selection considers both the independence and influence of the features under consideration. RFE focuses on eliminating features that are less influential. Both approaches yield good results on the German Credit Data.

For the classifier, we use logistic regression from scikit-learn. Unlike *linear* regression, where one can imagine two continuous variables and fitting a line to a scattered set of points, *logistic* regression assumes that the dependent variable is a category, for example, the zero and one of a binary classifier. The fitted line no longer predicts the value of the dependent variable, but rather the probability that the dependent variable will have a specific value. It is a well-established technique with a long pedigree [15].

To provide a benchmark, we apply logistic regression to the full feature set. Out of the box (i.e., without tuning), the logistic regression class from scikit-learn gives a 75% success rate on the German Credit Data. This is comparable to other methods reported in the literature, such as support vector machines (SVMs), decision trees, neural networks, k-Nearest Neighbors (k-NN) classification schemes, and so forth, as in Waad [12] or Huang [7].

Standing on the shoulders of giants, then, we will now approach the optimal feature selection problem in a perfectly ordinary way.

19.4 Formulation of the Credit Scoring and Classification Problem

Assume that we have some data on past credit applicants that we believe will be useful in predicting the creditworthiness of new applicants. The data is composed of *features*, where each feature may be:

- An integer, where the order (lower to higher) has potential meaning, for example, bank balances, years of education, and so on.
- A category, such as a geographic region code, where higher and lower values are arbitrary. It may also include value such as "missing" or "refused to answer."
- A decimal number, representing, for example, age, dollar amounts, interest rates, and so on, where the order has meaning. Data such as the latitude and longitude of the applicant's home would be better represented as a category.
- A Boolean (yes/no) value, which may be considered as an integer or a category.

A credit observation, or sample, consists of an observation for each feature, some means of identifying the applicant, and the outcome. The raw data may come from many sources.

In practice, we clean the raw data to create a vector of "feature variables" for each observation. For example, we may convert some or all of the categorical variables to binary indicators, a step described later, in Section 19.7. We may also scale the data and (in some cases) replace missing values with inferences. In the description that follows, we will assume that these steps have been taken and the data is in a form suitable for input to our feature selector and classifier.

For convenience, we organize the clean data as a matrix of m rows and n columns. Each column represents a feature, and each row represents the specific data values for a specific past credit applicant.

$$U = \begin{bmatrix} u_{11} & u_{12} & u_{13} & \cdots & u_{1n} \\ u_{21} & v_{22} & u_{23} & \cdots & u_{2n} \\ \vdots & \vdots & \vdots & \ddots & \vdots \\ u_{m1} & u_{m2} & u_{m3} & \cdots & u_{mn} \end{bmatrix}.$$

Our goal is to determine how the data on past applicants can inform us on the creditworthiness of new applicants. For this, we need a record of the decisions that were made. We represent these as the m-element vector

$$V = \begin{bmatrix} v_1 \\ v_2 \\ \vdots \\ v_m \end{bmatrix}.$$

The v_i will be constrained to take on the values 0 and 1, where 0 represents the acceptance and 1 represents the rejection of credit application i.

Conceptually, the classifier will be a "credit risk detector" that signals when an applicant should not be granted credit. This is consistent with the U.S. Fed data showing that rejections are less common than acceptances. Acceptance is the rule, and rejection is the exception. The credit risk detector has many analogues in other fields, and a well-established terminology exists in the literature [16].

19.5 Feature Selection

Assume that from the original set of n features, we want to select a subset of K features to use in making a credit decision. Data costs money, and we may want to experiment with different sources. We may be prevented from using certain data in certain jurisdictions. Or we may simply be curious, looking for the insights that we won't know until we find them.

We want to search broadly, without prejudging the data. However, the number of possible subsets for each K is given by the combinatorial function $C(n, K)$. Even if we can eliminate some candidates at the outset, the search space will typically be very large. We want to focus our search on areas where a good subset is likely to be found.

Mathematically, our goal will be to find the columns of U that are correlated with V, but not correlated with each other. We have not yet defined what correlation means here, but we assume that such a calculation is possible and that the value of the correlation coefficient can take on values from -1 to 1. Note that we can interpret "correlation" quite liberally: a "hard-won credit observation" might appear as a "hard requirement" that an attribute be present. Taking this a step further, the conversion of categorical variables to binary indicators (see below) can be expanded to include corroborations that the lender sees as being necessary.

Let ρ_{ij} represent the correlation between column i and column j of the matrix U, and let ρ_{V_j} represent the correlation between column j of U and the single column of V.

To find the "best" subset, we introduce n binary variables x_j, which have the property

$$x_j = \begin{cases} 1, & \text{if feature } j \text{ is in the subset} \\ 0, & \text{otherwise.} \end{cases}$$

We refer to these collectively as the vector X, where

$$X = \begin{bmatrix} x_1 \\ x_2 \\ \vdots \\ x_n \end{bmatrix}.$$

We will associate the best subset with the value of X that minimizes an objective function, which we construct from two components.

The first component of the objective function represents the influence that features have on the marked class, shown here in a form that *increases* as more terms are included:

$$\sum_{j=1}^{n} x_j |\rho_{Vj}|.$$

The second component of the objective function represents the independence of the features. The form shown below *increases* as more of the cross-correlated terms are included, which is the opposite of independence.

$$\sum_{j=1}^{n} \sum_{\substack{k=1, \\ k \neq j}}^{n} x_j x_k |\rho_{jk}|.$$

To obtain an objective function that is maximized at the optimum, we will need to subtract the second term from the first term.

We perform this subtraction with the aid of a parameter α $(0 \leq \alpha \leq 1)$, which represents the relative weighting of independence (greatest at $\alpha = 0$) and influence (greatest at $\alpha = 1$).

Finally, we negate the expression overall *to optimize at the minimum*, which yields the objective function

$$f(\mathbf{x}) = -\left[\alpha \sum_{j=1}^{n} x_j |\rho_{Vj}| - (1-\alpha) \sum_{j=1}^{n} \sum_{\substack{k=1,\\ k \neq j}}^{n} x_j x_k |\rho_{jk}|\right].$$

We can make use of the property that $x_j x_j = x_j$ for binary variables, which allows us to rewrite the summation as a vector product:

$$f(\mathbf{x}) = -\mathbf{x}^{\mathrm{T}} Q \mathbf{x}.$$

From this, we can express the problem in terms of the argmin operator, which returns the vector $\mathbf{x}^*$ for which its function argument is minimized:

$$\mathbf{x}^* = \underset{\mathbf{x}}{\operatorname{argmin}} \left[-\mathbf{x}^{\mathrm{T}} Q \mathbf{x} \right].$$

19.6 Classification

The classification problem may be stated as follows: given a row vector $\mathbf{u}$ of new observations from a new applicant, calculate whether the vector belongs to the creditworthy class. More specifically, find a function $f(\mathbf{u})$ that returns 0 for acceptance and 1 for rejection.

One of the premises of machine learning is that such a function can be derived from a programmatic analysis of existing data. The existing data is divided into a training set and a test set. A candidate function is derived from the training set, and its performance is measured on the test set. Much has been written on the best way to define such functions, the best way to divide the data, how to adapt to new data, and so on. For example, see the citation lists at the UCI Machine Learning Repository [13] or Chen [17]. For a cautionary note, however, the well-known Anscombe's Quartet is worth revisiting as a problem in dividing points as opposed to fitting lines through them [18].

In our example here, we use a simple classifier based on logistic regression. However, in the code examples we will see that other classifiers could easily be used in its place. One might also imagine a classifier with tunable parameters, and searching through these to obtain the settings for best overall performance. In the future, the speed of QUBO Feature Selection on a quantum annealer might enable searches on quite large spaces to be done interactively, as opposed to being spread out over hours or even days.

19.7 Binarizing, Scaling, and Correlating the German Credit Data

The German Credit Data under consideration was originally published in 1994 by Hans Hofmann at the Institute for Statistics and Econometrics, the University of Hamburg. It has been studied extensively.

The data consists of 20 features (7 numerical, 13 categorical) and a binary classification (good credit or bad credit). There are 1000 rows, of which 700 are "good" and 300 are "bad." The data is intended for use with a cost matrix, where giving credit to a bad applicant is five times as bad as not giving credit to a good applicant. In this example, however, we are concerned mainly with the relative "predictive power" of the feature subsets, so the cost matrix was not used.

We prepare the data as follows:

- The *german.data* file from UCI is imported into a Jupyter (iPython) notebook as a pandas DataFrame and given column headers with names from the accompanying *german.doc* file.
- The categorical variables are converted to "one-hot" binary indicators using the DictVectorizer class from scikit-learn.
- The first binary indicator in each group is removed (for k indicators, only $k-1$ are independent).
- All of the numerical features are scaled to mean zero and variance one.
- The classification variable is transformed to $0 =$ good, $1 =$ bad.

The subsequent correlation step is not so straightforward. In preparing this example, we looked at a variety of correlation methods. These led to a small difference in the feature subsets, but no real difference in the accuracy of the classifications. It was noted, however, that methods with a "smooth" distribution of coefficients (Spearman, Pearson, and so on) worked better with the quadratic objective function than correlation methods with sharper jumps, such as "mutual information" scores, as seen in Pedregosa [14] or Rosenberg [19]. In the end, the Spearman method was chosen as being simple and easy to reproduce. However, this is an area where more research is needed.

The binarization and scaling procedures transform the 20 features in the German Credit Data into 48 feature variables. For example, Attribute 1 ("Status of existing checking account"), which has four possible values in the original data, gets converted into three binary indicators. (In the resulting DataFrame, these appear as columns "ChqStat=A12," "ChqStat=A13," and "ChqStat=A14.") These 48 feature variables are the input to the feature selection algorithm. The feature subset at the output of the feature selector then forms the input to the classifier.

19.8 Coding the Feature Selector

The 1QBit SDK [20] provides a toolkit for solving QUBO problems. The details of the underlying quantum hardware are abstracted away, so that the code used by the analyst is no more complex than the code needed to use machine learning packages such as Weka [21] or scikit-learn [14].

19.8.1 An inspiring simplicity

The RFECV class in scikit-learn shows just how easy a good package can make things for the analyst. There are other ways to do RFE, but RFECV "wraps" RFE with a classifier chosen by the user, as shown below with Logistic Regression,

```
featureMatrix = U.values
classVector   = V.values
estimator     = LogisticRegression()
selector      = RFECV(estimator, step = 1, cv = 3)
selector      = selector.fit(featureMatrix, classVector)
indexList     = selector.get_support()
featureList   = np.where(indexList)[0]
```

At the end of this simple code block, the analyst has a feature list that can be used to select columns from the feature matrix. The accuracy scores for the classifier (a.k.a. the estimator) can then be computed using testing and scoring classes such as ShuffleSplit or StratifiedShuffleSplit.

Strictly speaking, RFECV is not directly comparable with QUBO Feature Selection, since the feature list is calculated independently of the accuracy scoring. The simpler variant of RFE with a cardinality target is closer in terms of program flow. A list of candidate subsets (of varying cardinalities) is returned by the selector and subsequently tested for performance. However, the simplicity of the RFECV code block is both an example and an inspiration.

19.8.2 QUBO feature selection in the 1QBit SDK

QUBO Feature Selection involves searching for the value of α that yields the feature subset with the highest accuracy score (or similar performance metric). Huang [7] has found that the QUBO method returns a list of candidate subsets, corresponding to an objective function that consists of flat regions and discrete jumps. This is a reasonable behavior for an objective function that should change as features are added or removed from a candidate subset.

The remarkable thing about QUBO Feature Selection is that the number of candidate subsets is roughly equivalent to the number of features overall, that is, there is usually only one subset at each possible cardinality. There is no *a priori* reason why this should be so, and the authors of this chapter are hopeful that the wider use of QUBO Feature Selection may lead to some deeper insights.

In the code example below, we construct a Q matrix using code that can be easily written from the equations shown in the previous section. We can assign our value of α at the outer level of a grid search, or at the core of a bisection search that looks for jumps in the objective function, but in each case we must somewhere solve the optimization problem to obtain a candidate feature subset.

To do this, we use the SDK's QuadraticBinaryPolynomialBuilder class, which returns a polynomial object representing the objective function seen earlier. The poly object is passed to a solver and the solution is returned as a list of ones and zeros, referred to as a configuration. We convert this to a list of integer indices that can be used to extract columns from a pandas DataFrame or NumPy array, which is typically how feature matrices and class vectors are passed to machine learning methods.

In the example below, the D-Wave solver is assigned explicitly.

```
builder           = qdk.QuadraticBinaryPolynomialBuilder()
##                ... Some code to construct the Q matrix...
poly              = builder.build_polynomial()
solver            = qdk.DWaveSolver(HWDWaveSolver(url, token))
solutionList      = solver.minimize(poly)
lowEnergySolution = solutionList.get_minimum_energy_solution()
config            = lowEnergySolution.configuration
featureList       = np.argwhere(config.values()).flatten().
                    tolist()
featureMatrix     = U.iloc[:, featureList].values
classVector       = V.values
estimator         = LogisticRegression()   # and so on...
```

In the experimental study that led to this chapter being written, we fixed the parameters for the classifier and focused on studying how the value of α affected the feature subset returned by QUBO Feature Selection. A full holistic optimization across all of the available parameters is a topic for future work.

19.8.3 What happens in the call to *minimize*()

The physical D-Wave machine is located at a URL. Access is controlled by a security token assigned to the calling program, as well as by various network management techniques.

The D-Wave machine is based on magnetic effects that occur at very low temperatures [22]. The chip containing the qubits is cooled to 15 mK (about 180 times colder than interstellar space). It is shielded from electric and magnetic fields inside a metal enclosure. To maintain the low temperature, the chip operates in a high vacuum environment at a pressure some 10 billion times lower than atmospheric pressure. Getting the signal from the outside world to the chip is an engineering accomplishment in itself.

The qubits may be thought of as small circulating currents governed by superconducting effects. The direction of the current may be thought of as representing a one or a zero. The qubits can interact with each other, with the exact degree of interaction controlled by electric and magnetic fields that can be externally applied. Collectively, the qubits form an ensemble whose energy is determined by the signs of the magnetic fields, and their coupling with the imposed fields. The overall arrangement corresponds to an Ising model, a concept from statistical mechanics that has been studied for close to a century.

The quantum annealing process consists of

- Initialization. The chip containing the qubits is prepared so that it represents a trivial problem, and is in the ground state of that problem, which is an equally weighted superposition of all possible states.
- Adiabatic Transformation. The system is then transformed continuously to the point that it represents the optimization problem that we want to solve. If this process is done slowly enough, the adiabatic theorem guarantees that the system will remain in the ground state, as long as external disturbances are absent.
- Readout. The state of the system is then read, and in the ideal case it would correspond to the optimal solution of the optimization problem we wish to solve.

The adiabatic transformation is accomplished by slowly applying electric and magnetic fields whose magnitudes correspond to the coefficients in the Q matrix. In a real device, however, external interference is always present, so the result is probabilistic, and annealing the same problem multiple times will increase the probability of finding the optimum. The number of annealings to perform is determined by the controlling software.

The internal structure of the D-Wave chip is outside the scope of this chapter. However, to give an idea of the challenges facing the abstraction layer, recall that the qubits have to be fabricated and connected using structures that can be formed on the surface of a semiconducting chip. At the time of writing, the physical layout is described by a square Chimera graph, composed of a lattice of bipartite unit cells containing eight qubits, as shown in Figure 19.1.

If we label the number of unit cells along an edge as s, then the total number of qubits is $q = 8s^2$. The hardware graph is sparse and in general does not match the problem graph, which is defined by the adjacency matrix of the problem matrix Q. In order to solve problems that are denser than the hardware graph, we identify multiple physical qubits with a single logical qubit (a problem known as "minor embedding" at the cost of using many more physical qubits).

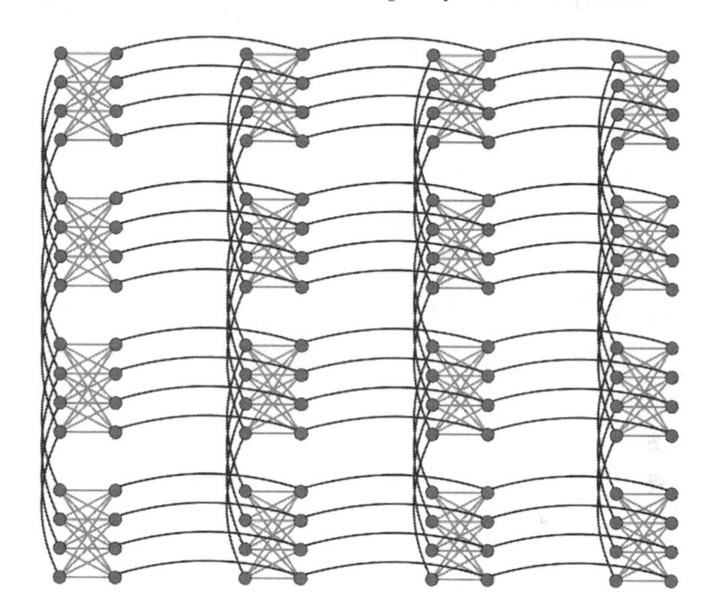

FIGURE 19.1: An example hardware graph, showing the connectivity of the qubits for a Chimera graph with $s = 4$ unit cells in each row/column, giving a total of $q = 128$ qubits.

For square Chimera hardware graphs, the size V of the largest fully dense problem that can be embedded on a chip with q qubits is $V = \sqrt{2q} + 1 = 4s + 1$, assuming no faulty qubits or couplers. For example, for a chip with $s = 12$ unit cells along each side (giving $q = 1152$ qubits), we find $V = 49$. Lower density problems of significantly larger size can be embedded. For example, experiments by 1QBit on annealers of this size have shown successful embeddings with $V_b \simeq 140$ and a density of $\simeq 0.1$.

It should be emphasized that users of the 1QBit SDK does not have to concern themselves with the size of the problem or the details of the embedding. For problems that exceed the capacity of the available D-Wave machine, the SDK can route the calculation to a simulated annealer (software) or a simulated Ising system (hardware).

The performance of simulated quantum annealers is now very good, and to anticipate our conclusions slightly, there is no real limit on the size of the feature selection problems that the SDK can handle, save that below a certain (but ever increasing) size, the solution will be passed through a quantum computer and be very fast as opposed to merely fast.

19.9 Evaluation Metrics

The evaluation of QUBO Feature Selection and RFE was performed by wrapping them with the LogisticRegression model from scikit-learn. The evaluation metric was defined as the unweighted accuracy, that is, the number of correct classifications divided by the total number of classifications made. Other metrics from scikit-learn were attempted, but they always led to the same optimal alpha or RFE feature set. Unweighted accuracy was kept for compatibility with other work and ease of understanding, as in Reference 7.

Testing and scoring were performed using the StratifiedShuffleSplit cross-validation class from scikit-learn. Given the feature matrix, this class returns sets of row indices that can be used to divide the matrix rows into a training set and a test set. The separation is done in folds, with the number of folds set by an argument. For example, a shuffle and split with 5 folds will take 80% of the matrix for the training set and 20% for the test set, and repeat this process until all 5 of the possible 20% folds have been used as test sets.

The accuracy score for each split is slightly different. Since these reflect a random selection of data for training and testing, it is conventional to report the mean of the individual accuracy scores. We follow the convention used by scikit-learn and calculate the error bars for the 95% confidence level.

19.10 Experimental Results

19.10.1 Establishing the zero-rule and other baseline properties

The German Credit Data has 700 class 0 samples ("good credit") and 300 class 1 samples ("bad credit"). A zero-rule classifier that assigns *all* of the samples to class 0 will therefore achieve a success rate of 70%.

We want our proposed feature selection and classification scheme to do better than the zero-rule. We want the feature selection component to choose subsets that are better than randomly selected subsets and (for that matter) better than no selection at all. We begin this section by establishing a baseline against which QUBO Feature Selection can be compared.

Feature selection is motivated by the intuitive concept that not all features are equally important. For example, in Figure 19.2, we see the feature variables from the binarized German Credit Data ranked by the Spearman correlation coefficient between the feature and the classification variable. There are four relatively important features at the left-hand end, followed by a gradual, almost linear decline. It turns out that the four features on the left are not enough to form a predictive subset on their own, so the problem for the feature selector is to find where on the line "enough is enough."

It also turns out that the smooth decline of the coefficients works well with the quadratic objective function used in QUBO Feature Selection, as was seen in comparisons of Pearson, Spearman, and Kendall correlations (readily available in the pandas package). In contrast, the use of a mutual information score in place of a correlation coefficient was not as successful. The integer features

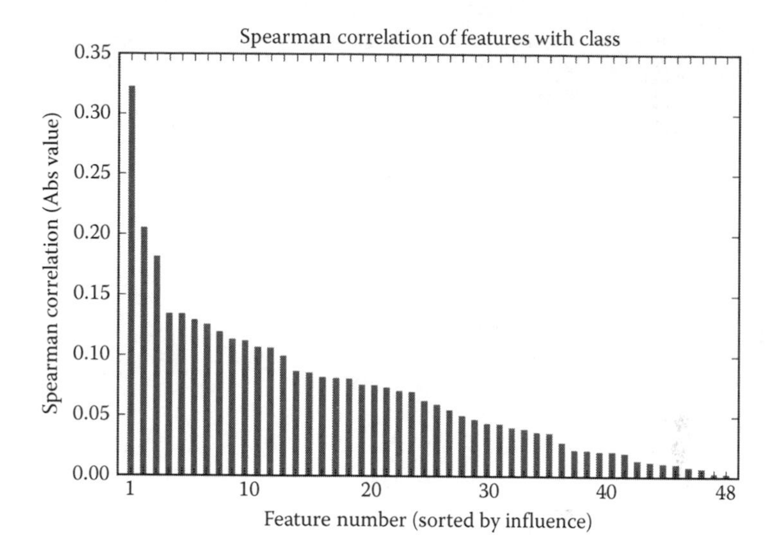

FIGURE 19.2: Spearman correlation. The most influential features involve the status of the applicant's checking account, savings account, and loans at other institutions, all of which are included in all practical feature subsets. As we move toward the right, however, the influence of each new feature declines.

were binned into ordered categories and generally behaved as expected. However, the more arbitrary binarized categories (e.g., loan purpose) had uniformly low mutual information scores, and the objective function tended to cycle among features. A plot of the ranked mutual information scores is shown in Figure 19.3.

The results shown in this chapter were all calculated using the Spearman correlation coefficient. As mentioned previously, this is an area where more research is needed, for example, using other data sets with a broader mix of feature variables.

Before we select any features, however, we first examine how well the logistic regression classifier performs on the full feature set, that is, all 48 feature variables. We do this using the StratifiedShuffleSplit() class from scikit-learn. Figure 19.4 shows that the mean accuracy depends on how many times the data is shuffled, and on how the data is split between the training set and the test set.

The combination of 1000 shuffles and 20% test share was chosen arbitrarily as the standard for initial performance comparisons, being much more convenient than the larger numbers. It avoids the fluctuations found below 500 shuffles, and is close to the converged scores at 3000 samples and above. For the definitive score comparison, however, the full 3000 shuffles were used. The results for 10%, 15%, and 20% share were always very close and typically in the median position. The 20% test share was therefore used throughout.

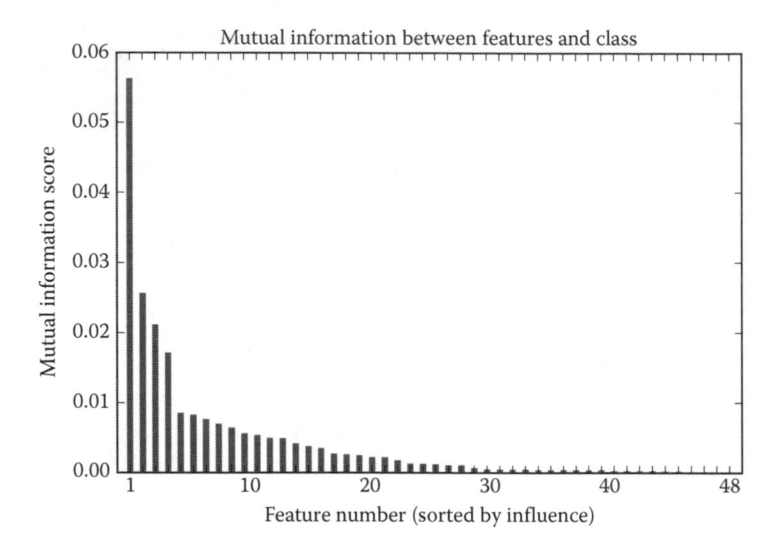

FIGURE 19.3: Mutual information scores. The age, term, and credit amount fields were binned into categories. However, the correlation matrix based on this technique led to fluctuations in the accuracy scores, a lower mean accuracy at the "best" feature subset, and a larger "best" feature subset cardinality of 34 elements.

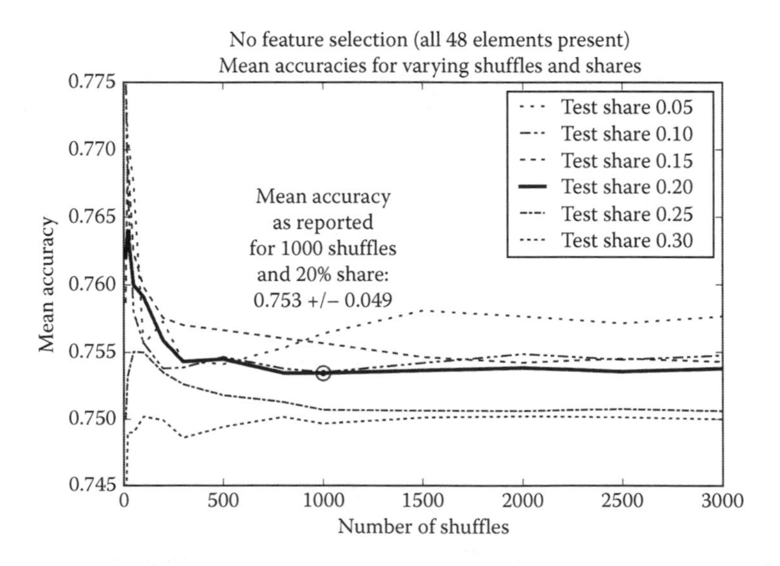

FIGURE 19.4: Mean accuracies measured for various numbers of shuffles (10–3000) and for different fractions of the data assigned to the test set.

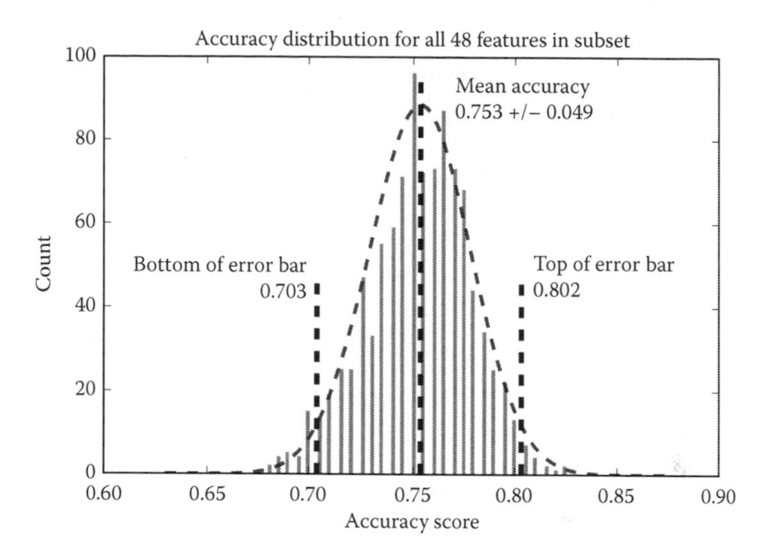

FIGURE 19.5: Accuracy scores for all 48 samples, using 1000 shuffles with 20% test share. In the work described in this chapter, this distribution was typical of the German Credit Data, regardless of the feature subset, number of shuffles, number of shares, classifier parameters, and the like.

It can be seen in Figure 19.4 that the mean accuracy increases as the test share is reduced, that is, a bigger training set yields a more accurate predictor. However, the difference is small in relation to the dispersion of scores from different shuffles, as can be seen in Figure 19.5.

Note that the 30% share curve does not fluctuate as much as the curves for smaller shares. It turns out that scikit-learn chose a threefold default for RFECV, and this may be one reason why it can attain a good (although not optimal) result in relatively little time. It is also important to keep track of absolute numbers as well as percentages: a 5% test share of 1000 samples consists of only 50 samples, which stratification on the German Credit Data will constrain to 35 good credit samples and 15 bad samples. It fluctuates widely at the outset and converges slowly.

In Figure 19.5, we see the distribution of accuracy scores for all 48 features at a 20% test share (800/200 train/test split) counted over 1000 shuffles. Stratification forces the training set and the test set to have the same 70/30 distribution of good and bad credit samples, so that a 200-sample test set will contain 60 bad credit samples chosen from 300 in the set overall. In a large number of shuffles, there will inevitably be some repetition (a point highlighted by scikit-learn in its documentation). Thus although the data looks "Gaussian" and fits into the Gaussian overlay, in practice there are certain scores that occur more frequently, and extreme values are not observed above a certain limit. Note that in comparison to the spread of accuracy scores from 0.7 to 0.8 seen in Figure 19.5, the (converged) spread of mean scores from

0.75 to 0.76 in Figure 19.4 is relatively small. So long as we avoid small test shares and low numbers of shuffles, the error bars from the accuracy scores will dominate the uncertainty overall.

We now take a moment to examine the behavior of logistic regression on feature subsets with fewer than 48 features.

The number of possible subsets is given by the combinatorial function $C(48, K)$, where K is the cardinality of the subset. The largest number of possible subsets occurs when 24 feature variables are selected, and is approximately 32 trillion. There are some 280 trillion subsets possible overall.

It is not possible to test these trillions of subsets systematically. However, we can gain an idea of how they behave from random sampling. For example, Figure 19.6 shows the accuracy of logistic regression for 10,000 randomly selected subsets at each of the 48 possible cardinalities, which examines 432,354 feature subsets out of 281,474,976,710,656 possibilities. If we record the best of the mean accuracies for each group of 10,000 subsets, and plot them separately (the triangle markers in Figure 19.6), we can identify a "best detected" subset at cardinality 35, with an accuracy of 0.76 ± 0.05. We then examine how this mean was calculated from the accuracy scores for 1000 shuffles with a 20% test share. In Figure 19.7, we see that the variance is large, and comparable to what we saw with all 48 features present. Feature selection looks to be a search for small improvements in collections of very noisy test results.

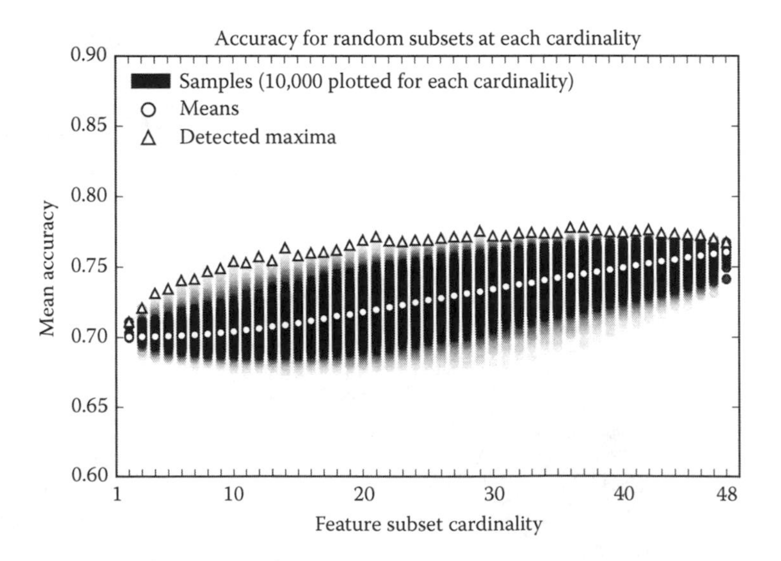

FIGURE 19.6: Sample means plotted as nearly transparent black circles. The largest mean found for each cardinality group is shown by the triangle markers. The "maximum" at cardinality 35 is somewhat arbitrary, but given the values of the maxima at the endpoints, there is clearly a maximum at least somewhere between 1 and 48.

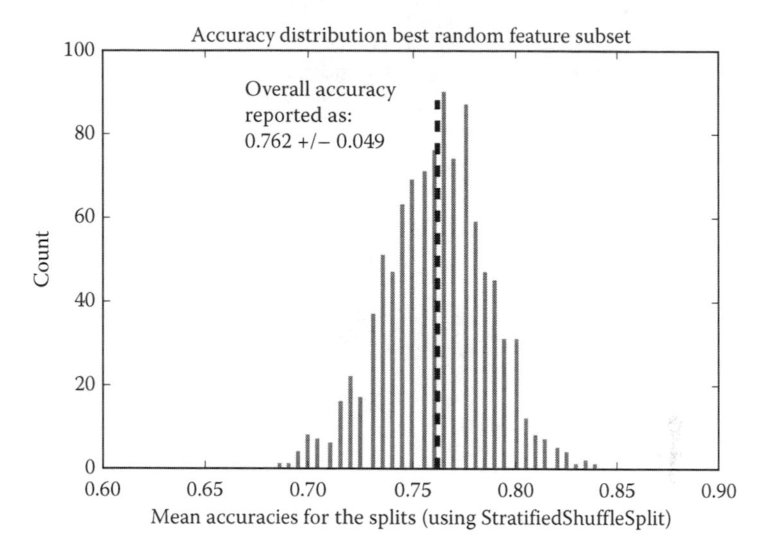

FIGURE 19.7: Distribution of accuracy scores for the best feature subset found through random search.

It must be stated at this point that accuracy scores reported by other researchers on the German Credit Data are typically in the 70%–75% range, with standard deviations around 5%. For example, see Chen [17], Rao [23], or Huang [7]. Our 0.76 ± 0.05 baseline accuracy puts QUBO Feature Selection squarely in the mainstream.

We can now summarize our baseline requirements as follows:

- Select a feature subset with 35 features or less.
- Deliver accuracy equal to or better than 0.76 ± 0.05.
- Calculate the feature subset in an efficient way that can scale to larger initial feature sets.

19.10.2 QUBO feature selection with logistic regression

QUBO Feature Selection and a logistic regression classifier were "wrapped" together. Practically speaking, this means that the feature selector and the classifier were placed together at the interior of the loops used to optimize the selection and classification parameters. For simplicity, the only optimization parameter was α, which determines the relative weighting of *independence* (greatest at $\alpha = 0$) and *influence* (greatest at $\alpha = 1$).

It was found that the cardinality of the feature subset tended to increase with α, as had been noted by other researchers [7,11]. However, an advantage

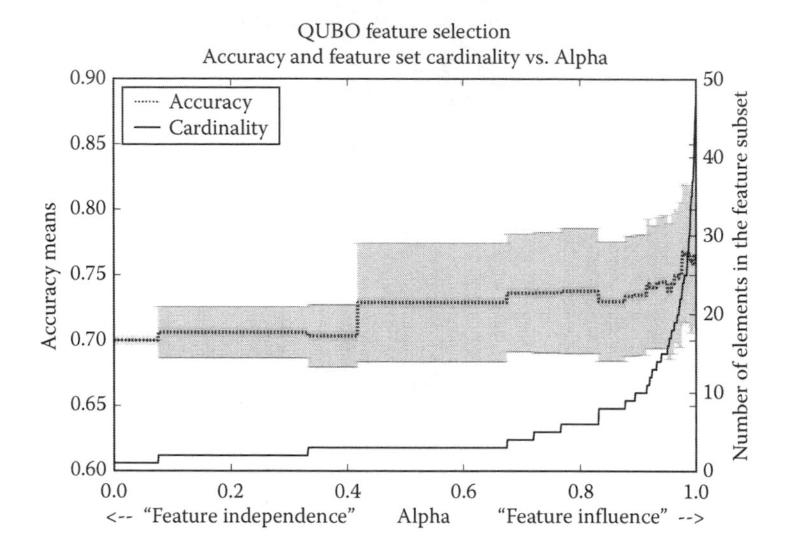

FIGURE 19.8: Increase in accuracy as influential features are chosen over independent ones.

of the wrapper model is that optimization can be done "at the α level" without looking at the details of the subsets.

Figure 19.8 shows the full range of α from 0 to 1. On the left-hand side, where α is close to zero, the emphasis is on feature independence. This favors small subsets, and since their regression coefficients are often not large enough to "push" the classifier across the cutoff point of $p \geq 0.5$, the predicted class is 0. They classify almost all of the samples as "good credit" and achieve the zero-rule's 70% success rate.

In Figure 19.9, we look more closely at the region between $\alpha = 0.9$ and $\alpha = 1$. Here, the emphasis is on feature influence, and the subsets eventually grow to include all 48 available feature variables.

It is interesting that accuracy increases with the size of the subset, reaches a peak at $\alpha = 0.977$ with 24 elements, and then declines gradually as more features are added. The drop in accuracy to the left of $\alpha = 0.977$ is quite sharp, and although it is encouraging to see a global maximum so clearly defined, this may be due to the data, and should be further investigated.

19.10.3 Recursive feature elimination with logistic regression

RFE is a technique for pruning features from a feature list. The procedure begins with a set of available features. It fits the logistic regression model and eliminates the feature with the lowest weight. The fitting and elimination is continued until the desired number of features is reached. In RFECV, the

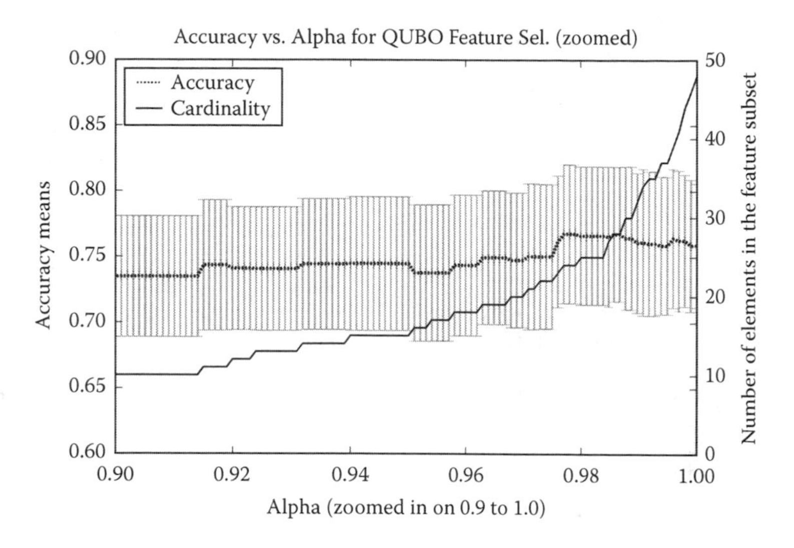

FIGURE 19.9: A closer look at QUBO Feature Selection in the $\alpha = 0.9$ to $\alpha = 1$ region.

fitting is accompanied by testing, using a training set and test set chosen according to a folding parameter.

Conceptually, RFE is like starting the QUBO objective function at $\alpha = 1$ and working downward toward $\alpha = 0$, except that RFE is recursive and treats each iteration as a new feature set.[1] Unlike QUBO Feature Selection, RFE does not test explicitly for feature independence, nor does it allow a feature to "come back" after it has been eliminated.

We began with the direct version of RFE, where we specified the desired number of features and then measured the performance of the returned feature subset.

The accuracies were measured with the same 1000 shuffles and 20% test share that was used with the other methods. The results are shown in Figure 19.10.

RFECV was very fast, although it converged to different feature subsets as the cross-validation settings and random seeds were varied. In practice, however, it was easy to search these (and faster than running RFE for thousands of shuffles). RFECV ultimately delivered a 31-element feature subset with an accuracy of 0.76 ± 0.05, comparable with the other methods.

[1] One could imagine the recursive elimination of features as α is iterated from 1 down to 0. A recursive version of QUBO Feature Selection would make an interesting topic for future work.

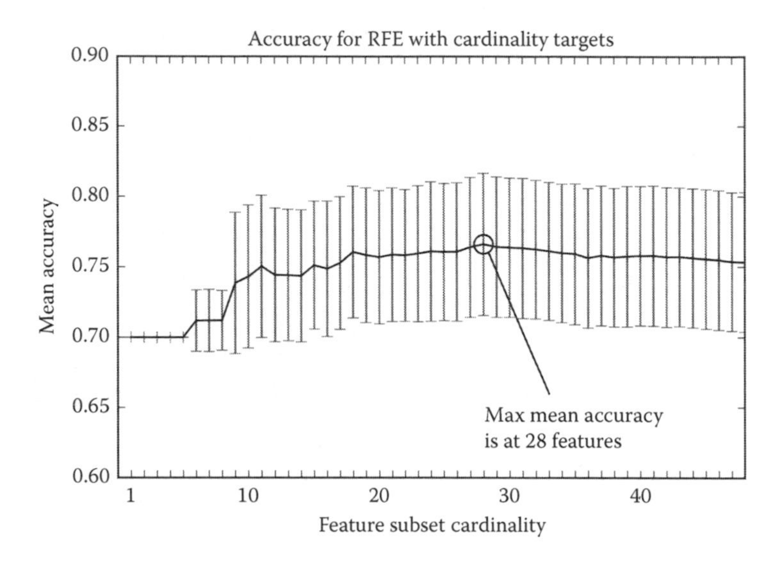

FIGURE 19.10: RFE for cardinality targets from 1 to 48, using 1000 shuffles with 20% test share. Error bars represent a confidence level of 95%. The maximum mean accuracy is 0.77 ± 0.05.

19.10.4 Comparison of QUBO feature selection and recursive feature elimination

Figure 19.11 shows the mean accuracies for QUBO Feature Selection and RFE (and, for comparison, a random search of 10,000 sample subsets). The differences between the methods are smaller than the error bars.

In Figure 19.12, a closeup view of the region between 20 features and 49 features shows that the results never differ by more than the spread of mean accuracy scores in Figure 19.4.

When the best subsets from each method are compared at 3000 shuffles, we obtain the accuracies shown in Table 19.1, ranked by the number of features in the subset.

These are equivalent accuracies, with QUBO Feature Selection giving the smallest feature subset. However, random feature selection with 10,000 samples per cardinality had the lowest score of all, which shows that 10,000 samples (out of several trillion) cannot tell us much about extreme values. There may be other subsets "out there" that were missed by both QUBO Feature Selection and RFE. To assess this possibility, we examine some of the feature subsets that are "near" the QUBO Feature Selection and RFE results, in the sense that they differ by one or two features.

The F1-Score was computed using the f1_score() method of the Logistic-Regression() class from scikit-learn. It is defined as the harmonic mean of the precision and the recall, where *precision* is the ratio of true positives to all predicted positives, and *recall* is the ratio of true positives to all actual

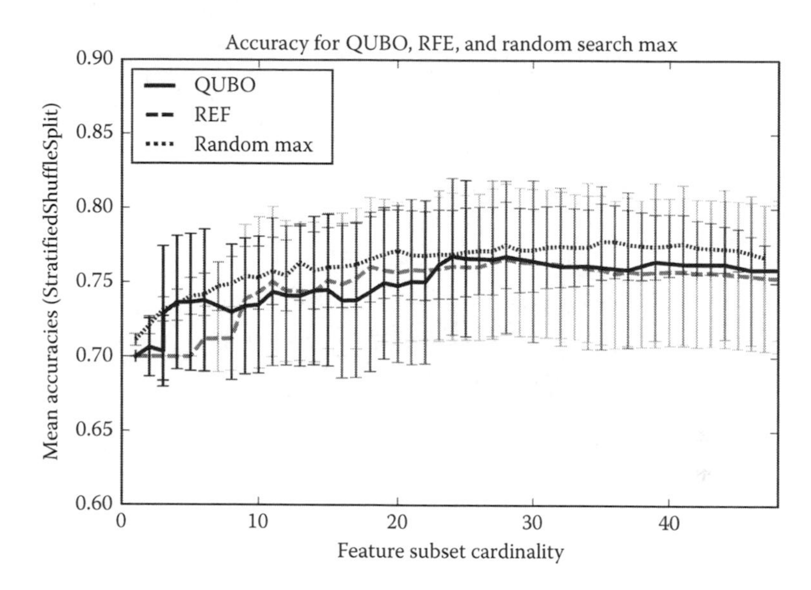

FIGURE 19.11: Comparison of QUBO Feature Selection, RFE, and random search over all 48 cardinalities.

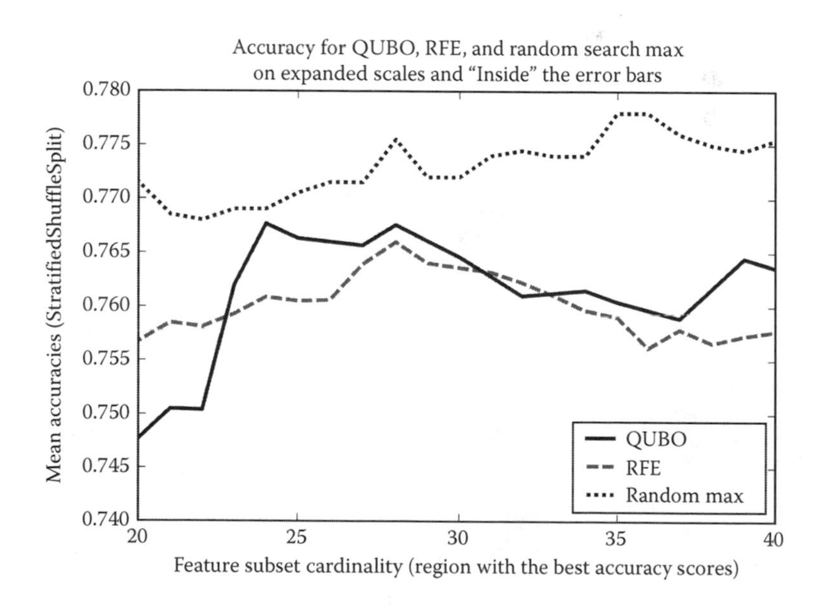

FIGURE 19.12: Comparison (using a different vertical scale) of QUBO Feature Selection, RFE, and random search.

TABLE 19.1: Comparison of accuracy scores at 3000 shuffles

Method	Accuracy	F1-Score	No. of features
QUBO Feature Selection	0.764 ± 0.049	0.54 ± 0.1	24
RFE28	0.767 ± 0.050	0.55 ± 0.1	28
RFECV	0.764 ± 0.050	0.54 ± 0.1	31
Rand10k	0.762 ± 0.050	0.54 ± 0.1	35

positives [16]. The above values reflect the classification of the German Credit Data without the application of a cost matrix and can be compared with results in the literature that were calculated in the same way.

19.10.5 Comparison with potentially missed subsets

In Figure 19.13, we see the mean accuracy results from feature subsets that have the same cardinality as the best subsets, but which differ in one feature, two features, and so on. Gaussian curves showing the distribution of the accuracy scores for the best subsets have been overlaid on the histograms.

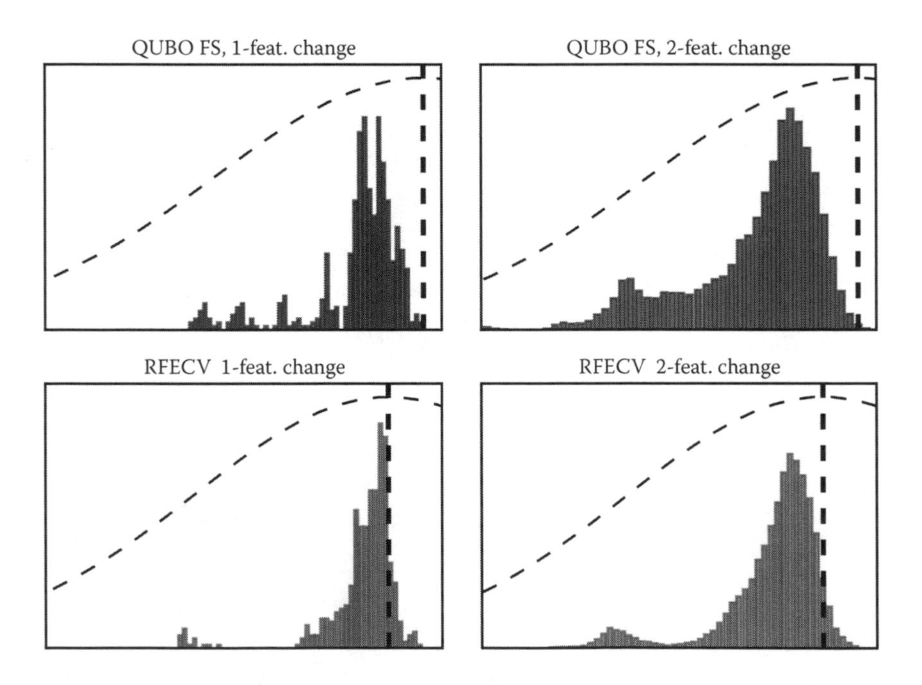

FIGURE 19.13: Comparison of mean accuracies from the "best" QUBO Feature Selection (dark gray) and "best" RFE (light gray) with mean accuracies from feature sets of the same cardinality but differing by one or two features. Gaussian distributions with the corresponding "best" standard deviations have been overlaid. The horizontal scale on all four graphs is the same, in both this figure and the next.

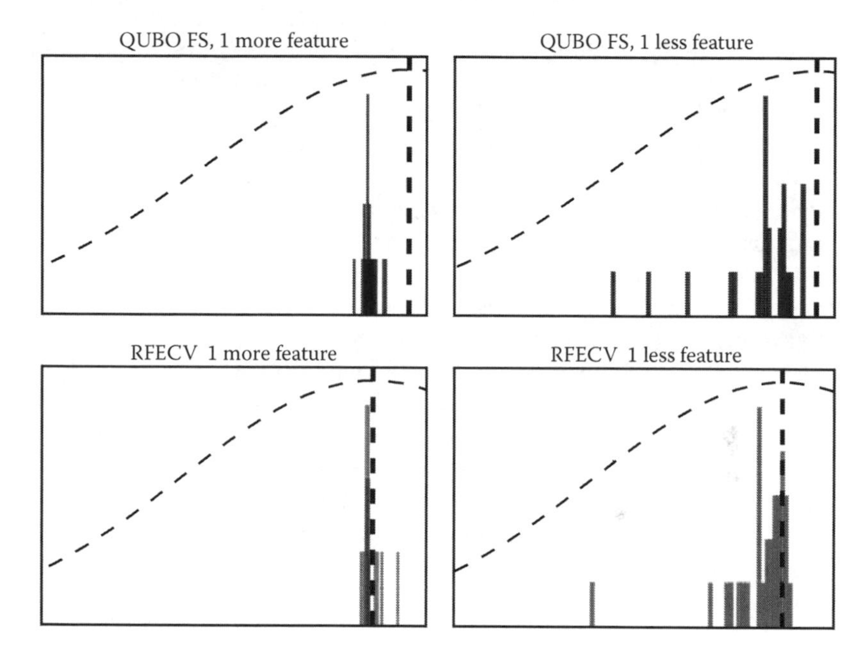

FIGURE 19.14: Comparison of mean accuracies from the "best" QUBO Feature Selection (dark gray) and "best" RFE (light gray) with mean accuracies from feature sets with one more feature and one less feature. Gaussian distributions with the corresponding "best" standard deviations have been overlaid.

The dashed vertical lines represent the means of the best subsets. The vertical bars represent the counted occurrences of mean accuracies for the perturbed feature subsets. QUBO Feature Selection does a better job than RFE in finding a subset that is better than its neighbors. This reflects the behavior of the quadratic objective function, which considers all of the feature sets simultaneously (especially in the quantum annealer implementation). In contrast, once a feature has been eliminated by RFE, there is no possibility of bringing it back on a later iteration.

The same behavior is observed when adding or subtracting a feature from the best subsets. In Figure 19.14, we see that the accuracy of the best QUBO subset is again better than the perturbed subsets.

19.11 Comparison with Previously Reported Results

The German Credit Data was originally published in 1994 by Hans Hofmann at the Institute for Statistics and Econometrics at the University of Hamburg. Since that time, it has been studied extensively. At the time of writing, the most thorough survey of how quadratic optimization compares with other methods is given by Waad [12]. Waad also used a publicly available machine learning package, Weka 3.7.0 [21], and created a "three-stage feature selection fusion" technique with QUBO Feature Selection as the first stage, which yielded very good accuracies. Waad's results are not directly comparable to the results in this chapter, since they were given in terms of precision and recall, which are affected by whether the 5:1 cost ratio prescribed by Hofmann has been applied in the training set. In this chapter, we did not use the cost ratio and Waad does not mention it. Also, the Weka package does not have a function like scikit-learn's StratifiedShuffleSplit(), and in Waad's reported experimental procedure, the division of the samples into a training set and a test set was done at an early stage with 10 folds but no shuffling.

Taken in total, however, Waad provides strong motivation for studying how QUBO Feature Selection might be used as part of a larger procedure. For example, Figure 19.12 shows that, although QUBO Feature Selection found a very good subset with 24 elements, there were other subsets nearby that were slightly better. Additional searching could uncover more.

For feature selection overall, Chen and Li [17] were able to achieve good results with 12 features from the original German Credit Data with its 20 categorical and integer (or fixed decimal) features. These were manually selected after comparing various correlation measures between the features and the classification. Chen and Li did not use the programmatic binarization of categorical variables that came into greater popularity between 1998 and the present day. Their results are primarily for SVMs and do not deliver notable accuracy, especially since Chen and Li also report performing only a single 10-fold cross-validation. The real lesson from this early work is that correlation makes a difference, and that automatic binarization might not always be a good idea.

19.12 Conclusion

Our objective in this chapter has been to show that quantum computing is now within the reach of everyone. We took an old problem and an old method that people used to think was slow, and implemented it using an SDK that can route the problem to either a quantum solver or an advanced classical solver. If the reader has forgotten the term "quantum annealer" by this point, that is perhaps a sign of success.

On the binarized German Credit Data, QUBO Feature Selection delivered a smaller feature set (24 features) than either RFE (28 features) or RFECV (31 features). All three methods showed comparable accuracy. A priority for future work is to study the behavior of the QUBO method on different data sets, with different correlation methods, and with a multistage approach that could improve its performance.

Also of interest is the unusually small number of candidate feature sets returned to the selector, and the possibility of applying the technique to much larger initial feature sets. The method performed best when the feature correlation coefficients fell off smoothly. One could imagine broadening the "wrapper" concept to include the selection of a correlation algorithm, and using the speed of the QUBO Feature Selector to explore this space more systematically.

The authors wishes that the availability of QUBO Feature Selection via 1QBit's quantum-ready SDK will make this method more accessible to researchers interested in studying its possibilities, and to practitioners wanting to add a new (and quantum-ready) tool to their metaphorical toolboxes.

Acknowledgments

The author would like to thank Majid Dadashi and the 1QBit Software Development Team for creating the SDK, and for their assistance with its use. Anna Levit provided useful comments on the draft. Jaspreet Oberoi contributed the idea that led to *recursive* QUBO Feature Selection, a concept whose possibilities we have yet to explore.

The author wishes to express his gratitude to Gili Rosenberg and his many collaborators on the 1QBit research paper, *Solving the Optimal Trading Trajectory Problem Using a Quantum Annealer*, and to 1QBit for permission to quote at length from their description of the D-Wave quantum annealer.

References

1. Organization for Economic Cooperation and Development (OECD). Household debt (indicator), December 2016.

2. World Bank. International Debt Statistics 2017, 2017.

3. U.S. Federal Deposit Insurance Corporation. Statistics at a Glance, 2017. From www.fdic.gov.

4. World Bank. Bank Non-Performing Loans to Gross Loans for United States, 2016. Data series DDSI02USA156NWDB retrieved from FRED, Federal Reserve Bank of St. Louis.

5. U.S. Federal Deposit Insurance Corporation. Bank Failures in Brief, 2017. From www.fdic.gov.

6. Center for Microeconomic Data U.S. Federal Reserve Bank of New York. SCE Credit Access Survey, 2016. From www.newyorkfed.org.

7. Huang, J. Feature Selection in Credit Scoring—A Quadratic Programming Approach Solving with Bisection Method Based on Tabu Search. PhD thesis, Texas A&M International University, 2014.

8. Harris, M. The Short History and Long Future of the Online Lending Industry. Forbes Valley Voices, 2017.

9. FICO (formerly the Fair Isaac Company). FICO Website www.fico.com, 2017.

10. Goel, V. Russian Cyberforgers steal millions a day with fake sites. *New York Times Online*, 2016.

11. Demirer, R. and B. Eksioglu. Subset Selection in Multiple Linear Regression: A New Mathematical Programming Approach (working paper). Technical report, University of Kansas, School of Business, 1998. A later version of the paper appeared in *Computers and Industrial Engineering*, vol. 49, August 2005.

12. N'Cir Waad, B. B. On Feature Selection Methods for Credit Scoring. PhD thesis, Université de Tunis, Institut Supérieur de Gestion, École Doctorale Sciences de Gestion LARODEC, 2016.

13. Lichman, M. Machine Learning Repository, School of Information and Computer Science, University of California, January 2017.

14. Pedregosa, F., Varoquaux, G., and Gramfort, A. Scikit-learn: Machine learning in python. *Journal of Machine Learning Research*, 12:2825–2830, 2011. Note: This is the citation requested at the scikit-learn.org website, accessed January 12, 2017.

15. Cox, D. The regression analysis of binary sequences (with discussion). *Journal of Royal Statistical Society B.*, 20:215–242, 1958.

16. Powers, D. M. W. Evaluation: From precision recall and F-Measure to ROC, informedness, markedness and correlation. *Journal of Machine Learning Technologies*, 2(1):37–63, 2011.

17. Chen, Fei-Long and Li, Feng-Chia. Combination of feature selection approaches with SVM in credit scoring. *Expert Systems with Applications*, 37:4902–4909, 2010.

18. Anscombe, F. J. Graphs in statistical analysis. *The American Statistician*, 27(1):17–21, 1973. See also the nicely illustrated Anscombe's Quartet article on Wikipedia.

19. Rosenberg, A. and Hirschberg, J. V-Measure: A conditional entropy-based external cluster evaluation measure. Technical report, Department of Computer Science, Columbia University, January 2007.

20. 1QBit Inc. www.1qbit.com, 2017.

21. Bouckaert, R. R., Frank, E., Hall, M., Kirkby, R., Reutemann, P., Seewald, A., and Scuse, D. Weka Manual (3.7.1), 2016. This is the citation given by Waad (see below) and describes the Weka features available at the time.

22. D-Wave Systems Inc., www.dwavesys.com, 2017.

23. Rao, M. *How to Evaluate Bank Credit Risk Prediction Accuracy based on SVM and Decision Tree Models*, Capgemini "Capping IT Off" Blog, November 2, 2016. www.capgemini.com, accessed January 18, 2017.

Index

Note: Page numbers followed by "n" indicate notes.

C

F

J

K

L

P